Advances in Diagnostic Analysis, Strategic Management, and Proactive Maintenance of Electrical Equipment

Advances in Diagnostic Analysis, Strategic Management, and Proactive Maintenance of Electrical Equipment

Guest Editors

Xiaozhen Zhao
Yiming Zang

MDPI

Basel • Beijing • Wuhan • Barcelona • Belgrade • Novi Sad • Cluj • Manchester

Guest Editors
Xiaozhen Zhao
School of Electrical Engineering
Hebei University of Technology
Tianjin
China

Yiming Zang
Department of Electrical Engineering
Shanghai Jiao Tong University
Shanghai
China

Editorial Office
MDPI AG
Grosspeteranlage 5
4052 Basel, Switzerland

This is a reprint of the Special Issue, published open access by the journal *Energies* (ISSN 1996-1073), freely accessible at: https://www.mdpi.com/journal/energies/special_issues/T9H8LO1KTR.

For citation purposes, cite each article independently as indicated on the article page online and as indicated below:

Lastname, A.A.; Lastname, B.B. Article Title. *Journal Name* **Year**, *Volume Number*, Page Range.

ISBN 978-3-7258-6740-0 (Hbk)
ISBN 978-3-7258-6741-7 (PDF)
https://doi.org/10.3390/books978-3-7258-6741-7

Contents

About the Editors

Xiaozhen Zhao

Xiaozhen Zhao is a researcher at the School of Electrical Engineering, Hebei University of Technology, Tianjin, China. His research focuses on the fault diagnosis, reliability assessment, and life-cycle management of electrical equipment, as well as pulsed power technology and its applications. He has participated in multiple national and provincial research projects, published over 30 SCI/EI-indexed papers, and holds several invention patents. Zhao completed a joint training Ph.D. program at Curtin University, Australia, under the China Scholarship Council, and worked at the College of Advanced Interdisciplinary Studies at the National University of Defense Technology. His contributions have been recognized with awards including the Second Prize in the Science and Technology Progress Award by the China Electrotechnical Society.

Yiming Zang

Yiming Zang is an assistant researcher in the Department of Electrical Engineering, Shanghai Jiaotong University, and a part-time researcher in Shanghai Key Laboratory of Emergency Management. He has undertaken/participated in more than ten major national scientific and technological projects, national key research and development programs, and projects for the National Natural Science Foundation of China. He has published more than 40 SCI/EI search papers and has been awarded 10+ national invention patents. He was awarded the Nomination Award for Excellent Doctoral Dissertation by China Energy Research Society and Shanghai Jiaotong University.

energies

MDPI

Article

Harmonic Current Effect on Vibration Characteristics of Oil-Immersed Transformers and Their Experimental Verification

Dexu Zou [1], Jian Hao [2,*], Weiju Dai [1], Guochao Qian [1], Haoruo Sun [1] and Jing Xu [2]

1 Electric Power Research Institute of Yunnan Power Grid Co., Ltd., Kunming 650217, China; zdxdb@126.com (D.Z.); 15911530479@163.com (W.D.); qianguochao_yndky@163.com (G.Q.); s_hr1997@163.com (H.S.)

2 State Key Laboratory of Power Transmission Equipment Technology, School of Electrical Engineering, Chongqing University, Chongqing 400044, China; xj12345620210531@163.com

* Correspondence: cquhaojian@126.com

Abstract: Harmonic currents can intensify transformer vibrations, seriously threatening their mechanical stability and safe operation. Drawing upon this foundation, the present paper undertakes a thorough simulation and experimental investigation into the vibration characteristics of transformers under diverse harmonic current scenarios. Initially, a multi-field coupling model incorporating both "electromagnetic and structural forces" was developed to simulate and analyze how the vibration acceleration of a transformer is distributed under varying harmonic currents. Subsequently, a specialized transformer harmonic loading and vibration measurement platform was constructed to validate the multi-physical-field vibration simulation. Finally, through a rigorous experimental analysis of transformer vibrations under harmonic currents, this research elucidates the variation patterns of characteristic vibration parameters of transformers under different harmonic currents. The results demonstrate that as the proportion of harmonic current grows, the mean winding vibration acceleration escalates following a power-function law. With increasing harmonic current frequency, the vibration acceleration augmentation in high-voltage (HV) windings exceeds that which is observed in low-voltage (LV) windings. Empirical validation confirms that the discrepancy between the measured and simulated acceleration increases remains within 5%, indicating the effectiveness and reliability of the simulation method. Experimental findings reveal that as the harmonic current content increases, six vibration characteristic parameters—including root mean square value, absolute average value, peak-to-peak value, and mean frequency—exhibit a pronounced upward trend. Furthermore, harmonic currents significantly increase the spectral dispersion and high-frequency components of the vibration signal. These research findings provide valuable references for transformer operation, maintenance, and anti-vibration design strategies.

Keywords: oil-immersed transformer; harmonic current; vibration characteristics; electromagnetic-structural force field coupling

1. Introduction

In modern power systems, the widespread proliferation of power electronic devices has significantly exacerbated harmonic pollution issues. This occurrence not only diminishes the quality of power but also has significant adverse effects on the crucial equipment throughout the entire power system [1–3]. As the core equipment of the power system, the transformers' secure and stable operation is essential to the overall reliability of the power

https://doi.org/10.3390/en18071673

grid. Under the action of complex harmonic currents, internal components of converter transformers experience varying degrees of vibration, leading to elevated noise and posing serious threats to the mechanical stability of the internal components [4–10]. Therefore, an in-depth exploration of the harmonic currents' impact on the vibration features of transformers holds profound significance for enhancing transformer reliability and extending operational lifespan.

Recent years have witnessed extensive scholarly research on transformer vibration characteristics across multiple dimensions. Regarding theoretical modeling, a multiphysics analytical framework was established in theoretical research by Jin Mingkai's team at Xi'an Jiaotong University [11], employing coupled electromechanical-field interactions through concurrent circuit-magnetic-structural integration to quantify short-circuit-induced alterations in stray flux profiles, Lorentz force generation, and mechanical displacement responses. Their findings revealed that during winding vibration, various fields interact dynamically. The short-circuit current is less affected by the change in reactance, while the distortion in the leakage magnetic flux intensifies the short-circuit force and the vibration intensity. Xu Jing et al. from Chongqing University [12] constructed an elaborate multi-physical field 2D simulation model to analyze winding stress distribution characteristics while accounting for harmonic current effects. The results demonstrated an exponential increase corresponding to harmonic current frequency elevation; furthermore, the relationship between maximum winding stress variations under different frequency harmonic currents and harmonic content follows a power function correlation. Xiong Jun et al. from Southwest Jiaotong University [13] established an electromagnetic–mechanical multi-physical-field model to examine internal transformer electromagnetic force distribution and its excitation effects on structural vibration. Their research indicated that increased load current amplifies vibration, with frequency components concentrated at the power supply frequency and its harmonics. As load rates escalate, vibration amplitudes across various frequencies intensify proportionally.

In experimental research, Jiang Peiyu et al. from Xi'an Jiaotong University [14] established a numerical simulation model to analyze harmonic-induced vibrations in transformer windings and cores. Through vibration testing experiments, they systematically examined the time-frequency domain vibration response characteristics corresponding to various valve-side windings configurations. Their findings revealed that harmonics amplify tank vibration amplitude and main frequency while broadening the frequency spectrum. Under harmonic currents conditions, the contribution of winding vibration to overall vibration exceeds that of the core, while windings connection methodology alters the primary vibration frequency distribution. Amir Esmaeili Nezhad et al. from the University of Tehran [15] employed the finite element method (FEM) and experimental modal analysis methods to investigate power transformer vibration. They classified vibration modes into three categories: core-related modes, winding-related modes, and coupling modes. Their investigation determined that transformers exhibit behavioral similarities to cylindrical structures with constrained ends under modal analysis, with winding vibration resembling cylindrical shell behavior—combining axial and radial vibrations. Excessive vibration can induce winding deformation, and most of the low-frequency modes are related to the core and coupling modes. Anthony John Moses et al. from Cardiff University [16] constructed an experimental system incorporating magnetization, vibration measurement, and noise measurement to evaluate single-phase and three-phase cores under different conditions. Their investigation revealed complex relationships between core surface vibration and noise generation, with out-of-plane vibration typically exceeding in-plane vibration. Additionally, the middle limb exhibits relatively high vibration, harmonics play crucial roles in noise and vibration generation, and joints constitute primary vibration sources. Although the existing research has explored transformer vibration characteristics extensively through theoretical modeling and experimental investigations across multiple dimensions, theoretical models predominantly focus on isolated factors, and experimental verification regarding harmonic current influence remains particularly scarce. There exists a significant research gap concerning transformer vibration characteristics under the combined influence of multiple characteristic harmonics.

Based on this, this paper focuses on the S20-M-100/10-0.4kV oil-immersed transformer as the subject of research and employs a methodology that integrates simulation analysis and experimental validation to examine the impact of varying harmonic currents. In Section 2, a multi-physical field simulation model of the electromagnetic–mechanical coupling of the transformer is constructed. Subsequently, in Section 3, the vibration acceleration response characteristics of the transformer under different harmonic currents are simulated and analyzed. In Section 4, a test platform for measuring the harmonic loading and vibration of the transformer is established. This platform serves to validate the accuracy of the simulation model and investigate the variation laws of the vibration characteristic quantities of the transformer under various harmonic currents. This research elucidates the intricate interaction mechanism between harmonic currents and the vibration characteristics of transformers, which is conducive to the development of more efficient operation and maintenance techniques. The research findings not only offer a solid theoretical foundation for the operation status monitoring and fault diagnosis of transformers in power systems, enabling the anticipation of potential fault risks in advance and ensuring the stability and reliability of power supply, but also play a pivotal role in the optimized design of transformers.

2. Theory and Modeling Method for Multi-Physical Field Simulation

2.1. Theory of Electromagnetic-Structural Force Field Coupling

Harmonic currents in the transformer windings generate a time-varying electromagnetic field, the dynamics of which are articulated by Maxwell's equations, as illustrated in Equation (1). This system of equations is constructed based on assumptions such as the homogeneity and isotropy of the medium, and the continuous and smooth variation in the electromagnetic field. The structure and electromagnetic properties of the winding influence the current density within them and the leakage magnetic field. This article integrates boundary conditions with the material properties to resolve Maxwell's equations and thereby determine the leakage magnetic field. Here, E represents the electric field strength; D signifies the electric displacement vector; H represents the magnetic field strength; B indicates the magnetic induction intensity; and J refers to the eddy current density. Meanwhile, the Maxwell stress tensor is an important physical quantity that describes the stress distribution in an electromagnetic field, and its expression is shown in Equation (2). This equation is used to describe the stress distribution of the electromagnetic field and relies on the assumptions of linear electromagnetic media and an inertial system. Here, c_0 is the permittivity of free space, m_0 is the permeability of free space, E_i and B_i are the components of the electric field strength and magnetic induction intensity in the i-direction, respectively, and d_{ij} is the Kronecker delta. Under the assumptions of a closed surface and the continuity of the tensor, the electromagnetic force on the winding is obtained through integral calculation [17], as shown in Equation (3). Here, F is the electromagnetic force vector, S is the surface of the winding, and dS is the surface element vector.

$$\begin{cases} \nabla \cdot \vec{D} = \rho \\ \nabla \cdot \vec{B} = 0 \\ \nabla \times \vec{E} = -\frac{\partial \vec{B}}{\partial t} \\ \nabla \times \vec{H} = \vec{J} + \frac{\partial \vec{D}}{\partial t} \end{cases} \quad (1)$$

$$\begin{cases} \overline{\overline{T}} = \begin{pmatrix} Txx & Txy & Txz \\ Tyx & Tyy & Tyz \\ Tzx & Tzy & Tzz \end{pmatrix} \\ T_{ij} = c_0(E_iE_j - \frac{1}{2}\delta_{ij}E^2) + \frac{1}{\mu_0}(B_iB_j - \frac{1}{2}\delta_{ij}B^2) \end{cases} \quad (2)$$

$$\vec{F} = \oint_S \overline{\overline{T}} \cdot d\vec{S} \quad (3)$$

The transformer windings are stimulated by electromagnetic forces due to harmonic currents, resulting in vibrations. Therefore, it is essential to establish a winding vibration displacement response model utilizing the elastodynamic equation and the mode superposition technique. In the Cartesian coordinate system, the elastodynamic equation can be expressed by Equation (4). The elastodynamic equation is premised on the continuum medium assumption, treating the transformer windings as continuous and homogeneous bodies, enabling their physical properties to be described by continuous functions. This paper presents the mode superposition method. Assume that the displacement response of the structure can be articulated as a linear combination of each mode order, as delineated in Equation (5). Incorporate this assumption into the elastodynamic equation, and utilizing the orthogonality of the modes, a set of decoupled equations about $q_n(t)$ can be obtained. Solving this set of decoupled equations allows for the determination of the vibration displacement response of the structure. In this context, r represents the material density, u_i denotes the displacement component in the i-direction, s_{ij} refers to the stress tensor component, f_i signifies the volume force component in the i-direction, f_n(x) represents the n-th order mode shape function that characterizes the deformation of the structure in the n-th order mode, and q_n(t) indicates the n-th order mode coordinate, reflecting the temporal variation in the n-th order mode. In addition, the ferromagnetic materials in the transformer core undergo magnetostriction under the action of a magnetic field, which has a nonnegligible impact on the vibration characteristics of the transformer [18]. The magnetostrictive strain can be expressed by Equation (6). Where l is the magnetostrictive strain, l_S is the saturation magnetostriction coefficient, B is the magnetic induction intensity, and B_S is the saturation magnetic induction intensity.

$$\rho \frac{\partial^2 u_i}{\partial t^2} = \sum_{j=1}^{3} \frac{\partial \sigma_{ij}}{\partial x_j} + f_i,\ i = 1,2,3 \quad (4)$$

$$u(x,t) = \sum_{n=1}^{\infty} \phi_n(x) q_n(t) \quad (5)$$

$$\lambda = \lambda_S \frac{3}{2} (\frac{B}{B_S})^2 - \frac{1}{2}\lambda_S \quad (6)$$

2.2. Development of the Three-Dimensional Model of the Oil-Immersed Transformer

This study primarily focuses on the S20-M-100/10-0.4kV oil-immersed transformer as the subject of investigation. Based on this, a 3D transformer model is established, as illustrated in Figure 1, with its key parameters detailed in Table 1. The transformer employs a three-phase three-limb core design, where both LV windings and HV coils of each phase are concentrically arranged on identical core limbs. The key simulation parameters for the electromagnetic–solid mechanics fields of this transformer are shown in Table 2.

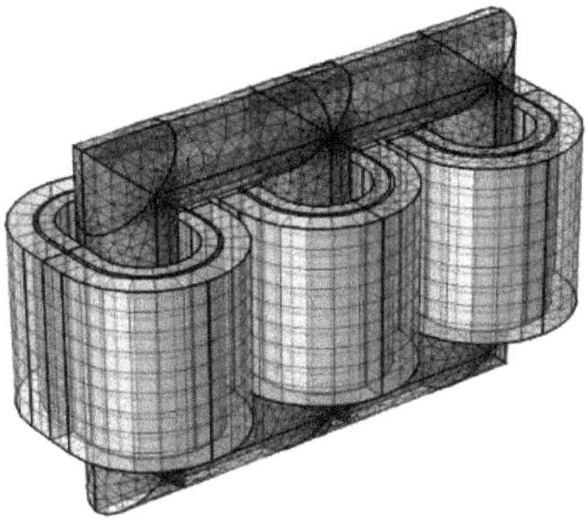

Figure 1. Three-dimensional simulation model of the S20-M-100/10-0.4kV transformer.

Table 1. Main parameters of the S20-M-100/10-0.4kV transformer.

Parameter	Rated Voltage (V)	Phase Current (A)	Number of Coil Turns
LV Winding	400	144.3	104
HV Winding	10000	3.3	2364

Table 2. Key simulation parameters of the S20 - M - 100/10 - 0.4kV transformer.

Parameter	Magnetic Permeability (H/m)	Electrical Conductivity (S/m)	Density (kg/m^3)	Elastic Modulus (Pa)	Thermal Conductivity (W/(m·K))	Poisson's Ratio
Core	5000	2×10^6	7500	1.8×10^{11}	45	0.24
Winding	/	5.8×10^7	8900	1.2×10^{11}	380	0.33

This paper simplifies the geometric model of the transformer to lower the computational demands of the simulation, considering the transformer's intricate construction, with simplification from the following aspect:

(1) The air gap between the laminated silicon steel sheets of the transformer core is minimal, and core vibration primarily arises from the magnetostrictive action. Consequently, the core is seen as a singular entity, and the interdependence among the silicon steel sheets is disregarded. Simultaneously, to more effectively simulate and analyze the comprehensive dynamic response of the transformer under harmonic stimulation, the HV and LV windings are each represented as a singular entity. The size of the core and the winding correspond accurately to the actual measurements of the transformer.

(2) Structures including clamping pieces, cushion blocks, and cooling devices are excluded. Fixed constraints are applied to the iron core and windings to simulate the fixing effects of the clamping pieces and buffer components on the iron core and windings.

During the simulation process, in view of the complex electromagnetic and structural characteristics of the transformer, we placed significant emphasis on optimizing the mesh generation. In the winding regions and at the core-winding interfaces where the electromagnetic fields and stresses vary drastically, triangular meshes with a minimum element size of 0.01 m were employed. This approach was adopted to ensure the accuracy of the simulation results and enhance the reliability of the simulation. This paper simulates and analyzes the distribution laws of the vibration characteristics of the transformer, utilizing a 3D simulation model. The analysis focuses on harmonic currents with frequencies of 250Hz, 350Hz, and 550Hz, with content levels ranging from 0% to 50%.

3. Simulation Analysis of Transformer Vibration Characteristics Under Harmonic Currents

3.1. Analysis of Transformer Vibration Characteristics Under Power Frequency Current

Under the influence of pure power-frequency sinusoidal current, the vibration acceleration of the transformer core attains its peak value at t = 0.012 s. Figure 2 illustrates the core deformation and vibration acceleration distribution. The core deformation distribution indicates that regions with significant core vibration are mostly located in the central core limb and at the interface of the yoke and core limb, with the average acceleration of the entire transformer core reaching 0.002 m/s^2.

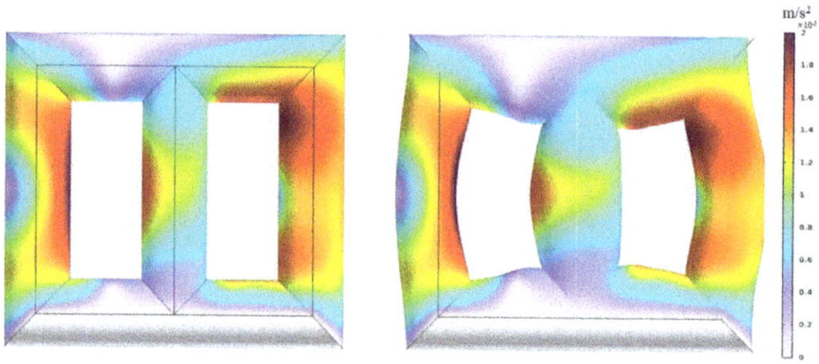

(a) Before the iron core is deformed (b) After the iron core is deformed

Figure 2. Nephogram depicting the acceleration of the transformer core under power frequency current.

The vibration distribution nephogram of the transformer's LV winding is illustrated in Figure 3a. The vibration of the LV winding of this transformer is concentrated at the center of the B-phase winding, with the vibration amplitude diminishing towards both ends. The vibration of the A-phase winding is predominantly localized at the end of the winding, whereas the vibration amplitude of the C-phase winding is comparatively minimal. The time-domain stress values at the inner and outer radii as well as different heights of the winding are solved through the finite element method. After collecting the acceleration data at each point, the data are aggregated and averaged to calculate that the overall average acceleration value of the low-voltage winding is 0.0175 m/s^2. The vibration distribution nephogram of the transformer's HV winding is depicted in Figure 3b.

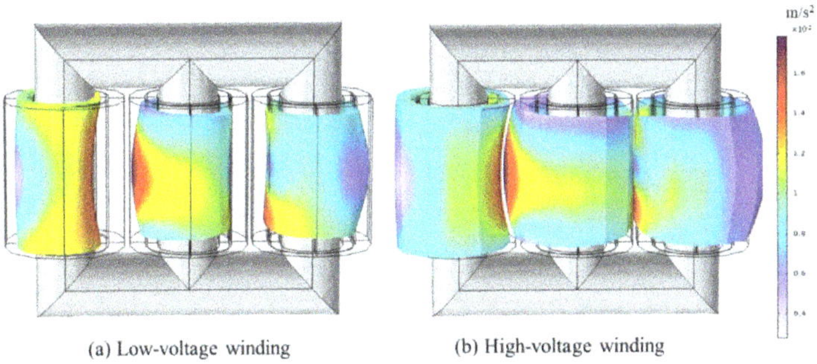

Figure 3. Nephogram depicting the vibration acceleration of the transformer windings under power frequency current.

The vibration of the high-voltage winding in this transformer model is mostly centered at the terminal of the A-phase winding, whereas the vibrations of the B-phase and C-phase winding are predominantly located in the central region of the winding. The average acceleration of the entire HV winding is 0.0125 m/s^2. Comprehensive analysis reveals that the vibration of the transformer is mostly attributed to the windings, with its acceleration amplitude being an order of magnitude more than that of the core. This paper focuses on the transformer windings as the primary subject of investigation to thoroughly examine their vibrational properties under varying operational conditions.

3.2. Analysis of Transformer Vibration Characteristics Under Different Harmonic Currents

Using the harmonic current with a composition of 25% as an illustration, as the harmonic current frequency escalates from 250 Hz to 550 Hz, the vibration distribution nephograms of the transformer windings are depicted in Figure 4. The acceleration distribution nephograms of the HV and LV windings reveal that as the harmonic current frequency increases, significant distribution disparities in acceleration manifest between the two windings, with the area of acceleration concentration migrating from the center to the ends of the windings. The trend of change is as indicated by the black arrow, and a stress peak appears at the end part, with the position of the peak shown as the red dot in the figure.

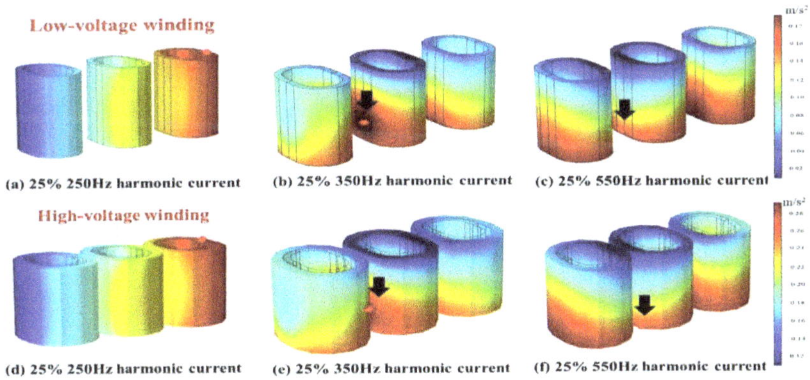

Figure 4. Distribution diagram of the vibration acceleration of the transformer windings under the harmonic current content of 25% at different frequencies of 250 Hz, 350 Hz, and 550 Hz.

The variations in the average values of the overall vibration accelerations for both windings at different harmonic current frequencies are presented in Figure 5. When the harmonic current frequency goes up in frequency from 220 Hz to 550 Hz, the average value of the acceleration of the LV winding rises from 0.157 m/s^2 to 0.179 m/s^2, reflecting an increase of 14.01%; the average acceleration of the HV winding ascends from 0.213 m/s^2 to 0.286 m/s^2, indicating a growth of 34.27%. Furthermore, Figure 5 illustrates that the augmentation of the vibration acceleration in the HV winding is more significant than that in the LV winding.

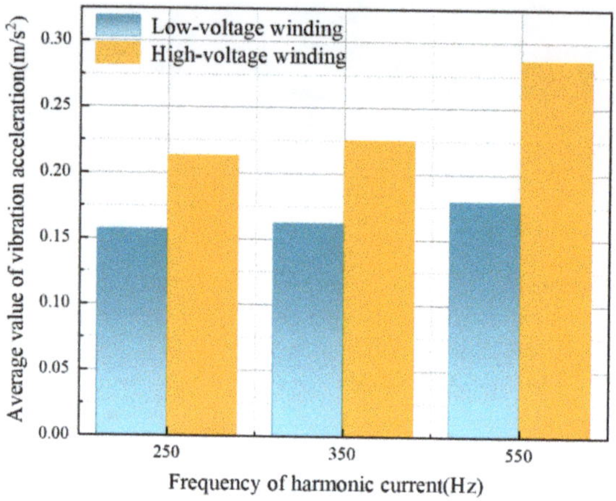

Figure 5. Graph of the variation in the average vibration acceleration of the transformer windings under the action of harmonic currents with frequencies of 250 Hz, 350 Hz, and 550 Hz.

Taking the harmonic current with a frequency of 250 Hz as an example, when the harmonic current content rises gradually from 0% to 50%, the vibration nephew-grams of the transformer windings are illustrated in Figure 6. The acceleration distribution nephograms of the HV and LV windings indicate that under pure power frequency conditions, the acceleration of the windings is predominantly concentrated in their central regions. The introduction of harmonic current excitation causes the region of concentrated acceleration to migrate from the middle part of the winding to the ends. The migration trend is depicted by the black arrow. Moreover, a stress peak emerges at the winding ends, with the peak location denoted by the red dot in the figure. Simultaneously, the average vibration acceleration increases in proportion to the growth of harmonic content.

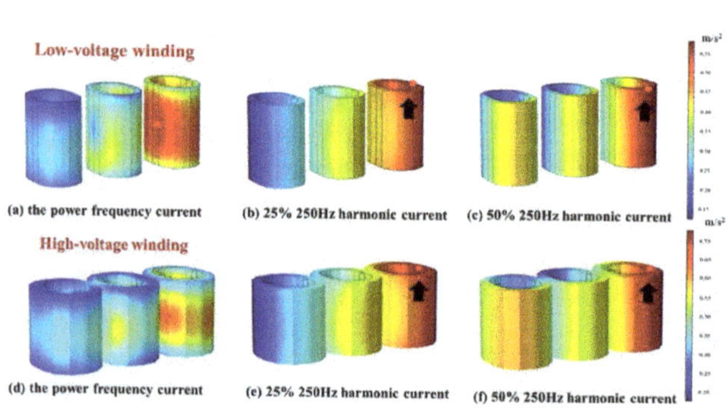

Figure 6. Diagrams of the vibration acceleration distribution of the LV winding and the HV winding under the harmonic current of 250 Hz with the contents of 25% and 50%. Note: The black arrow serves to demonstrate the migration trend of the acceleration concentration area, showing the direction from the middle part of the winding to the ends as the harmonic current excitation is applied. The red dot is used to pinpoint the location of the stress peak at the winding ends, which appears due to the change in acceleration distribution caused by the harmonic current. This makes it easier to visually identify the position with the highest stress value at the winding ends.

Figure 7 illustrates the variation curves of the overall average vibration values for both the HV and LV windings under varying harmonic current levels. When the quantity of harmonic current rises from 0% to 50%, the average acceleration of the LV winding ascends from 0.0205 m/s^2 to 0.543 m/s^2, and the average value of the acceleration of the HV winding surges from 0.0213 m/s^2 to 0.754 m/s^2. This study derived Equation (7) by fitting the average vibration acceleration values of the high-voltage (HV) and low-voltage (LV) windings presented in Figure 7. Specifically, the fitting formula for the HV winding is indicated by the yellow arrow, while that for the LV winding is denoted by the blue arrow. Based on the fitting results, as the harmonic current content increases, the average values of the vibration accelerations of the HV and LV windings rise corresponding with a power function law.

$$\begin{cases} a_l = 0.0205 + 2.09\alpha^2 \\ a_h = 0.0205 + 2.934\alpha^2 \end{cases} \quad (7)$$

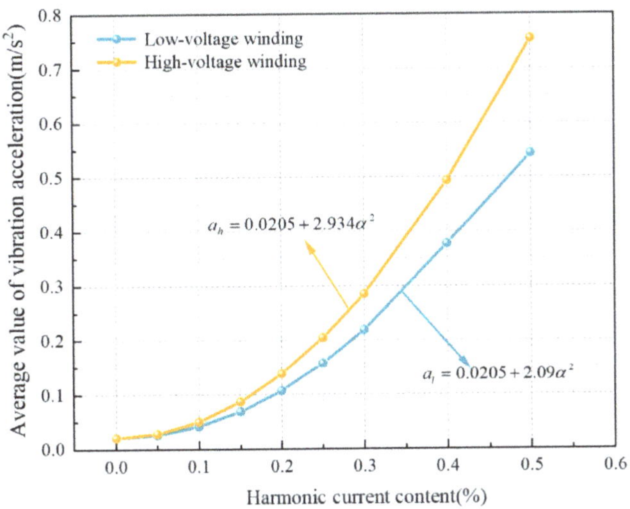

Figure 7. Variation curve of the average vibration acceleration of the transformer windings under the harmonic current content ranging from 0 to 50% at a frequency of 250 Hz.

4. Experimental Design and Validation

4.1. Vibration Testing Platform for the Oil-Immersed Transformer

This study presents the construction of a transformer harmonic loading and vibration monitoring platform based on the S20-M-100/10-0.4kV transformer. Figure 8 illustrates this test platform. The platform is composed of an adjustable harmonic power supply, an intermediate frequency transformer, a test transformer, sensors, and a vibration test platform. The adjustable harmonic power supply has a capacity of 180 kVA and can produce harmonic currents at up to 10 distinct frequencies, including the fundamental power frequency. The amplitude and phase of each frequency component are continually adjustable and thereafter are transmitted to the tested transformer via the intermediate frequency transformer. The specifications of the intermediate frequency transformer are presented in Table 3. Based on this, this paper conducts the winding vibration tests under the conditions that the frequencies of the harmonic currents are 250 Hz, 350 Hz, and 550 Hz, and the harmonic content ranges from 0 to 50%. This project conducts a vibration test on the winding of phase A of the transformer under no-load conditions. Considering that the winding vibration contains axial and radial components, four A25G01 piezoelectric acceleration sensors are strategically positioned around the phase A winding for vibration measurement. The positions of the four measuring points are indicated by the four colored numbers in Figure 8. Measuring point 1 is located in the middle of the oil tank on the side of the phase A winding of the transformer, measuring point 2 is situated in the middle of the phase A winding on the front of the transformer oil tank, measuring point 3 is positioned externally on the outside of the tank above the phase A winding of the transformer, and measuring point 4 is installed on the rear of the transformer oil tank and is symmetrically positioned relative to measuring point 2. The signal is ultimately gathered using the automated vibration signal collecting method. First, white noise was collected, and, subsequently, the white noise component was removed from the measured noise to effectively control the impact of background noise on the experimental results. The indoor temperature was maintained at 25 °C ± 1 °C to minimize the influence of temperature-induced changes in material properties. The relative humidity was controlled within the range of RH = 40 ± 5% to minimize the impact of humidity-related changes in material properties. Finally, the experimental setup was placed in a magnetic-shielding chamber, and all electrical circuits and components were shielded and grounded to strictly control the environmental conditions.

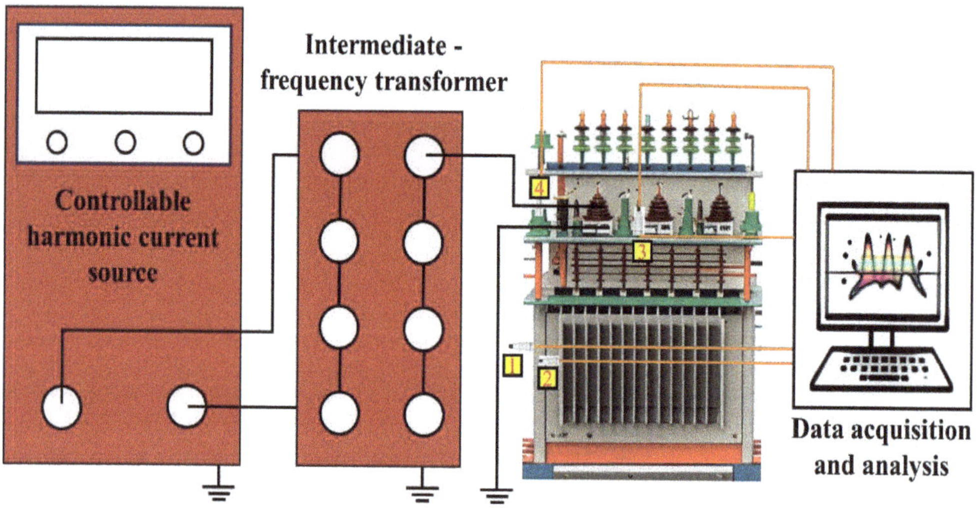

Figure 8. Platform for measuring transformer harmonic loading and vibration.

Table 3. Parameters of the intermediate frequency transformer.

Parameter Name	Value
Rated Capacity (kVA)	240
Rated Input Voltage (V)	340
Rated Input Current (A)	705.88
Rated Output Voltage (V)	12 × 600
Rated Output Current (A)	12 × 34
Operating Frequency (Hz)	50–2500

4.2. Comparative Analysis Between Simulation and Experimentation

This article conducts verification of multi-physical field simulations based on the vibration characteristic testing platform of the transformer. The field test is undertaken just in phase A. Given that measuring point 3 is positioned directly above the phase A winding, the test data from measuring point 3 are utilized in this paper for the verification and interpretation of the test and simulation data. Firstly, through simulation calculation, the location of maximum vibration intensity at the terminal of the phase A winding is precisely identified, and to investigate the changes in the vibration acceleration of this point under the influence of harmonic current, a point probe is placed at this location. Figure 9a displays the change curve of the point probe's vibration acceleration when power frequency current is applied. The data obtained from the point probe are compared with the acceleration recorded at measuring point 3 under the action of pure power frequency current, as illustrated in Figure 9. The trends in vibration change are generally consistent, with the acceleration peaks occurring simultaneously at t = 0.175 s, as shown by the red dot in Figure 9. The highest vibration value of the point probe is 0.00398 m/s^2, while the peak value of the measured vibration signal at measuring point 3 is 0.00149 m/s^2. These two values are of the same order of magnitude. There exists a time lag between the simulated and tested vibration peaks, and meanwhile, the vibration amplitude decreases. This can be attributed to the complex vibration transmission path involving the transformer oil. The viscosity of the oil impedes the propagation of vibration energy, not only causing a time lag in the occurrence of the vibration peaks but also dissipating energy, thus reducing the vibration amplitude. In addition, since the viscosity of the oil varies with temperature, changes in the oil temperature during the test will affect the transmission and dissipation rates of vibration. This further exacerbates the time lag of the vibration peaks and the decrease in the vibration amplitude, ultimately leading to the differences between the simulation and the measurement.

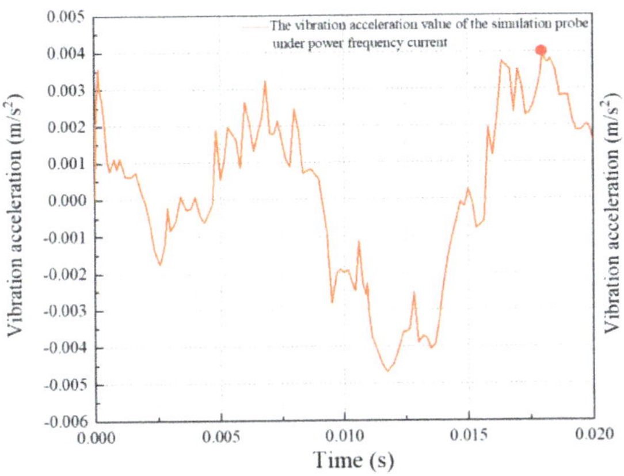

(a) The acceleration of the simulation probe

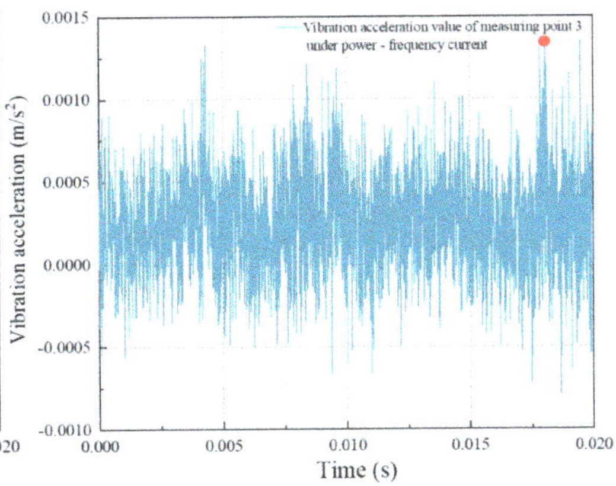

(b) The vibration acceleration measured at the measuring point 3

Figure 9. Comparative diagram of simulation probe acceleration and measured acceleration at test measuring point 3 under power frequency current.

Therefore, in this study, the effectiveness verification was carried out by comparing the peak amplitude increases of the vibration acceleration at measuring point 3 and that of the point probe under different harmonic content compositions. Given that the vibration amplitude of the winding under power frequency current is relatively small, a combination of 75% power frequency current and 25% 350 Hz harmonic current was selected as the benchmark for comparison. The comparison of the amplitude increases between the simulated and measured values is presented in Table 4. Through comparative analysis, the error between the vibration acceleration amplitude increase at measuring point 3 and that of the point probe on the winding was found to be within 5%. This indicates the effectiveness and reliability of the proposed multi-physical-field vibration simulation method.

Table 4. Comparison of amplitude increase for simulated and measured peak values of vibration acceleration across various harmonic compositions.

Current Composition	Simulated Value (m/s^2)	Increase Amplitude (%)	Test Value (m/s^2)	Increase Amplitude (%)	Error (%)
75% 50 Hz/25% 350 Hz	0.589	/	0.218	/	/
75% 50 Hz/25% 550 Hz	0.653	10.87%	0.239	9.63%	1.24%
75% 50 Hz/25% 250 Hz	0.664	12.73%	0.256	12.8%	0.07%
60% 50 Hz/40% 250 Hz	0.805	36.67%	0.298	36.7%	0.03%
50% 50 Hz/50% 250 Hz	0.902	53.14%	0.334	53.2%	0.06%

5. Analysis of the Characteristic Parameters of Transformer Vibration Under Harmonic Currents

5.1. Analysis of Transformer Vibration Properties at Various Harmonic Current Frequencies

This paper comprehensively analyzes the vibration characteristics of the transformer, grounded on the principle of maintaining a constant total effective value of the current, under various combinations of harmonic current contents and frequencies. Figure 10 illustrates the vibration acceleration distributions of the transformer when the harmonic currents with frequencies of 250 Hz, 350 Hz, and 550 Hz, with harmonic current contents of 25%, 30%, and 40%, are superimposed on the power frequency current. Figure 10 illustrates that as the contents of the superimposed multiple harmonic currents increase, the peak of the transformer vibration acceleration also rises. When the content of the multiple harmonic currents is increased from 25% to 40%, the peak vibration acceleration at measuring point 3 escalates from 0.183 m/s^2 to 0.325 m/s^2, reflecting an increase rate of 77.6%. This clearly indicates that the overall vibration intensity of the transformer has been significantly enhanced.

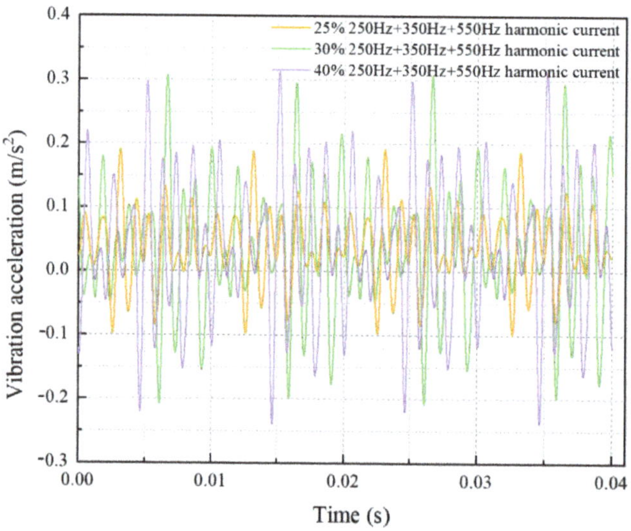

Figure 10. Variation in vibration acceleration under the action of harmonic currents at different frequencies.

This study analyzes the statistical properties of different harmonic frequencies in both the time and frequency domains and defines six key characteristics for quantitative analysis. In the time domain, the root mean square value reflects the vibration energy, the absolute average value shows the degree to which the signal deviates from zero, and the peak—to—peak value represents the maximum vibration range within one period. In the frequency domain, the mean frequency determines the average distribution position of energy, the random coefficient quantifies the randomness of the signal, and the centroid frequency describes the energy—related frequency characteristics. Their calculation formulas are shown in Table 5. In the equations: x_i represents the time-series vibration signal, whereas N means the total sample points. The frequency-domain signal subsequent to the FFT (Fast Fourier Transform) is defined as X_i, and the frequency value corresponding to the i-th frequency-domain point is defined as f_i.

Table 5. Vibration Characteristics in Time Domain/Frequency Domain.

	Feature Name	Calculation Formula
Time Domain Feature	Root Mean Square Value	$X_{RMS} = \sqrt{\frac{1}{N}\sum_{i=1}^{N} x_i^2}$
	Absolute Average Value	$X_{am} = \frac{1}{N}\sum_{i=1}^{N} \lvert x_i \rvert$
	Peak-to-Peak Value	$X_{pp} = \max(x_i) - \min(x_i)$
Frequency Domain Feature	Mean Frequency	$MF = \sum_{i=1}^{N} X_i / N$
	Random Coefficient	$RC = \sum_{i=1}^{N} (X_i - MF)^4 / \sum_{i=1}^{N} X_i \times VF^2$
	Centroid Frequency	$F_c = \sum_{i=1}^{N} f_i \times X_i / \sum_{i=1}^{N} X_i$

Table 6 illustrates the modification of the transformer's vibration characteristic values in response to harmonics of differing frequencies. When the harmonic frequency increases from 250 Hz to 350 Hz and subsequently to 550 Hz, both the quadratic mean value and the absolute average value exhibit a progressively increasing trend. Among them, the properties affected by the 550 Hz harmonic stand out the most. The influence of the 250 Hz harmonic is the next most significant, while the influence of the 350 Hz harmonic is relatively small. The peak-to-peak value indicates that the 250 Hz harmonic exhibits the smallest action, while the 550 Hz harmonic is the largest. The random coefficient characteristics under the 350 Hz and 550 Hz harmonics are nearly identical, but the random coefficient value under the 250 Hz harmonic is relatively large, indicating a heightened dispersion of the signal spectrum at this frequency. The centroid frequency first rises and then decreases, with the maximum centroid frequency occurring under the influence of the 350 Hz harmonic, during which the high-frequency components in the spectrum are relatively large at this time.

Table 6. Vibration Characteristic Values under the Action of Harmonics of Different Frequencies.

	250 Hz	350 Hz	550 Hz
RMS Value	0.085	0.074	0.091
Absolute Average Value	0.072	0.057158443	0.097
Peak-to-Peak Value	0.374176	0.397119	0.498
Mean Frequency	8.18×10^{-5}	7.51×10^{-5}	9.14×10^{-5}
Random Coefficient	1390.010	1203.685	1204.277
Centroid Frequency	2422.362	2569.912	2382.367

5.2. Analysis of Transformer Vibration Characteristics Under Different Harmonic Current Contents

This research examines the variance in the oscillatory behaviors of the transformer under power frequency currents with varying compositions. The experiment quantifies the variations in vibration acceleration at measuring point 3 under power frequency currents with contents of 15%, 45%, and 70%, respectively, as seen in Figure 11. Figure 11 illustrates that under power frequency current, the amplitude of its vibration acceleration minimal, with the order of magnitude at 10^{-3}.

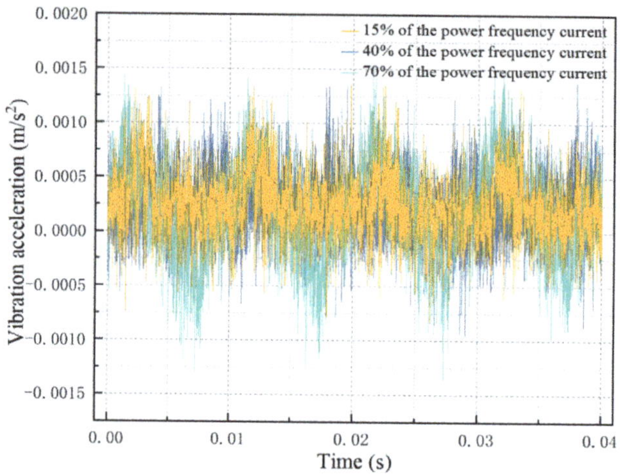

Figure 11. Variations in the Transformer's Vibration Acceleration under Power Frequency Currents with Different Contents.

As the quantity of the fundamental—frequency current rises, the intensity of the vibration acceleration correspondingly strengthens. As the power frequency current increases from 15% to 70%, the peak of the acceleration at measuring point 3 increases from 0.00128 m/s^2 to 0.00162 m/s^2, reflecting an increase of 26.56%. However, the overall vibration under the pure power frequency current is minimal, and the noise at the test site is also rather low. This indicates that the influence of the unaltered, basic power—frequency current on the transformer's vibration is relatively restricted, leading to both low vibration amplitude and decreased noise levels.

This work examines the vibration characteristics of the transformer by analyzing pure power frequency current superimposed with 250 Hz harmonic current as a research sample under varying harmonic current. As the proportion of the 250 Hz harmonic current increases from 25% to 40%, Figure 12 shows the pattern of vibration acceleration values at measurement point 3. Figure 12 illustrates that Figure 12 shows that the magnitude of the transformer's vibration acceleration grows remarkably, attaining an order of magnitude of 10^{-1}. Its amplitude exceeds the acceleration by two orders of magnitude when only fundamental—frequency current is present. Moreover, as the intensity of the combined harmonic current rises, the amplitude of the vibration acceleration correspondingly intensifies. The maximum vibration acceleration at measuring point 3 rises from 0.135 m/s^2 to 0.334 m/s^2, reflecting an increase of 147.41%. Upon superimposing the harmonic current, the overall oscillatory motion of the transformer is markedly more intense than that under the power frequency current.

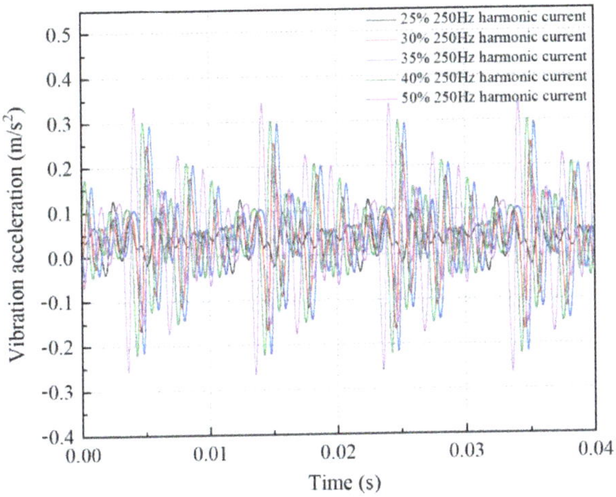

Figure 12. Variations in the Transformer's Vibration Acceleration under the Action of 250 Hz Currents.

Table 7 illustrates the variations in the transformer's vibration characteristic values under the action of harmonics with different contents. As the harmonic content increases from 25% to 50%, all six vibration characteristic values exhibit an upward trend. The changes in four characteristics—root mean square value, absolute average value, peak-to-peak value, and mean frequency—are substantial as the harmonic content increases from 25% to 30%. When the harmonic content reaches 50%, the centroid frequency escalates from 2000 Hz to beyond 4000 Hz, indicating a significant alteration. The random coefficient nearly exhibits a linear increase with the rise in harmonic content, accompanied by a progressive intensification of dispersion and an incremental rise in high-frequency components.

Table 7. Vibration characteristic values under the action of harmonics of different contents.

	25%	30%	35%	40%	50%
RMS Value	0.031	0.074	0.089	0.091	0.107
Absolute Average Value	0.023	0.056	0.068	0.070	0.082
Peak-to-Peak Value	0.167	0.412	0.494	0.498	0.593
Mean Frequency	3.81×10^{-7}	2.18×10^{-6}	3.16×10^{-6}	3.31×10^{-6}	4.6×10^{-6}
Random Coefficient	914.628	1091.872	1180.066	1204.277	1244.019
Centroid Frequency	2152.368	2382.367	2391.157	2586.288	4571.565

6. Conclusions

This study focuses on the S20-M-100/10-0.4kV oil-immersed transformer, analyzing the effects of various harmonic currents on its vibration characteristics. By utilizing an "electromagnetic structural force" simulation of multi-field coupling, the transformer's vibration acceleration distribution under various harmonic currents is derived. At power frequency, the winding vibration prevails, with its acceleration amplitude exceeding that of the iron core by one order of magnitude. As the harmonic current content increases, the average vibration accelerations of the windings vary according to a power function. With increasing frequency, the acceleration distributions of the two windings diverge markedly, with the concentrated area relocating from the center to the terminus of the winding, and the acceleration increase in the high-voltage winding becoming more pronounced. To validate the simulation, a platform for harmonic loading and vibration measurement is constructed. The discrepancy between the acceleration increments at the measured points and the simulation probe points is within 5%, demonstrating the effectiveness and dependability of the multi-physical field vibration simulation methodology.

The transformer vibration experiment conducted under harmonic currents reveals the variation laws of vibration characteristics at various harmonic current contents and frequencies. The vibration acceleration increases significantly as the harmonic current frequency rises from 250 Hz to 550 Hz, reaching a maximum of 0.298 m/s^2 at 550 Hz. When the content increases from 25% to 40%, the peak value surges by 77.6%, and vibration characteristic values, such as the root mean square value, show a significant increase, indicating that the augmentation in harmonic content intensifies the vibration. Compared to the pure power frequency current, the amplitude of vibration acceleration increases by two orders of magnitude upon the superposition of harmonic currents, and the random coefficient exhibits a linear increase with the harmonic content, suggesting that harmonic currents substantially augment the spectrum dispersion and high-frequency components of the vibration signal.

Furthermore, transformer oil characteristics, environmental conditions, and aging significantly impact transformer vibration behavior. Oil viscosity affects winding-core damping, with high-viscosity oil-enhancing damping, reducing vibrations, and altering the transmission path. The dielectric constant influences the electric field, thus affecting electromagnetic forces and vibration. Environmentally, temperature changes winding and core material properties, and humidity can create extra vibration sources due to insulation degradation. Transformer aging, like winding insulation aging and core magnetic property changes, weakens the structure and modifies electromagnetic forces, affecting vibration. In future research, integrating these factors is crucial for better transformer condition monitoring, fault prediction, and power system reliability.

Author Contributions: Conceptualization, D.Z.; methodology, W.D. and J.H.; software, H.S.; validation, J.X.; formal analysis, G.Q.; investigation, J.H.; writing—original draft preparation, G.Q.; writing—review and editing, W.D.; supervision, H.S. All authors have read and agreed to the published version of the manuscript.

Funding: This research was funded by the Science and Technology Project of China Southern Power Grid Co., Ltd. (YNKJXM20210186). "Research on insulation life prediction of key components of equipment under the effect of harmonic coil in power system and the transformer trial scale test device production with typical defects" of the Electric Power Research Institute of Yunnan Power Grid Co, Ltd. We express our gratitude to Xi'an Jiaotong University for supplying the test platform for harmonic vibration research, which is crucial to the success of our work.

Data Availability Statement: Data are contained within the article.

Conflicts of Interest: Authors Dexu Zou, Weiju Dai, Guochao Qian and Haoruo Sun were employed by the company Electric Power Research Institute of Yunnan Power Grid Co., Ltd. The remaining authors declare that the research was conducted in the absence of any commercial or financial relationships that could be construed as a potential conflict of interest.

References

1. Alawasa, K.M.; Al-Badi, A.H. Investigation and Analysis of the Power Quality in an Academic Institution's Electrical Distribution System. *Energies* **2024**, *17*, 3998. [CrossRef]
2. Arranz-Gimon, A.; Zorita-Lamadrid, A.; Morinigo-Sotelo, D.; Duque-Perez, O. A review of total harmonic distortion factors for the measurement of harmonic and interharmonic pollution in modern power systems. *Energies* **2021**, *14*, 6467. [CrossRef]
3. Taghvaie, A.; Warnakulasuriya, T.; Kumar, D.; Zare, F.; Sharma, R.; Vilathgamuwa, D.M. A comprehensive review of harmonic issues and estimation techniques in power system networks based on traditional and artificial intelligence/machine learning. *IEEE Access* **2023**, *11*, 31417–31442. [CrossRef]
4. Qian, M.; Yin, F.; Yuan, Y.; Jiang, P.; Wang, L.; Zhao, L. Vibration characteristics of ± 800 kV converter transformers part I: Under no-load conditions. *Int. J. Electr. Power Energy Syst.* **2024**, *159*, 110060. [CrossRef]

5. Qian, M.; Yin, F.; Yuan, Y.; Jiang, P.; Wang, L.; Zhao, L. Vibration characteristics of ± 800 kV converter transformers part II: Under load conditions. *Int. J. Electr. Power Energy Syst.* **2024**, *159*, 110026. [CrossRef]
6. Liu, C.; Hao, J.; Liao, R.; Yang, F.; Li, W.; Li, Z. High proportion and large value harmonic current influence on the magnetic field, loss and temperature distribution for ultrahigh voltage converter transformer. *IET Electr. Power Appl.* **2024**, *18*, 208–225. [CrossRef]
7. Dao, T.; Phung, B.T. Effects of voltage harmonic on losses and temperature rise in distribution transformers. *IET Gener. Transm. Distrib.* **2018**, *12*, 347–354. [CrossRef]
8. Yaghoobi, J.; Alduraibi, A.; Martin, D.; Zare, F.; Eghbal, D.; Memisevic, R. Impact of high-frequency harmonics (0–9 kHz) generated by grid-connected inverters on distribution transformers. *Int. J. Electr. Power Energy Syst.* **2020**, *122*, 106177. [CrossRef]
9. Kumar, I.; Bakshi, A. Effects of axial sticks and proof stress of conductor material on the buckling strength of transformer inner winding. *IEEE Trans. Power Deliv.* **2022**, *11*, 5465–5468. [CrossRef]
10. Hu, Z.; Han, Y.; Zalhaf, A.S.; Zhou, S.; Zhao, E.; Yang, P. Harmonic sources modeling and characterization in modern power systems: A comprehensive overview. *Electr. Power Syst. Res.* **2023**, *218*, 109234. [CrossRef]
11. Jiang, M.; Sha, Z.; Zhao, Y.; Chen, W.; Zhang, Q.; Wen, T. Strongly Coupled Circuit-Magnetic-Structural Effect on Winding Vibration Process of Power Transformer. *IEEJ Trans. Electr. Electron. Eng.* **2024**, *19*, 1147–1155. [CrossRef]
12. Xu, J.; Hao, J.; Zhang, N.; Liao, R.; Feng, Y.; Liao, W.; Cheng, H. Simulation study on converter transformer windings stress characteristics under harmonic current and temperature rise effect. *Int. J. Electr. Power Energy Syst.* **2025**, *165*, 110505. [CrossRef]
13. Xiong, J.; Luo, Y.; Ren., S.; Lu, S.; Pei, X.; Zha, X.; Zhou, W. Simulation and Experimental Study on Vibration and Noise of Power Transformer Under Load. In Proceedings of the 2023 4th International Symposium on Insulation and Discharge Computation for Power Equipment (IDCOMPU 2023), Wuhan, China, 26–28 May 2023. [CrossRef]
14. Jiang, P.; Zhang, Z.; Dong, Z.; Yang, Y.; Pan, Z.; Deng, J. Axial and radial electromagnetic-vibration characteristics of converter transformer windings under current harmonics. *High Volt.* **2023**, *8*, 477–491. [CrossRef]
15. Amir Esmaeili, M.; Hamed Nezhad, S. Analysis of Power Transformer Vibration Modes. *J. Power Eng.* **2024**, *10*, 45–55.
16. Anthony, J.; Philip, A.; Teeraphon, P. Localized Surface Vibration and Acoustic Noise Emitted from Laboratory-Scale Transformer Cores Assembled From Grain-Oriented Electrical Steel. *IEEE Trans. Magn.* **2016**, *52*, 1–15. [CrossRef]
17. Jin, M.; Chen, W.; Zhao, Y.; Wen, T.; Wu, J.; Wu, X.; Zhang, Q. Coupled magnetic-structural modeling of power transformer for axial vibration analysis under short-circuit condition. *IEEE Trans. Magn.* **2022**, *58*, 1–9. [CrossRef]
18. Sun, C.; Zhu, L.; Qian, G.; Dai, W.; Hong, Z.; Wang, S.; Xu, J.; Wang, J. Study on the Effect of Harmonic Current on the Vibration Characteristics of Oil-Immersed Transformer. In Proceedings of the 2023 IEEE International Conference on Power Science and Technology (ICPST), Kunming, China, 5–7 May 2023. [CrossRef]

energies

MDPI

Article

Risk Assessment Model for Converter Transformers Based on Entropy-Weight Analytic Hierarchy Process

Guochao Qian [1], Weiju Dai [1], Dexu Zou [1], Haoruo Sun [1], Hanting Zhang [2] and Jian Hao [2,*]

1 Electric Power Research Institute Yunnan Power Grid Co., Ltd., Kunming 650217, China; qianguochao_yndky@163.com (G.Q.); 15911530479@163.com (W.D.); zdxdb@126.com (D.Z.); s_hr1997@163.com (H.S.)

2 State Key Laboratory of Power Transmission Equipment Technology, School of Electrical Engineering, Chongqing University, Chongqing 400044, China; zht689986@163.com

* Correspondence: cquhaojian@126.com

Abstract: As a critical component in voltage–current conversion and power transmission within HVDC systems, the risk assessment of converter transformers plays a significant role in ensuring their operational safety and enhancing the reliability of the power supply. To address the issues of the incomplete characteristic parameters and limited fault data used for model training in existing transformer evaluation models, this paper develops a risk assessment model for converter transformers based on the entropy-weighted analytic hierarchy process (AHP). Firstly, in accordance with relevant standards in the power industry and existing experimental research, 14 'electrical–thermal–mechanical' multi-dimensional characteristic parameters, including partial discharge, dissolved gases in oil, and hot spot temperature rise, are selected to effectively reflect the risk state of converter transformers. The risk state is then categorized into four levels. Next, the AHP, which uses a subjective weighting method, is combined with the entropy-weight method, an objective weighting approach, to construct the risk assessment model for converter transformers based on the entropy-weighted AHP. Finally, the effectiveness of the model is validated through four case studies of converter transformers. The results indicate that the risk assessment model proposed in this paper can accurately and effectively reflect the risk state of transformers at different levels, providing valuable guidance for the development of maintenance strategies for converter transformers.

Keywords: converter transformer; analytic hierarchy process; entropy-weight method; risk assessment

1. Introduction

As a core component in the voltage–current conversion and power transmission of HVDC systems, maintaining the converter transformer in good health is crucial for ensuring the safe operation of the power system and a stable power supply [1–3]. During actual operation, the insulation components of the valve-side winding, outlet device, and internal wiring experience irreversible degradation due to factors such as electrical stress, thermal stress, and mechanical stress [4–6]. A failure of the converter transformer can severely disrupt the normal operation of the power system and lead to significant economic losses [7]. It is of great importance to extend the transformer's service life and ensure the safety of the power grid by scientifically acquiring operational parameters and comprehensively evaluating its risk level, detecting anomalies in time, and implementing appropriate maintenance measures [8–10].

Currently, maintenance methods in China for transformer risk states are still based on a combination of scheduled and post-maintenance. However, there is a lack of real-time monitoring of the transformer's risk state, leading to issues such as 'under-repair' or 'over-repair', which result in insufficient transformer reliability or the excessive waste of human and material resources [11,12]. In recent years, with the rise and development of wireless sensor technology, condition-based maintenance methods for transformers, based on condition monitoring technology, are gradually being implemented in power systems. By monitoring characteristic parameters such as dissolved gas in oil, top oil temperature, and vibration signals, the transformer's state can be assessed in real time, preventing failures and maintenance risks due to improper maintenance [13–16]. As a necessary prerequisite for condition-based maintenance, the condition assessment of the converter transformer enables the determination of its operational capacity by evaluating its risk state, thereby helping to formulate a reasonable maintenance plan to ensure its stable operation.

Due to its unique operating characteristics and numerous indicators, the converter transformer requires the development of a comprehensive and scientific index system to ensure accurate risk assessment [17]. As a typical oil-immersed transformer, the converter transformer has been the subject of extensive research, both domestically and internationally, on oil-immersed transformer state evaluation methods. Jiang Tao et al. used characteristic parameters, such as transformer load coefficient, dissolved gas in oil, and partial discharge, combined with health index theory, to construct a three-level health evaluation model based on ontological data, service data, and correction factors, achieving a scientific evaluation of the power transformer's insulation state [18]. Zou Yang et al. extracted characteristic parameters such as the polarization spectrum peak voltage, insulation resistance, and geometric capacitance, employing a combination weighting method that integrated gray correlation analysis and the improved analytic hierarchy process (AHP). They used the cloud model's atomization characteristics to account for the fuzziness and randomness of the classification boundaries in the evaluation index, constructing a model for evaluating the transformer oil–paper insulation state, addressing the issue of sensitivity differences in insulation state reflection, and avoiding data loss [19]. Zhou Jian fused state characteristic parameters such as dissolved gas, insulation resistance, and micro-water in oil to build a transformer state evaluation model based on a support vector classification mechanism, enhancing the evaluation accuracy [20]. Based on dissolved gas in oil data, Ming Tao et al. measured the relative proximity of data in different transformer states according to the point density criterion and big data clustering principles, constructing a transformer state evaluation model based on a time series, which could effectively assess transformer state changes over time [21]. P. Zhang et al., using dissolved gas data, constructed the phase space of the model using the delayed coordinate method and calculated phase space parameters through the C-C method. By analyzing the convergence of the correlation integral, they developed an optimal data-mining-based transformer state evaluation model that improved the evaluation speed [22]. T. Manoj et al. combined AHP and gray relational analysis to construct a transformer health status assessment model based on parameters such as dissolved gas in oil, oleic acid value, and breakdown voltage, with high accuracy in the results [10]. Z. Hamed et al. developed transformer health status assessment models using multiple linear regression, adaptive neural networks, and artificial neural networks, based on parameters such as dissolved gas in oil, degree of polymerization, and dielectric loss factor. Actual case verification demonstrated the high evaluation accuracy of all three models [23]. In general, existing evaluation models can provide more accurate assessments of oil-immersed transformers, but the number of characteristic parameters used is limited, and comprehensive evaluation remains difficult with inherent uncertainty.

Computationally intensive machine learning approaches fail to meet real-time monitoring requirements. Mathematically driven models over-rely on expert knowledge and are prone to misjudgment. Transformer health status assessment methods based on data-driven approaches and machine learning require sufficient training data. However, due to the complex structure, harsh onsite environment, and strong magnetic fields of converter transformers, it is challenging to collect data from different states, leading to lower accuracy. Furthermore, due to factors such as sensor aging, transformer monitoring systems may exhibit data distribution shifts during prolonged operation, requiring continuous model updates to maintain accuracy.

To address the above issues, this paper integrates the 'electrical–thermal–mechanical' multi-dimensional characteristic parameters of electricity, heat, and mechanics, proposing a risk assessment model for converter transformers based on the entropy-weight AHP. Firstly, 14 characteristic parameters that effectively reflect the risk state of the transformer are selected based on relevant standards and experimental studies, and the risk states of the transformer are classified. Then, using the AHP, a judgment matrix is constructed to determine the subjective weights of the index layer. Following the entropy-weight method, the risk decision matrix is developed, and the evaluation model is objectively weighted. Finally, by combining the AHP with the entropy-weight method, the comprehensive weight of the evaluation model is obtained, revealing the internal correlations and importance rankings of the state parameters of the converter transformer. Based on the established evaluation model, the risk state of the transformer is assessed.

2. Model Development and Validation Framework

Due to the varying effects of different 'electric–heat–machine' characteristic parameters on reflecting the risk state of converter transformers, it is necessary to assign weights to these parameters to ensure the rationality of the risk assessment model. Based on the weight distribution and assignment methods, existing weighting techniques can be categorized into subjective and objective methods. Common subjective methods include the AHP [24,25], expert evaluation, and Delphi method. These techniques take into account the evolution mechanism and law of each characteristic parameter; however, these methods heavily depend on expert experience, which can lead to issues of contingency and poor stability [26,27]. On the other hand, objective weighting methods include the entropy-weight method [28,29], coefficient of variation method, and principal component analysis. These methods have a solid data and theoretical foundation; however, they rely entirely on existing statistical data and are prone to misjudging index weights [30,31].

To address these issues, this study first establishes classification thresholds for different risk levels of the 'electrical–thermal–mechanical' multi-dimensional characteristic parameters in converter transformers based on relevant standards from the power industry and existing experimental research. Subsequently, the analytic hierarchy process is employed to construct a hierarchical judgment matrix based on the relative importance of these multi-dimensional parameters, deriving the maximum eigenvalue λ_{max} and the corresponding normalized eigenvector ω_j'. Concurrently, using the entropy-weight method and the classification thresholds for different risk levels of multi-dimensional characteristic parameters, a risk decision matrix is developed to obtain the normalized deviation matrix μ_j. Then, the AHP and entropy-weight method are integrated to create a risk assessment matrix model for converter transformers. Finally, the model's effectiveness is validated using actual characteristic parameter monitoring data from converter transformers. The model development and validation framework are illustrated in Figure 1.

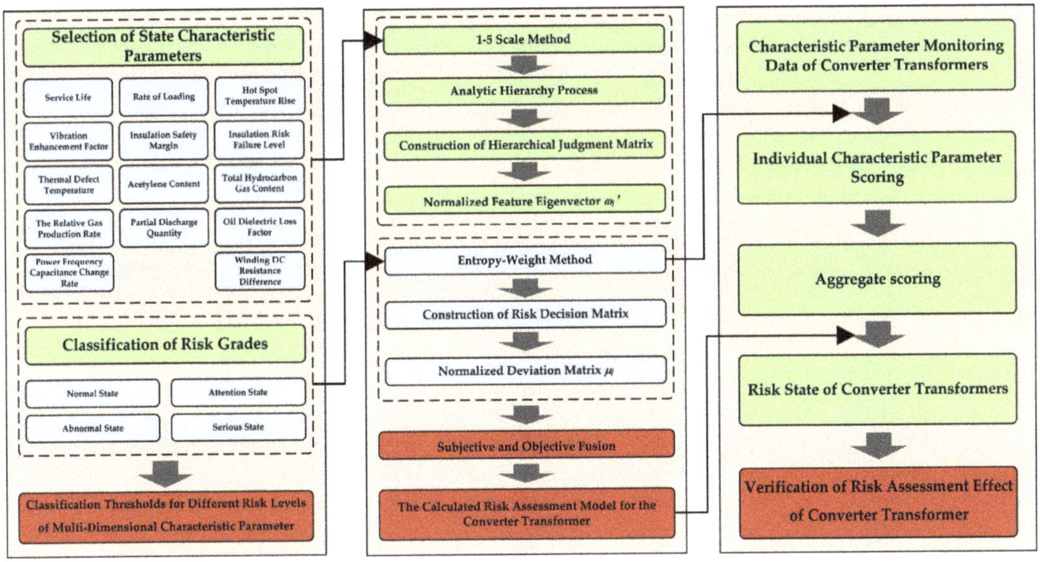

Figure 1. Model development and validation framework.

This paper combines the AHP from the subjective weighting method with the entropy-weight method from the objective weighting method to establish a converter transformer risk assessment model, integrating both expert experience and existing test data to comprehensively assess the risk state of transformers and improve the reliability and accuracy of the model.

3. Selection of State Characteristic Parameters and Classification of Risk Grades

The risk state of the converter transformer is influenced by electrical, thermal, mechanical, and other factors, each of which exerts a different impact on the risk state of the transformer. To conduct a more scientific and comprehensive analysis of the risk level of the converter transformer, this study selects 14 'electrical–thermal–mechanical' multi-dimensional characteristic parameters, including partial discharge, power frequency capacitance change rate, dissolved gas in oil, relative gas production rate, hot spot temperature rise, and thermal defect temperature, which effectively reflect the risk state of the converter transformer. These parameters are based on relevant standards from the power industry and existing experimental research [32–35]. The service life and rate of loading effectively indicate the aging degree and current operational intensity of converter transformers. Hot spot temperature rise and thermal defect temperature serve as critical indicators for identifying overheating faults and cooling system failures. The vibration enhancement factor reflects the mechanical stability of the transformer structure. Insulation safety margin and insulation risk failure level provide comprehensive evaluations of the insulation system's endurance capabilities. Acetylene content, total hydrocarbon gas content, and the relative gas generation rate are essential parameters for detecting faults and assessing oil–paper insulation degradation. Partial discharge quantity is a vital indicator for internal insulation defects. The oil dielectric loss factor and power frequency capacitance change rate directly reflect the deterioration of the insulation medium's performance. Winding DC resistance differences can reveal poor electrical contact and winding strand breakage issues. The selected parameters comprehensively cover the characteristics of various indices of the converter transformer under different risk states, making them highly significant for the accurate and thorough assessment of the risk state.

The proposed risk assessment model for converter transformers requires only four operational parameters: service life, vibration enhancement factor, insulation safety margin, and insulation failure risk level, which are to be obtained through testing and threshold classification. The remaining parameters are derived from industry standards, with their risk thresholds determined based on practical transformer maintenance experience. The model minimizes data requirements, significantly reducing acquisition costs.

Given that the influence of different electrical, thermal, and mechanical characteristic parameters on the risk state of the converter transformer varies, and their sensitivity in diagnosing different defective transformers differs, this paper performs a comprehensive evaluation of the weighting values of these parameters. The evaluation is conducted in accordance with the "Guidelines for State Evaluation of Oil-immersed Transformers (Reactors)" [32]. The risk state of the transformer is classified into four levels: normal, attention, abnormal, and serious. The threshold values for the multi-dimensional characteristic parameters of the converter transformer under different risk levels are provided in Table 1. The specified thresholds for characteristic parameters used in transformer risk assessment differ between domestic [32–35] and international standards [36–38]. Without modifying the classification thresholds for different risk levels of multi-dimensional characteristic parameters, the model cannot directly evaluate the risk state of converter transformers in different countries. Therefore, to enable the assessment of converter transformer risk state across various countries, it is necessary to redefine the classification thresholds for different risk levels of multi-dimensional characteristic parameters according to the relevant standards.

Table 1. Classification thresholds for different risk levels of multi-dimensional characteristic parameters.

Characteristic Parameter Index	Normal State	Attention State	Abnormal State	Serious State
Service Life K_1	700	500	250	150
Rate of Loading K_2	1	1.1	1.3	1.5
Hot Spot Temperature Rise K_3/K	60	68	78	90
Vibration Enhancement Factor K_4	1	1.2	1.4	1.5
Insulation Safety Margin K_5	0.8	0.7	0.6	0.5
Insulation Risk Failure Level K_6	1	2	3	4
Thermal Defect Temperature K_7/°C	150	300	500	700
Acetylene Content K_8/μL/L	2	5	20	50
Total Hydrocarbon Gas Content K_9/μL/L	100	150	350	500
The Relative Gas Production Rate K_{10}/%/month	7	10	20	30
Partial Discharge Quantity K_{11}/pC	100	250	700	1000
Oil Dielectric Loss Factor tanδ K_{12}/%	0.5	2	3	4
Power Frequency Capacitance Change Rate K_{13}/%	2	3	4	5
Winding DC Resistance Difference K_{14}/%	1	2	4	5

4. The Converter Transformer Risk Assessment Model Based on the Entropy-Weight AHP

4.1. Construction of Hierarchical Judgment Matrix Based on AHP

Firstly, the hierarchical structure of the evaluation system is established. When analyzing objects with multiple characteristics, it is essential to first construct the logical relationships for evaluating these objects. In the case of a redundant model without embedded features, the hierarchy is typically divided into three levels: the target level, the index level, and the decision level. Based on this logic, the hierarchical structure of the converter transformer risk assessment model is constructed, as shown in Figure 2.

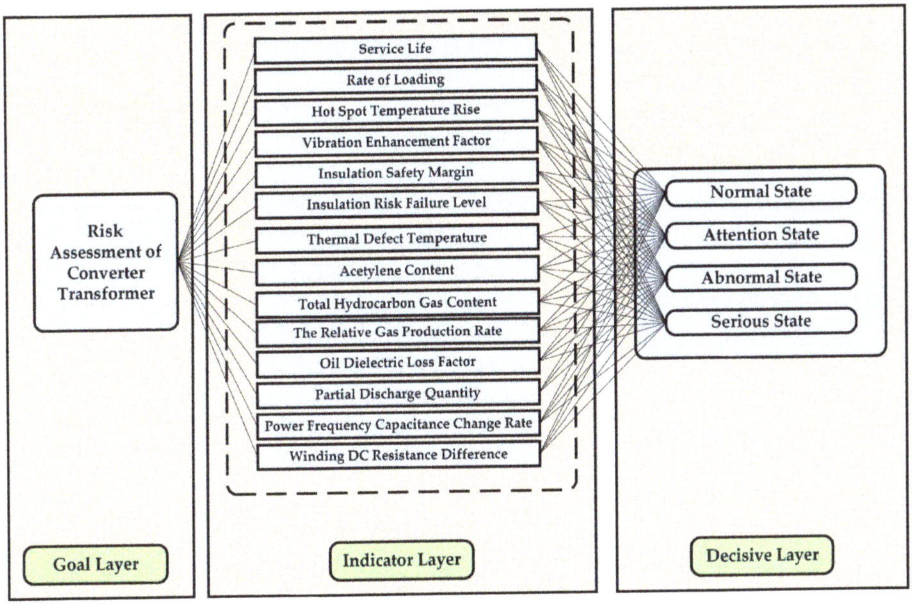

Figure 2. Hierarchy relationship of the 'Goal–Indicator–Decision' layers.

The hierarchical judgment matrix A is then constructed. Based on expert experience, the relative importance and correlation between the characteristic parameters in the index layer are assessed, and the judgment matrix A is formed, as shown in Equation (1).

$$A = \begin{bmatrix} a_{11} & a_{12} & \cdots & a_{1j} \\ a_{21} & \cdots & \cdots & \cdots \\ \vdots & \vdots & \ddots & \vdots \\ a_{i1} & \cdots & \cdots & a_{ij} \end{bmatrix} \quad (1)$$

Here, i denotes the feature vector in the ith row, j represents the feature vector in the jth column, and a_{ij} indicates the relative importance of the feature vector in the ith row compared to the one in the jth column.

Since the evaluation model established in this study involves 14 characteristic parameters, $i = j = 14$. The 1–5 scale method, similar to the 1–9 scale method, is used to indicate the relative importance between different feature vectors, thereby constructing the judgment matrix A. The meaning of each scale is outlined in Table 2. Pairwise comparisons are made for each feature vector to effectively reflect the importance of different indicators.

Table 2. Meaning of each scale.

Scale	Implication
1	The two feature vectors have the same importance
2	The ith row feature vector is more important than the jth column feature vector
3	The ith row feature vector is much more important than the jth column feature vector
4	The ith row feature vector is extremely important compared to the jth column feature vector
5	The ith row feature vector is the most important compared to the jth column feature vector
$1/a_{ij}$	The feature vector of column j is more important than the feature vector of row i

The eigenvalues of the judgment matrix A are first calculated. Using Equation (2), the maximum eigenvalue λ_{max} of matrix A and its corresponding eigenvector ω_j are determined and normalized, obtaining the normalized eigenvector ω_j'.

$$A = \lambda_{max}\omega_j \tag{2}$$

Subsequently, the consistency test of matrix A is performed. The consistency index CI and consistency ratio CR of matrix A are calculated according to Equations (3) and (4).

$$CI = \frac{\lambda_{max} - n}{n-1} \tag{3}$$

$$CR = \frac{CI}{RI} \tag{4}$$

Here, RI denotes the average random consistency index, and n represents the order of the matrix, which corresponds to the number of feature vectors. The values of RI for different values of n are shown in Table 3 [39]. A smaller CR indicates a higher reliability of matrix A. When $CR < 0.1$, the CR value of matrix A is considered acceptable, suggesting that the logical structure of matrix A meets the required standards. At this point, the initial subjective weight of the matrix is represented by ω after normalization. If $CR > 0.1$, the logical structure of matrix A is deemed unsatisfactory, and the matrix must be revised until $CR < 0.1$.

Table 3. Classification thresholds for different risk levels of 'electrical–thermal–mechanical' multi-dimensional eigenparameters.

n	1	2	3	4	5	6	7	8	9	10	11	12	13	14
RI	0	0	0.52	0.89	1.12	1.24	1.36	1.41	1.46	1.49	1.52	1.54	1.56	1.58

4.2. Construction of Risk Decision Matrix Based on Entropy-Weight Method

To assess the variability of different characteristic parameters in the evaluation model, the degree of influence of each characteristic parameter on the overall outcome is evaluated. First, the primary risk decision matrix $Y = x_{mj}$ is constructed, as shown in Equation (5).

$$Y = \begin{pmatrix} x_{11} & x_{12} & \cdots & x_{1j} \\ x_{21} & x_{22} & \cdots & x_{2j} \\ \vdots & \vdots & \ddots & \vdots \\ x_{m1} & x_{m2} & \cdots & x_{mj} \end{pmatrix} \tag{5}$$

where m represents the risk level.

Next, the optimal value for each characteristic parameter is defined, and the risk characteristic decision matrix $Z = x_{mj}'$ is constructed. The data for each characteristic parameter in matrix Z is normalized. For parameters where higher values are better, such as the insulation safety margin, $x_{mj}^* = \max x_{mj}$, and $x_{mj}' = x_{mj}/x_{mj}^*$; for parameters where smaller values are better, such as total hydrocarbon gas content, $x_{mj}^* = \min x_{mj}$, and $x_{mj}' = x_{mj}^*/x_{mj}$. The normalized values of each characteristic parameter in matrix Z fall within the range [0, 1]. The higher the value of a characteristic parameter, the better it is, with the optimal value being equal to 1.

Subsequently, the risk characteristic decision matrix $Z = x_{mj}'$ is standardized to eliminate any ambiguity in the weight assignment of each characteristic parameter and to improve computational efficiency. Without altering the correlation of the characteristic parameters, the data are transformed into real numbers, b_{mj}, within the range [0, 1], to reflect the relative importance of each characteristic parameter. The standardized risk characteristic decision matrix B is shown in Equation (6).

$$B = \begin{pmatrix} b_{11} & b_{12} & \cdots & b_{1j} \\ b_{21} & b_{22} & \cdots & b_{2j} \\ \vdots & \vdots & \ddots & \vdots \\ b_{m1} & b_{m2} & \cdots & b_{mj} \end{pmatrix} \quad (6)$$

After the construction of the risk decision matrix is completed, the information entropy matrix E_j for each characteristic parameter is computed, as shown in Equation (7), to determine the influence of different characteristic parameters on the risk state of the converter transformer.

$$E_j = \frac{\sum_{m=1}^{4} b_{mj} \ln b_{mj}}{\ln m} \quad (7)$$

Here, when $b_{mj} = 0$ or $b_{mj} = 1$, it is defined that $b_{mj}\ln b_{mj} = 0$.

Next, the deviation matrix d_j for each characteristic parameter is calculated, as shown in Equation (8).

$$d_j = 1 - E_j \quad (8)$$

Finally, the deviation matrix d_j for each feature parameter is normalized, resulting in the normalized deviation matrix μ_j, as shown in Equation (9). At this point, μ_j represents the objective weight of the decision matrix constructed using the entropy-weight method.

$$\mu_j = \frac{d_j}{\sum_{j=1}^{n} d_j} \quad (9)$$

4.3. Comprehensive Weight Calculation Based on Subjective and Objective Fusion

Firstly, the μ_j values of the risk decision matrix B in the entropy-weight method are corrected using the λ_{max} values from the hierarchical judgment matrix A in the AHP, along with their corresponding ω_j'. The correction deviation θ_j is then computed, as shown in Equation (10).

$$\theta_j = \frac{\mu_j \omega_j'}{\sum_{j=1}^{n} \mu_j \omega_j'} \quad (10)$$

Subsequently, the weight assignment model is further optimized, and the comprehensive weight ξ_j for the converter transformer risk assessment model using the entropy-weight AHP is derived, as shown in Equation (11).

$$\xi_j = \frac{\lambda \theta_j \times (1-\lambda)\mu_j}{\sum_{i=1}^{n} \lambda \theta_j \times (1-\lambda)\mu_j} \quad (11)$$

Since the entropy-weight method is sensitive to the extreme values of characteristic parameters, while the analytic hierarchy process over-relies on expert judgment, $\lambda = 0.5$ is set to counteract the respective biases of both methods and ensure their balanced contributions in the integrated assessment model.

5. Research on Risk Assessment of Converter Transformers

Firstly, based on Equation (1), the hierarchical structure of the converter transformer risk assessment system, as depicted in Figure 1, is established. Using the 1–5 scale method, 14 'electrical–thermal–mechanical' multi-dimensional characteristic parameters, including partial discharge, power frequency capacitance change rate, dissolved gas in oil, relative gas production rate, hot spot temperature rise, and thermal defect temperature, which effectively reflect the risk state of the converter transformer, are compared and analyzed. The judgment matrix A is constructed, as shown in Equation (12). In Equation (12), the characteristic parameters K_1–K_{14} are arranged from left to right and top to bottom. The intersection of the horizontal and vertical parameters indicates the relative importance of the ith row parameter compared to the jth column parameter. For example, the matrix values at the 1st row and 2nd column, 3rd row and 4th column, and 6th row and 1st column illustrate this relationship. Specifically, a value of 1/2 indicates that the rate of loading K_2 is more important than the service life K_1, a value of 3 signifies that the hot spot temperature rise K_3 is much more critical than the vibration enhancement factor K_4, and a value of 5 demonstrates that the insulation risk failure level K_6 is the most important compared to the service life K_1.

$$A = \begin{pmatrix} 1 & 1/2 & 1/3 & 1 & 1/2 & 1/5 & 1/5 & 1/5 & 1/4 & 1/4 & 1/5 & 1 & 1/3 & 1/3 \\ 2 & 1 & 1/2 & 2 & 1 & 1/4 & 1/4 & 1/4 & 1/3 & 1/3 & 1/4 & 2 & 1/2 & 1/2 \\ 3 & 2 & 1 & 3 & 2 & 1/3 & 1/3 & 1/3 & 1/2 & 1/2 & 1/3 & 3 & 1 & 1 \\ 1 & 1/2 & 1/3 & 1 & 1/2 & 1/5 & 1/5 & 1/5 & 1/4 & 1/4 & 1/5 & 1 & 1/3 & 1/3 \\ 2 & 1 & 1/2 & 2 & 1 & 1/4 & 1/4 & 1/4 & 1/3 & 1/4 & 1/4 & 3 & 1/2 & 1/2 \\ 5 & 4 & 3 & 5 & 4 & 1 & 1 & 1 & 2 & 2 & 1 & 5 & 3 & 3 \\ 5 & 4 & 3 & 5 & 4 & 1 & 1 & 1 & 2 & 2 & 1 & 5 & 3 & 3 \\ 5 & 4 & 3 & 5 & 4 & 1 & 1 & 1 & 2 & 2 & 1 & 5 & 3 & 3 \\ 4 & 3 & 2 & 4 & 3 & 1/2 & 1/2 & 1/2 & 1 & 1 & 1/2 & 4 & 3 & 3 \\ 4 & 3 & 2 & 4 & 3 & 1/2 & 1/2 & 1/2 & 1 & 1 & 1/2 & 4 & 2 & 2 \\ 5 & 4 & 3 & 5 & 4 & 1 & 1 & 1 & 2 & 2 & 1 & 5 & 3 & 3 \\ 1 & 1/2 & 1/3 & 1 & 1/2 & 1/5 & 1/5 & 1/5 & 1/4 & 1/4 & 1/5 & 1 & 1/3 & 1/3 \\ 3 & 2 & 1 & 3 & 2 & 1/3 & 1/3 & 1/3 & 1/2 & 1/2 & 1/3 & 3 & 1 & 1 \\ 3 & 2 & 1 & 3 & 2 & 1/3 & 1/3 & 1/3 & 1/2 & 1/2 & 1/3 & 3 & 1 & 1 \end{pmatrix} \quad (12)$$

The maximum eigenvalue, λ_{max}, of matrix A is 14.3328, and the corresponding eigenvector is ω_j = (0.0661, 0.1021, 0.1641, 0.0661, 0.1067, 0.4272, 0.4272, 0.4272, 0.2890, 0.2661, 0.4272, 0.0661, 0.1641, 0.1641)^{T}. The eigenvector is normalized, and the normalized feature eigenvector, $\omega_j{}'$ = (0.0046, 0.0071, 0.0114, 0.0046, 0.0074, 0.0298, 0.0298, 0.0298, 0.0202, 0.0186, 0.0298, 0.0046, 0.0114, 0.0114)^{T}, is obtained.

The consistency test of matrix A is then conducted. Given that n = 14, it follows from Table 3 that RI = 1.58 and CR = 0.0162 < 0.1, as calculated using Equations (3) and (4). Therefore, the logical consistency of the judgment matrix A meets the necessary requirements, and $\omega_j{}'$ can be adopted as the initial weight in the evaluation model.

Next, based on Table 1 and employing the entropy-weight method, the primary risk decision matrix Y is constructed, as shown in Equation (13). Subsequently, the data for each characteristic parameter in matrix Y is processed on a per-unit value basis, resulting in the establishment of the risk characteristic decision matrix Z, as shown in Equation (14). Finally, matrix Z is standardized, producing the standardized risk characteristic decision matrix B, as shown in Equation (15).

$$Y = \begin{pmatrix} 0.0014 & 1.0 & 60 & 1.0 & 1.2500 & 1 & 150 & 2 & 100 & 7 & 100 & 0.0050 & 2 & 1 \\ 0.0020 & 1.1 & 68 & 1.2 & 1.4286 & 2 & 300 & 5 & 150 & 10 & 250 & 0.0200 & 3 & 2 \\ 0.0040 & 1.3 & 78 & 1.4 & 1.6667 & 3 & 500 & 20 & 350 & 20 & 700 & 0.0300 & 4 & 4 \\ 0.0067 & 1.5 & 90 & 1.5 & 2.0000 & 4 & 700 & 50 & 500 & 30 & 1000 & 0.0400 & 5 & 5 \end{pmatrix} \tag{13}$$

$$Z = \begin{pmatrix} 1 & 1 & 1 & 1 & 1 & 1 & 1 & 1 & 1 & 1 & 1 & 1 & 1 & 1 \\ 0.7143 & 0.9091 & 0.8823 & 0.8333 & 0.8750 & 0.5000 & 0.5000 & 0.4000 & 0.6667 & 0.7000 & 0.4000 & 0.2500 & 0.6667 & 0.5000 \\ 0.3571 & 0.7292 & 0.7692 & 0.7143 & 0.7500 & 0.3333 & 0.3000 & 0.1000 & 0.2857 & 0.3500 & 0.1429 & 0.1667 & 0.5000 & 0.2500 \\ 0.2143 & 0.6667 & 0.6667 & 0.6667 & 0.6250 & 0.2500 & 0.2143 & 0.0400 & 0.2000 & 0.2333 & 0.1000 & 0.1250 & 0.4000 & 0.2000 \end{pmatrix} \tag{14}$$

$$B = \begin{pmatrix} 0.0714 & 0.0714 & 0.0714 & 0.0714 & 0.0714 & 0.0714 & 0.0714 & 0.0714 & 0.0714 & 0.0714 & 0.0714 & 0.0714 & 0.0714 & 0.0714 \\ 0.0812 & 0.1033 & 0.1003 & 0.0947 & 0.0995 & 0.0568 & 0.0568 & 0.0455 & 0.0758 & 0.0796 & 0.0455 & 0.0284 & 0.0758 & 0.0568 \\ 0.0617 & 0.1329 & 0.1329 & 0.1234 & 0.1210 & 0.0576 & 0.0518 & 0.0173 & 0.0494 & 0.0605 & 0.0247 & 0.0288 & 0.0864 & 0.0432 \\ 0.0466 & 0.1449 & 0.1449 & 0.1449 & 0.1358 & 0.0543 & 0.0466 & 0.0087 & 0.0434 & 0.0507 & 0.0217 & 0.0272 & 0.0869 & 0.0435 \end{pmatrix} \tag{15}$$

The values of E_j, d_j, and μ_j for each characteristic parameter, as calculated using matrix B, are shown in Equation (16).

$$E_j = \begin{pmatrix} -0.5100 \\ -0.7005 \\ -0.6977 \\ -0.6851 \\ -0.6882 \\ -0.4863 \\ -0.4672 \\ -0.3177 \\ -0.4824 \\ -0.5127 \\ -0.3627 \\ -0.3533 \\ -0.5828 \\ -0.4497 \end{pmatrix} \quad d_j = \begin{pmatrix} 1.5100 \\ 1.7005 \\ 1.6977 \\ 1.6851 \\ 1.6882 \\ 1.4863 \\ 1.4672 \\ 1.3177 \\ 1.4824 \\ 1.5127 \\ 1.3627 \\ 1.3533 \\ 1.5828 \\ 1.4497 \end{pmatrix} \quad \mu_j = \begin{pmatrix} 0.0709 \\ 0.0798 \\ 0.0797 \\ 0.7913 \\ 0.7927 \\ 0.0698 \\ 0.0689 \\ 0.0619 \\ 0.0696 \\ 0.0710 \\ 0.0640 \\ 0.0635 \\ 0.0743 \\ 0.0681 \end{pmatrix} \tag{16}$$

Subsequently, the AHP and the entropy-weight method are combined to compute ξ_j and θ_j, as shown in Equation (17).

$$\theta_j = \begin{pmatrix} 0.0214 \\ 0.0372 \\ 0.0597 \\ 0.0238 \\ 0.0386 \\ 0.1360 \\ 0.1343 \\ 0.1206 \\ 0.0918 \\ 0.0862 \\ 0.1247 \\ 0.0192 \\ 0.0556 \\ 0.0510 \end{pmatrix} \quad \xi_j = \begin{pmatrix} 0.0046 \\ 0.0123 \\ 0.0317 \\ 0.0051 \\ 0.0133 \\ 0.1879 \\ 0.1855 \\ 0.0167 \\ 0.0858 \\ 0.0742 \\ 0.1723 \\ 0.0041 \\ 0.0292 \\ 0.0270 \end{pmatrix} \tag{17}$$

Finally, based on the risk characteristic decision matrix Z and the comprehensive weight ξ_j, the risk assessment matrix model Q of the converter transformer, incorporating all characteristic parameters, is calculated, as shown in (18).

$$Q = \sum_{j=1}^{n} Z \cdot \xi_j \tag{18}$$

The calculated risk assessment model for the converter transformer is $Q = (1,\ 0.5239,\ 0.2813,\ 0.2048)^{\mathrm{T}}$. The four values of matrix Q categorize the risk state of the converter transformer into four levels: normal, attention, abnormal, and critical. The values in the matrix Q are [0, 1]. A value of 0 indicates that the characteristic parameters of the converter transformer fall below the critical threshold, implying that the transformer is at a serious risk level and should be immediately shut down for maintenance. A value of 1 indicates that the characteristic parameters of the converter transformer exceed the normal threshold, meaning the transformer is in a completely normal state and can continue normal operations.

6. Verification of Risk Assessment Effects of Converter Transformers

To verify the effectiveness of the risk assessment model for converter transformers based on the entropy-weight analytic hierarchy process proposed in Section 4, four representative converter transformers were selected from converter stations across multiple provincial grids, with their operational data serving as validation case studies. The monitoring data for the characteristic parameters are presented in Table 4.

Table 4. Characteristic parameter monitoring data.

Parameter	Transformer A	Transformer B	Transformer C	Transformer D
K_1	800	667	476	118
K_2	1	1	1.2	1.5
K_3	53	62	75	98
K_4	1	1	1.1	1.3
K_5	0.80	0.78	0.63	0.47
K_6	1	1	2	4
K_7	0	87	462	885
K_8	0.53	1.38	15.26	187.05
K_9	87.04	134.20	267.42	961.99
K_{10}	4	6	13	45
K_{11}	27.47	74.88	296.57	1332.45
K_{12}	0.5	0.5	2	3
K_{13}	2	2	3	5
K_{14}	1	1	2	5

The evaluation criteria of the risk assessment model proposed in this study were applied to score the individual characteristic parameters of the selected transformers. Following the risk characteristic decision matrix value calculation method described in Section 4.2, the normal state values of each characteristic parameter from Table 1 were selected as optimal values. The characteristic parameter monitoring data of the four converter transformers in Table 4 were normalized, yielding the individual characteristic parameter scoring presented in Table 5.

By combining Equations (17) and (18) and Table 5, the comprehensive scores of the four converter transformers are calculated, as presented in Table 6.

The aggregate scoring of the four converter transformers from Table 6 were input into the risk assessment model $Q = (1, 0.5239, 0.2813, 0.2048)^T$ developed in Section 4, enabling the determination of their current risk states. The comprehensive score results of different converter transformers reveal that Transformer A, which had recently been put into operation, had a score of 1, indicating a completely normal risk state. Transformer B, which had been in operation for several years, showed signs of internal insulation degradation, with its score falling between normal and attention, warranting enhanced monitoring. Transformer C, with winding defects, had a score between attention and seriousness, requiring timely testing and repair. Transformer D, exhibiting severe winding deformation, had a score indicating a state below serious risk, necessitating immediate overhauling. The risk assessment model aligns with the actual conditions of converter transformers at varying severity levels. It not only retains expert knowledge but also incorporates data-driven theory, ensuring high accuracy.

Table 5. Individual characteristic parameter scoring.

Parameter	Transformer A	Transformer B	Transformer C	Transformer D
K_1	1	0.9524	0.6803	0.1681
K_2	1	1	0.8333	0.6667
K_3	1	0.9677	0.8333	0.6122
K_4	1	1	0.9091	0.7692
K_5	1	0.9690	0.7911	0.5952
K_6	1	1	0.5000	0.2500
K_7	1	0.58	0.3247	0.1695
K_8	1	1	0.1311	0.0107
K_9	1	0.7452	0.3739	0.1040
K_{10}	1	1	0.5385	0.1556
K_{11}	1	1	0.3372	0.0750
K_{12}	1	1	0.2500	0.1667
K_{13}	1	1	0.6667	0.4000
K_{14}	1	1	0.5000	0.2000

Table 6. Aggregate scoring.

Transformer Type	Transformer A	Transformer B	Transformer C	Transformer D
Comprehensive score	1	0.8986	0.3953	0.1717

7. Conclusions

Considering the limitations of the current risk assessment model for converter transformers, which is based on a relatively simple set of characteristic parameters and limited training data, a comprehensive evaluation of the state of the transformer remains unachieved. Building on existing standards and research, this study analyzes and selects 14 multi-dimensional 'electrical–thermal–mechanical' characteristic parameters that effectively reflect the risk state of converter transformers. Using the entropy-weight AHP, a risk assessment model for converter transformers is developed. The main conclusions are as follows:

1. This paper selects 14 'electrical–thermal–mechanical' multi-dimensional characteristic parameters, such as partial discharge, power frequency capacitance change rate, dissolved gas in oil, relative gas production rate, hot spot temperature rise, and thermal defect temperature, to comprehensively evaluate the risk state of the converter transformer, ensuring the comprehensiveness of the assessment model.
2. By combining the subjective weighting method from the AHP with the objective weighting method from the entropy-weight method, the risk assessment model based on the entropy-weight AHP effectively addresses the issues of over-reliance on expert experience or statistical data in existing models, thereby enhancing the reliability and accuracy of the model, while simultaneously satisfying the requirements for both real-time online monitoring and continuous model updating.
3. The effectiveness of the model risk assessment is validated using actual operational data from the converter transformer. The results indicate that the proposed risk assessment model can accurately assess the risk state of the transformer by providing comprehensive protection for HVDC systems, spanning from equipment-level anomaly detection to system-wide stability maintenance. The model exhibits particularly robust capabilities in predicting and preventing risks under complex operating conditions, such as commutation failures and harmonic resonance, enabling the proactive mitigation of potential failures.

Author Contributions: Conceptualization, G.Q. and J.H.; methodology, H.Z.; software, W.D.; validation, D.Z.; formal analysis, H.Z.; investigation, H.S.; writing—original draft preparation, G.Q. and H.Z.; writing—review and editing, J.H. and H.Z.; supervision, W.D. All authors have read and agreed to the published version of the manuscript.

Funding: This research was funded by the Science and Technology Project of China Southern Power Grid Co., Ltd. (YNKJXM20210186).

Data Availability Statement: Data available upon request.

Conflicts of Interest: The authors declare that this study received funding from China Southern Power Grid Co., Ltd. Authors Guochao Qian, Weiju Dai, Dexu Zou, Haoruo Sun were employed by the Electric Power Research Institute Yunnan Power Grid Co., Ltd. The remaining authors declare that the research was conducted in the absence of any commercial or financial relationships that could be construed as a potential conflict of interest. The funder was not involved in the study design, collection, analysis, interpretation of data, the writing of this article or the decision to submit it for publication.

References

1. He, W.; Zhang, Z.; Yang, Y.; Gao, J.; Pei, X.; Deng, J.; Pan, Z.; Lv, J.; Yao, H. Risk Assessment Study of Oil Flow Under Inrush Current on the Misoperation of Converter Transformer Gas Relay. *Appl. Sci.* **2025**, *15*, 2235. [CrossRef]
2. Yu, Y.; Zhang, Z.; He, W.; Chen, J.; Li, D. Simulation Analysis of Inrush Current and Electromagnetic-Force Characteristics of Converter Transformer During No-Load Closing. In Proceedings of the 2023 2nd International Conference on Power Systems and Electrical Technology (PSET), Milan, Italy, 25–27 August 2023; pp. 137–143.
3. Liu, Z.; Yin, X.; Pan, Y.; Xi, W.; Yin, X.; Liu, B. Analysis of zero-mode inrush current characteristics of converter transformers. In Proceedings of the 2021 56th International Universities Power Engineering Conference (UPEC), Middlesbrough, UK, 31 August–3 September 2021; pp. 1–6.
4. Li, X.; Chen, B.; Ma, Y.; Liu, J.; Yang, M. Differential analysis of gas generation between gas to liquid and mineral oil under electric-thermal combined fault based on molecular dynamics. In *Electrical Engineering*; Springer: Berlin/Heidelberg, Germany, 2025; pp. 1–9.
5. Amalanathan, A.J.; Sarathi, R.; Zdanowski, M.; Vinu, R.; Nadolny, Z. Review on Gassing Tendency of Different Insulating Fluids Towards Transformer Applications. *Energies* **2023**, *16*, 488. [CrossRef]
6. Manea, A.; Gorjanu, T.; Lazeanu, A.; Dumitran, L.M. Effect of Electrical Accelerated Aging on DC Resistivity of Mineral Oil Used in Power Transformers. *Energies* **2023**, *16*, 294.
7. Zhang, S.; Zhou, H. Transformer Fault Diagnosis Based on Multi-Strategy Enhanced Dung Beetle Algorithm and Optimized SVM. *Energies* **2024**, *17*, 6296. [CrossRef]
8. Saha, T.K. Review of modern diagnostic techniques for assessing insulation condition in aged transformers. *IEEE Trans. Dielectr. Electr. Insul.* **2003**, *10*, 903–917.
9. Jin, L.; Kim, D.; Abu-Siada, A.; Kumar, S. Oil-Immersed Power Transformer Condition Monitoring Methodologies: A Review. *Energies* **2022**, *15*, 3379. [CrossRef]
10. Manoj, T.; Ranga, C.; Abu-Siada, A.; Ghoneim, S.S.M. Analytic Hierarchy Processed Grey Relational Fuzzy Approach for Health Assessment of Power Transformers. *IEEE Trans. Dielectr. Electr. Insul.* **2024**, *31*, 1480–1489.
11. Žarković, M.; Stojković, Z. Analysis of artificial intelligence expert systems for power transformer condition monitoring and diagnostics. *Electr. Power Syst. Res.* **2017**, *149*, 25–136.
12. Yang, J.; Wu, H.; Yang, Y.; Zhao, X.; Xun, H.; Wei, X.; Guo, Z. A Transformer Maintenance Interval Optimization Method Considering Imperfect Maintenance and Dynamic Maintenance Costs. *Appl. Sci.* **2024**, *14*, 6845. [CrossRef]
13. Luo, C.; Zhu, Y.; Li, Z.; Yu, P.; Zhou, Z.; Yang, X.; Liao, M. Diffusion Characteristics of Dissolved Gases in Oil Under Different Oil Flow Circulations. *Energies* **2025**, *18*, 432. [CrossRef]
14. Zhang, C.; He, Y.; Du, B.; Li, B.; Jiang, S. Transformer fault diagnosis method using IoT based monitoring system and ensemble machine learning. *Future Gener. Comput. Syst.* **2020**, *108*, 533–545.
15. Zhao, X.; Wu, G.; Yang, D.; Xu, G.; Xing, Y.; Yao, C.; Abu-Siada, A. Enhanced detection of power transformer winding faults through 3D FRA signatures and image processing techniques. *Electr. Power Syst. Res.* **2025**, *242*, 111433.
16. Farhadian, A.H.; Mikhak-Beyranvand, M.; Fard, S.S.M. Aged Transformer Oil Analysis Through Laser-Induced Breakdown Spectroscopy. *IEEE Trans. Dielectr. Electr. Insul.* **2024**, *31*, 1988–1994.
17. Velásquez, R.M.A.; Lara, J.V.M. Root cause analysis improved with machine learning for failure analysis in power transformers. *Eng. Fail. Anal.* **2020**, *115*, 104684.

18. Jiang, T.; Yang, C.; Wang, J.; Huang, G.; Cao, L.; Shi, Y.; State Grid Shaanxi Electric Power Company; State Grid Shaanxi Electric Power Research Institute. Research on Insulation State Assessment Method of Power Transformer based on Health Index. *High Volt. Appar.* **2024**, *60*, 124–129.
19. Zou, Y.; Lin, J.; Li, A.; Zhang, Y. Evaluation of Transformer Oil—Paper Insulation Status based on Grey Relational Analysis and a Cluster Cloud Model. *Power Syst. Prot. Control* **2023**, *51*, 35–43.
20. Zhou, J. *Transformer Fault Diagnosis and Condition Assessment Based on Neural Networks and Support Vector Machines*; Kunming University of Science and Technology: Kunming, China, 2022. (In Chinese)
21. Ming, T.; Zhang, L.; Wang, Q. Research on Transformer State Assessment based on Clustering and Time Series Analysis. *Autom. Devices Equip.* **2021**, *43*, 108–111. (In Chinese)
22. Zhang, P.; Qi, B.; Shao, M.; Li, C.; Rong, Z.; Chen, J.; Wang, H. Optimal data selection rule mining for transformer condition assessment. *IEEE Access* **2021**, *9*, 156962–156972.
23. Zeinoddini-Meymand, H.; Kamel, S.; Khan, B. An efficient approach with application of linear and nonlinear models for evaluation of power transformer health index. *IEEE Access* **2021**, *9*, 150172–150186.
24. Sangwook, L. Determination of Priority Weights under Multiattribute Decision-Making Situations: AHP Versus Fuzzy AHP. *J. Constr. Eng. Manag.* **2015**, *141*, 05014015.
25. Bidisha, B.; Mahapatro, J. Performance Evaluation of Priority-based Scheduling in Hybrid VANETs for Different Criteria Weights Using AHP-AHP and AHP-TOPSIS. *IETE J. Res.* **2024**, *70*, 5759–5770.
26. Sui, L.; Huang, Q.; Liu, X. Research on safety investment decision evaluation and optimization of network booking taxi platform enterprise based on subjective-objective assessment method. *Teh. Vjesn.* **2023**, *30*, 1201–1208.
27. Sun, D.; Hu, X.; Liu, B. Comprehensive evaluation for the sustainable development of fresh agricultural products logistics enterprises based on combination empowerment-TOPSIS method. *PeerJ Comput. Sci.* **2023**, *9*, 1719.
28. Zhu, C. The Local Variational Principle of Weighted Entropy and Its Applications. *J. Dyn. Differ. Equ.* **2024**, *36*, 797–831.
29. Wu, J.; Zhang, Q. Multicriteria decision making method based on intuitionistic fuzzy weighted entropy. *Expert Syst. Appl.* **2011**, *38*, 916–922.
30. Xu, X. A note on the subjective and objective integrated approach to determine attribute weights. *Eur. J. Oper. Res.* **2004**, *156*, 530–532.
31. Liu, H.; Hu, Y. An Evaluating Method with Combined Assigning-Weight Based on Maximizing Variance. *Sci. Program.* **2015**, *1*, 8.
32. *DL/T 1685-2017*; Guideline for Condition Evaluation of Oil-Immersed Transformers (Reactors). China Electric Power Press: Beijing, China, 2017.
33. *DL/T 984-2018*; Guideline for Judging Insulation Aging of Oil-Immersed Transformers. China Electricity Power Press: Beijing, China, 2018.
34. *DL/T 2002-2019*; Converter Transformer Operation Procedure. China Electricity Power Press: Beijing, China, 2019.
35. *GB/T 1094.2-2013*; Power Transformer Part 2: Temperature Rise of Liquid-Immersed Transformer. China Standards Press: Beijing, China, 2013.
36. *IEC 60076-7-2018*; Power Transformers-Part 7: Loading Guide for Mineral-Oil-Immersed Power Transformers. Multiple. Distributed through American National Standards Institute (ANSI): New York, NY, USA, 2018.
37. *IEEE C57.91-2011*; IEEE Guide for Loading Mineral-Oil-Immersed Transformers and Step-Voltage Regulators. IEEE: New York, NY, USA, 2011.
38. *IEC 60599-2022*; Mineral Oil-Filled Electrical Equipment in Service-Guidance on the Interpretation of Dissolved and Free Gases Analysis. International Electrotechnical Commission: Geneva, Switzerland, 2022.
39. Franek, J.; Kresta, A. Judgment Scales and Consistency Measure in AHP. *Procedia Econ. Financ.* **2014**, *12*, 164–173.

energies

MDPI

Article

Evaluation on the Long-Term Operational Reliability of Closing Springs in High-Voltage Circuit Breakers

Mingkun Yang [1], Liangliang Wei [2], Pengfeng Qiu [1], Guangfu Hu [1], Kun Yang [1], Xiaohui He [1], Zhaoyu Peng [1], Fangrong Zhou [1], Yun Zhang [3], Jie Luo [3] and Xuetong Zhao [2,*]

1 Electric Power Research Institute of Yunnan Power Grid Co., Ltd., Kunming 650200, China; xgllymk@163.com (M.Y.); university20241009@163.com (P.Q.); yyt12110427@163.com (G.H.); zju20241010@163.com (K.Y.); nzgsmfj123@163.com (X.H.); 13206263732@163.com (Z.P.); 15070551243@163.com (F.Z.)

2 State Key Laboratory of Power Transmission Equipment Technology, School of Electrical Engineering, Chongqing University, Chongqing 400044, China; liangliangwei0208@gmail.com

3 Yuxi Power Supply Bureau of Yunnan Power Grid, Yuxi 653100, China; shuike20241009@163.com (Y.Z.); 19132052895@163.com (J.L.)

* Correspondence: zxt201314@cqu.edu.cn

Abstract: As a key energy storage component in high-voltage circuit breakers, closing springs are susceptible to stress relaxation, resulting in a decline in closing performance due to high operational loads, prolonged usage, and environmental factors. In this work, the 60Si2CrVA alloy steel springs used in 110 kV high-voltage circuit breakers were utilized to study their mechanical behaviors under various temperatures, salt spray corrosion, and repeated closing operations. It is found that the conditions of salt spray corrosion, and repeated closing operations demonstrate a slight impact on its stress loss, while the operation temperature above 70 °C will result in an apparent increase in the stress loss rate. A threshold for closing failure related to the spring's stress loss rate was established, and a life prediction method based on an improved Arrhenius acceleration model was proposed. The results indicate that the calculated service life of the spring is approximately 27.09 years based on the stress loss rate threshold of 4.5%. This work provides a novel method to evaluate the long-term operational state and service life of closing springs in high-voltage circuit breakers.

Keywords: springs; high-voltage circuit breaker; stress loss; Arrhenius acceleration model

1. Introduction

High-voltage circuit breakers are utilized to control electric current breaking of electrical circuits in power systems, and their performance reliability directly affects the stability and safety of the entire power grid [1–3]. Under normal operating conditions, circuit breakers typically remain in a static state, but they must switch swiftly to connect or disconnect power sources during operations or fault events. Currently, spring-operated mechanisms have become the main priority for high-voltage circuit breakers in power systems due to their simple structure, rapid response, low pollution, and compact size. These mechanisms mainly contain hydraulic-operated springs, pneumatic-operated springs, and fully operated springs. The performance of the closing and opening springs directly determines the response speed and reliability of the circuit breaker during operation.

Faults of high-voltage circuit breakers are generally categorized into insulation breakdown, electricity interruption, and mechanical failure. Specifically, over 60% of circuit breaker failures in China stem from the mechanical failure during their operating process [4,5]. Instability of opening and closing operations often signifies the potential failure risk for circuit breakers. Under the rated operating conditions, the closing springs of circuit breakers generally work in a compression state for a long period, inevitably leading to issues such as strength degradation and stress relaxation [6]. These problems will affect the closing performance of the circuit breaker, and even result in spring damage, potentially inducing a power grid accident.

Recently, the characteristics of springs in high-voltage circuit breakers have been widely studied. Wang et al. proposed the non-contact detection of the spring deformation based on the computer vision technology, and evaluated the energy-storing property of the spring [7]. Li et al. [8] established a simulation model of high-voltage circuit breakers with the dynamic simulation software of ADAMS, and analyzed the impact of stress relaxation in springs on the dynamic contact time and displacement of high-voltage circuit breakers.

Huang et al. [9] proposed a detection method for anomalies in the operating mechanism of circuit breaker by combining local mean decomposition with a support vector machine (SVM), which effectively identifies fault types caused by abnormal energy states in the springs. Ding et al. [10] analyzed the fatigue life of the spring operating mechanism in high-voltage circuit breakers in detail by using the local stress strain method combined with an impact dynamics experiment. Pal et al. [11] found that surface defects, particularly those resembling 'crow's foot' shapes in hydraulic valves, can easily lead to fatigue fractures of springs.

In practical applications, high-voltage circuit breakers are typically installed in outdoor substations, which are generally exposed to fluctuating temperatures, humidity, salt spray (especially in coastal areas), and occasional mechanical vibrations. The closing springs usually operate under long-term compression, particularly in standby conditions, which may lead to stress relaxation and mechanical fatigue over time. These environmental and operational factors collectively contribute to the degradation of spring performance and directly affect the long-term reliability of the circuit breaker mechanism. However, most existing studies have focused on fault feature analysis and state detection, while the degradation of closing springs under realistic operational environments remains insufficiently explored. Therefore, in this work, the degradation characteristics of closing springs in high-voltage circuit breakers were investigated from two perspectives: (1) performance degradation and lifespan prediction under long-term energy storage conditions, and (2) fatigue characteristics during prolonged closing operations. This study provides a new method to evaluate the long-term operational reliability of circuit breaker springs under complex environmental conditions.

2. Materials and Methods

2.1. Test Object

In this study, the energy-storage spring (using for closing) used in the 110 kV CT-26 operating mechanism (ZF-126, Taikai High voltage switchgear Co., Ltd., Tai'an, China) was selected. The spring is made of 60Si2CrVA alloy steel and manufactured using a hot-coiling process, followed by heavy compression and surface shot peening treatments. Detailed parameters of the spring are provided in Table 1.

Table 1. Parameters of 110 kV closing spring.

Parameter	Value
Material	60Si2CrVA
Free Height (mm)	380
Wire Diameter (mm)	22
Active Coils	7.5
Total Coils	9.5
Maximum Compression (mm)	135
Standard Working Height (mm)	$H_1 = 347$, $H_2 = 245$
Closing Time (ms)	50–80
Operating Temperature	−30 °C–70 °C

2.2. *Test and Equipment*

The energy storage spring is an essential component for storing elastic potential energy in circuit breakers. Under normal operating conditions, the spring remains tightly compressed within the circuit breaker, ready for activation. When a trigger signal is received, the spring quickly releases its stored potential energy, allowing the circuit breaker to open or close rapidly and perform its switching function. The performance degradation analysis of high-voltage circuit breaker springs can be approached from two main perspectives: the degradation occurring under long-term energy storage conditions or an extended closing operation.

To investigate the degradation of the closing spring under long-term energy storage, the spring is compressed at the preset stress and fixed through a set of pressure plates. Afterwards, the setup is placed in an environmental simulation chamber, as shown in Figure 1. This equipment can accurately simulate temperature conditions ranging from −20 °C to 150 °C and relative humidity levels from 5% to 90%.

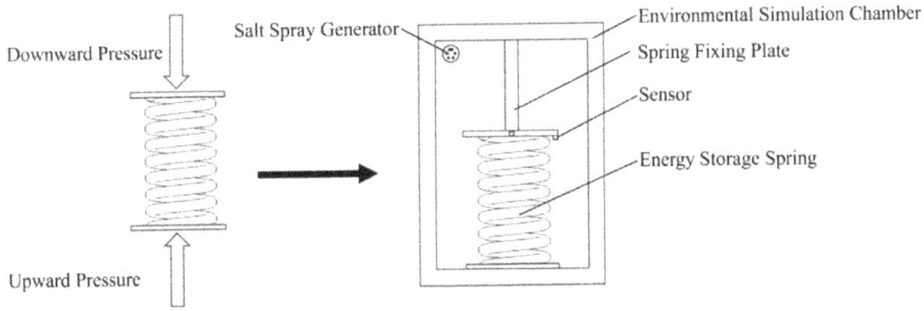

Figure 1. Spring environment simulation chamber.

Circuit breaker equipment in coastal areas may also be affected by salt spray corrosion. Therefore, the test chamber is equipped with a salt spray generator, which allows for adjustable salt spray concentration (1–10% NaCl) and spray frequency. The selection of this concentration range is based on the industry standard GB/T 2423.17-2008 [12] for salt spray testing of power equipment, which specifies 3–5% NaCl as the typical concentration for simulating coastal environmental conditions. To assess the corrosion effects under both moderate and extreme conditions, we extended the testing range to 1–10% NaCl. During the testing period, the spring's stress and closing behaviors are measured at a regular interval (every half-month or month) to monitor its performance degradation.

In order to analyze the mechanical response of the spring under the long-term working condition, the repeated closing and opening operation is generally required to evaluate the spring's elastic recovery capability and mechanical properties under varying operation cycles. In Figure 2, an accelerated test was carried out using the tension compression testing machine (Model: SDS-1000, Changchun Institute of Mechanical Science Co., Ltd., Changchun, China) at a fixed frequency of 0.1 Hz (i.e., 360 cycles per hour). The machine was operated continuously for 17 h per day over a 5-day period, reaching a total of approximately 30,000 cycles. In every cycle, the spring was compressed to its standard working height (from H_1 to H_2) and then released, simulating the complete energy storage and release process that occurs during a typical closing operation of the circuit breaker. Meanwhile, the displacement and pressure sensors were also installed in the accelerated test machine to ensure the accuracy and reliability of the measurements.

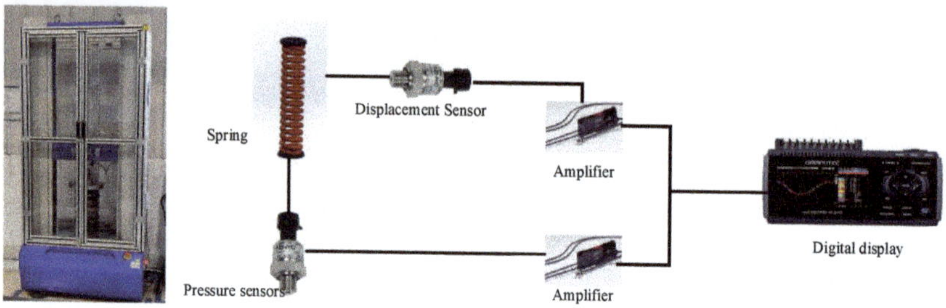

Figure 2. The tension–compression testing machine.

2.3. *Stress Loss Rate and Failure Thresholds of the Spring*

The stress loss rate of the spring refers to the loss of its original load-bearing capacity under a long-term or repeated loading, which represents the decrease in the spring's initial stiffness. The stress loss rate (δ) of the spring is defined using Equation (1) [13,14]:

$$\delta = \frac{F - F_{\mathrm{i}}}{F} \times 100\% \quad (1)$$

where F is the original load of a spring and F_{i} is the load of a used or aged spring.

In this work, a stress loss rate of 4.5% is adopted as a conservative threshold for assessing the degradation of spring performance in high-voltage circuit breakers. This value was selected based on operational requirements, and the acceptable closing time was suggested to be maintained within 50–80 ms (Table 1). When the spring's stress loss exceeds 4.5%, the closing time tends to surpass this limit, indicating that the spring is severely degraded. The experimental validation is provided in Section 5, where the relationship between closing time and stress loss rate is further discussed.

3. Degradation of Springs in the Long-Term Energy Storage State

3.1. *Temperature Factor*

The degradation of seven groups of spring samples was studied at different temperatures of −10 °C, 10 °C, 30 °C, 50 °C, 70 °C, 90 °C, and 110 °C. The spring samples were compressed and fixed using pressure plates to maintain their energy-storage state, and then placed inside an environmental simulation chamber with a temperature control device. In the initial experimental stage (the initial 4 months), the mechanical parameters such as spring stress, stress loss rate, and closing time of the springs were tested every half month, while these parameters were tested every month in the second stage (4 to 11 months).

As shown in Figure 3, the stress relaxation process of high-voltage circuit breaker springs at different temperatures can be divided into two stages. In the first stage (the initial 4 months), the slope of the stress curve is relatively large, indicating a rapid decay in stress, while in the second stage (4 to 11 months), the slope of the curve gradually decreases, suggesting a slower rate of stress decay. Moreover, when the temperature is between −10 °C and 70 °C, the effect of temperature on the stress change of the spring is very small, but when the temperature exceeds this range, it can be clearly seen that the stress change of the spring becomes larger with the increase in temperature.

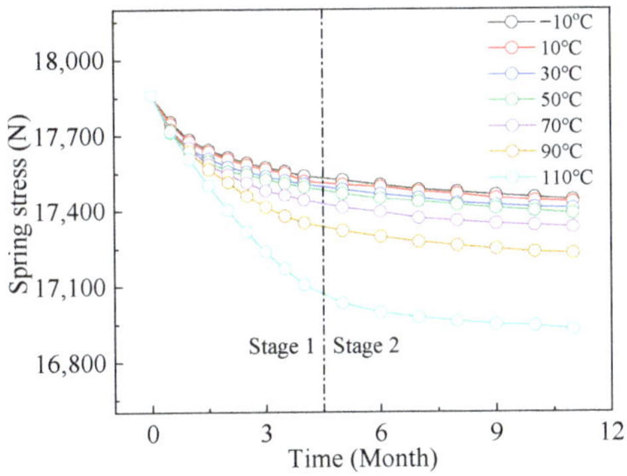

Figure 3. Spring stress curves at different temperatures.

Figure 4 presents the variation in stress loss rate of the circuit breaker spring operating in the temperature range of −10 °C to 110 °C. It can be observed that, within the temperature range of −10 °C to 50 °C, the stress loss rate of the spring after 11 months of degradation falls within the interval of 2.33% to 2.64%. When the temperature increases to the extreme condition of 110 °C, the stress loss rate of the spring reaches 5.21%.

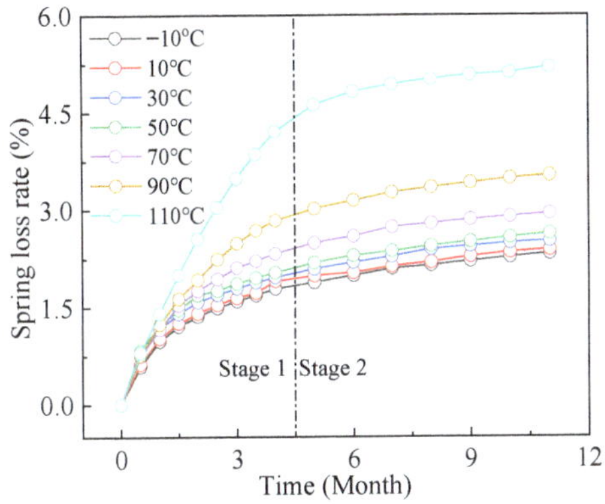

Figure 4. Spring stress loss rates at different temperatures.

3.2. *Effect of Salt Spray Corrosion*

For 60Si2CrVA alloy steel springs, prolonged exposure to an environment with varying humidity or salt spray concentrations can lead to surface and internal corrosion, subsequently affecting their mechanical properties. In this study, the springs were subjected to varying concentrations of salt spray (1 wt%–10 wt% NaCl solution) at 30 °C for a period of 11 months, after which their final rate of stress loss was determined. The experiment involved two groups of different springs. In Group 1, springs were produced under normal manufacturing conditions, while in Group 2, the used springs with damaged electrophoretic paint on the surface were adopted. The results of the experiment are shown in Figures 5 and 6.

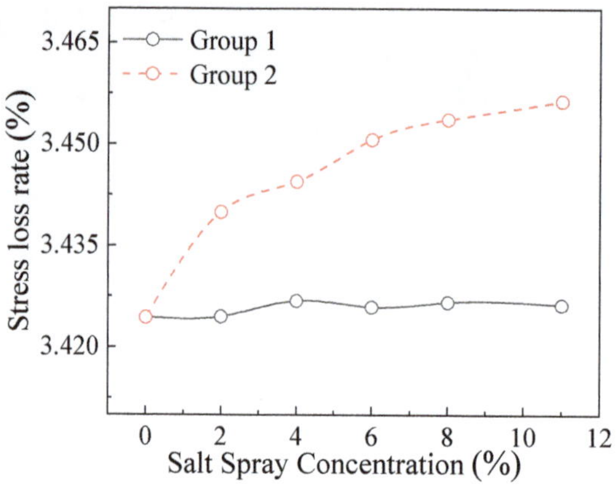

Figure 5. Variation in stress loss rate of two groups of springs after 11 months in different salt spray environments.

Figure 6. Surface corrosion of springs in a 4% NaCl environment.

The surface of the springs in Group 1 coated with electrophoretic paint remained intact and presented effective corrosion resistance. Consequently, the stress loss rate of these springs does not show any obvious variation across the different salt spray concentrations. In contrast, the surface of the springs in Group 2 was significantly corroded after exposing to the salt spray environment, and the stress loss rate was also increased with the increase in the salt spray concentration. However, the stress loss rate of the corroded spring samples was only marginally higher (by approximately 0.01–0.03) than that in Group 1, which would not affect the mechanical properties of the springs. The reason for this minimal impact can be attributed to a specific corrosion process for the springs in Group 2. In the initial corrosion stage, the salt spray is in direct contact with the spring surface, resulting in a chemical reaction that produces a layer of red rust. This rust covers the spring surface to create a protective layer, which hinders further reaction between the salt spray solution and the metal substrate and thereby slows down the penetration of corrosion into the internal structure of the spring [15,16]. Moreover, the results showed that stress loss rate of springs of Group 2 was obviously corroded by the salt spray condition. However, in real working conditions, normal springs are almost fully coated with electrophoretic paint, and are hardly affected by salt spray corrosion [17].

4. Characteristics of Springs Under Long-Term Closing Operation

In general, the operational spring of a high-voltage circuit breaker possesses a service life spanning for several decades, during which it undergoes numerous opening and closing operations. During the closing process, the operational spring is initially in an energy-storage state. Upon receiving the closing command, the spring rapidly releases its stored energy, driving the operating mechanism and completing the closing operation. Then, the spring returns to its uncompressed state, ready for the next energy-storage cycle.

Therefore, the closing operation of the spring refers to compressing the spring for store energy and subsequently releasing the energy. In Figure 7, the effect of the closing operations on spring performance was studied. A spring tension–compression testing machine was used to drive the closing spring for 30,000 cycles of compression over 5 days. The compression range was set between the spring's standard operating heights, H_1 and H_2, which are the spring closing parameters listed in Table 1, to simulate the energy storage and release process during the closing operation.

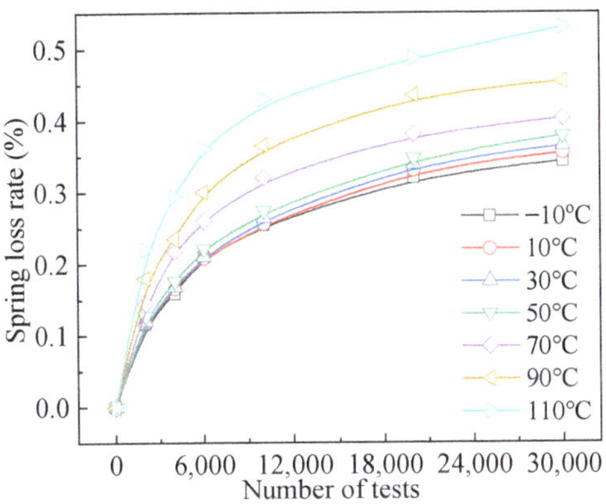

Figure 7. Effect of closing operation on spring stress loss rate.

As shown in Figure 7, the power function curve illustrates the relationship between the stress loss rate of the spring and the number of closing operations. With rising temperature, the stress loss rate exhibits an upward trend. Especially, when the temperature exceeds 70 °C, the stress loss rate of the spring increases significantly, and the curve displays a more pronounced upward slope. After 30,000 simulated closing operations, the stress loss rates of the springs at temperatures of −10 °C, 10 °C, 30 °C, 50 °C, 70 °C, 90 °C, and 110 °C reach 0.342%, 0.354%, 0.36%, 0.387%, 0.414%, 0.454%, and 0.53%, respectively, which are all lower than the 4.5% stress loss rate threshold. This indicates that the stress loss rate during long-term spring closing operations exhibits minimal variation across different operating temperatures. By fitting the spring stress loss rate using the data tested at 30 °C, the following equation was derived:

$$\delta = 0.008 \cdot x^{0.389} \tag{2}$$

where δ is the stress loss rate of the spring, and x is the number of closing operations. The fitting results are illustrated in Figure 8.

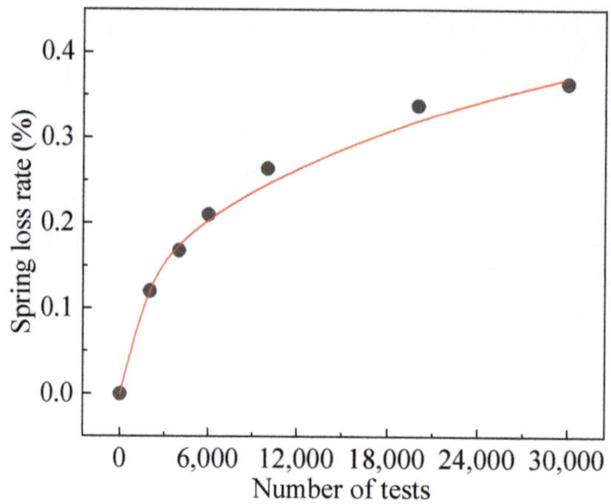

Figure 8. Fitting relationship chart between spring stress loss rate and number of closing tests.

When the number of closing operations reaches 200,000 cycles, the calculated stress loss rate is only 0.92%, which is still below the 4.5% safety threshold. Under actual working conditions, 110 kV circuit breakers are primarily used for the protection and control of transmission systems. They are just operated during system maintenance, fault isolation, or load regulation. Most of the time, the circuit breaker remains in the closed position. Based on the aforementioned test results and analysis, it can be concluded that the closing spring maintains good performance under long-term closing operation and meets the requirements for stable long-term operation of high-voltage circuit breakers.

5. Performance Degradation Assessment and Lifetime Prediction

From the above experiments, it is evident that the stress loss of high-voltage circuit breaker springs is primarily influenced by the duration of the energy storage state. The long-term energy storage state at different temperatures leads to varying rates of stress decay, thereby affecting the closing speed. Figure 9 illustrates the relationship between stress loss rate and closing time. The results show that the closing time of the circuit breaker exhibits an exponential increase with the increased stress loss rate. When the spring's stress loss rate exceeds 4.5%, the closing time fails to meet the operational requirements of the operating mechanism (50–80 ms). Furthermore, when the spring's stress loss rate surpasses 6%, the mechanism is unable to complete the closing operation due to the insufficient pre-tightening force in the closing spring [18]. To ensure the safe operation of circuit breaker spring closing, a stress loss rate threshold of 4.5% can be conservatively established to assess the spring's closing performance.

In engineering practice, the Arrhenius acceleration model was widely used to describe the degradation process of the spring [19–22] and evaluate the effect of temperature on lifespan of materials. The Arrhenius equation is given in Equation (3).

$$k = \mathrm{A}\exp(-\frac{E_a}{RT}) \tag{3}$$

where k the spring stress loss rate, A is the Arrhenius constant, E_a is the activation energy, T is the absolute temperature, and R is the universal gas constant.

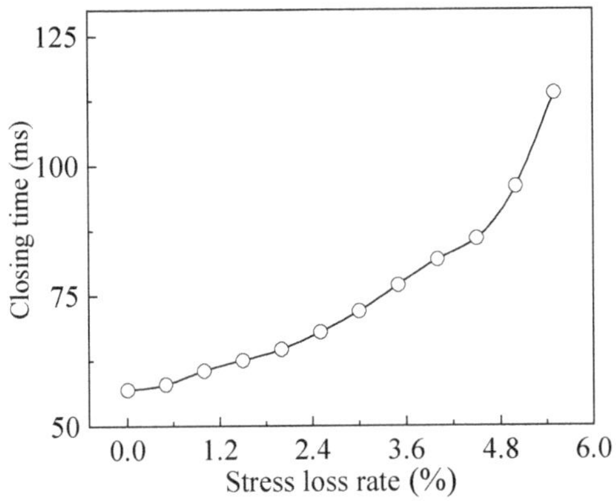

Figure 9. The relationship between spring stress loss and closing time.

The Arrhenius Equation (3) can be transformed into a linear Equation (4) using the natural logarithm, as shown below:

$$\ln(k) = \ln(A) - \frac{E_a}{R} \cdot \frac{1}{T} \tag{4}$$

In addition, as shown in Figure 4, a nonlinear relationship can be found between the spring's stress loss rate (δ, in %) and time (t), which can be expressed as

$$\delta = k \cdot \ln t + c \tag{5}$$

In Equation (5), k is the spring stress loss rate, and c is a constant. This equation shows that the stress loss rate δ is linearly dependent on time t under different temperatures. The linear relationships under various temperatures are obtained as summarized in Table 2.

Table 2. Linear regression equation between stress loss rate and logarithm of time at different temperatures.

Temperature (°C)	Regression Equation	Rate of Stress Loss (k)	Coefficient of Determination (r^2)
−10	$\delta = 0.563 \cdot \ln(t) + 0.981$	0.563	0.983
10	$\delta = 0.569 \cdot \ln(t) + 1.025$	0.569	0.990
30	$\delta = 0.571 \cdot \ln(t) + 1.176$	0.571	0.987
50	$\delta = 0.592 \cdot \ln(t) + 1.226$	0.592	0.986
70	$\delta = 0.741 \cdot \ln(t) + 1.254$	0.741	0.995
90	$\delta = 0.958 \cdot \ln(t) + 1.367$	0.958	0.988
110	$\delta = 0.957 \cdot \ln(t) + 1.907$	0.957	0.921

In Table 2, the coefficient of determination ($0 \leq r^2 \leq 1$) is an important statistical indicator that was used to evaluate the fitting goodness of the regression model. A high R^2 value indicates good linear fitting between the stress loss rate and the logarithmic time. In practical engineering applications, the operating temperature of high-voltage circuit breaker springs does not exceed 70 °C. According to Equation (4), the slope $e = -E_a/R$ and the intercept $b = ln(A)$ of the linear equation can be obtained by fitting ln(k) and $1/T$ at this temperature. The fitting results are illustrated in Figure 10.

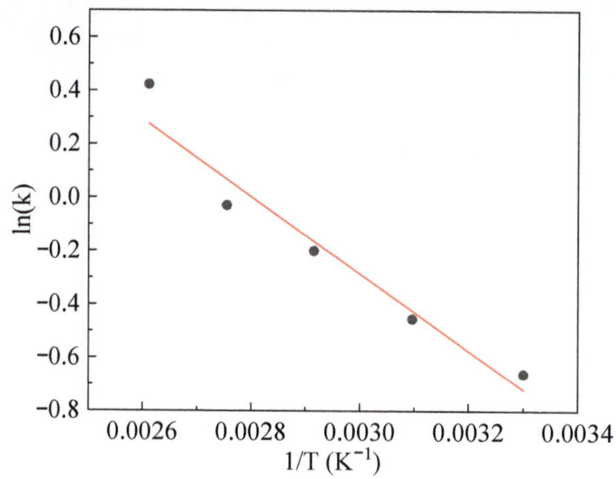

Figure 10. Fitted relationship between logarithm of stress loss rate and inverse of temperature.

The fitting result is given by Equation (6).

$$\ln(k) = 4.3 - 1441.83 \cdot \frac{1}{T} \quad (6)$$

For Equation (5), when t = 1 month, $\delta = c$. This means that c represents the stress loss rate after one month. By substituting c into Equation (3), the following relationship is obtained:

$$c = \int_0^1 A \exp(-\frac{E_a}{RT})dt = A \exp(-\frac{E_a}{RT}) \quad (7)$$

Taking the natural logarithm of both sides,

$$\ln(c) = \ln(A) - \frac{E_a}{RT} = e - f \cdot \frac{1}{T} \quad (8)$$

Here, e and f are constants. Using the data in Table 2, a linear regression of Equation (8) is performed and the fitting result is shown in Figure 11.

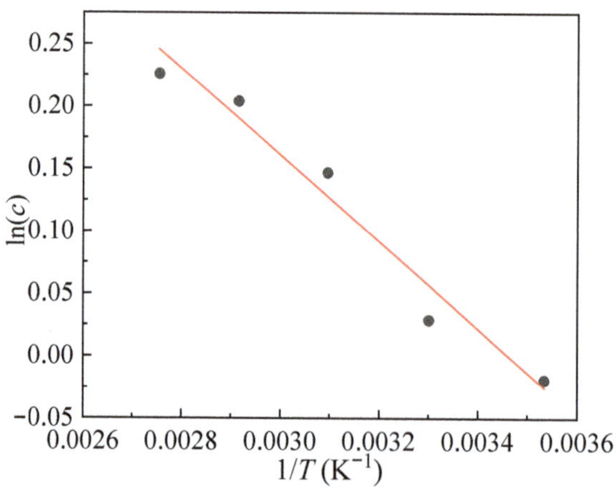

Figure 11. Fitting relationship between logarithm of the constant of integration and the temperature inverse.

The fitting result is given by Equation (9).

$$\ln(c) = 1.14 - 305.03 \cdot \frac{1}{T} \quad (9)$$

In the classical Arrhenius acceleration model, a constant degradation rate k is adopted under isothermal conditions, while the improved model introduces time-dependent degradation behavior expressed as $\delta = kt + c$, where the constant c represents the initial nonlinear degradation at the early stage of spring stress relaxation. This improvement allows the model to better describe the full degradation profile over time and provides a more accurate estimation for the service life of the spring under varying thermal environments.

With the spring stress loss rate threshold set at 4.5% and the room temperature of 25 °C, the stress loss rate k can be calculated using Equations (6) and (9). Substituting these into the linear regression model of Equation (5), the estimation of the service life of the spring under long-term energy storage conditions can be described as Equation (10).

$$4.5 = 0.584 \cdot \ln(t) + 1.122 \quad (10)$$

By solving this equation, the maximum stress loss rate of 4.5% under the condition of 25 °C corresponds to a service life of the spring of t = 325.14 months, or approximately 27.09 years.

6. Conclusions

In summary, the degradation characteristics of 60Si2CrVA alloy steel springs used in 110 kV high-voltage circuit breakers were studied under long-term energy storage and repeated closing operations. It was found that the stress loss rate of the spring is significantly enhanced when the operation temperature exceeds 70 °C, while other factors such as the closing operations and salt spray corrosion show a minimal impact on its stress loss. The threshold of stress loss rate related with closing failure was determined, and a lifetime prediction method for the springs was proposed based on the Arrhenius accelerated model. At room temperature, the calculated service life of the spring is approximately 27.09 years. This study provides an experimental basis and theoretical support for the long-term operational reliability of closing springs in high-voltage circuit breakers under complex working conditions, which is of great significance for ensuring the stable operation of power systems.

Author Contributions: Conceptualization, M.Y. and P.Q.; data curation, M.Y. and K.Y.; methodology, L.W. and X.H.; validation, Z.P. and F.Z.; writing—original draft, L.W., G.H. and Y.Z.; writing—review and editing, X.Z. and J.L. All authors have read and agreed to the published version of the manuscript.

Funding: This work was supported by the Innovation Project of China Southern Power Grid Co., Ltd. (Grant No. YNKJXM20222318).

Data Availability Statement: The data presented in this study are available on request from the corresponding author.

Conflicts of Interest: Authors Mingkun Yang, Pengfeng Qiu, Guangfu Hu, Kun Yang, Xiaohui He, Zhaoyu Peng and, Fangrong Zhou were employed by the company Yunnan Electric Power Research Institute. Authors Yun Zhang and Jie Luo were employed by the company Yunnan Power Grid. The remaining authors declare that the research was conducted in the absence of any commercial or financial relationships that could be construed as a potential conflict of interest. The authors declare that this study received funding from China Southern Power Grid. The funder was not involved in the study design, collection, analysis, interpretation of data, the writing of this article or the decision to submit it for publication.

References

1. Liu, Y.; Zhang, G.; Zhao, C.; Lei, S.; Qin, H.; Yang, J. Mechanical condition identification and prediction of spring operating mechanism of high voltage circuit breaker. *IEEE Access* **2020**, *8*, 210328–210338. [CrossRef]
2. Jiang, J.; Lin, H.; Fang, S. Multi-Objective Optimization of a Permanent Magnet Actuator for High Voltage Vacuum Circuit Breaker Based on Adaptive Surrogate Modeling Technique. *Energies* **2019**, *12*, 4695. [CrossRef]
3. Ma, H.; Qian, G.; Zhang, J.; Chen, J.; Zhou, F.; Qiu, P.; Zhang, A.; Wang, T.; Yao, X.; Liu, Z. Fatigue Life Estimation of Critical Components in a Motor-Energized Spring Operating Mechanism Based on Theory of Reliability. *Energies* **2024**, *17*, 1623. [CrossRef]
4. Fu, R.; Zhao, L.; Rong, Q. Study on mechanical characteristics of operating mechanism of high voltage circuit breaker. *High Volt. Appar.* **2017**, *53*, 56–62.
5. Tahvilzadeh, M.; Aliyari-Shoorehdeli, M.; Razi-Kazemi, A.A. Model-Aided Approach for Intelligent Fault Detection System for SF_6 High-Voltage Circuit Breaker with Spring Operating Mechanism. *IEEE Trans. Power Deliv.* **2023**, *38*, 3356–3365.
6. Sun, T. Intelligent Diagnosis Technology for Mechanical Faults of High Voltage Circuit Breakers Based on Weighted Evidence Theory. Master's Thesis, Southeast University, Nanjing, China, 2017.
7. Wang, K. Research on Evaluation Method of Performance State of Circuit Breaker Energy Storage Spring. Master's Thesis, North China Electric Power University, Beijing, China, 2020.
8. Li, X.; Wu, S.; Li, X.; Zhao, D.; Li, Q.; Li, X. Kinetic characteristics of high voltage circuit breakers based on spring stress relaxation. *J. Cent. South Univ. Sci. Technol.* **2019**, *7*, 1575–1583.
9. Huang, H.; Wang, F.; Su, Y.; Lu, Y.; Xia, X. Research on detection method for abnormal state of spring operating mechanism in high voltage circuit breaker based on LMD and SVM. *High Volt. Appar.* **2020**, *5*, 243–248.
10. Ding, G.; Zhu, Y.; Wu, J.; Guan, T.; Dong, H. Fatigue life analysis of key components of spring operating mechanism based on local stress-strain method. *CSEE J. Power Energy Syst.* **2023**, *20*, 8161–8170.
11. Pal, U.; Mukhopadhyay, G.; Bhattacharya, S. Failure analysis of spring of hydraulic operated valve. *Eng. Fail. Anal.* **2019**, *95*, 191–198.
12. *GB/T 2423.17-2008*; Environmental Testing for Electric and Electronic Products-Part 2: Test Method-Test Ka: Salt Mist. Standards Press of China: Beijing, China, 2009.
13. Yang, C.; Gu, X.; Pan, S. Reliability evaluation of coil spring for fuse with multiple performance parameters. *J. Natl. Univ. Def. Technol.* **2023**, *45*, 200–207.
14. Zhang, Y.; Liu, H.; Wang, D. (Eds.) *Spring Handbook*, 3rd ed.; Mechanical Industry Press: South Norwalk, CT, USA, 2017; pp. 297–333.
15. Zuo, M.; Chen, Y.; Mi, Z.; Wang, Y.; Jiang, H. Effects of Cr content on corrosion behaviour and corrosion products of spring steels. *J. Iron Steel Res. Int.* **2019**, *26*, 1000–1010. [CrossRef]
16. Wang, X.; Chen, Y.; Niu, G. The study on corrosion resistance of high-strength spring steel. *Corros. Eng. Sci. Technol.* **2018**, *53*, 54–64. [CrossRef]
17. Huang, J. Study on Fatigue Crack Propagation Rate and Corrosion Mechanism of 60Si2Mn Spring Steel Under Salt Spray Environment. Master's Thesis, Kunming University of Science and Technology, Kunming, China, 2023.
18. Wang, K.; Shi, J.; Jin, P. Experimental analysis and application of stress relaxation characteristics of helical compression spring. *Chin. J. Appl. Mech.* **2021**, *38*, 158–165.
19. Su, Y.; Xie, Z.; Wang, J.; Lu, Y.; Jiang, Z. Fatigue characteristics test and life prediction on closing spring of 12 kV vacuum circuit breaker. *High Volt. Appar.* **2022**, *58*, 165–171.
20. Xu, Z.; Cheng, Y.; Weng, D.; Pan, Q.; Xie, P.; Chen, K.; Chen, S.; Lin, J.; Chen, J.; Wang, W. Research on the failure and fracture factors of spring in high voltage circuit breakers. In Proceedings of the Fourth International Conference on Mechanical Engineering, Intelligent Manufacturing, and Automation Technology (MEMAT 2023), Guilin, China, 1–3 December 2023.
21. Zhang, X.; Liu, H.; Ma, S.; Yang, P. Thermal air aging and lifespan prediction of PVC-P geomembranes: An Arrhenius equation-based approach. *Power Eng. Eng. Thermophys.* **2023**, *2*, 199–211.
22. Qin, Z.; Jiang, H.; Yuan, H.; Yao, J.; Zhou, B.; Nong, W.; Liu, L. Stress relaxation characteristics analysis and storage life prediction of butt separation pusher spring. In Proceedings of the 13th International Workshop of Advanced Manufacturing and Automation, Shanghai, China, 16–17 October 2023.

energies

MDPI

Article

Numerical Analysis and Experimental Verification of Optical Fiber Composite Overhead Ground Wire (OPGW) Direct Current (DC) Ice Melting Dynamic Process Considering Gap Convection Heat Transfer

Shuang Wang [1,2,*], Long Cheng [2], Bo Tang [1,2], Wangsheng Xu [3] and Zheng Wang [3]

1 Hubei Provincial Engineering Technology Research Center for Power Transmission Line, China Three Gorges University, Yichang 443002, China; tangboemail@sina.com
2 College of Electrical Engineering and New Energy, China Three Gorges University, Yichang 443002, China; ultramarine920@outlook.com
3 China Southern Power Grid Extra High Voltage Power Transmission Company Guiyang Branch, Guiyang 550081, China; xws202409@163.com (W.X.); wangzheng202403@126.com (Z.W.)
* Correspondence: wangshuang@ctgu.edu.cn

Abstract: An accurate analysis of the dynamic process of ice melting in an optical fiber composite overhead ground wire (OPGW) is of great reference significance for the selection of an ice melting current and the formulation of an ice melting strategy. Existing analytical models for the dynamic process of DC ice melting in an OPGW ignore the gap convective heat transfer after the formation of the air gap between the ground wire and the ice layer, and lack the study of the dynamic process of the phase transition of the ice layer. To this end, a finite element model of the DC ice melting process of OPGW was established by introducing the mushy zone constant to consider the influence of the convective heat transfer in the gap, and at the same time, the apparent heat capacity method was used to simulate the changes of the physical property parameters of the melted ice layer. The dynamic process of the ice layer phase transition and OPGW temperature rise during ice melting are calculated, and the effects of the half-width of phase transition interval dT and the mushy zone constant A_m on the DC ice melting process are summarized and analyzed. The accuracy of the OPGW DC ice melting model is verified by conducting DC ice melting experiments. The results show that during the ice melting process, the gap convection heat transfer mainly affects the temperature distribution of the air gap between the ice layer and the OPGW as well as the location of the phase transition interface, and the width of the air gap at the same height below the OPGW increases by about 3 mm after considering the gap convection; the half-width of phase transition interval, dT, mainly affects the location of the phase transition interface and the temperature rise of the modeled heat source, OPGW, while the mushy zone constant, A_m, mainly affects the temperature distribution in the mushy zone, the air gap region. The elliptical phase transition cross-section formed by the OPGW DC ice melting experiment is consistent with the shape of the ice melting simulation model results, and the measured temperature rise curves of the OPGW during DC ice melting are in good agreement with the simulation results, with a maximum difference of about 3.5 K in temperature and 10 min in ice melting time, but the overall trend is consistent, all showing as increasing first and then decreasing.

Keywords: OPGW; DC ice melting; numerical simulation; apparent heat capacity method; gap convection heat transfer

1. Introduction

Ground wire ice covering refers to the phenomenon that the supercooled water in the air condenses on the overhead ground wire when it gets cold, so that the ground wire is covered by ice over a large area. Ground wire ice cover is prone to cause serious defects, such as broken strands, broken wires, and damage to transmission line fittings and towers, which are a great threat to the safe and stable operation of the power system [1,2]. Existing thermal ice melting research mainly focuses on transmission conductors, and there are relatively few studies on ground wire ice melting methods and techniques. Optical fiber composite overhead ground wire (OPGW) is currently widely used in engineering; under the influence of ice melting working temperature, the difference in the elastic modulus of the optical fiber and the coating will cause the optical fiber to be deformed in many micro deformations, which will result in larger optical loss, and the long-term accumulation will inevitably lead to the damage of the optical fiber due to thermal aging and the reduction of its strength, life, and reliability [3]. So, it is particularly important to select the appropriate melting current and melting time, reasonably control the melting temperature, and reduce the impact on the OPGW as much as possible while guaranteeing the melting effect when melting the ice on the OPGW. The accurate simulation and analysis of the dynamic process of OPGW ice melting has an important reference role in the selection of ice melting current and the formulation of ice melting strategy. By constructing the numerical simulation model of the dynamic process of OPGW ice melting, the physical process of ice melting is finely simulated to obtain the rule of change, and then provides technical support for the melting of overhead ground wire, which is of great significance to theoretical research and engineering application value.

Over the years, scholars have proposed ice melting dynamic models, elliptical ice melting models, etc. for the simulation calculation of the thermal ice melting method [4–7]. Jiang Xingliang [8] from Chongqing University relied on the long-term observation of the ice-cover melting process of the line by the artificial climate laboratory and the natural ice covering test station, and through the analysis of the thermal equilibrium process of ice melting, they established the physical-mathematical model of the critical ice melting conditions of the DC, put forward the method of calculating the ice surface temperature and the critical ice melting current, and analyzed the factors affecting the ice surface temperature and the critical ice melting current. Meng Zhigao [9] established a dynamic numerical calculation model for DC ice melting of OPGW, considered the distribution of ice melting current inside the OPGW and the dynamic change in heat exchange between the ice layer and the environment in the process of ice melting, and analyzed the growth process of the air gap, and obtained the change rule of the DC ice melting time of OPGW and the temperature of the optical fiber inside the OPGW in the process of DC ice melting. Yang Guolin [10] analyzed and compared the DC ice melting time for single conductor circular and wing shaped ice cover based on the fact that the shape of the ice cover varies across the conductor. Zhou Yushen [11,12] from Changsha University of Science and Technology established a high-frequency excitation ice melting model for transmission conductors by considering local convective heat transfer, and obtained the effects of local convective heat transfer on the temperature field of the high-frequency ice melting model and its windward and leeward sides. He Zhu [13] constructed a mathematical model of the melting ice vibration characteristics based on the temperature distribution of the ground wire melting process and the characteristic matrix of the ground wire structure. Hu Qin [14] established a dynamic numerical calculation model for DC ice melting, and analyzed the influence of the temperature change in the optical fiber in terms of ambient temperature, wind speed, and ice melting current for the temperature distribution of the OPGW during ice melting and after complete ice melting. Nidhal Ben Khedher [15,16] offered in-depth modeling and experimental evaluation methods for heat transfer improvement with curved fins in latent heat systems and presented comprehensive 3D numerical simulations of melting processes, which combine fin design and heat propagation to benchmark phase change and heat transfer models for this research, while closely aligning these approaches with the thermal modeling of melting phenomena in OPGW systems.

As there are both heat transfer and ice phase change in the ice melting process, the current study has less simulation for the ice phase change process and does not consider the convective heat transfer in the air gap generated in the ice melting process, which affects the accuracy of the dynamic simulation model of ice melting and leads to a discrepancy in the calculation results of the ice melting current and ice melting time with the actual situation. Therefore, considering the gap convective heat transfer, the finite element analysis model of OPGW DC ice melting dynamic process is established in this paper. The phase change problem of the ice layer is solved by using the apparent heat capacity method, and the influence of the half-width of phase transition interval dT and the mushy zone constant A_{m} on the phase change simulation of ice layer is analyzed. Finally, the accuracy of the ice melting model and calculation results in this paper are verified by OPGW DC ice melting experiments.

The structure of this paper is as follows: Section 2 explains the three stages of the physical process of OPGW DC ice melting. Section 3 gives the boundary conditions and control equations of the dynamic numerical simulation of DC ice melting and establishes the geometric model of OPGW DC ice melting. Section 4 analyzes whether to consider the influence of gap convection, half-width of phase transition interval dT, and mushy zone constant A_{m} on the simulation of OPGW DC ice melting. In the Section 5, the OPGW DC ice melting experiment was carried out and compared with the simulation results.

2. Physical Process of OPGW DC Ice Melting

When DC ice melting starts, the surface temperature of the OPGW increases under the effect of Joule heat. When the surface temperature of the OPGW reaches 0 °C, the ice in contact with the OPGW starts to melt. The ice melting process can be divided into three stages as follows:

① After the current flows through the OPGW, the Joule heat heats the OPGW and the ice, and the heat is transferred to the outer surface of the ice to exchange heat with the surroundings by radiation and convection. The surface temperature of the OPGW does not reach 0 °C at this stage, and the overlying ice does not undergo phase change.

② When the surface temperature reaches 0 °C, the ice-covered ground wire is in the critical ice melting state, and the ice in the grooves on the surface of OPGW absorbs heat and melts. After the ice begins to melt, the melted water is lost through the ice gap infiltration, and an air gap is formed between the OPGW and the ice, and the ice gradually moves down by gravity, and the upper surface of the OPGW is in contact with the ice, and the distance between the lower surface and the ice gap increases, so the melting speed of the inner surface of the ice layer is different at each place, and the air gap is in the shape of an ellipse.

③ As the ice moves down, when the thickness of the ice cover on the upper surface of the OPGW reaches a certain level, the upper ice cover cannot withstand its own weight and is cut off by the OPGW, and the OPGW DC ice melting process ends.

3. Numerical Simulation of OPGW DC Ice Melting Process

3.1. Numerical Calculation Modeling of OPGW Ice Melting

3.1.1. Control Equations and Boundary Conditions for Electrical and Thermal Fields

The Joule heat generated by the OPGW flux during the DC ice melting process is partly used to heat the OPGW and the ice and provide heat for the latent heat of phase change; the other part is transferred to the outer surface of the ice by heat conduction through the OPGW and the ice, and is finally lost in the external environment through convective and radiative heat exchanges. Therefore, the energy equation during ice melting can be expressed as [17]:

$$I^2Rt = \int h(T_{\mathrm{i}} - T_{\mathrm{a}})dl \cdot \Delta t + \iint \rho_\theta C_\theta dSdT + \rho_{\mathrm{ice}} L_{\mathrm{f}} \Delta S \tag{1}$$

where I represents the ice melting current, R represents the OPGW DC resistance, h represents the heat transfer coefficient between the outer surface of the ice layer and the external environment, T_{i} and T_{a} represent the temperature of the outer surface of the ice layer and the temperature of the external environment, respectively, ρ_θ represents the density of the microelement of the area of the solution area, C_θ represents the specific heat capacity of the microelement of the area of the solution area, ρ_{ice} represents the density of the ice layer, L_{f} represents the enthalpy of the phase change of the ice, and ΔS represents the area of ice melting in the Δt area of the melted cross-section.

The OPGW through-current region during DC ice melting is the aluminum layer, steel core, and stainless-steel tube, and the current distribution in the OPGW is assumed to be I for the ice melting current:

$$\begin{cases} I_{\mathrm{al}} = \frac{\sigma_{\mathrm{al}} S_{\mathrm{al}}}{\sigma_{\mathrm{al}} S_{\mathrm{al}} + \sigma_{\mathrm{st}} S_{\mathrm{st}}} I \\ I_{\mathrm{st}} = \frac{\sigma_{\mathrm{st}} S_{\mathrm{st}}}{\sigma_{\mathrm{al}} S_{\mathrm{al}} + \sigma_{\mathrm{st}} S_{\mathrm{st}}} I - I'_{\mathrm{st}} \\ I'_{\mathrm{st}} = \frac{S'_{\mathrm{st}}}{S_{\mathrm{st}}} I_{\mathrm{st}} \end{cases} \tag{2}$$

where I_{al}, I_{st}, I'_{st} represent the current in the region of the aluminum layer, steel core, and stainless steel tube, respectively, σ_{al}, σ_{st} represent the conductivity of aluminum and steel, respectively, and S_{al}, S_{st}, S'_{st} represent the cross-sectional area of the aluminum layer, steel core, and stainless steel tube, respectively.

The inner region of the ice layer is mainly for heat conduction and satisfies the following differential equation:

$$\rho c \frac{\partial T}{\partial t} = \nabla \cdot (\lambda \nabla T) + Q \tag{3}$$

where ρ represents the density of the material, c represents the specific heat capacity of the material, λ represents the heat transfer coefficient of the material, and Q represents the intensity of the internal heat source, which can be expressed as follows:

$$Q = \begin{cases} \frac{I_{\mathrm{al}}{}^2 R_{\mathrm{al}}}{S_{\mathrm{al}}} \\ \frac{I_{\mathrm{st}}{}^2 R_{\mathrm{st}}}{S_{\mathrm{st}}} \end{cases} \tag{4}$$

There are convective and radiative heat exchanges between the outer surface of the ice layer and the external environment, so the third type of boundary conditions should be satisfied [18]:

$$-\lambda \frac{\partial T}{\partial n} = h(T_{\mathrm{i}} - T_{\mathrm{a}}) \tag{5}$$

where h represents the sum of the convective heat transfer coefficient and the radiative heat transfer coefficient; the convective heat transfer coefficient includes the natural convective heat transfer coefficient and the forced convective heat transfer coefficient, expressed as follows:

$$\begin{cases} h = h_{\mathrm{D}} + h_{\mathrm{R}} \\ h_{\mathrm{D}} = h_{\mathrm{N}} + h_{\mathrm{F}} \\ h_{\mathrm{N}} = \frac{\lambda_{\mathrm{air}} B(GrPr)^{b}}{D} \\ h_{\mathrm{F}} = \frac{\lambda_{\mathrm{air}} C(Re)^{n}(Pr)^{1/3}}{D} \end{cases} \tag{6}$$

where h_{D}, h_{R}, h_{N}, h_{F} represent convective heat transfer coefficient, radiative heat transfer coefficient, natural convective heat transfer coefficient, and forced convective heat transfer coefficient, respectively, λ_{air} represents the air heat transfer rate, D represents the geometrical diameter of the ice-covered area, and Re, Gr, Pr represent the Reynolds number, the Prandlt number, and the Grashof number, respectively:

$$\begin{cases} Re = \frac{V_{\mathrm{a}} D \rho_{\mathrm{air}}}{\mu_{\mathrm{air}}} \\ Gr = \frac{2g(T-T_{\mathrm{a}})D^{3}}{v_{\mathrm{air}}^{2}(T+T_{\mathrm{a}})} \\ Pr = \frac{\mu_{\mathrm{air}} C_{\mathrm{air}}}{\lambda_{\mathrm{air}}} \end{cases} \tag{7}$$

where g represents the gravitational constant, v represents the air kinematic viscosity, μ represents the coefficient of kinetic viscosity of the air, C_{air} represents the specific heat capacity of the air, V_{a} represents the external ambient wind speed, ρ_{air} represents the air density, and B, b, C, and n represent the coefficients determined by the Grashof number and Reynolds number, which can be obtained by looking up the table [18].

The radiative heat transfer coefficient h_{R} is expressed as follows:

$$h_{\mathrm{R}} = \varepsilon \sigma_{\mathrm{r}} (T^{2} + {T_{a}}^{2})(T + T_{a}) \tag{8}$$

where ε represents the emissivity of the outer surface of the ice layer, $\varepsilon = 0.9$. σ_{r} represents the radiation constant, $\sigma_{\mathrm{r}} = 5.67 \times 10^{-8}\ \mathrm{W/(m^{2} \cdot K^{4})}$.

3.1.2. Characterization of Ice Layer Material Properties by Apparent Heat Capacity Method

During the physical process of ice melting, there are three types of regions in the phase change distribution of the ice: solid phase, mushy zone, and liquid phase. The paste region is a semi-solid region that exists as an interface between the melted and unmelted regions of the phase change material.

Based on the relative magnitudes of the actual temperature T in the solution region with respect to the melting point T_{m} of the ice layer and the width of the phase transition interval ΔT, the phase transition process of the ice layer during the DC ice melting process of the OPGW is solved by using the apparent heat capacity method, and the three types of regions can be characterized by the segmented function liquid phase fraction $\varphi(T)$ [19]:

$$\varphi(T) = \begin{cases} 0 & T < T_{\mathrm{m}} - \frac{\Delta T}{2} \\ \frac{T-(T_{\mathrm{m}}-\frac{\Delta T}{2})}{\Delta T} & T_{\mathrm{m}} - \frac{\Delta T}{2} \leq T \leq T_{\mathrm{m}} + \frac{\Delta T}{2} \\ 1 & T > T_{\mathrm{m}} + \frac{\Delta T}{2} \end{cases} \tag{9}$$

where T_{m} represents the melting point of the ice layer and ΔT represents the width of the phase transition interval.

The material properties during the phase transition of the ice layer in each of the three regions: solid phase, mushy zone, and liquid phase can be expressed as [20]:

① Equivalent heat capacity $C_e(T)$

$$C_e(T) = \begin{cases} C_{fs}(T) & T < T_m - \frac{\Delta T}{2} \\ C_{fs}(T) + [C_{fg}(T) - C_{fs}(T)]\varphi(T) + L_f \cdot D(T) & T_m - \frac{\Delta T}{2} \le T \le T_m + \frac{\Delta T}{2} \\ C_{fg}(T) & T > T_m + \frac{\Delta T}{2} \end{cases} \quad (10)$$

where C_{fs} and C_{fg} represent the specific heat capacity of ice and air, respectively, and L_f represents the latent heat of phase transition; the contribution in the equivalent specific heat capacity is represented by a Gaussian function $D(T)$ centered on T_m [21]:

$$D(T) = \frac{e^{-\frac{(T-T_m)^2}{(\Delta T/4)^2}}}{\sqrt{\pi(\Delta T/4)^2}} \quad (11)$$

② Equivalent heat transfer coefficient $k_e(T)$

$$k_e(T) = \begin{cases} k_{fs}(T) & T < T_m - \frac{\Delta T}{2} \\ k_{fs}(T) + [k_{fg}(T) - k_{fs}(T)]\varphi(T) & T_m - \frac{\Delta T}{2} \le T \le T_m + \frac{\Delta T}{2} \\ k_{fg}(T) & T > T_m + \Delta T/2 \end{cases} \quad (12)$$

where k_{fs} and k_{fg} represent the heat transfer coefficients for ice and air, respectively.

③ Equivalent density $\rho_e(T)$

$$\rho_e(T) = \begin{cases} \rho_{fs}(T) & T < T_m - \frac{\Delta T}{2} \\ \rho_f(T) & T_m - \frac{\Delta T}{2} \le T \le T_m + \frac{\Delta T}{2} \\ \rho_{fg}(T) & T > T_m + \frac{\Delta T}{2} \end{cases} \quad (13)$$

where ρ_{fs} and ρ_{fg} represent the densities of ice and air, respectively.

$\rho_f(T)$ is related to the solid phase rate φ_{fs}, which is 1 in the solid phase and 0 in the liquid phase for phase change materials, and linearly correlates in the paste region between the solid phase rate and the temperature distribution in the paste region, expressed as follows:

$$\begin{cases} \rho_f(T) = \phi_{fg}\rho_{fg}(T) + \phi_{fs}\rho_{fs}(T) \\ \phi_{fs} = \frac{(T_m + \Delta T/2) - T}{\Delta T} \\ \phi_{fs} + \phi_{fg} = 1 \end{cases} \quad (14)$$

where φ_{fs} represents the solid phase rate and φ_{fg} represents the liquid phase rate.

3.1.3. Convective Flow in Fiber Optic Paste, Paste Region, Air Gap

The fiber optic fiber paste as well as the air gap solution region from the ice melting is an incompressible flow, which can be represented by the Navier–Stokes equations [22,23]:

$$\begin{cases} \rho\nabla\cdot\mathbf{u} = 0 \\ \rho(\frac{\partial\mathbf{u}}{\partial t} + \mathbf{u}\cdot\nabla\mathbf{u}) = \nabla\cdot[-p\mathbf{I} + \mu(\nabla\mathbf{u} + (\nabla\mathbf{u})^T)] - S(T)\cdot\mathbf{u} \end{cases} \quad (15)$$

where ρ represents the fluid density, $\mathbf{u}$ represents the fluid velocity vector, p represents the fluid pressure, $\mathbf{I}$ represents the unit matrix, μ represents the hydrodynamic viscosity, and $S(T)$ represents the damping term.

In order to describe the convective flow after a phase change in the ice layer, while preventing the solid from generating flow, a Darcy's law damping term $S(T)$ is added to the momentum equation, incorporating the Carman–Kozeny relation, which has a large value when the liquid phase fraction tends to 0, forcing the solid-phase velocity field to be 0, and eliminating the term when the liquid-phase fraction region is 1. It can be expressed as follows:

$$S(T) = A_{\mathrm{m}} \frac{(1-\varphi(T))^2}{\varphi(T)^3 + \varepsilon} \tag{16}$$

where A_{m} denotes the mushy zone constant, for incompressible flow, $A_{\mathrm{m}} \geq 1 \times 10^4$ kg/(m^3·s), and the constant ε, $\varepsilon = 0.001$, is added to avoid de-zeroing in the case of $\varphi(T) = 0$.

3.1.4. Dynamic Mesh Following at the Ice Layer Phase Transition Interface

In the ice layer melting model, there are thermal and mass balances at the phase transition interface, from which the Stefan condition, which the interface velocity V_{s} should satisfy, can be derived [24]:

$$\begin{cases} V_{\mathrm{s}} = \frac{Q_{\mathrm{s}}}{\rho_{\mathrm{ice}}(T) L_{\mathrm{f}}} \\ Q_{\mathrm{s}} = \lambda_{\mathrm{ice}} \frac{\partial T}{\partial n} \end{cases} \tag{17}$$

where Q_{s} represents the sudden change in normal heat flux at the interface, ρ_{ice} is the density of ice, and λ_{ice} represents the heat transfer coefficient of ice.

With the OPGW DC ice melting process, the air gap between the ice layer and the OPGW gradually increases, and the ice layer will gradually move downward after melting due to the influence of gravity, and the process is simulated in the model by a moving mesh, and the mesh moving speed is the phase transition interface velocity V_{s}, which can be determined by Equation (17).

3.2. Geometric Modeling

Fiber optic composite overhead ground wire OPGW-24B1-98 [124.2; 51.8] was selected to establish its DC ice melting two-dimensional model, and the corresponding parameters are shown in Tables 1 and 2. The density, specific heat capacity, and heat transfer coefficient of the ice layer are determined according to the apparent heat capacity method in Section 3.1.2. In addition, the initial conditions of ice melting are ambient temperature $T_{\mathrm{a}} = 268.15$ K, wind speed $V_{\mathrm{a}} = 5$ m/s, ice melting current $I = 500$ A, and the thickness of ice cover $d = 15$ mm.

Table 1. OPGW technical parameters.

Ground Type	Calculated Cross-Sectional Area (mm^2)		Outer Diameter (mm)	20 °C DC Resistance (Ω/km)
	Aluminum	Steel		
OPGW-24B1-98 [124.2; 51.8]	26.069	79.377	13.20	0.884

Table 2. OPGW electric heating parameters.

Material	Density (kg·m^{-3})	Thermal Conductivity (W·(m·K)$^{-1}$)	Specific Heat Capacity (J·(kg·K)$^{-1}$)	Conductivity (S/m)
Steel	7850	45.00	460	7.69×10^6
Aluminum	2700	238	900	3.77×10^7

The ice cover OPGW simulation model established in this paper is glaze icing. Due to the complexity of the actual ice cover morphology and ice melting conditions of OPGW, the following assumptions and simplifications are made for the convenience of DC ice melting modeling and calculations while taking into account its accuracy [25–27]:

① Ice-covered cross-section is circular: Due to the small diameter of OPGW, the torsional stiffness is small; when the ice cover is piled up on the windward side, it is more likely to torsion under the action of the gravity of the ice layer to make the shape of the ice-covered cross-section close to circular.

② Neglecting the axial heat transfer of OPGW: When the OPGW is de-icing in the same period or the de-icing part of the wire is long, the axial heat transfer of OPGW can be neglected, so the two-dimensional field is used to analyze and solve the temperature distribution of OPGW.

③ Neglect de-icing shear: the process of ice melting on the ground wire is the process of expanding the air gap, and when the thickness of the ice layer on the upper surface of the conductor is zero, the ice layer is dislodged from the surface of the conductor.

④ Neglecting the effect of OPGW strand gap on heat transfer: the OPGW diameter is relatively small, and the strand gap is also small, which has less influence on the heat transfer process of ice melting. In order to improve the quality of the dissecting mesh and the convergence of the solution, the geometry of the optical fiber is taken into account, and the geometrical models of the corresponding aluminum and steel parts are established with the equivalent aluminum and steel cross-sectional areas as a whole, respectively, so that the overall geometrical model of the OPGW is finally formed, as shown in Figure 1.

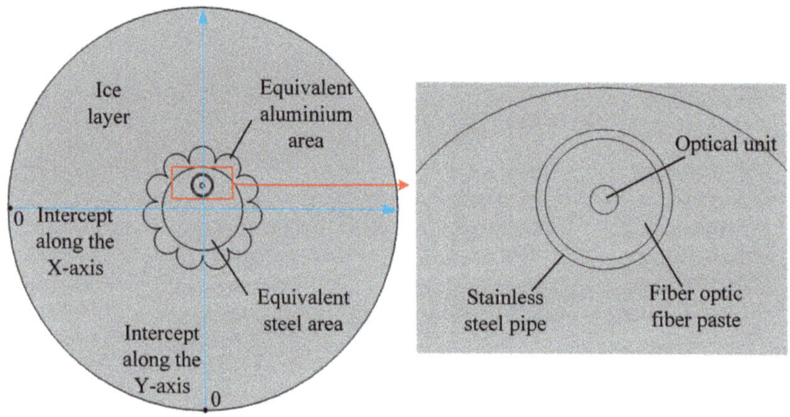

Figure 1. OPGW geometric model.

3.3. Numerical Calculation of the Dynamic Process of Ice Melting in OPGW

By establishing a numerical analysis model of OPGW DC ice melting and using the finite element method, the laws of the phase change process and temperature distribution of the OPGW ice layer considering the interstitial convective heat transfer are calculated, and the calculation flow is shown in Figure 2.

In the simulation of the dynamic process of ice melting, the difficulty lies in the tracking of the dynamic mesh following the phase transition interface; therefore, the first step is to determine the location of the phase transition interface. After solving for the heat flow density inside the model, the heat and mass balances existing at the phase transition interface can be determined, and then the interface velocity, i.e., the rate of phase transition at the top of the inner surface of the ice layer, can be derived from Equation (17) and used as the following velocity of the moving mesh. Equation (9) is used as a judgement condition for whether or not to perform moving mesh tracking; when the liquid phase fraction is less than 1, the mesh does not move, i.e., no tracking is performed. When the liquid phase fraction is equal to 1, the mesh starts to move, and the mesh tracking is performed.

As the mesh moves following the phase transition interface, the mesh cells are stretched or squeezed, which may lead to non-convergence or inaccurate results, and there must also be an automatic updating of the mesh, which uses the minimum relative cell volume as the mesh redrawing condition, and the mesh is redrawn when it is less than a threshold value of 0.5. Otherwise, the mesh continues to move until it rejoins the outer boundary of the ice layer.

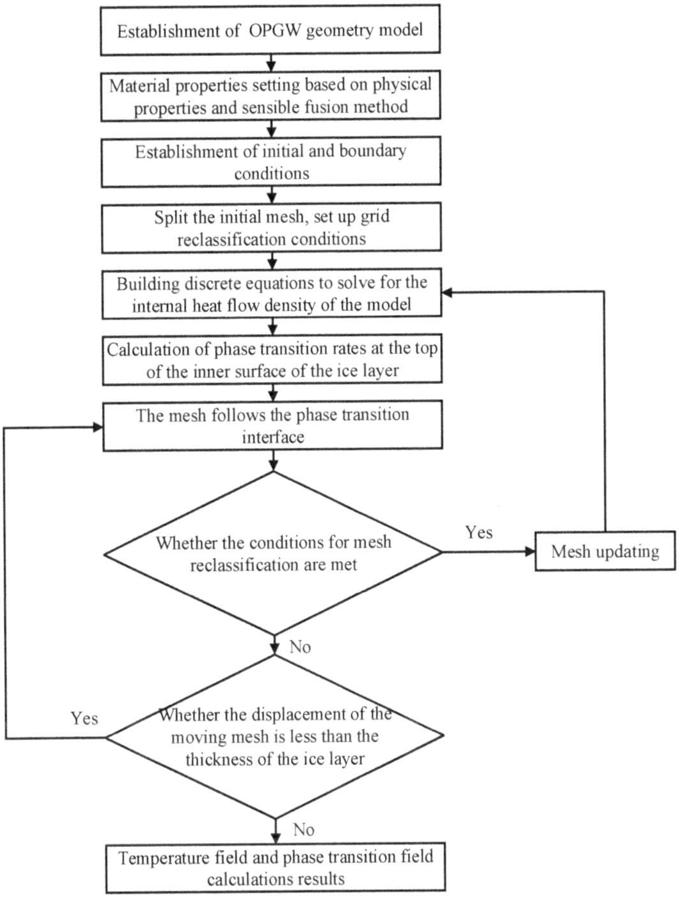

Figure 2. Flow chart of OPGW DC ice melting simulation calculation.

After the initial mesh dissection, in addition to the domain boundaries, the computational domain mesh is a free triangular mesh, the domain boundaries are surrounded by a quadrilateral boundary layer mesh, and corner refinement is used in the corner region, and the overall mesh delineation is shown in Figure 3, with the largest cell size of 1.61 mm and the number of divided meshes of 15,862. The outer boundary of the OPGW is a moving mesh boundary, which is used to simulate the downward movement of the ice layer in the process of ice-melting, and the movement speed is determined according to Equation (17).

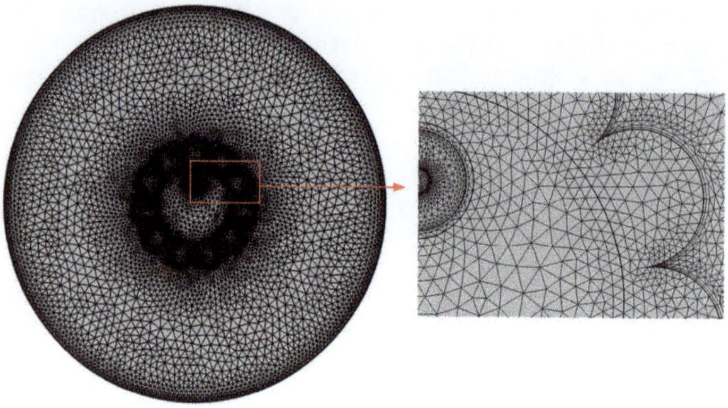

Figure 3. Computational domain meshing.

3.4. Mesh Independence Verification

In order to minimize the influence of the mesh density on the accuracy of the numerical calculation results and to ensure the reliability of the results and the efficiency of the calculation. Three mesh scales with mesh numbers of 11,266, 15,862, and 21,240 were selected. By choosing the truncated path along the Y-axis as shown in Figure 1, the temperature profiles and phase transition profiles for different mesh densities were obtained, as shown in Figure 4.

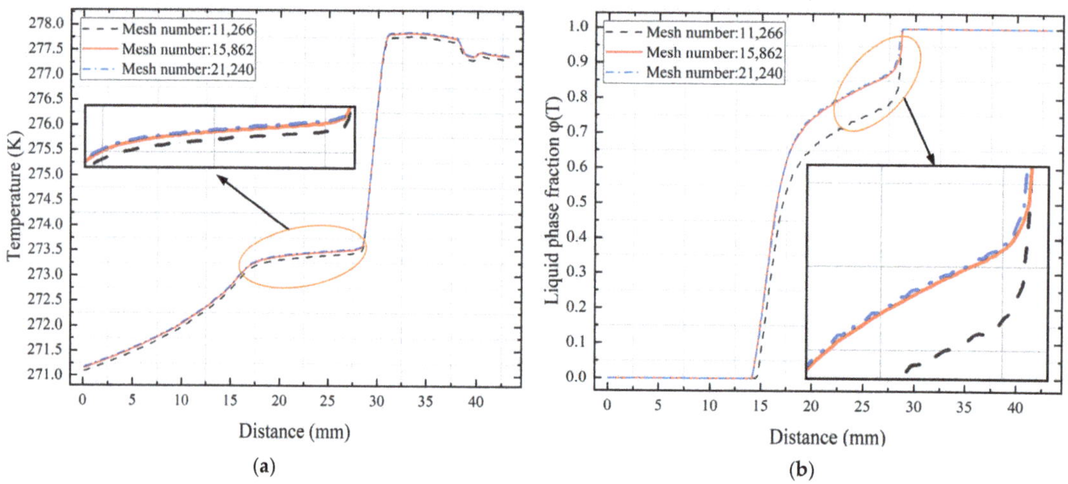

Figure 4. Temperature profiles and phase transition profiles for different mesh densities. (**a**) Temperature profile along the Y-axis intercept. (**b**) Phase transition curve along the Y-axis intercept.

It can be seen that as the mesh density increases, its impact on the accuracy of the calculation results decreases, and the number of meshes of 15,862, 21,240, the two temperature curves and phase transition curves are basically the same. Therefore, considering the calculation accuracy and cost, the mesh scale of 15,862 is selected for numerical simulation in the subsequent study.

4. Simulation Analysis of OPGW DC Ice Melting

4.1. *Influence of Interstitial Convective Heat Transfer on the Temperature and Phase Change Distribution of the Dynamic Process of Ice Melting*

Considering the interstitial convective heat transfer, the temperature and phase change distributions of the dynamic process of OPGW ice melting were obtained by applying the calculation method shown in Figure 2 to the established OPGW DC ice melting model. Figure 5a,b show the curves of the temperature and convective heat transfer coefficient of the outer surface of the top and bottom of the ice layer with time during ice melting, respectively.

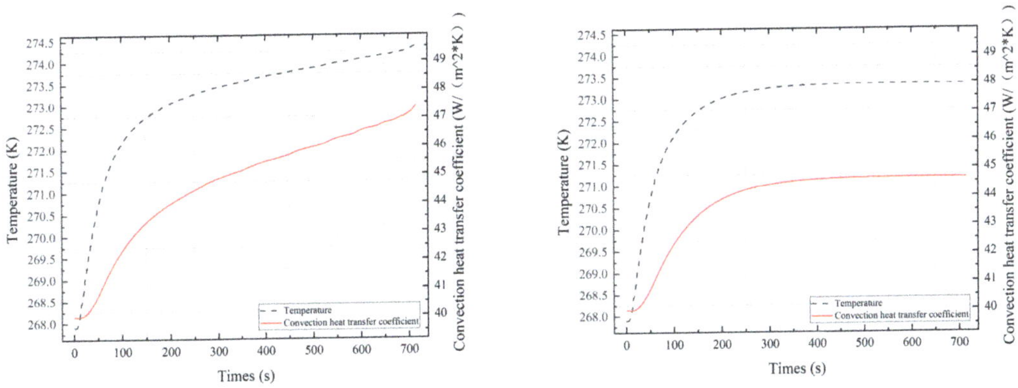

(a) (b)

Figure 5. Changes in the outer surface temperature and convective heat transfer coefficient of the ice layer during the ice melting process. (**a**) Top outer surface of the ice layer. (**b**) Outer surface at the base of the ice layer.

Prior to the start of DC ice melting, there is no temperature difference between the outer surface of the ice layer and the outside environment, and there is no heat exchange. After the ice melting current is applied, the convective heat transfer coefficient increases as the temperature of the OPGW and the ice layer increases under the action of Joule heat, and there is a temperature difference between the ice layer and the external environment.

From Figure 6, it can be seen that in the initial stage of ice melting, the temperature rise of the outer surface of the ice layer is faster and then gradually tends to the melting point, so the convective heat transfer coefficient shows the same growth trend; but, as the ice layer continues to melt and move downward, the thermal resistance of the ice layer at the top decreases, and at the bottom, due to the continuous growth of air gaps hindering the transfer of heat, the temperature rise of the ice layer as well as the convective heat transfer coefficient grows more gently, so that in the process of melting, the OPGW temperature. Therefore, the maximum value of the OPGW temperature during the ice melting process occurs at the bottom outer surface.

The temperature distributions and phase transition distributions of the OPGW ice melting model with or without considering the gap convection were obtained by selecting three typical time points in the pre-melting, mid-melting, and deglazing processes, as shown in Figure 6a–d, respectively.

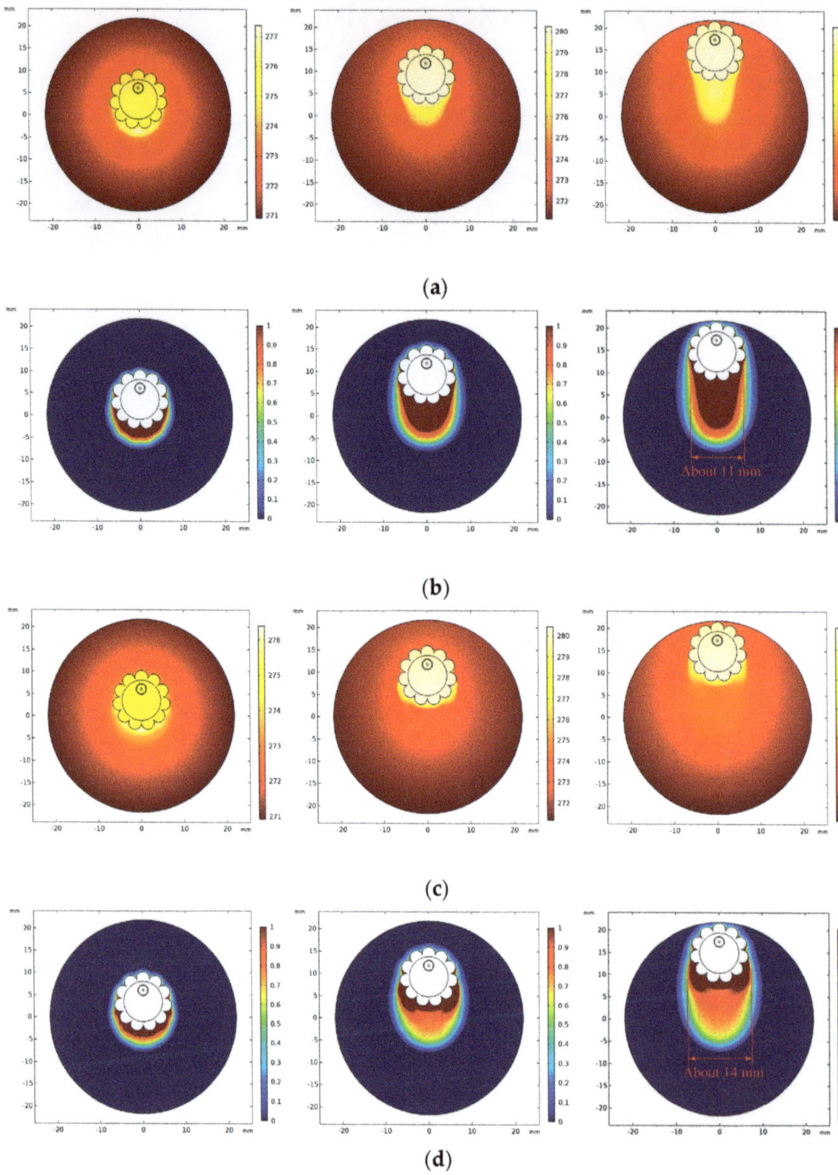

Figure 6. Temperature distribution and phase transition distribution during ice melting process. (**a**) Temperature distribution without considering gap convection. (**b**) Distribution of phase transitions without considering gap convection. (**c**) Temperature distribution considering gap convection. (**d**) Distribution of phase transitions considering gap convection.

In Figure 6b,d, the phase transition distribution is characterized by Equation (9), where the dark blue region is 0, indicating that no phase transition has occurred, i.e., the ice layer, and the dark red region is 1, indicating that the ice layer has undergone a phase transition, i.e., the air gap, and the region between 0 and 1 indicates the phase transition interval, i.e., the mushy zone, of the ice layer.

Comparing and analyzing the temperature distribution at the same time point in Figure 6a,c, it can be found that there is a significant difference in the temperature distributions in the air gap region between the two, with the latter having a lower temperature in the air gap region, suggesting that it is necessary to consider interstitial convection in the numerical simulation of DC ice melting. In addition, near the lower part of the OPGW, the maximum difference in the width of the air gap at the same phase change interface location of the ice layer is about 3 mm after considering the gap convection, and the phase transition cross-section shape of the ice layer in Figure 6b,d is elliptical.

Selecting the paths along the X-axis intercept and along the Y-axis intercept as shown in Figure 1, the temperature and phase transition calculations for the corresponding locations at the moment of OPGW de-icing are extracted with or without consideration of interstitial convection, as shown in Figure 7.

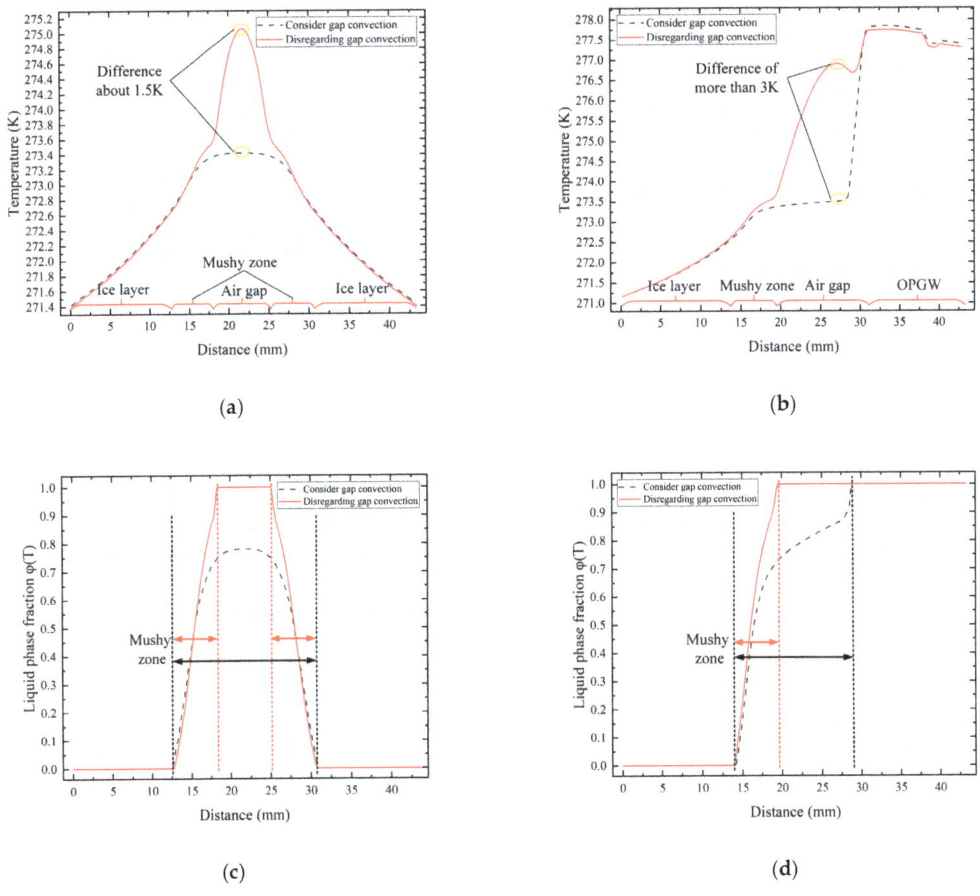

Figure 7. Temperature profiles and phase transition profiles during ice melting with or without considering gap convection. (**a**) Temperature profile along the X-axis intercept. (**b**) Temperature profile along the Y-axis intercept. (**c**) Phase transition curve along the X-axis intercept. (**d**) Phase transition curve along the Y-axis intercept.

As an example, the range of intercept distances for the ice layer, mushy zone, air gap, and OPGW without considering gap convection is indicated as shown in Figure 7a,b. The location of the mushy zone for the case of whether or not to consider gap convection is shown in Figure 7c,d.

From Figure 7a,b, it can be seen that the temperature curve is almost fit in the range of the ice layer, mushy zone, and OPGW truncation distance, while in the range of air gap truncation distance. The temperature difference along the X-axis intercept is about 1.5 K, and the temperature difference along the Y-axis intercept is more than 3 K. From Figure 7c,d, it can be seen that whether the gap convection is considered, the width of the mushy zone is different and the phase transition distribution is also different, but the starting point of the mushy zone overlaps.

The above analysis illustrates that during the ice melting process, whether or not to consider interstitial convection mainly affects the temperature distribution in the air gap region and the location of the ice layer phase transition interface, and has no significant effect on the temperature change of the OPGW. This is due to the fact that after considering gap convection, there is no longer only a heat conduction process in the air gap region, but it also contains the convective heat transfer process, which makes the heat in the air gap transfer to the inner surface of the ice layer better and accelerates the melting of the ice layer, which results in the difference of the air gap width with a maximum difference of about 3 mm as shown in Figure 6b,d; the melting time of ice melting without considering the gap convection is 4776.87 s, i.e., about 80 min, while the melting time after considering the gap convection is about 4563.34 s, i.e., about 76 min. Therefore, it is necessary to consider the gap convection in the numerical simulation of DC ice melting.

4.2. Influence of the Half-Width of the Phase Transition Interval and Mushy Zone Constants on the Dynamic Process of Ice Melting

When performing the phase change analysis of the OPGW DC ice melting process, since the volume and shape of the air gap during the ice melting process directly affects the convective motion and convective strength of the air, which in turn, also affects the heat transfer process within the gap, and thus in turn, affects the volume and shape of the air gap, determining the location of the phase change interface and its changes, and simulating the volume and shape of the air gap that agrees with the actual situation have a significant impact on the ice melting dynamic physical process simulation results.

In this paper, the apparent heat capacity method is used to characterize the properties of phase change materials, and the half-width of the phase transition interval dT (dT = $\Delta T/2$) and the constant A_m of mushy zone are used to determine and simulate the position of phase change interface.

From Equation (9), the half-width of phase transition interval dT is used to define the temperature interval in which the material undergoes a phase transition, i.e., the start temperature of the phase transition and the end temperature of the phase transition, which determines the size of the mushy zone. The mushy zone constant A_m is used to indicate the amount of change in fluid velocity as the ice melting; the larger the value, the faster the velocity decreases. It has been shown that the values of dT and A_m have an important influence on the accuracy of the simulation calculation of the phase transition process [28], and reasonable values of the parameters should be selected for different phase transition materials and different geometries [22].

According to the literature [20], three groups of typical half-widths of phase transition intervals and mushy zone constants were selected, respectively, to analyze their influence laws on the dynamic process of ice melting and to provide reference for the corresponding parameter selection in the numerical analysis of ice melting. The half-width dT values of the three groups of phase transition intervals are 0.25 K, 0.5 K, and 0.75 K, and the values of the three groups of mushy zone constants A_m are 10^4, 10^5, and 10^6.

In this paper, the influence of the half width dT of the phase transition interval in Equation (9) and the constant A_m of the mushy zone in Equation (16) on the phase transition interface is compared and analyzed. The X-axis and Y-axis cross-section paths shown in

Figure 1 are selected to extract the temperature and phase transition calculation results of the corresponding position at the time of OPGW de-icing, as shown in Figures 8 and 9.

4.2.1. Simulation Analysis of the Dynamic Process of Ice Melting for Different Phase Transition Interval Half-Widths

Taking dT = 0.5 K as an example, the range of cut-off distance of the ice layer, mushy zone, air gap, and OPGW is indicated, as shown in Figure 8a,b. The position of the half-wide mushy zone of the three groups of phase transition intervals is shown in Figure 8c,d.

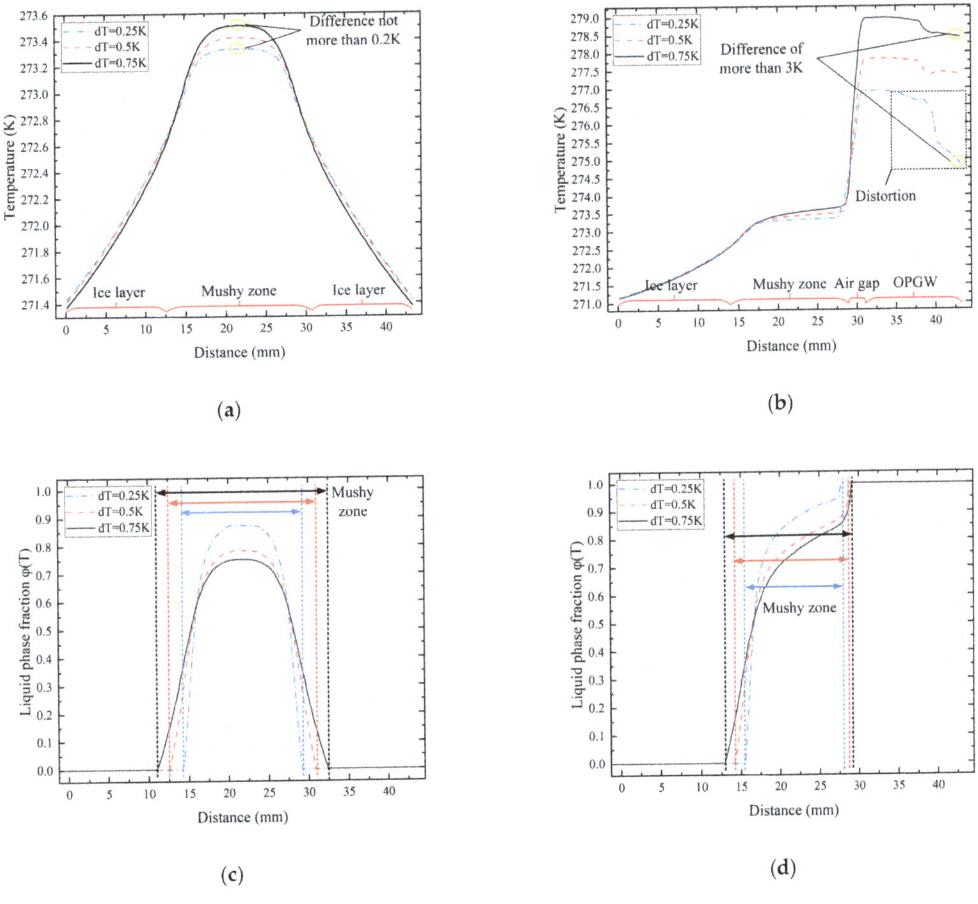

Figure 8. Temperature profiles and phase transition profiles during ice melting at different half-width of phase transition intervals. (**a**) Temperature profile along the X-axis intercept. (**b**) Temperature profile along the Y-axis intercept. (**c**) Phase transition curve along the X-axis intercept. (**d**) Phase transition curve along the Y-axis intercept.

From Figure 8a,b, it can be seen that the temperature difference does not exceed 0.2 K in the mushy zone as well as in the air gap within the truncation distance, whereas in Figure 8b, the temperature difference can be up to more than 3 K within the OPGW truncation distance. As can be seen from Figure 8c,d, the distribution of phase transitions within the mushy zone is different for three sets of values of the half-width of the phase transition interval, dT, with different starting points and widths of the mushy zone.

The above analysis illustrates that different values of the half-width of phase transition interval, dT, have an impact on the phase transition simulation, mainly in terms of the location of the phase transition interface and the temperature rise of the heat source, OPGW, in the model, due to the fact that the temperature interval of the phase transition, as defined by dT, determines the magnitude of the heat required to be absorbed by the melting process of the phase-transitioned material. From Figure 8c,d, it can be seen that when dT is larger, the phase transition region, i.e., the mushy zone, is also wider and the overall temperature in the computational domain is higher, and when dT is too small, the abrupt change in material properties due to phase transition will be distorted, as shown in Figure 8b with dT = 0.25 K.

4.2.2. Simulation Analysis of the Dynamic Process of Ice Melting for Different Mushy Zone Constants

Taking $A_m = 10^5$ as an example, the range of cut-off distance of ice layer, mushy zone, air gap, and OPGW is indicated, as shown in Figure 9a,b. The mushy zone positions of the three groups of mushy zone constants are shown in Figure 9c,d.

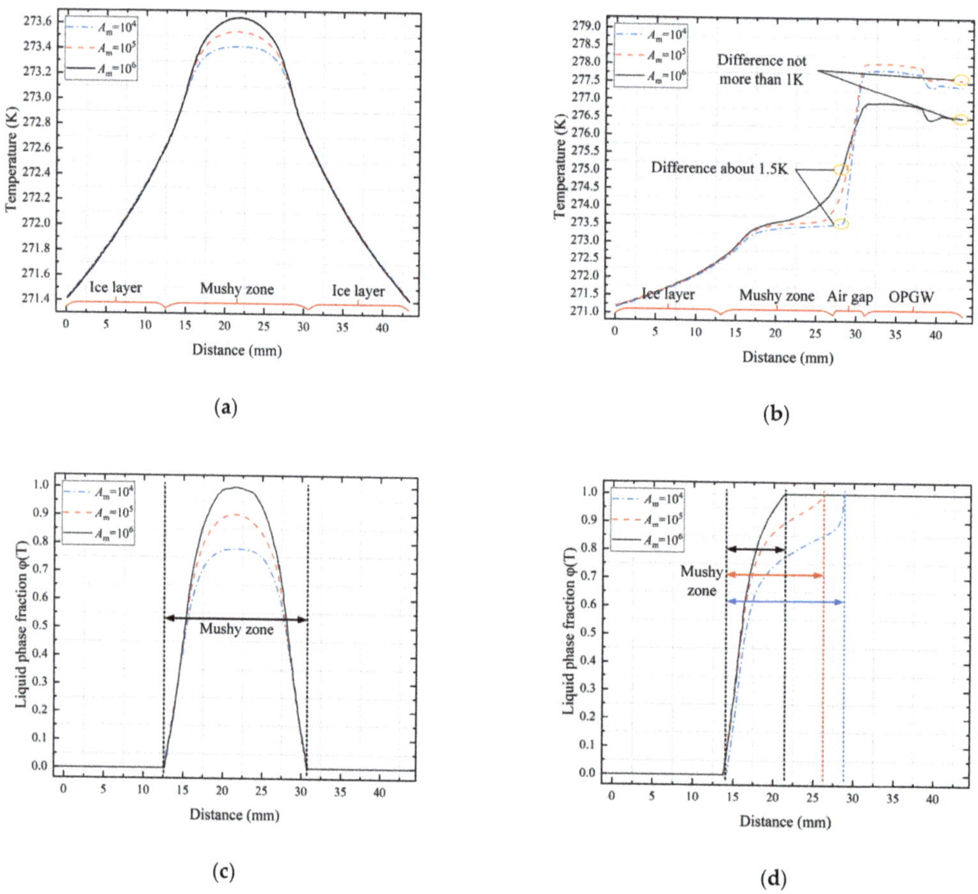

Figure 9. Temperature profiles and phase transition profiles during ice melting with different mushy zone constants. (**a**) Temperature profile along the X-axis intercept. (**b**) Temperature profile along the Y-axis intercept. (**c**) Phase transition curve along the X-axis intercept. (**d**) Phase transition curve along the Y-axis intercept.

As can be seen from Figure 9a,b, compared to Figure 8a,b, the temperature difference increases to a maximum of about 1.5 K in the mushy zone as well as in the air gap in the range of the cut-off distances, whereas the temperature difference decreases to within 1 K in the range of the OPGW cut-off distances. As can be seen from Figure 9c,d, the onset of the mushy zone overlaps for the three sets of values of the mushy zone constant A_m compared to Figure 8c,d, even though the distribution of phase transitions within the mushy zone is also different.

The above analysis illustrates that the effect of different mushy zone constants A_m on the phase transition simulation is mainly reflected in the temperature distribution in the mushy zone, air gap region. This is because A_m is characterizing the state of convective motion of the fluid and affects the convective heat transfer process of the fluid in the mushy zone, air gap region. At A_m = 10^4, 10^5, the OPGW temperature difference is within 0.25 K; when A_m is too large, the existence of damping term makes it difficult for convective heat transfer to occur in the air gap, which is closer to the simulation process of ice melting without considering the convection in the gap, such as the Y-axis truncated phase transition curve, shown by A_m = 10^6 in Figure 9d, which shows the step trend. Although the effect of the mushy zone constant A_m on the position of the phase transition interface is small, it can be seen from Figure 6 that, without considering the gap convection, i.e., when the mushy zone constant is not introduced, there is a 3 mm difference in the size of the air gap for the phase transition interface at the same height below the OPGW, so this parameter should not be neglected in the numerical simulation of the phase transition.

5. Experiment Verification of OPGW DC Ice Melting

The artificial icing of OPGW was carried out in the laboratory, and the DC ice melting method was used to carry out the ice melting experiment. The temperature variation law of OPGW during the ice melting process was measured, and the melting process of ice melting was observed and compared with the simulation results of the dynamic process of ice melting.

The OPGW sample used in the experiment is 1000 mm long and the model is the same as that in the simulation model. The PT-100 patch leaded platinum resistor is used as a temperature sensor, which is pre-buried on the outer surface of the OPGW and close to the ground side, as shown in Figure 10.

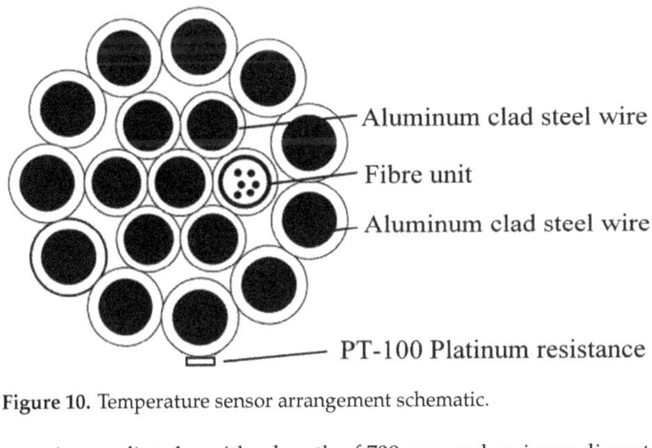

Figure 10. Temperature sensor arrangement schematic.

An acrylic tube with a length of 700 mm and an inner diameter of 43.2 mm was used as the OPGW ice-covering device, and the OPGW sample with a pre-buried temperature sensor was placed in the center of the acrylic tube, which was then filled with water and placed in the cooler. After the outside of the OPGW in the acrylic tube was completely frozen, it was removed from the cooler, the acrylic tube was removed, and DC ice melting experiments were conducted on the ice-covered OPGW. Figure 11 shows the prepared ice-covered OPGW sample.

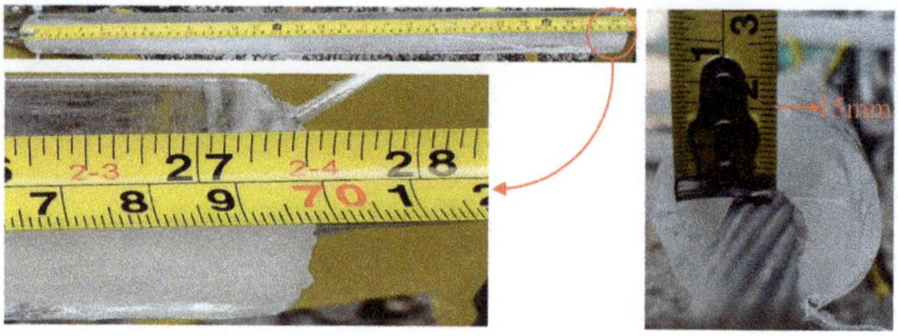

Figure 11. Ice-covered OPGW sample.

The experimental wiring is shown on Figure 12. In order to ensure that the temperature of the ice melting experiment is consistent with the ambient temperature of the simulation model, the ice melting experiment is carried out in a temperature-settable constant-temperature freezer, and the temperature of the freezer is set to 268.15 K, i.e., −5 °C, during the ice melting process. The effect of ambient wind speed on ice melting was simulated by an adjustable axial fan, and the wind speed of the fan was set to 5 m/s. The experimental site layout is shown in Figure 13.

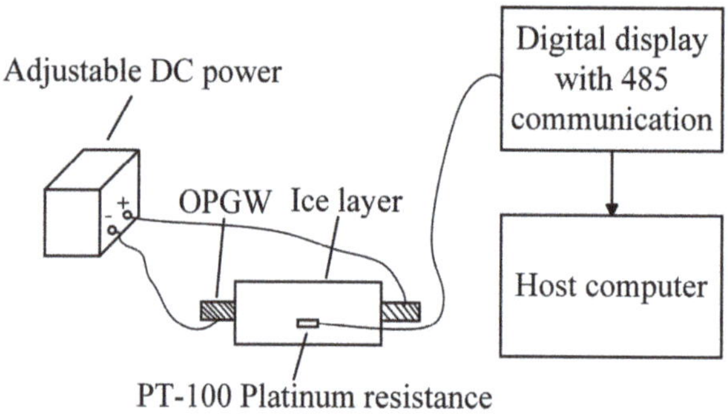

Figure 12. OPGW DC ice melting wiring diagram.

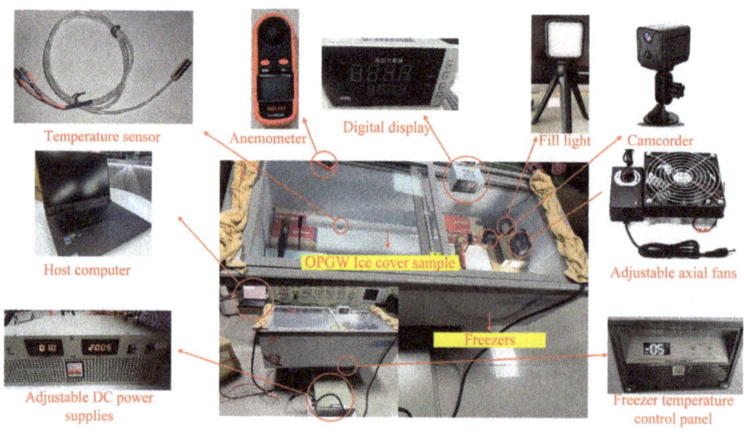

Figure 13. OPGW DC ice melting experiment site arrangement.

The adjustable DC power supply is used as the ice melting power supply, and the ice melting current is set at 200 A. In order to observe the dynamic process of ice melting in OPGW in real time, a high-definition camera is used to film the whole process of ice melting and to observe the changes of the ice layer. The experimentally observed process of OPGW ice melt dynamics is shown in Figure 14. The platinum resistance temperature sensor was placed at the bottom of the outer surface of the OPGW; the measured temperature data were first processed by the digital display instrument, and then through the 485 communication busy line, and through the host computer real-time monitoring and display every 1 min to record the surface temperature value of the OPGW. The measured temperature rise curve of the OPGW DC ice melting is shown as the red solid line in Figure 15.

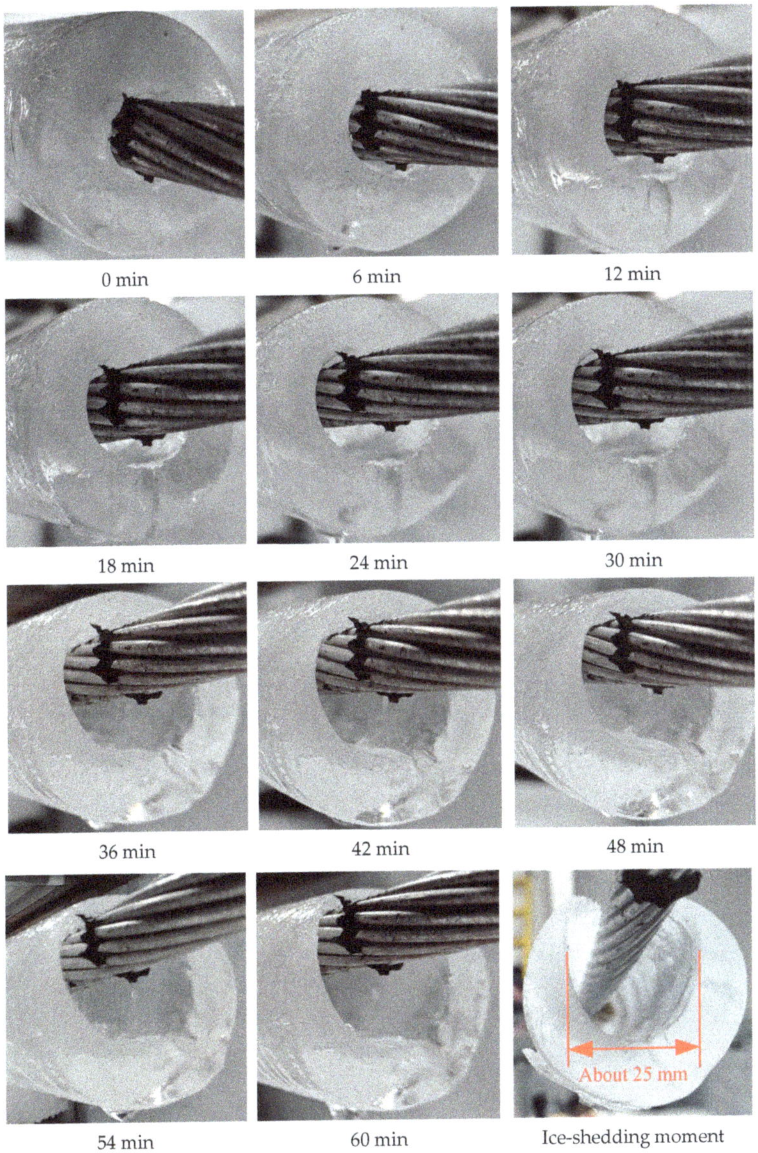

Figure 14. OPGW DC ice melting experiment ice melting process.

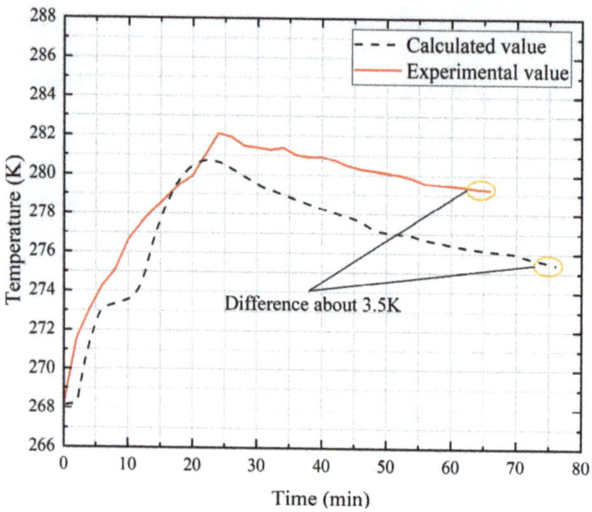

Figure 15. OPGW DC ice melting temperature rise curve.

As shown in Figure 14, during the ice melting process, due to the effect of gravity, the ice layer gradually moves downward after forming an air gap, gradually forming an elliptical air gap, which makes the distance between the outer surface of the OPGW and each specific point on the inner surface of the ice layer inconsistent. The thermal resistance of the air gap in different heat transfer directions is also inconsistent, thus making the heat absorbed by the ice layer in different heat transfer directions, different, which leads to different melting rates of ice layers at different locations. Meanwhile, considering the gap convection heat transfer, the airflow velocity at the angle between the two sides of the OPGW and the inner surface of the ice layer is the highest, and the convective heat transfer coefficient is also large. The faster the heat transfer rate, the faster the ice layer melting rate, and the melting rate of the inner surface of the ice layer farthest from the OPGW surface is the slowest. Therefore, the shape of the phase transition interface obtained from simulation and experiment is elliptical.

This phenomenon is consistent with the shape of the phase change cross-section demonstrated by the simulation results in Figure 6b,d, and also with the results of the ice melting test carried out by Jiang Xingliang of Chongqing University at the Xuefengshan natural ice-covered test station in the literature [29], which shows that it is necessary to consider the interstitial convection during the numerical simulation of ice melting, and verifies the accuracy of the OPGW ice melting model and the computational results in this paper.

The temperature rise curve of OPGW measured in the DC ice melting experiment is consistent with the calculated temperature rise curve, which shows the trend of temperature increase and then decrease. This is due to the fact that with the downward movement of the ice layer, the thickness of the OPGW to the top upper surface of the ice layer is gradually reduced, resulting in the reduction of the ice layer's thermal resistance, and at the same time, the air gap is also gradually increased, the loss of heat to the external environment is increased, and the temperature is gradually lowered.

The measured OPGW DC melting ice temperature rise curves are in good agreement with simulation results, and the relative standard deviation is 5.45%, with a maximum difference of about 3.5 K in temperature and about 10 min in ice melting time, but the overall trend is consistent, which further verifies the accuracy of the OPGW ice melting model established in this paper and the calculation results.

The error between the two is because the ground wire span often reaches several hundred meters in practical engineering. Therefore, the influence of OPGW axial heat transfer is neglected in the simulation model, and the OPGW geometric model is simplified accordingly. In this experiment, the axial length of OPGW is short, and the geometric shape is woven by strands. At the same time, there are uncontrollable differences in wind speed and wind direction on the ice surface in the experiment, which leads to certain errors.

6. Conclusions

Aiming at the dynamic physical process of OPGW DC ice melting, considering the gap convection heat transfer, the corresponding numerical analysis model is established, and the dynamic process of ice phase transition and OPGW temperature rise during ice melting is simulated. The following conclusions are drawn:

(1) Whether or not to consider gap convective heat transfer during ice melting mainly affects the temperature distribution in the air gap region and has no significant effect on the temperature change in the OPGW. After considering the gap convection heat transfer, the temperature in the air gap area is lower, and at the same time, near the bottom of the OPGW, the maximum difference in the width of the air gap at the same phase change interface position of the ice layer is about 3 mm, and the phase change cross-section of the ice layer is elliptical. This is in line with the results of the actual melting test, and it shows that it is necessary to consider the gap convection heat transfer in the process of DC ice melting.

(2) In the numerical simulation of the phase transition process, the values of the half width of the phase transition interval dT and the constant A_m of the mushy zone have a direct influence on its correctness, the former mainly reflects the position of the phase transition interface and the temperature rise of the heat source OPGW in the model, while the latter reflects the temperature distribution in the mushy zone and the air gap region, which has a smaller influence on the determination of the position of the phase transition interface and the temperature rise of the OPGW, but it can be seen that this parameter is still not neglected in the numerical simulation of phase transition. From Conclusion (1), it can be seen that this parameter is still not negligible in the numerical simulation of phase change.

(3) When the half-width of phase transition interval dT is too small, the sudden change in material properties caused by phase transition will be distorted or even the singularity will not converge; when the mushy zone constant A_m is too large, the convective heat transfer in the mushy zone and the air gap is difficult to occur due to the existence of the damping term. At this time, it will approach the ice melting simulation process without considering the gap convection. Combined with the calculation and analysis results, the half-width of phase transition interval dT and the mushy zone constant A_m can be selected as 0.5 K and 10^4, respectively.

(4) The shape of the elliptical phase transition cross-section formed during the OPGW DC ice melting experiment is consistent with that of the ice melting simulation model, and is also consistent with the results of the ice melting experiment carried out by the literature [26] at the Xuefengshan natural ice-covered experimental station. Due to the assumption and simplification of the model, as well as the uncontrollable factors, such as wind speed and wind direction in the experiment, the temperature rise curves of OPGW DC ice melting measured in the experiment have some errors with the simulation results, and the relative standard deviation is 5.45%, with a maximum difference of about 3.5 K in temperature and 10 min in ice melting time. But, the overall

trend is the identical, and the temperature rise curves are all shown as increasing first and then decreasing, which verifies the accuracy of the OPGW DC ice melting model and the calculation results.

The refined numerical analysis model of OPGW DC ice melting constructed in this paper can provide technical support for evaluating or predicting the real-time ice melting status of overhead ground lines, and then provide reference for the reasonable selection of ice melting current and ice melting time, which is of important engineering practical value.

Author Contributions: Conceptualization, S.W. and L.C.; methodology, S.W., B.T. and L.C.; software, L.C.; validation, S.W., L.C., W.X. and Z.W.; formal analysis, L.C.; investigation, W.X. and Z.W.; resources, S.W., W.X. and Z.W.; data curation, S.W. and L.C.; writing—original draft preparation, S.W. and L.C.; writing—review and editing, S.W., B.T., W.X. and Z.W.; supervision, B.T.; project administration, S.W.; funding acquisition, S.W. and B.T. All authors have read and agreed to the published version of the manuscript.

Funding: This research received no external funding.

Data Availability Statement: The original contributions presented in the study are included in the article; further inquiries can be directed to the corresponding author.

Conflicts of Interest: Authors Wangsheng Xu and Zheng Wang were employed by the company China Southern Power Grid Extra High Voltage Power Transmission Company Guiyang Branch. The remaining authors declare that the research was conducted in the absence of any commercial or financial relationships that could be construed as a potential conflict of interest.

References

1. Zhang, Z.; Zhang, H.; Yue, S.; Zeng, W. A Review of Icing and Anti-Icing Technology for Transmission Lines. *Energies* **2023**, *16*, 601. [CrossRef]
2. Zhao, X.; Wu, G.; Yang, D.; Xu, G.; Xing, Y.; Yao, C.; Abu-Siada, A. Enhanced Detection of Power Transformer Winding Faults through 3D FRA Signatures and Image Processing Techniques. *Electr. Power Syst. Res.* **2025**, *242*, 111433. [CrossRef]
3. Li, M. The influence of DC Ice-Melting Process of OPGW on the Communicationof Optical Fibre. *Electr. Power ICT* **2013**, *11*, 124–128.
4. Sadov, S.Y.; Shivakumar, P.N.; Firsov, D.; Lui, S.H.; Thulasiram, R. Mathematical Model of Ice Melting on Transmission Lines. *J. Math. Model. Algorithms* **2006**, *6*, 273–286. [CrossRef]
5. Fu, P.; Farzaneh, M. A CFD Approach for Modeling the Rime-Ice Accretion Process on a Horizontal-Axis Wind Turbine. *J. Wind. Eng. Ind. Aerodyn.* **2010**, *98*, 181–188. [CrossRef]
6. Tan, L.; Liu, Y.; Yuan, Z.; Li, R. Linear Modeling Analysis of the Heat Balance of the Transmission Line in High Frequency Critical Ice Melting State. *Int. J. Low-Carbon Technol.* **2024**, *19*, 508–516. [CrossRef]
7. Jiang, X.L.; Fan, S.H.; Zhang, Z.J.; Sun, C.; Shu, L. Simulation and Experimental Investigation of DC Ice-Melting Process on an Ice Conductor. *IEEE Trans. Power Deliv.* **2010**, *25*, 919–929. [CrossRef]
8. Jiang, X.L.; Fan, S.H.; Hu, J.L.; Zhang, Z.J.; Sun, C.X. Critical Current Analysis of DC Short Circuit Ice Melting in Transmission Lines. *Chin. J. Electr. Eng.* **2010**, *30*, 111–116.
9. Meng, Z.G. Research on DC Ice Melting Process and Modeling of Fiber Optic Composite Overhead Ground Wire (OPGW). Ph.D. Thesis, Chongqing University, Chongqing, China, 2017.
10. Yang, G.L.; Jiang, X.L.; Wang, M.Z.; Hu, J.L.; Zhang, Z.J. Influence of the Shape of Single Conductor Ice Cover on the Melting Time of DC High Current on Transmission Lines. *J. Electrotechnol.* **2024**, *39*, 2916–2924.
11. Peng, Z.Y.; Zhou, Y.S.; He, Y.; Zhou, K.; Zeng, C.; Tang, C. Analysis of Critical Current for High-Frequency Excitation Melting of Ice on Transmission Conductors. *High Volt. Electr. Appar.* **2021**, *57*, 108–114.
12. Yuan, X.L.; Zhou, Y.S.; Wang, Y.A.; Xiong, Q.; Wang, X.J.; Ma, X.H.; Dai, Z. Influence of Local Convective Heat Transfer on High-Frequency Excitation Ice Melting on Transmission Lines. *High Volt. Electr. Appar.* **2020**, *56*, 224–229.
13. Zhu, H.; Han, Z.B.; Xing, H.C.; Xu, D.; Zhang, R.; Liu, C.; Chen, W. DC Ice-Melting Operation of the Ground Wire Based on the Characteristic Investigation of the Thermal Structure Coupling Effect. *Electr. Power Syst. Res.* **2023**, *218*, 109181. [CrossRef]
14. Hu, Q.; Zhang, Y.; Chen, X.Y.; Lin, H.X.; He, J.; Wang, Z.G. DC Ice-Melting Temperature Rise Characteristics of OPGW and Its Influencing Factors. *High Volt. Electr. Appar.* **2024**. [CrossRef]

15. Khedher, N.B.; Mohammed, H.I.; Ahed, A.M.; Biswas, N.; Togun, H.; Hammoodi, K.A.; Mahdi, J.M.; Talebizadehsardari, P. Maximizing Charging/Discharging Capabilities of Horizontal Shell-and-tube Latent Heat Storage Systems with Innovative Curved Fin Inserts. *Int. J. Heat Mass Transf.* **2025**, *236*, 126289. [CrossRef]
16. Khedher, N.B.; Togun, H.; Aljibori, H.S.S.; Mohammed, H.I.; Khosravi, K.; Abed, A.M.; Boujelbene, M.; Mahdi, J.M. Accelerated Melting Dynamics in Latent-heat storage Systems via Longitudinal and Circular Fins: A Comprehensive 3D Analysis. *Int. Commun. Heat Mass Transf.* **2024**, *156*, 107602. [CrossRef]
17. Yang, G.L.; Jiang, X.L.; Liao, Y.; Li, T.; Deng, Y.; Wang, M.; Hu, J.; Ren, X.; Zhang, Z. Research on Load Transfer Melt-Icing Technology of Transmission Lines: Its Critical Melt-Icing Thickness and Experimental Validation. *Electr. Power Syst. Res.* **2023**, *221*, 109409. [CrossRef]
18. Donald, P.; Leighton, S. *Heat Transfer*; McGraw Hill: New York, NY, USA, 2011.
19. Ramin, K.; Babak, K. Investigation of the Effect of Inclination Angle on the Melting Enhancement of Phase Change Material in Finned Latent Heat Thermal Storage Units. *Appl. Therm. Eng.* **2019**, *146*, 45–60.
20. Ali, C.; Kheirabadi, D.G. Simulating Phase Change Heat Transfer Using COMSOL and Fluent: Effect of the mushy-zone Constant. *Comput. Therm. Sci.* **2015**, *7*, 427–440.
21. Moench, S.; Dittrich, R. Influence of Natural Convection and Volume Change on Numerical Simulation of Phase Change Materials for Latent Heat Storage. *Energies* **2022**, *15*, 2746. [CrossRef]
22. Zhan, T.; Huo, D.X.; Wang, C.Y.; Shi, Z. Review of the Modeling Approaches of Phase Change Processes. *Renew. Sustain. Energy Rev.* **2023**, *187*, 1–28.
23. Milad-Tajik, J.; Cortes, C.; Pallarés, J.; Gil, A.; Arauzo, I. Comparison of Different Numerical Models for Solidification of Paraffin-Based PCMs and Evaluation of the Effect of Variable MushyZoneConstant. *Energies* **2024**, *17*, 5982. [CrossRef]
24. Claus, S.; Bigot, S.; Kerfriden, P. Cutfem Method for Stefan-Signorini Problems With Application in Pulsed Laser Ablation. *Siam J. Sci. Comput.* **2018**, *40*, 1444–1469. [CrossRef]
25. Jiang, X.L.; Meng, Z.G.; Zhang, Z.J.; Hu, J.; Lei, Y. DC Ice-Melting and Temperature Variation of Optical Fibre for Ice-Covered Overhead Ground Wire. *IET Gener. Transm. Distrib.* **2016**, *10*, 352–358. [CrossRef]
26. Zhu, Y.C.; Huang, X.B.; Zhao, L.; Tian, Y.; Mu, J.-Y.; Gao, H. Thermodynamic Model of Critical Ice-Meleting Current on Iced Transmission Lines. *Therm. Sci.* **2019**, *23*, 3187–3198. [CrossRef]
27. Fan, S.H.; Jiang, X.L.; Shu, L.C.; He, P.; Liu, P.; Nie, H. DC Ice-Melting Model for Elliptic Glaze Iced Conductor. *IEEE Trans. Power Deliv.* **2010**, *26*, 2697–2704. [CrossRef]
28. Yang, B.; Raza, A.; Bai, F.; Zhang, T.; Wang, Z. Microstructural Evolution Within Mushy Zone during Paraffin's Melting and Solidification. *Int. J. Heat Mass Transf.* **2019**, *141*, 769–778. [CrossRef]
29. Fan, S.H.; Jiang, X.L.; Sun, C.X.; Zhang, Z.; Shu, L. Temperature Characteristic of DC Ice-Melting Conductor. *Cold Reg. Sci. Technol.* **2011**, *65*, 29–38. [CrossRef]

energies

MDPI

Article

Fault Diagnosis Method for Transformer Winding Based on Differentiated M-Training Classification Optimized by White Shark Optimization Algorithm

Guochao Qian [1,*], Kun Yang [1], Jin Hu [1], Hongwen Liu [1], Shun He [1], Dexu Zou [1], Weiju Dai [1], Haozhou Wang [1] and Dongyang Wang [2]

[1] Electric Power Research Institute of Yunnan Power Grid, Kunming 650214, China
[2] School of Electrical Engineering, Southwest Jiaotong University, Chengdu 611756, China; jiulingthebest@163.com
* Correspondence: qianguochao_1981@163.com

Abstract: Transformers, serving as critical components in power systems, are predominantly affected by winding faults that compromise their operational safety and reliability. Frequency Response Analysis (FRA) has emerged as the prevailing methodology for the status assessment of transformer windings in contemporary power engineering practice. To mitigate the accuracy limitations of single-classifier approaches in winding status assessment, this paper proposes a differentiated M-training classification algorithm based on White Shark Optimization (WSO). The principal contributions are threefold: First, building upon the fundamental principles of the M-training algorithm, we establish a classification model incorporating diversified classifiers. For each base classifier, a parameter optimization method leveraging WSO is developed to enhance diagnostic precision. Second, an experimental platform for transformer fault simulation is constructed, capable of replicating various fault types with programmable severity levels. Through controlled experiments, frequency response curves and associated characteristic parameters are systematically acquired under diverse winding statuses. Finally, the model undergoes comprehensive training and validation using experimental datasets, and the model is verified and analyzed by the actual transformer test results. The experimental findings demonstrate that implementing WSO for base classifier optimization enhances the M-training algorithm's diagnostic precision by 8.92% in fault-type identification and 8.17% in severity-level recognition. The proposed differentiated M-training architecture achieves classification accuracies of 98.33% for fault-type discrimination and 97.17% for severity quantification, representing statistically significant improvements over standalone classifiers.

Keywords: transformer; frequency response method; fault identification; white shark optimizer; M-training algorithm

1. Introduction

Power transformers, functioning as pivotal infrastructure in electrical power systems, undertake critical responsibilities in energy generation, transmission, and distribution. Their operational integrity directly governs the quality of transmitted electrical energy across the grid [1,2]. Statistical analyses of failure patterns reveal that approximately 30% of transformer faults originate from winding-related anomalies [3]. Therefore, an accurate assessment of the transformer winding status is of great significance in enhancing the stability and reliability of the power system.

At present, the main detection methods for transformer winding status are the short circuit impedance method, vibration signal method and FRA. FRA has higher sensitivity to transformer winding faults than the other methods [4], and it is also one of the most widely used methods for on-site detection. For the frequency response method, standards such as IEC 60076-18-2012 and C57.149-2012 have been developed, which state that the correlation coefficient can be used to determine the status of the winding [5,6]. In addition to the correlation coefficients mentioned in the standard, scholars at home and abroad have carried out a lot of research on feature extraction and fault identification methods for FRA results. In terms of feature extraction, scholars at home and abroad have carried out research from various aspects such as statistical features, waveform features and image features, for example: scholars at the University of Toronto proposed the vector matching method [7], which proved that the transformer status can be recognized by the frequency response curve obtained by fitting; scholars at the University of Tehran adopted the mean square deviation and standard deviation of amplitude-frequency curves for the judgment of winding state [8], and the results showed that this feature has a strong anti-interference ability against the external noise; scholars from the University of Queensland convert the frequency response curve into a Nyquist diagram [9], which explores and verifies the existence of a certain correspondence between the transformer winding state and the Nyquist diagram; scholars from the Southwest University binarize the image of the frequency response curve [10], and extract the feature coefficients by using the similarity of the image matrix to achieve the identification of the three kinds of common winding faults. In terms of fault diagnosis methods, the threshold comparison of eigenvalues was mainly used for evaluation and diagnosis at the beginning, and in recent years, with the development of artificial intelligence, machine learning methods have been used in the identification of transformer winding faults to improve accuracy, for example, scholars at the Southwest Jiaotong University used support vector machine (SVM) to analyze the area of the transformer crossover band as well as the centre-of-mass deviation, and to identify the state of the windings [11]; scholars at the Northwest A&F University used the Kernel Extreme Learning Machine (KELM) to identify the type and degree of faults in transformer winding deformation [12]. The scholars of the Amirkabir University of Technology proposed a model using artificial neural networks (ANN) for the identification of the type and extent of winding faults [13], but the model performed poorly in the identification of the extent of winding faults and the identification of axial displacement faults. In addition, the scholars of the Northwest A&F University and Amirkabir University of Technology used Random Forest (RF) [14] and K-Nearest Neighbors (KNN) [15] classifiers to identify the winding faults of the transformer, respectively. All of the above scholars use a single classifier to identify winding faults, and each classifier involves the selection of key parameters. The randomness of the key parameters will lead to the uncertainty of the model identification effect, which ultimately affects the accuracy and reliability of the algorithm in identifying winding faults. Meanwhile, the application of optimization algorithms in the field of power systems has made significant progress. For example, the decentralized stochastic recursive gradient method (DSRG) [16] achieves an efficient parallel solution in the optimized tidal current computation of multi-area power systems through the distributed stochastic gradient updating mechanism; the beluga whale optimization algorithm (BWO) [17] demonstrates an excellent global search capability in energy management and battery scheduling of microgrids by simulating the beluga whales' group collaborative behaviors. However, the applicability of these methods in transformer fault diagnosis has not been fully explored: DSRG relies on the gradient information of the objective function, which makes it difficult to be directly applied to discrete parameter optimization; and

BWO, although it excels in continuous-space optimization, is weakly adapted dynamically to high-dimensional, nonlinear parameter spaces.

To address the above problems, this paper applies the M-training algorithm to the fault identification of transformer windings by combining multiple classifiers to reduce the impact of parameter selection and using WSO to solve the parameter optimization problem of the original classifiers to improve the classification effect of each classifier. Secondly, the transformer fault simulation experimental platform is constructed to obtain the frequency response data under different winding fault types, and the M-training algorithm is trained and tested with the obtained frequency response data, comparing the differences between the M-training algorithm and the WSO-RF, SVM and KNN algorithm, and the results show that WSO differentiated M-training classification algorithm proposed in this paper works well.

2. M-Training Algorithm for Differentiated Classifiers

2.1. M-Training Algorithm Based on Differentiated Classifiers

2.1.1. Fundamentals of the M-Training Algorithm

There are N base classifiers in the M-training algorithm, denoted as c_i, $i = 1, 2, 3, \ldots N$, where N is a positive integer and $N > 3$. Let L denote all the labeled samples in the initial sample set, and $|L|$ denote the total number of labeled samples. Let U denote all unlabeled samples in the initial sample set, and $|U|$ denote the total number of unlabeled sample sets.

Assuming a training dataset containing m samples and a noise rate of η, the worst-case error rate ξ of the classifier satisfies the following equation:

$$m = \frac{c}{\xi^2(1-2\eta)^2} \quad (1)$$

where c is a constant, Equation (1) can be transformed to:

$$u = \frac{c}{\xi^2} = m(1-2\eta)^2 \quad (2)$$

The sample complexity m in Equation (1) derives from the Probably Approximately Correct learning framework under label noise [18], where the term $(1-2\eta)^2$ accounts for the degradation of learning efficiency due to mislabeled samples. Equation (2) reformulates this relationship to define the normalized constant u, which encapsulates the interplay between η and ξ.

Use $L_i(t-1)$ and $L_i(t)$ to denote the set of samples labeled as completed by the auxiliary classifier c_i for the main classifier in the $t-1$-th and t-th cycles, respectively, and the total number of main classifier training sample sets U in the $t-1$-th and t-th cycles are $|L \cup L_i(t-1)|$ and $|L \cup L_i(t)|$, respectively.

Let $e_i(t)$ denote the maximum value of the classification error rate of the auxiliary classifier c_i at the t-th cycle; each training refines the main classifier c_i, and each cycle also changes the labeled and unlabeled sample sets. Let η denote the classification noise rate of L. The number of mislabeled samples in L is $\eta_L|L|$. The classification noise rate for the t-th round of the loop is:

$$\eta_i(t) = \frac{\eta_L|L| + e_i(t)|L_i(t)|}{|L \cup L_i(t)|} \quad (3)$$

Equation (3) computes the effective noise rate $\eta_i(t)$ by combining the original noise and the auxiliary classifier's noise. This weighted average ensures that the algorithm adapts to both historical and newly introduced label errors. For example, if the auxiliary classifier improves ($e_i(t) < \eta_L$), the overall noise rate decreases, enabling more aggressive data expansion. Conversely, if $e_i(t) > \eta_L$, subsampling is triggered to maintain robustness.

Substituting Equation (3) into Equation (2) yields:

$$u_i(t) = m_i(t)(1-2\eta_i(t))^2 = \left| L \cup L_i(t)\left(1 - 2\frac{\eta_L|L|+e_i(t)|L_i(t)|}{|L \cup L_i(t)|}\right)\right| \tag{4}$$

According to Equation (2), in order to make $e_i(t) < e_i(t-1)$, $u_i(t) > u_i(t-1)$ needs to be satisfied:

$$\begin{aligned} &\left| L \cup L_i(t)\left(1 - 2\tfrac{\eta_L|L|+e_i(t)|L_i(t)|}{|L \cup L_i(t)|}\right)\right| > \\ &\left| L \cup L_i(t-1)\left(1 - 2\tfrac{\eta_L|L|+e_i(t-1)|L_i(t-1)|}{|L \cup L_i(t-1)|}\right)\right| \end{aligned} \tag{5}$$

Assuming $0 \leq e_i(t-1), e_i(t) \leq 0.5$, Equation (5) can be simplified as:

$$0 < \frac{e_i(t)}{e_i(t-1)} < \frac{|L_i(t-1)|}{|L_i(t)|} < 1 \tag{6}$$

The transition from Equation (5) to (6) formalizes the conditions for iterative refinement in M-training. By requiring the auxiliary classifier's error rate to decrease ($e_i(t) < e_i(t-1)$) while its labeled dataset grows ($|L_i(t)| > |L_i(t-1)|$), Equation (6) ensures that each training cycle reduces noise and enhances model accuracy.

Since $|L_i(t)|$ may be much larger than $|L_i(t-1)|$, even if $e_i(t)$ is smaller than $e_i(t-1)$, there will still exist $e_i(t)|L_i(t)|$ larger than $e_i(t-1)|L_i(t-1)|$, and the sub-sampling method proposed by the paper [19] is used to randomly sample $L_i(t)$. Define the integer S_i to denote the number of $L_i(t)$ after sampling, and be able to let $e_i(t)|L_i(t)|$ be greater than $e_i(t-1)|L_i(t-1)|$ hold when S_i satisfies the condition of the following equation:

$$S_i = \left\lceil \frac{e_i(t-1)|L_i(t-1)|}{e_i(t)} - 1 \right\rceil \tag{7}$$

In addition, $L_i(t-1)$ needs to satisfy the following condition to ensure that the size of $L_i(t)$ is still larger than $|L_i(t-1)|$ after subsampling:

$$|L_i(t-1)| > \frac{e_i(t)}{e_i(t-1) - e_i(t)} \tag{8}$$

Equations (7) and (8) formalize the subsampling mechanism. Equation (7) ensures that the retained samples S_i do not amplify label noise, while Equation (8) guarantees dataset stability. For Equation (7), if $S_i < 0$, subsampling is skipped, and all samples from $L_i(t-1)$ are retained. Equation (8) requires that the classifier's error rate improvement ($e_i(t) < e_i(t-1)$is sufficient to justify expanding the labeled set. When $S_i \leq 0$, the algorithm retains all samples in $L_i(t-1)$ to avoid data loss but imposes a confidence threshold ($\tau = 0.9$) to filter low-probability predictions. This balances stability and noise tolerance. Skipping subsampling increases the overfitting by 4.8%, which highlights the importance of Equations (7) and (8) in maintaining robustness.

When the above conditions (6), (7), and (8) are satisfied, the unlabeled sample U is decided by the auxiliary classifier and added to the main classifier for the classifier adjustment work. Separately, the remaining classifiers are assumed to be the same classification method when cycling the above process for the main classifier until all the classifiers are no longer changed, and finally, the voting method will be taken for decision fusion. Each base classifier independently predicts the class label for a test sample. The majority class among the four classifiers' predictions is selected as the final output. In cases of a tie, the algorithm prioritizes the prediction from the RF classifier with higher accuracy. This prioritization is based on empirical validation of individual classifier performance on the training set. Building on the foundational M-training framework [20], our method introduces differentiated classifiers and WSO-based optimization. We combine RF, SVM, and KNN to enhance robustness and adaptability. Further, WSO replaces manual parameter tuning, ensuring optimal performance across diverse fault scenarios.

2.1.2. Diversity of Base Classifiers

In order to eliminate the problem of lack of variability in the base classifiers of the traditional M-training algorithm, this paper uses multiple classifiers with large differences to be combined, and the specific reasons for the selection of classifiers are as follows:

In the original feature space, if the data is not linearly separable, i.e., it is not possible to completely separate different classes of data points by a linear hyperplane, the SVM can map the data to a higher dimensional space through a kernel function to make the original data linearly separable in this higher dimensional space [21]. Therefore, using SVM as a base classifier can improve the M-training algorithm's ability to handle nonlinear data. However, SVM classifies the dataset by means of support vectors, which also leads to the fact that SVM is more sensitive to noise and outliers. RF is more robust to noise and outliers due to the integration of multiple decision trees [22], and RF is more suitable for dealing with large-scale datasets compared to SVM. Therefore, combining RF with SVM not only improves the algorithm's ability to handle noise in data but also improves the M-training algorithm's ability to handle data of different sizes.

SVM is, in principle, trying to find a globally optimal solution from a global point of view, which will ignore the local features of the data, whereas KNN is based on samples from the nearest neighbors, so it can adapt well to local changes in the data. Therefore, combining KNN with SVM allows the algorithm to deal with the problem from both a global perspective and observe the local information of the data.

In summary, there are large differences between RF, SVM, and KNN, and the advantages and disadvantages of the algorithms are complementary. Combining them can effectively improve the differences between the base classifiers of the M-training algorithm. Since RF has better results in transformer fault identification compared to SVM and KNN [23], this paper uses two RFs, one SVM and one KNN, as the base classifiers of the M-training algorithm, as shown in Figure 1.

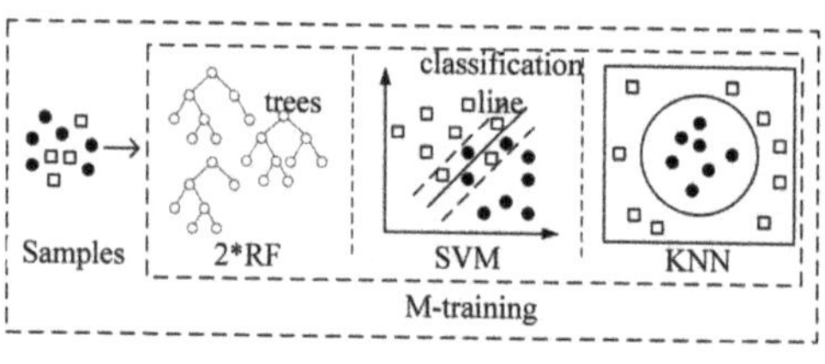

Figure 1. Structure of the M-training algorithm.

2.2. *Base Classifier Based on WSO*

When SVM deals with linearly indivisible problems, it maps the data from the original space to the high-dimensional space through the kernel function and constructs the optimal classification surface in this space to realize data classification. Therefore, its kernel parameter δ directly affects the performance of the kernel function as well as the performance of SVM. The penalty factor C controls the confidence and error balance of the SVM in the feature space, thus optimizing the generalization ability of the model and affecting the classification results.

Random forest algorithm is an integrated learning method based on decision trees, which improves the accuracy and robustness of the model by constructing multiple decision trees and combining their predictions; the main parameters include mtry and ntrees, mtry represents the number of tree forks, too many forks may learn noise and outliers in the data and reduce the generalization ability of the model. ntrees represents the number of decision trees. Too large ntrees will lead to model redundancy and reduce the model estimation efficiency, while too small will reduce the model accuracy.

KNN is a classification method based on a distance metric, where the closest K samples are selected for classification by calculating the distance between test samples and training samples. The value of K determines the number of nearest neighbors that are considered when making a decision [24]. Smaller values of K make the model more sensitive to local features and may lead to overfitting; larger values of K may smooth the model but may ignore important local information.

For the key parameters in each of the above base classifiers, this paper adopts WSO, the core idea of which is to mimic the special hearing, olfactory and fish behavior of white sharks when tracking and foraging and mathematically model the hunting characteristics of the white sharks, searching for and exploiting each potential region in the space in order to achieve parameter optimization. The specific hyperparameters of the algorithm are configured as follows: population size of 50, maximum iterations of 100, and olfactory sensitivity coefficient of 0.8. These parameter settings are referenced from the convergence experiments in [24]. Detailed mathematical models and formulations of the algorithm can be found in [24], with the specific workflow illustrated in Figure 2.

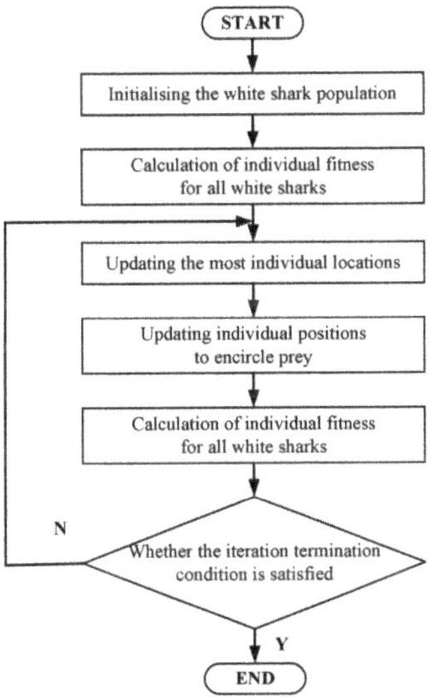

Figure 2. Flowchart of white shark optimization algorithm.

While the WSO algorithm incurs a computational cost proportional to population size and iterations, its rapid convergence and parallelizable fitness evaluations mitigate practical overhead. In our experiments, WSO achieved stable optimization within 20 iterations, outperforming PSO and GA in both speed and solution quality. Further efficiency gains are possible through early termination and distributed computing.

The differences between the proposed method in this paper and the existing integrated learning methods are mainly reflected in the following two aspects: Unlike the stacking of homogeneous classifiers in traditional integrated methods, this study selects significantly different classifiers, such as globally optimized SVMH and locally sensitive KNN, which improves the robustness of the model through complementarity, whereas in traditional methods, the base classifiers are trained independently, and are weighted at the output layer only by the meta-classifiers or the stacking for simple weighting, which may lead to insufficient diversity of base classifiers. WSO is firstly applied to the tuning of the base classifier parameters, which solves the performance fluctuation problem caused by the randomness of parameters in the traditional integration method.

2.3. Feature Extraction

According to the circuit equivalence theory, the transformer winding can be equivalent to a two-port network consisting of resistive, capacitive and inductive elements at frequencies of 1 kHz and above. When the internal structure of the transformer winding changes, the electrical parameters of the equivalent circuit model will also change, which is manifested in the amplitude change of the resonance point on the frequency response curve and the frequency shift. In this paper, the correlation coefficient, mathematical statistical features and waveform statistical features in the IEC 60076-18-2012 standard [5], which are commonly used in the frequency response method to diagnose winding deformation, are used as the characteristic information of the frequency response curve.

Let *X*(*k*) and *Y*(*k*) be the data $k = 0, 1, \ldots, N-1$ under the amplitude-frequency response curves of the normal transformer and test transformer at 1~1000 kHz, respectively.

(1) The correlation coefficient *R* is calculated as follows:

Calculate the standardized variance *D* of the two series:

$$D_x = \frac{1}{N}\sum_{k=0}^{N-1}\left[X(k) - \frac{1}{N}\sum_{k=0}^{N-1}X(k)\right]^2 \tag{9}$$

Calculate the covariance c_{xy} of the two series:

$$c_{xy} = \frac{1}{N}\sum_{k=0}^{N-1}[X(k) - \frac{1}{N}\sum_{k=0}^{N-1}X(k)] \times [Y(k) - \frac{1}{N}\sum_{k=0}^{N-1}Y(k)] \tag{10}$$

Calculate the normalized covariance coefficient LR_{xy} of the two series:

$$LR_{xy} = \frac{c_{xy}}{\sqrt{D_x D_y}} \tag{11}$$

where D_x, D_y are the standardized variance of *X*(*k*), *Y*(*k*).

Calculate the correlation coefficient R_{xy}:

$$R_{xy} = \begin{cases} 10 & 1 - LR_{xy} < 10^{-10} \\ -\lg(1 - LR_{xy}) & \text{else} \end{cases} \tag{12}$$

(2) Mathematical and statistical characteristics

Standard Deviation *SD*:

$$SD_{(X,Y)} = \sqrt{\frac{\sum_{i=1}^{N}(X_i - Y_i)^2}{N-1}} \tag{13}$$

Absolute Sum of Logarithmic Errors *ALSE*:

$$ALSE_{(X,Y)} = \frac{1}{N}\left(\sum_{i=1}^{N}\left|20\log_{10}|X_i| - 20\log_{10}|Y_i|\right|\right) \tag{14}$$

Mean Absolute Errors *DABS*:

$$DABS_{(X,Y)} = \frac{1}{N}(\sum_{i=1}^{N}|X_i - Y_i|) \tag{15}$$

Sum of Squared Relative Errors *SSRE*:

$$SSRE_{(X,Y)} = \frac{1}{N}(\sum_{i=1}^{N}(\frac{Y_i}{X_i} - 1)^2) \tag{16}$$

(3) Waveform statistical characteristics

Kurtosis (*Ku*):

$$Ku = \frac{\frac{1}{N}\sum_{i=1}^{N}(Z_i - \overline{Z}_i)^4}{\left(\frac{1}{N}\sum_{i=1}^{N}(Z_i - \overline{Z}_i)^2\right)^2} \tag{17}$$

Crest Factor (*Cf*):

$$Cf = \frac{Z_{\max}}{RMS} \tag{18}$$

Skewness (*Sk*):

$$Sk = \frac{\frac{1}{N}\sum_{i=1}^{N}(Z_i - \overline{Z}_i)^3}{\left(\frac{1}{N}\sum_{i=1}^{N}(Z_i - \overline{Z}_i)^2\right)^{\frac{3}{2}}} \tag{19}$$

where Z_i is the amplitude value at frequency point i of the amplitude-frequency response curve, $\overline{Z}_i$ is the average value of amplitude-frequency characteristic curve amplitude, *RMS* is the effective value of amplitude-frequency response curve amplitude, and Z_{max} is the maximum value of amplitude-frequency response curve amplitude.

2.4. Diagnostic Process Based on Differentiated M-Training Algorithm

The M-training algorithm based on different classifiers in this paper is shown in Figure 3, which includes two RF classifiers, one SVM classifier and one KNN classifier and the key parameters in the base classifiers are optimized by WSO, the pseudocode of the algorithm is shown in Appendix A.

Step1: Obtain the frequency response curves under different winding fault types to extract the characteristic information of the samples;

Step2: Use labeled samples to train the base classifier to identify unlabeled samples by auxiliary classifiers; when the classification results obtained by the auxiliary classifiers are highly consistent, the samples will be added to the set of labeled samples; if there are no unlabeled samples that need to be added to the set of labeled samples, the set of labeled samples will be updated by random sampling;

Step3: Train the main classifier using the updated labeled sample set;

Step4: Each base classifier acts as the main classifier in turn, repeat Step2 and stop training when all base classifiers are not changing;

Step5: The test sample is recognized, and the voting result of the base classifier is used as the final output.

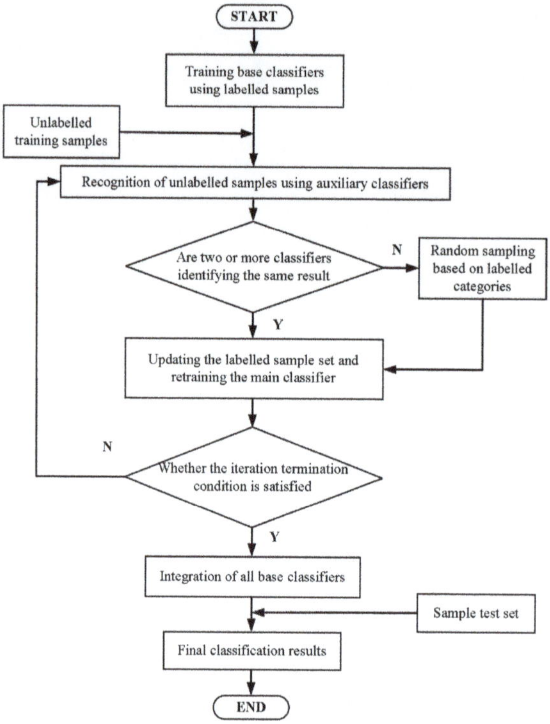

Figure 3. Flowchart of M-training algorithm.

3. Experimental Platform Construction and Model Training

In order to study the transformer winding fault identification method based on the frequency response method, this paper utilizes an experimental transformer with a voltage level of 10 kV, and the internal structure of the transformer is shown in Figure 4. From outside to inside, there are 16 cakes of high-voltage windings, 16 cakes of low-voltage windings and the iron core, and each cake of the high-voltage and low-voltage windings consists of 32 turns and 16 turns of coils, respectively, which are separated by pads to separate the wires from the cakes.

(**a**) Overall

(**b**) Single-phase

Figure 4. Internal structure of the transformer.

3.1. Fault Simulation

(1) Axial displacement faults (AD)

The axial displacement fault is specifically characterized by an overall shift in the height of the windings at the fault location, which is simulated in this paper by adding pads between the windings, as shown in Figure 5a. The degree of failure of axial displacement of transformer windings can be quantified by Equation (20) [25].

$$DF = \frac{\Delta h}{H} \times 100\% \tag{20}$$

where Δh is the height of the axial displacement of the winding, i.e., the thickness of the pads when performing the fault simulation.

(**a**) AD (**b**) SC (**c**) SCV

Figure 5. Transformer winding fault simulation.

The transformer windings were simulated with AD of 1~5% with different fault levels, and the faults were set at three locations: the top (between 1 and 2 cakes), middle (between 8 and 9 cakes) and bottom (between 15 and 16 cakes) of the windings in the way shown in Table 1.

Table 1. Fault settings for axial displacement of windings.

Ordinal Number	Thread Cake Position	Degree of Displacement/%	Ordinal Number	Thread Cake Position	Degree of Displacement/%
1	1, 2	1	9	8, 9	4
2	1, 2	2	10	8, 9	5
3	1, 2	3	11	15, 16	1
4	1, 2	4	12	15, 16	2
5	1, 2	5	13	15, 16	3
6	8, 9	1	14	15, 16	4
7	8, 9	2	15	15, 16	5
8	8, 9	3			

(2) Short circuit fault between cakes (SC)

For the fault simulation of SC, this paper uses alligator clip wires to directly connect the copper noses of different cakes and simulates different degrees of SC between cakes by shorting different numbers of cakes, which is realized as shown in Figure 5b, and different degrees of SC are simulated at the top, middle, and bottom of the winding, which is set up in the way as shown in Table 2.

Table 2. Fault settings for short circuits between cakes of the windings.

Ordinal Number	Thread Cake Position	Ordinal Number	Thread Cake Position
1	1, 2	6	6, 9
2	1, 3	7	16, 15
3	1, 4	8	16, 14
4	6, 7	9	16, 13
5	6, 8		

(3) Bulging and warping faults (SCV)

When the transformer windings are twisted, bulging, and other deformations occur, the spacing between the wires and cakes is reduced, resulting in a larger longitudinal equivalent capacitance between the winding cakes [26]. The longitudinal equivalent capacitance of the model transformer can be obtained at the level of 100 pF based on the calculation method of the literature [26]. In this paper, different degrees of SCV are simulated by connecting capacitors ranging from 220 pF to 680 pF in parallel between the winding wires and cakes, which are implemented as shown in Figure 5c, and different degrees of SCV are simulated at the top, middle, and bottom of the windings, which are set up as shown in Table 3.

Table 3. Fault settings for bulging and warping of windings.

Ordinal Number	Thread Cake Position	Capacitance/pF	Ordinal Number	Thread Cake Position	Capacitance/pF
1	1, 2	100	9	8, 9	470
2	1, 2	220	10	8, 9	680
3	1, 2	330	11	15, 16	330
4	1, 2	470	12	15, 16	470
5	1, 2	680	13	15, 16	680
6	8, 9	100	14	15, 16	100
7	8, 9	220	15	15, 16	220
8	8, 9	330			

3.2. Test Results

Figure 6 shows the frequency response curve of AD under different fault degrees and fault locations; the amplitude of the curve in the frequency band 1~100 kHz is basically not fluctuating; the curve in the frequency band 100~500 kHz is shifted downward as a whole, and the degree of shifting becomes smaller with the increase of frequency; the curve in the frequency band 800~1000 kHz fluctuates markedly, which is specifically reflected in the decrease of the amplitude of the peaks and valleys of the high-frequency band; and it can be seen that with the aggravation of the degree of AD, the curve also tends to move to the low-frequency band. The detailed figure shows that with the increase in AD, the curve also shows a tendency to move to the low-frequency band.

Figure 7 shows the frequency response curves under different fault degrees and fault locations of SC, and the curves are shifted in the whole frequency range; the curve in the frequency band 1~100 kHz shows an upward shift, and the valley of the band at 10 kHz as well as the peak of the band at 80 kHz are shifted to the high-frequency band; the curve in the frequency band 100~500 kHz is upwardly shifted overall, and the extent of the shift increases with the increase of the degree of SC. The curve in the frequency band 100~500 kHz is shifted upward, and the degree of shift increases with the SC degree, which has a better resolution of the fault degree; the curve in the frequency band 500~1000 kHz fluctuates a lot, and there is a tendency to shift to the high-frequency band on the basis of the upward shift.

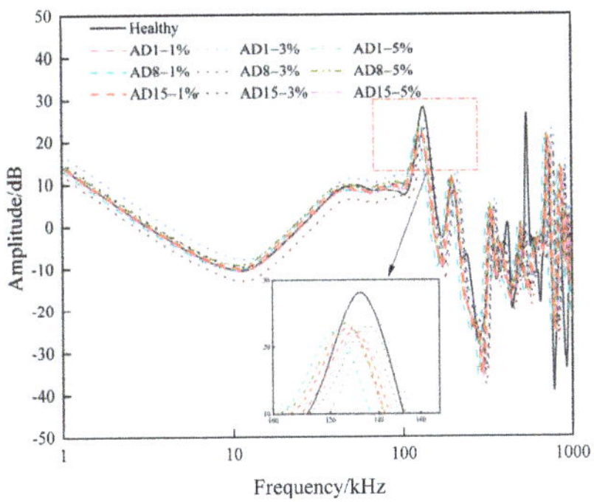

Figure 6. Frequency response curve of transformer winding axial displacement faults.

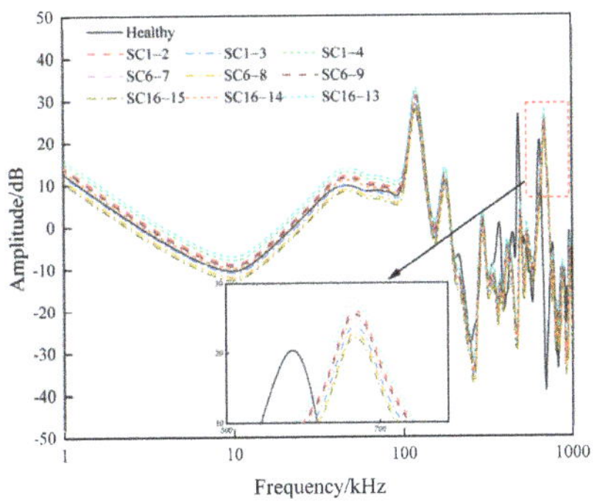

Figure 7. Frequency response curve of transformer winding short circuit fault between cakes.

Figure 8 shows the frequency response curves under different fault degrees and fault locations of the winding SCV; when SCV occurs, the overall amplitude of the frequency response curve increases, the upward translation, the frequency response curve moves toward the low-frequency band, and with the increasing degree of fault, the amplitude of the move toward the low-frequency band increases.

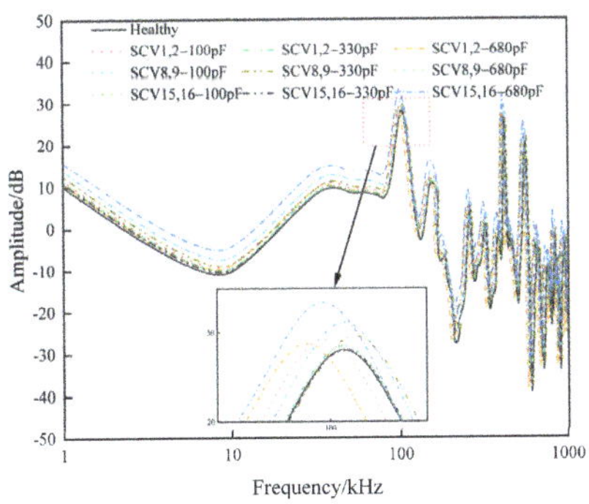

Figure 8. Frequency response curve of transformer winding bulging and warping faults.

The distinct impacts of AS, SC, and SCV on frequency response curves arise from their unique modifications to the winding's equivalent electrical parameters. AD primarily alters longitudinal capacitance and mutual inductance, manifesting as a downward shift in mid-frequency resonance. SC reduces inter-turn impedance, causing global upward shifts and high-frequency noise. SCV increases localized capacitance, leading to amplitude elevation and resonance damping. In summary, the frequency response curve presents a certain degree of differentiation for different fault types and fault degrees, but when applied in the field, it is impossible to accurately identify transformer winding faults only through the trend of waveform changes. Therefore, it is necessary to quantify the changes in the frequency response curve, analyze the subtle changes in the frequency response curve through machine learning methods, and accurately identify the winding operation status.

3.3. Model Training Results

In this paper, a total of 1200 sample data were obtained for different fault types, fault locations, and fault levels. According to the failure type, there are 30 normal state samples, 450 AD samples, 270 SC samples and 450 SCV samples in the failure samples. According to the degree of failure, there are 30 normal state samples, 270 samples with 1~3% failure degree, and 180 samples with 4% and 5% failure degree in the fault samples. In order to reduce the complexity of the base classifiers and improve the generalization ability of the algorithms, this paper uses two M-training algorithms to identify the fault types in terms of degree, respectively. The different types of samples under both divisions are divided into training and test sets in the ratio of 7:3; 3/7 samples in the training set are labeled samples, and 4/7 samples are unlabeled samples.

WSO is used to optimize each base classifier to obtain the best key parameters, as shown in Table 4. Meanwhile, the convergence speed and final optimization effect of WSO with Particle Swarm Optimization (PSO) and Genetic Algorithm (GA) are compared, as shown in Figure 9. It can be seen that compared with PSO and GA, the convergence of WSO for both fault type identification base classifiers and fault degree identification base classifiers reaches a lower error rate within 20 iterations, and the classification error rates are lower than those of the comparison algorithms by 2.97–15.41%. This indicates that WSO has better convergence speed and optimization effect for each base classifier.

Table 4. Parameters of the base classifier after optimization.

Model	Base Classifier	Optimization Parameters	
Fault Recognition M-training	RF1	ntrees	49
		mtry	7
	RF2	ntrees	58
		mtry	7
	SVM	δ	1.3779
		C	1.4793
	KNN	K	23
Degree Recognition M-training	RF1	ntrees	51
		mtry	7
	RF2	ntrees	47
		mtry	7
	SVM	δ	3.2069
		C	10.3954
	KNN	K	27

Tables 5 and 6 demonstrate the recognition effectiveness of the M-training algorithm with or without optimization of the base classifier using WSO, including the recognition effectiveness for fault type and fault degree. In terms of fault type and degree recognition, optimizing the base classifier can improve the total recognition accuracy by 8.92% and 8.17%, respectively. To validate the statistical significance of the accuracy improvements, we conducted paired t-tests and McNemar's tests across 10 independent runs [27]. For fault type recognition, the mean improvement of 8.92% yielded $t(9) = 12.74$, $p < 0.001$, with a 95% confidence interval of 7.21–10.63%. Similarly, for severity recognition, the

8.17% improvement showed $t(9) = 9.85$, $p < 0.001$, and a confidence interval of 6.05–10.29%. McNemar's tests further confirmed the robustness of these gains ($\chi^2 > 14.2$, $p < 0.001$). These results conclusively demonstrate that the WSO-optimized M-training algorithm achieves statistically and practically significant enhancements in diagnostic accuracy.

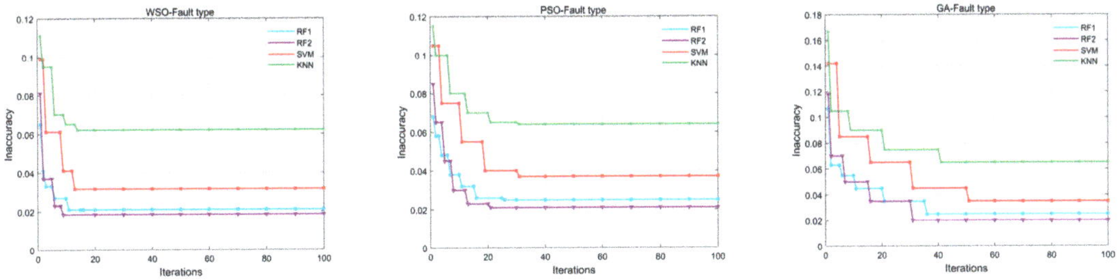

(**a**) Fault type recognition base classifiers

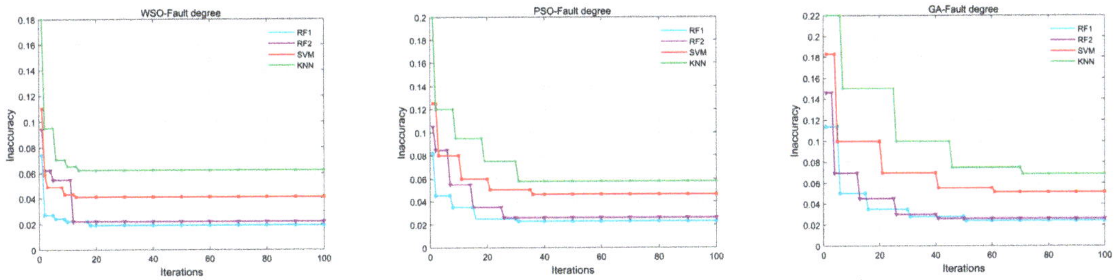

(**b**) Fault degree recognition base classifier

Figure 9. Iterative curves of error rate for different algorithms optimized base classifiers.

Table 5. Fault type recognition accuracy.

M-Training	Accuracy/%				
	Normal	AD	SC	SCV	Average
Pre-optimization	100	91.55	86.29	88.44	89.41
Post-optimization	100	98.00	98.51	98.44	98.33

Table 6. Fault level recognition accuracy.

M-Training	Accuracy/%						
	Normal	1%	2%	3%	4%	5%	Average
Pre-optimization	100	87.78	85.19	91.11	89.44	91.11	89.00
Post-optimization	100	95.93	97.41	98.15	96.11	97.78	97.17

In order to verify whether the M-training algorithm has the ability to synthesize the advantages of each base classifier, the M-training algorithm and each individual base classifier are trained using the training set respectively, and the training set is labeled samples when training the individual base classifiers, and the effect of each classifier is tested through the test set, and the accuracies of the fault type recognition and fault degree recognition are shown in Tables 7 and 8.

Table 7. Fault type recognition accuracy for each classifier.

Sorter	Accuracy/%				
	Normal	AD	SC	SCV	Average
M-training	100	98.00	98.51	98.44	98.33
RF	100	93.55	93.33	91.33	92.83
RFs	100	94.89	94.27	93.16	94.11
SVM	100	89.55	91.85	86.22	89.08
KNN	100	83.11	91.48	79.33	83.91

Table 8. Fault level recognition accuracy for each classifier.

Sorter	Accuracy/%						
	Normal	1%	2%	3%	4%	5%	Average
M-training	100	95.93	97.41	98.15	96.11	97.78	97.17
RF	100	91.85	94.44	93.70	90.56	91.67	92.83
RFs	100	93.78	95.16	94.59	93.06	92.28	93.77
SVM	100	90.37	89.63	87.04	89.44	88.33	89.25
KNN	100	81.85	80.74	84.81	83.89	80.56	82.83

Redundancy from two RF classifiers reduces the risk of overfitting to noise or outliers. By aggregating predictions from slightly divergent RF models, the ensemble becomes less sensitive to perturbations in specific subsets of training data. The RF configuration improves fault-type accuracy by 1.28%, and fault degree accuracy by 0.94% improves fault-type accuracy compared to the RF classifiers. The optimized RFs exhibit complementary error patterns, ensuring that misclassifications from one RF are often corrected by the other.

The accuracy of the M-training algorithm for fault type recognition is above 98%, and the accuracy of fault degree recognition is above 97%, which is higher than that of the base classifiers, which shows that the M-training algorithm is able to make the ability of multiple classifiers complement each other so that the base classifiers can effectively play their respective advantages, which makes the integrated model have higher fault recognition accuracy than a single model. The M-training algorithm is able to complement the abilities of multiple classifiers, allowing the base classifiers to effectively utilize their respective advantages, resulting in a higher fault recognition accuracy of the combined model than a single model.

3.4. Live Case Study

This paper takes a 110 kV power transformer of Yunnan Power Grid as the main research object, as shown in Figure 10, and the specific parameters of this transformer are shown in Table 9. Due to the presence of a substation in the vicinity, the SNR for ambient noise at the site is approximately 15 dB. The high and medium voltage windings are in star wiring mode, and the low voltage windings are in triangular wiring mode. The high, medium and low voltage windings are tested separately, and the test program for the low voltage side is A injection B test, B injection C test and C injection A test; the test program for the high and medium windings is: take the O point as the signal injection point, and collect signals in the three phases of A, B and C, and name the collected signals through the file naming provisions in the IEC 60076-18-2012 [5].

A total of nine signals are obtained from the field test, and in this paper, the M-training algorithm is verified using the signals from the MV side of the three-phase windings of A, B and C. The test results are shown in Figure 11.

Figure 10. 110 kV transformer field test.

Table 9. 110 kV power transformer parameters.

Parameters	Data		
Rated Capacity/kVA		40,000	
Linkage Group Labeling		YNyn0d11	
Rated Voltage/kV	110	35	10
Rated Current/A	363	1142	4000

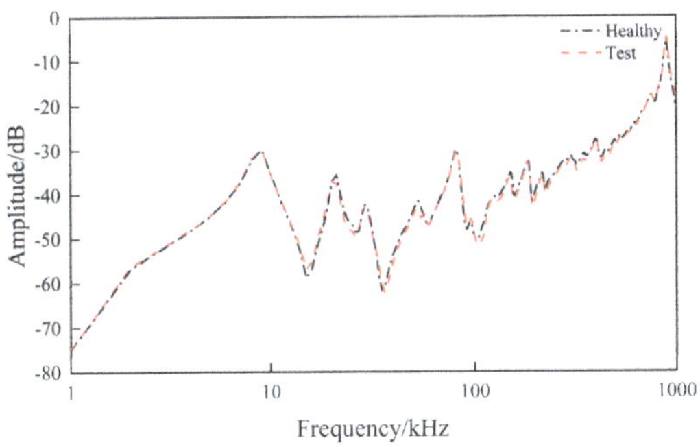

(**a**) A−phase

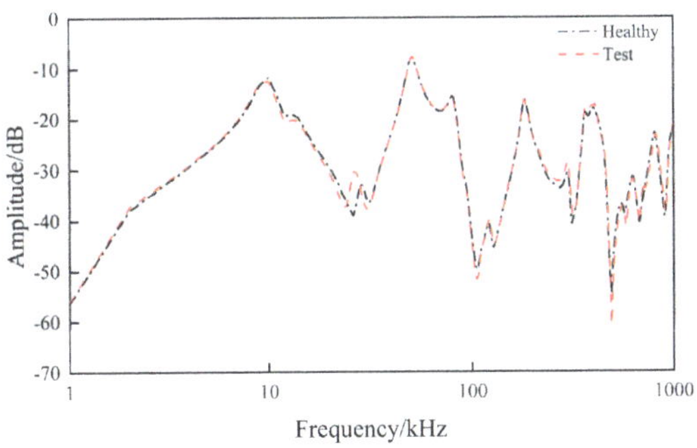

(**b**) B−phase

Figure 11. *Cont.*

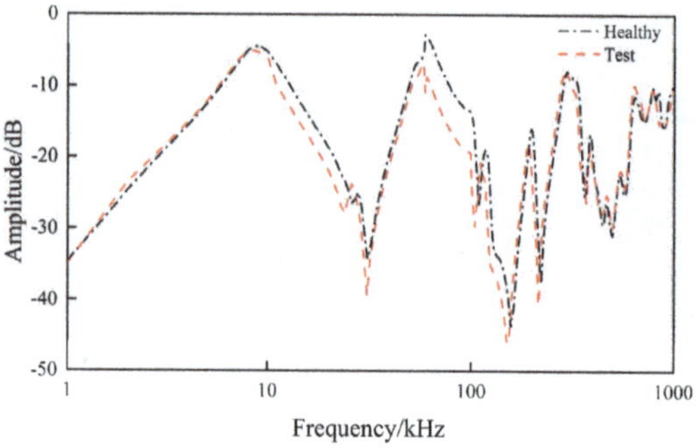

(**c**) C–phase

Figure 11. Frequency response curve of the medium voltage side of the transformer.

In order to determine the specific fault condition of the winding, the characteristic quantity mentioned before was calculated for the frequency response curve of the winding, which was inputted into the M-training algorithm, and the result showed that 5% AD appeared in the medium-voltage side of the winding in the C-phase of the winding.

In order to verify the validity of the above analysis method, the transformer lifting cover overhaul was carried out on-site. Although ambient noise (SNR $\approx$ 15 dB) and load cycles (40–85%), The overhaul results still show that the upper part of the C-phase winding of the power transformer in the medium voltage side has a heavy AD in 5 cakes of windings, as shown in Figure 12. The overhaul results are consistent with the analysis of the M-training algorithm, which verifies the effectiveness of the model in this paper.

Figure 12. Hanger overhaul results.

4. Conclusions

This paper carries out research on transformer winding fault identification, proposes a transformer fault identification method based on the WSO-differentiated M-training algorithm, acquires data based on a laboratory platform, trains the established evaluation model, and, finally, applies the proposed fault identification method to the field transformer for analysis and validation. The conclusions are obtained as follows:

1. Optimization of the base classifiers using WSO can improve the accuracy of the M-training algorithm by 8.92% and 8.17% in terms of fault type and degree identification.
2. The WSO-differentiated M-training algorithm proposed in this paper has an accuracy of 98.33% and 97.17% for the identification of transformer fault type and fault degree, respectively, which is a significant improvement compared to the identification effect of a single classifier, indicating that the method proposed in this paper is able to accurately diagnose transformer winding faults.

3. In the field test of the 110 kV transformer, under the actual noise and load variation, it was accurately judged that 5% AD appeared in the medium voltage side of winding phase C, and the overhaul results were consistent with the analysis results of the M-training algorithm, which verified the validity of the model in this paper.

Author Contributions: Conceptualization, G.Q. and K.Y.; methodology, J.H. and D.W.; validation, K.Y., J.H. and H.L.; formal analysis, S.H. and D.W.; investigation, D.Z.; resources, G.Q.; data curation, K.Y. and D.W.; writing—original draft preparation, W.D.; writing—review and editing, W.D.; visualization, H.W.; supervision, H.W.; project administration, K.Y.; funding acquisition, G.Q. All authors have read and agreed to the published version of the manuscript.

Funding: This research was funded by the Electric Power Research Institute of Yunnan Power Grid (Principle and system development of online monitoring of transformer winding faults under load-side multi-signal source excitation,·YNKJXM20222300).

Data Availability Statement: The raw data supporting the conclusions of this article will be made available by the authors on reasonable request.

Conflicts of Interest: Guochao Qian, Kun Yang, Jin Hu, Hongwen Liu, Shun He, Dexu Zou, Weiju Dai, and Haozhou Wang are affiliated with Electric Power Research Institute of Yunnan Power Grid. The authors declare no conflicts of interest.

Appendix A

Pseudocode for the differentiated M-training classification optimized by the WSO algorithm:

Input: Labeled dataset L; Unlabeled dataset U; Base classifiers: RF1, RF2, SVM, KNN; WSO parameters: population size, max iterations; Voting threshold τ

Output: Final fault type/severity prediction for test samples

1. WSO-based Parameter Optimization

For each base classifier $c_i \in$ { RF1, RF2, SVM, KNN }:

a. Initialize the WSO population with random hyperparameters.

b. While iteration < max_iter: i. Evaluate fitness; ii. Update WSO positions using hunting and foraging mechanisms.

c. Select optimal hyperparameters θ_i^* for c_i.

2. Differentiated M-training Process

Initialize: L_pool = L, U_pool = U; Trained classifiers $C = \{c_1\theta_1^*, c_2\theta_2^*, c_3\theta_3^*, c_4\theta_4^*\}$

Repeat until all c_i stabilize (no changes in the labeled set):

For each $c_i \in C$ acting as the main classifier:

a. Label Propagation: i. Use auxiliary classifiers $C \backslash \{c_i\}$ to predict labels for U_pool; ii. Select samples where $\geq \tau$ auxiliary classifiers agree; iii. Add these samples to L_pool; remove them from U_pool.

b. Subsampling (if no new labels are added): Randomly select S_i samples from L_pool.

c. Retrain c_i using updated L_pool.

3. Final Voting Mechanism

For a test sample x:

a. Obtain predictions $\{y_1, y_2, y_3, y_4\}$ from C.

b. Compute majority vote: If a tie occurs, Prioritize predictions from RF1/RF2 (higher accuracy).

c. Return final prediction y_final.

References

1. Hamzeh, M.; Vahidi, B. The Impact of Cyber Network Configuration on the Dynamic-thermal Failure of Transformers Considering Distributed Generator Controller. *Int. J. Electr. Power Energy Syst.* **2022**, *137*, 107786. [CrossRef]
2. Yao, H.; Xu, Y.; Guo, Q.; Chen, S.; Lu, B.; Huang, Y. Study on Transformer Fault Diagnosisbased on Improved Deep Residual Shrinkage Network and Optimized Residual Variational Autoencoder. *Energy Rep.* **2025**, *13*, 1608–1619. [CrossRef]
3. Zhao, X.; Yao, C.; Zhang, C.; Abu-Siada, A. Toward Reliable Interpretation of Power Transformer Sweep Frequency Impedance Signatures: Experimental Analysis. *IEEE Electr. Insul. Mag.* **2018**, *34*, 40–51. [CrossRef]
4. Zhang, H.; Zhang, H.; Ma, Q.; Han, H.; Wang, S. Frequency Response Signature Analysis for Winding Mechanical Fault Detection of Power Transformer Using Sensitivity Method. *Int. J. Appl. Electromagn. Mech.* **2019**, *61*, 593–603. [CrossRef]
5. *IEC 60076-18:2012*; Power Transformers—Part 18: Measurement of Frequency Response. IEC: Geneva, Switzerland, 2012.
6. *IEEE Standard C57.149-2012*; IEEE Guide for the Application and Interpretation of Frequency Response Analysis for Oil-Immersed Transformers. IEEE: New York, NY, USA, 2013.
7. Gustavsen, B.; Semlyen, A. Rational Approximation of Frequency Domain Responses by Vector Fitting. *IEEE Trans. Power Deliv.* **1999**, *14*, 1052–1061. [CrossRef]
8. Samimi, M.; Tenbohlen, S.; Akmal, A.; Mohseni, H. Dismissing Uncertainties in the FRA Interpretation. *IEEE Trans. Power Deliv.* **2018**, *33*, 2041–2043. [CrossRef]
9. Yousof, M.; Ekanayake, C.; Saha, T. Frequency Response Analysis to Investigate Deformation of Transformer Winding. *IEEE Trans. Dielectr. Electr. Insul.* **2015**, *22*, 2359–2367. [CrossRef]
10. Zhao, Z.; Yao, C.; Saha, T.; Li, C.; Islam, S. Detection of Power Transformer Winding Deformation Using Improved FRA Based on Binary Morphology and Extreme Point Variation. *IEEE Trans. Ind. Electron.* **2018**, *65*, 3509–3519. [CrossRef]
11. Zhou, L.; Jiang, J.; Zhou, X.; Wu, Z.; Lin, T.; Wang, D. Detection of Transformer Winding Faults Using FRA and Image Features. *IET Electr. Power Appl.* **2020**, *14*, 972–980. [CrossRef]
12. Wang, G.; Qiu, S.; Xie, F.; Luo, T.; Song, Y.; Wang, S. Diagnosing Fault Types and Degrees of Transformer Winding Combining FRA Method With SOA-KELM. *IEEE Access* **2024**, *12*, 50287–50299. [CrossRef]
13. Ghanizadeh, A.J.; Gharehpetian, G.B. ANN and Cross-correlation Based Features for Discrimination between Electrical and Mchanical Defects and Their Localization in Transformer Winding. *IEEE Trans. Dielectr. Electr. Insul.* **2014**, *21*, 2374–2382. [CrossRef]
14. Wang, S.; Qiu, S.; Xie, F.; Yang, S.; Yu, K.; Li, T. Diagnosis of AD and DSV Winding Faults Based on FRA Method and Random Forest Algorithm. In Proceedings of the 4th IEEE International Conference on Electrical Materials and Power Equipment (ICEMPE), Shanghai, China, 7–10 May 2023. [CrossRef]
15. Behkam, R.; Moradzadeh, A.; Karami, H.; Naderi, M.S.; Mohammadi-Ivatloo, B.; Gharehpetian, G.B.; Tenbohlen, S. Mechanical Fault Types Detection in Transformer Windings Using Interpretation of Frequency Responses via Multilayer Perceptron. *J. Oper. Autom. Power Eng.* **2023**, *11*, 11–21. [CrossRef]
16. Hussan, U.; Wang, H.; Ayub, M.A.; Rasheed, H.; Majeed, M.A.; Peng, J.; Jiang, H. Decentralized Stochastic Recursive Gradient Method for Fully Decentralized OPF in Multi-Area Power Systems. *Mathematics* **2024**, *12*, 3064. [CrossRef]
17. Ayub, M.A.; Hussan, U.; Rasheed, H.; Liu, Y.; Peng, J. Optimal energy management of MG for cost-effective operations and battery scheduling using BWO. *Energy Rep.* **2024**, *12*, 294–304. [CrossRef]
18. Kearns, M.; Li, M. Learning in the Presence of Malicious Errors. *Siam. J. Comput.* **1993**, *22*, 807–837. [CrossRef]
19. Sally, G.; Yan, Z. Enhancing Supervised Learning with Unlabeled Data. In Proceedings of the Seventeenth International Conference on Machine learning (ECML 2000), Stanford, CA, USA, 29 June–2 July 2000.
20. Pengfei, J.; Tailai, H.; Shukai, D.; Lingpu, G.; Jia, Y.; Lidan, W. A Novel Semi-Supervised Electronic Nose Learning Technique: M-Training. *Sensors* **2016**, *16*, 370. [CrossRef]
21. Efeoglu, E.; Tuna, G. Machine Learning for Predictive Maintenance: Support Vector Machines and Different Kernel Functions. *J. Mach. Manuf. Reliab.* **2022**, *51*, 447–456. [CrossRef]
22. Xiong, W.; Yu, G.; Ma, J.; Liu, S. A Novel Robust Adaptive Subspace Learning Framework for Dimensionality Reduction. *Appl. Intell.* **2024**, *54*, 8939–8967. [CrossRef]
23. Shabani, S.; Samadianfard, S.; Sattari, M.T.; Mosavi, A.; Shamshirband, S.; Kmet, T.; Varkonyi-Koczy, A.R. Modeling Pan Evaporation Using Gaussian Process Regression K-Nearest Neighbors Random Forest and Support Vector Machines; Comparative Analysis. *J. Mach. Manuf. Reliab.* **2020**, *11*, 66. [CrossRef]
24. Ali, A.K. An Optimal Design for An Automatic Voltage Regulation System Using a Multivariable PID Controller Based on Hybrid Simulated Annealing-White Shark Optimization. *Sci. Rep.* **2024**, *14*, 30218. [CrossRef]
25. Hashemnia, N.; Abu-Siada, A.; Islam, S. Improved Power Transformer Winding Fault Detection Using FRA Diagnostics? Part 1: Axial Displacement Simulation. *IEEE Trans. Dielectr. Electr. Insul.* **2015**, *22*, 556–563. [CrossRef]

26. Abu-Siada, A.; Hashemnia, N.; Islam, S.; Masoum, M.A.S. Understanding Power Transformer Frequency Response Analysis Signatures. *IEEE Electri. Insul. Mag.* **2013**, *29*, 48–56. [CrossRef]
27. Kebalepile, M.M.; Chakane, P.M. Commonly Used Statistical Tests and Their Application. *S. Afr. J. Anaesth* **2022**, *28*, S80–S84. [CrossRef]

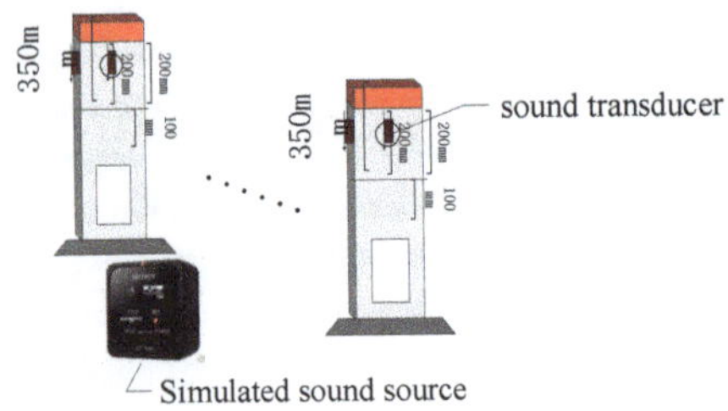

Figure 1. System framework schematic diagram.

The system deploys four acoustic sensors at the cardinal directions of each cable marker to capture signals from a common source. Additionally, multiple markers equipped with sensor arrays are distributed around potential damage sites, allowing for collaborative signal analysis and precise localization through sensor fusion.

2.2. Sound Signal Characteristics Analysis

This section presents an analysis of the distinctive characteristics of cable damage sounds, which are typically generated by construction machinery such as excavators and pile drivers. In the time domain, the amplitude of these signals reflects sound intensity, with higher amplitudes indicating closer sound sources. Figures 2 and 3 respectively utilize image processing techniques to illustrate the time-domain waveforms captured at distances of 5 m and 30 m from the sound source [21].

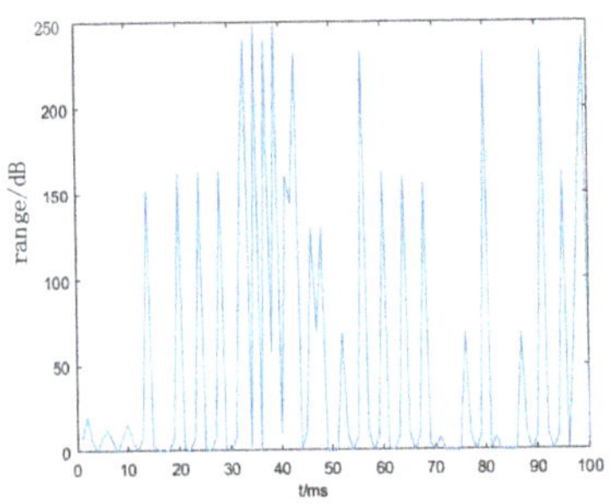

Figure 2. Sound signal at 5 m distance.

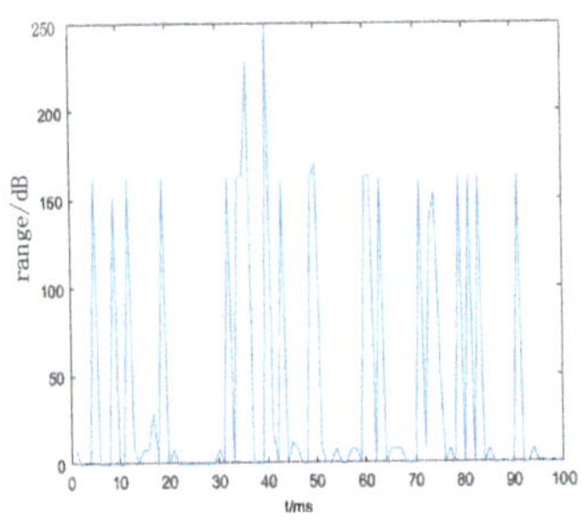

Figure 3. Sound signal at 30 m distance.

As illustrated in Figures 2 and 3, as the sound source approaches the sensor, the amplitude of the sound signal becomes increasingly concentrated in regions of higher decibel levels. This observation suggests that significant mechanical operations are occurring in proximity to the cable marker pile, thereby informing the positioning research of external cable rupture sound sources.

2.3. Sound Propagation Modeling

In real-world environments, sound propagation is influenced by various factors, including obstructions and echoes. Consequently, it is essential to develop a refined propagation model. In this system, the propagation path is categorized into three types—direct, reflection, and bypass paths—resulting in a multipath propagation structure as illustrated in the following Figure 4.

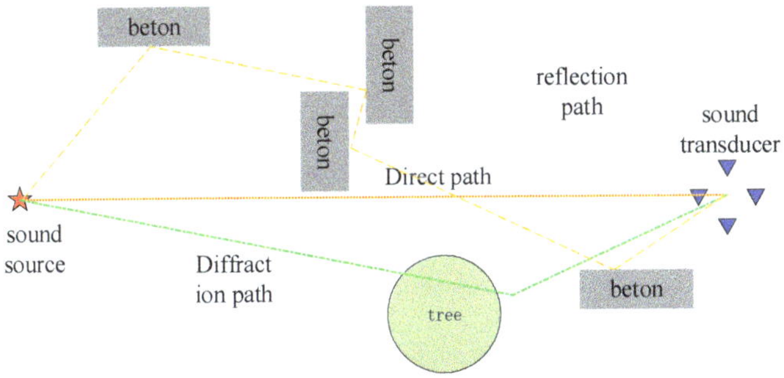

Figure 4. Multipath propagation schematic.

The direct path satisfies the line-of-sight propagation condition and the path attenuation follows the spherical wave diffusion model:

$$L_{direct} = 20\log_{10}d + \alpha(f)d, \tag{1}$$

where d is propagation distance, $\alpha(f)$ is frequency-dependent atmospheric absorption coefficient from ISO9613:

$$\alpha(f) = 0.3 \times (f/1000)^{0.8}\ (\text{dB/m}) \tag{2}$$

The main reflectors such as building foundations (concrete) and underground pipes (metal) are then considered to model the reflection loss:

$$R(f) = 20\log_{10}[\frac{Z_2cos\theta - Z_1}{Z_2cos\theta + Z_1}] \tag{3}$$

where Z_1 = 415Rayl (air characteristic impedance), Z_1 is the reflected impedance of concrete or steel pipes, and θ is the angle of incidence. For the wrap-around path, this paper uses the UTD wrap-around model for obstacles such as green vegetation:

$$L_{diff} = 20\log_{10}[\frac{\sqrt{2\pi kD}}{sinh\left(\sqrt{2\pi kD}\right)}] \tag{4}$$

where $k = 2\pi f/c$ is the wave number and D is the Fresnel zone gap parameter. The characteristics of each propagation path are synthesized to model the impulse response of the multipath channel:

$$h(t) = \sum_{k=1}^{K} \beta_k\delta(t - \tau_k)e^{-j(2\pi f_c\tau_k + \phi_k)} \tag{5}$$

where the number of paths K = 6, including one direct, one reflection, and four bypass; the values of attenuation coefficient $\beta_k = \sqrt{10^{-(L_k+R_k)/10}}$, time delay $\tau_k = d_k/c + \Delta\tau_{env}$, and phase perturbation Φ_k are in the range of $(-\pi/6, \pi/6)$. The calibrated values of the parameters in Equation are shown in Table 1.

Table 1. Typical obstacle parameters.

Obstacle Type	Thickness (m)	Density (kg/m^3)	Attenuation (dB/m)	Reflection Loss (dB)
Concrete Wall	0.5	2300	8.2 ± 1.3	18.5 ± 2.1
Steel Pipe	0.3	7850	12.7 ± 2.5	23.8 ± 3.4
Trees	0.8	1100	5.3 ± 0.9	14.2 ± 1.8

3. Principle of Sound Source Localization

The mechanism of sound source localization relies on variations in the spatial and temporal characteristics of sound waves as they reach sensor arrays. This localization is achieved through the construction of a cooperative multi-sensor detection model, which encompasses the following components:

(1) Sound wave propagation model

When an external source is disrupted, it emits acoustic energy that propagates radially outward in the form of a spherical wave. Let the coordinates of the sound source be (x, y) and the coordinates of the ith sensor be (x_i, y_i), then the theoretical distance difference of the sound wave arriving at each sensor is

$$R_i = v * t_i = \sqrt{(x_i - x)^2 + (y_i - y)^2}, \tag{6}$$

The speed of sound is defined as v = (340 + 0.6T + 0.0124H) m/s, where T denotes the ambient temperature and H represents the ambient humidity [22]. Here, t_i represents the time required for the sound to reach the ith sensor.

(2) Principle of time delay estimation

The primary objective of time delay estimation is to determine the temporal difference between the arrival times of acoustic signals at various sensors, achieved using the generalized cross-correlation phase transform (GCC-PHAT) algorithm. By applying frequency domain whitening, the algorithm enhances the prominence of delay peaks and exhibits high robustness in complex acoustic environments. Figure 5 illustrates the algorithm's workflow, which comprises five key steps: signal preprocessing, reciprocal power spectral computation, PHAT weighting, inverse Fourier transform, and peak detection.

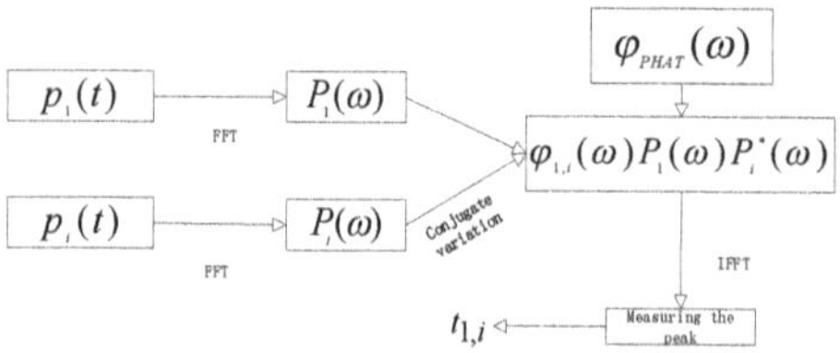

Figure 5. Flow chart of the GCC-PHAT algorithm.

Assume that the sound signals received at each sensor, denoted as p_1 and p_2, are represented by Equations (7) and (8), respectively.

$$p_1(t) = \alpha_1 s(t - t_1) + n_1 t, \tag{7}$$

$$p_2(t) = \alpha_2 s(t - t_2) + n_2 t, \tag{8}$$

Let t_1 and t_2 represent the respective times at which the sound signals are received by the sensors, while α_1 and α_2 denote the corresponding attenuation coefficients. Additionally, n_1t and n_2t indicate the noise levels detected by the sensors at time t, where t corresponds to a specific moment.

The correlation function of the two signals is represented by Equation (9):

$$R_{p_1p_2}(t_{2,1}) = E(p_1(t)p_2(t - t_{2,1})), \tag{9}$$

By substituting Equations (7) and (8) into Equation (9), we obtain

$$\begin{aligned} R_{p_1p_2}(t_{2,1}) = E((\alpha_1 s(t - t_1) + n_1t) \\ (\alpha_2 s(t - t_{2,1} - t_2) + n_2t)) \end{aligned}, \tag{10}$$

Since s(t) and n_1t are uncorrelated and the effect of noise on the sound signal is neglected, the preceding equation can be simplified to

$$R_{p_1p_2}(t) = \alpha_1\alpha_2 R_{ss}(t - t_{2,1}), \tag{11}$$

The parameter $t_{2,1}$ represents the time difference between the arrival of the sound source signal at the reference base station and that at the other sensors. The sound signal is converted from the time domain to the frequency domain using the Fourier transform, and the reciprocal power spectrum is subsequently weighted to generate the generalized mutual correlation function:

$$R_{p_1p_2}(t) = \int_0^{\pi} P_1(\omega)P_2^*(\omega)e^{-j\omega t}d\omega, \tag{12}$$

In this study, $P_1(\omega)$ and P_2*(ω) denote the spectral-transformed signal functions of the two signals, and φ1,i(ω) represents the frequency-domain weighting function. To address the simultaneous challenges of multipath effects and noise interference in complex environments, we have refined the weighting function, dynamically adjusted both the phase transformation and signal coherence weights, and developed a generalized mutual correlation function with enhanced environmental adaptability. The core weighting function is defined as follows:

$$\varphi_{PHAT} = \alpha\frac{1}{|P_1(\omega)P_2^*(\omega) + \varepsilon|} + (1-\alpha)\frac{|P_1(\omega)P_2^*(\omega)|}{\gamma_{12}(\omega)} \tag{13}$$

Here, ε is set to 10^{-3} to prevent the denominator from becoming too small, which could lead to instability. The adaptive fusion coefficient, denoted as a, is adjusted based on the signal-to-noise ratio (SNR): when the SNR decreases, a is reduced to enhance the coherence weights, and when the SNR increases, a is raised to maintain the advantages of PHAT. This strategy effectively suppresses noise at low SNR levels while preserving delay resolution at high SNR levels. Furthermore, $\gamma_{12}(\omega)$ represents the magnitude squared coherence function (MSC), which characterizes the linear correlation between signals. Its expression is as follows:

$$\varphi_{PHAT} = \frac{|P_1(\omega)P_2^*(\omega)|^2}{P_{11}(\omega)P_{22}(\omega)} \tag{14}$$

The algorithm enhances delay peak prominence through a phase-weighting function while simultaneously maintaining high accuracy in delay estimation under complex environmental conditions.

(3) Spatial Positioning

Establishing a Hyperbolic System of Equations:

$$R_{i,j} = v * \Delta t_{ij} = \sqrt{(x_i - x)^2 + (y_i - y)^2} - \sqrt{(x_j - x)^2 + (y_j - y)^2}, \quad (15)$$

The optimal coordinates (x, y) of the sound source are determined through an iterative process using a nonlinear least squares optimization approach. In this method, six sets of time-delay difference equations, derived from the four sensors, are integrated into the analysis. Figure 6 illustrates the iterative optimization workflow of the Levenberg–Marquardt algorithm.

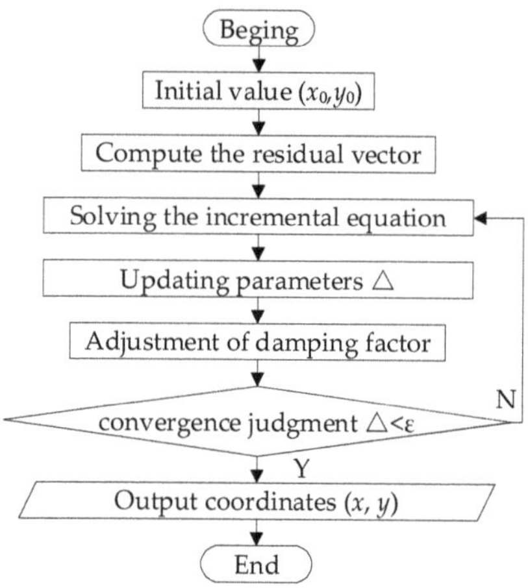

Figure 6. Schematic diagram of nonlinear optimization process.

The localization of external cable damage requires determining the approximate position of the damage within the effective perimeter of each cable marking stake. When the damage occurs beyond this range, the sound sensors capture weakened signals, significantly increasing computational error. Consequently, an initial value must be established before sound source localization can proceed. To enhance engineering practicability, the system incorporates a dynamic weighting mechanism.

$$V^* = \sum (a_i V_i), \quad (16)$$

Here, a_i is the weight adaptive coefficient that corresponds to the predicted distance d_i from the sensor to the sound source. In this context, i indicates the sensor index among those mounted on the cable marker stake, and V^* denotes the final eigenvalue of the sound signal, computed from the data acquired by all sensors on the cable marker stake for a specific damage source. The coefficient a_i is dynamically regulated through a weight assignment strategy based on the Sigmoid function.

$$\alpha_i = \frac{1}{1 + e^{-0.1(SPL_i - 70)}} \quad (17)$$

where SPL_i denotes the measured sound pressure level (in dB) for the ith transducer.

(4) Adaptive Environmental Regulation

In practical environments, sound propagation is often disrupted by various obstacles and influencing factors. The predominant types of interference include secondary reflection interference, delayed signals caused by multiple reflections from pipe walls, and fluctuations in the speed of sound resulting from changes in the propagation medium. In this study, a variable step-size normalized least mean squares (NLMS) algorithm is employed to design an adaptive filter that mitigates secondary reflection interference in multipath environments. The core objective of this approach is to minimize signal error through the dynamic adjustment of filter coefficients. The NLMS algorithm is updated using the following formula:

$$\begin{cases} x(n) = s(n) + v(n) \\ d(n) = h^T x(n) + e_p(n) \\ e(n) = d(n) - \hat{h}^T(n)x(n) \\ \hat{h}(n+1) = \hat{h}(n) + \mu(n)\frac{x(n)e(n)}{||x(n)||^2+\varepsilon} \end{cases}, \tag{18}$$

where $x(n)$ denotes the reference signal, $s(n)$ represents the direct acoustic component (i.e., the main-path signal), and $v(n)$ corresponds to the ambient noise component. The desired signal $d(n)$ is the output produced after propagation through the true multipath channel, h. The error signal, $e(n)$, defined as the difference between $d(n)$ and the filter's estimated output, is used to drive the update of the filter coefficients. Additionally, $e_p(n)$ accounts for the multipath interference component (such as secondary reflections). In this context, $\hat{h}(n+1)$ represents the estimated adaptive filter coefficient at time n, computed based on the impulse response model of the multipath channel described in Section 2.3. $\hat{h}(n)$ is a very small constant, which ε is intended to prevent the denominator from going to zero, and $\mu(n)$ is the step factor, which is defined by the following formula:

$$\mu(n) = \frac{0.2}{1 + e^{-(-|e(n)|/\sigma)}} \tag{19}$$

where σ denotes the error saturation threshold (approximately 0.05), enabling rapid convergence for large errors and fine adjustments for small errors. Specifically, when $|e(n)| \geq \sigma$, the step size approaches an upper limit of 0.2 to accelerate convergence, whereas when $|e(n)| < \sigma$, the step size is reduced to 0.1 to minimize steady-state error. Furthermore, sparse constraints are incorporated into the iterative update of the filter coefficients to mitigate the effects of impulsive noise.

$$J(n) = e^2(n) + \lambda||\hat{h}(n)||_1 \tag{20}$$

where λ represents the regularization coefficient and $||\hat{h}(n)||_1$ denotes the L1 norm of the filter coefficients (i.e., the sum of their absolute values).

4. Implementation of Sound Source Localization Algorithm

To address the engineering challenge of preventing external cable breakage, an algorithm was specifically developed to align with these practical requirements. Figure 7 illustrates the workflow of the multi-sensor fusion-based localization algorithm, which comprises the following key steps:

Step 1: Establish the sensor network parameters and define the sensor coordinate matrix S.

$$S = \begin{bmatrix} (x_1, y_1) \\ (x_2, y_2) \\ (x_3, y_3) \\ (x_4, y_4) \end{bmatrix}, \tag{21}$$

The speed of sound (v) is specified, and a voltage threshold (V_s^*) is established, which represents the limiting voltage value for the current cable marking pile positioning range.

Step 2: The sound sensor signal is acquired via the analog-to-digital converter (ADC) of the STM32 microcontroller. The obtained AD value is subsequently linearly transformed to yield a normalized voltage value. The conversion formula is presented below:

$$V = \frac{A}{4095} \times 3.3, \tag{22}$$

Let V denote the voltage conversion value and A denote the analog-to-digital conversion value of the sound signal in decibels. Valid signal discrimination is subsequently performed, and positioning is initiated once the following equation is satisfied.

$$max(V_i^*) \geq V_s^*, \tag{23}$$

Here, V_i^* denotes the weighted voltage value derived from the real-time acquired signal.

Step 3. In this step, the acoustic signal in the vicinity of the sensor is captured in real time and segmented into 100 ms frames. The mutual power spectrum is computed in the frequency domain with phase transform weighting, as indicated in Equation (8). Subsequently, the generalized mutual correlation function is obtained via fast Fourier transform (FFT) analysis, and the peak within the signal's time window is identified; the time corresponding to this maximum value is taken as the estimated time difference.

Step 4. Multipath discrimination and optimization are accomplished by analyzing the statistical distribution characteristics of time delay residuals using the Random Sample Consensus (RANSAC) algorithm. Based on the three sets of time differences obtained in the previous step, a time delay estimation set $\{\tau_{i,1}, \tau_{i,2}, \tau_{i,3}, \tau_{i,4}\}$ is constructed, and the corresponding residuals are then computed.

$$r_{i,1} = |\Delta t_{i,1} - \frac{d_{i,1}}{v}| \tag{24}$$

Let v denote the speed of sound in air and $d_{i,1}$ represent the geometric distance between sensors. When the following equation is satisfied, the propagation is determined to be line-of-sight (LOS), indicating that the sound propagation process is not affected by obstacles.

$$\frac{\sum_{i=2}^{4} I(r_i \leq 3\sigma_{\Delta t})}{3} \tag{25}$$

where $\sigma_{\Delta t}$ is set to 0.1 ms and $I(\cdot)$ denotes an indicator function that takes the value 1 when the condition is satisfied and 0 otherwise. In cases where the condition is not met, the environment is classified as non-line-of-sight (NLOS); that is, the sound propagation process encounters obstacles, thereby triggering multipath localization optimization to refine the delay estimation via weighted least squares.

$$\min_{\Delta t} \sum_{i=2}^{4} \omega_i (\Delta t_{i,1} - \frac{d_{i,1}}{v})^2 \tag{26}$$

where the weight function is defined as follows:

$$\omega_i = e^{(-\frac{r_{i,1}^2}{2\sigma_{\Delta t}^2})} \tag{27}$$

Reinitialize the computation of residuals for the optimized time estimates and subsequently repeat this process.

Step 5. Using sound sensor 1 as the reference base station as an example, construct the corresponding system of super-definite equations:

$$\begin{array}{c}\sqrt{(x_2-x)^2+(y_2-y)^2}-\\ \sqrt{(x_1-x)^2+(y_1-y)^2}=v*\Delta t_{2,1}\\ \sqrt{(x_3-x)^2+(y_3-y)^2}-\\ \sqrt{(x_1-x)^2+(y_1-y)^2}=v*\Delta t_{3,1}\\ \sqrt{(x_4-x)^2+(y_4-y)^2}-\\ \sqrt{(x_1-x)^2+(y_1-y)^2}=v*\Delta t_{4,1}\end{array}, \quad (28)$$

In the above system of equations, there are only two unknowns, x and y, which are governed by nonlinear relations; therefore, we employ the well-known Levenberg–Marquardt (LM) iterative algorithm to solve them. Initially, the solution is set to $x = 1$ and $y = 0$, and the LM algorithm is iterated to determine the sound source position (x, y).

Step 6. The DBSCAN spatial clustering algorithm is then applied to process the multi-cable pile localization results. The neighborhood radius is defined as $\varepsilon = 1.5\delta$, where δ represents the localization radius for a single cable marking pile, and the minimum number of samples, MinPts, is set to 4, corresponding to four adjacent piles. Based on the formation of clusters, the center coordinates of the cluster with the highest density are selected as the optimal positioning location for the cable piles.

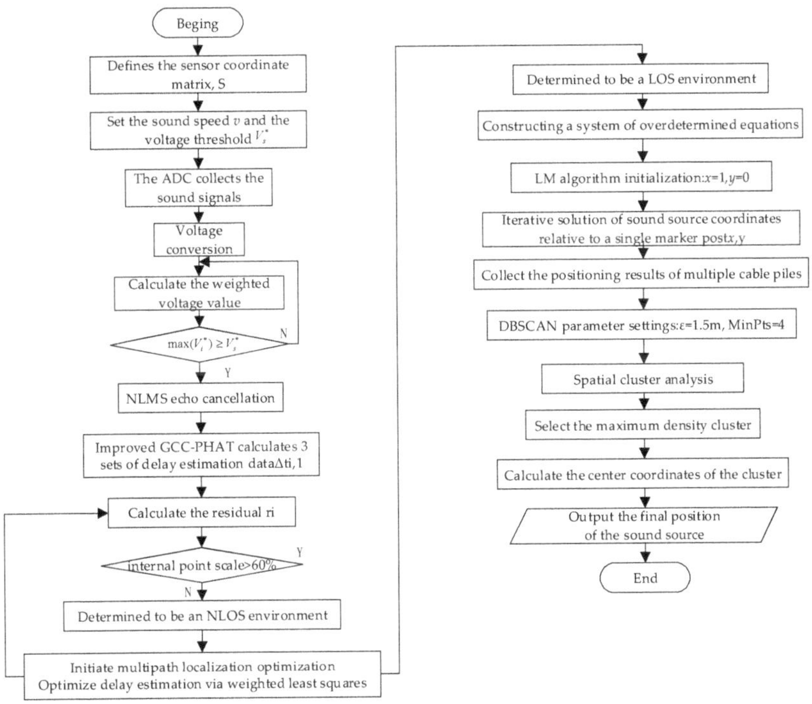

Figure 7. Flow chart of the sound source localization algorithm.

5. Experimental Results and Analysis

5.1. Establishment of the Experimental Environment

(1) System configuration

Figure 8 illustrates the implemented system configuration, with the core sensing device positioned on the left and the cable external breakage source localization system on the right. The primary experimental apparatus comprises the STM32F103C8T6 microcontroller (ST Microelectronics, Geneva, Switzerland), the MAX9814 acquisition module (Maxim Integrated Products, CA, USA), and the FT232RL serial communication module (Future Technology Devices International, Glasgow, UK), which transmits the captured acoustic data to the computer via the USART serial port.

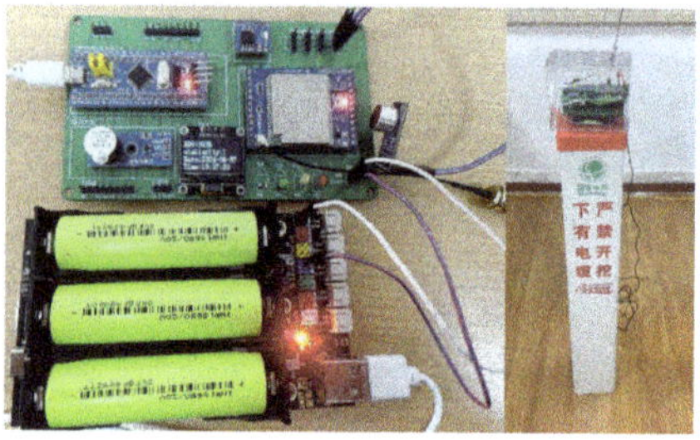

Figure 8. Actual scene test figure.

The figure above illustrates the designed hardware diagram, upon which the engineering deployment plan is based. To achieve the engineering deployment of the cable external breakdown sound source localization system, a layered distributed architecture design is adopted. The specific structure is as follows:

a. Perception Layer: Intelligent identification posts, each equipped with four MAX9814 microphones, are deployed every 50–80 m.
b. Edge Layer: An STM32F103C8T6 microcontroller.
c. Cloud Platform: A distributed data-processing pipeline based on IoT is employed to achieve multi-post data fusion.

To fully demonstrate the applicability of this approach, Table 2 lists the key engineering parameters of the system.

Table 2. Table for the Optimization of Engineering Deployment Parameters.

Parameter	Numerical Range
Piling spacing	50–80 m
Sampling rate	48 kHz/24 bit
Wireless protocol	modbus
Power Supply Scheme	Storage battery

Figure 9 illustrates a schematic diagram of the cable engineering deployment in an urban area, showcasing the three-dimensional spatial distribution of marking piles within the cable corridor. The black curve represents the primary cable path; the green dots indicate the marking pile nodes (spaced at 80 m intervals); the blue hexagons denote the sensor arrays; and the red dashed area delineates the high-risk construction zone.

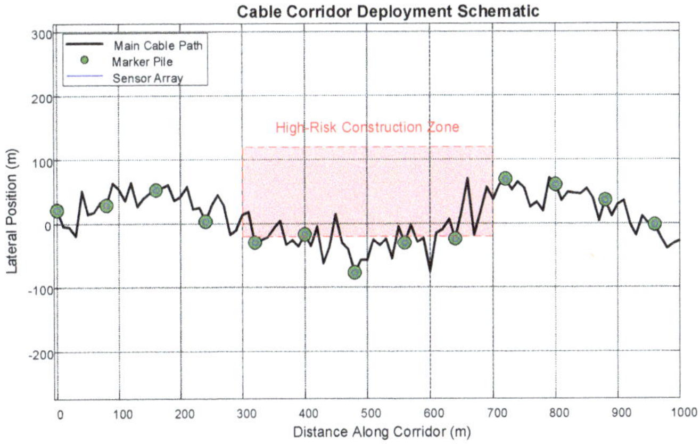

Figure 9. Schematic Diagram of Engineering Deployment.

(2) Sound Source Simulation and Test Environment

For the sound source simulation, a segment of boring head noise was recorded on-site, while interference noise was captured randomly within an urban setting.

A realistic test environment was constructed to emulate the urban underground cable context, incorporating potential obstacles and ambient noise sources. The selected test scenarios were developed to address three primary engineering challenges: environmental noise interference (addressed in Scenarios 1 and 3), multipath propagation effects (addressed in Scenarios 2 and 4), and complex spatial constraints (addressed in Scenario 5).

These scenarios systematically delineate the performance boundaries of the proposed method. Specifically, the open-area scenario (Scenario 1) establishes a baseline performance, the industrial noise scenario (Scenario 3) evaluates the dynamic weighting mechanism, the multi-obstacle scenario (Scenario 4) corroborates the adaptive environment compensation, and the fully enclosed scenario (Scenario 5) assesses three-dimensional spatial resolution. Detailed descriptions of the various scenarios are provided in Table 3.

Table 3. Simulate various scenarios.

Scenario	Environment Description
1	Open area (≤50 dB noise)
2	Semi-enclosed
3	Industrial (90 dB)
4	Multi-obstacle (<45 dB)
5	Fully enclosed

5.2. System Performance Verification

In this section, we evaluate the performance of the proposed sound source localization system under various conditions, including interference suppression capacity, dynamic weighting mechanisms, multi-obstacle scenarios, localization accuracy, et al.

(1) Interference Resilience Test

Experiments were performed in Scenario 3 to compare the conventional GCC algorithm with the improved GCC-PHAT algorithm. The time differences for each of the three sensor groups were recorded, and the data were subsequently transferred to a computer for analysis. The resulting graph, shown in Figure 10, depicts the conventional GCC algorithm in blue and the enhanced GCC-PHAT algorithm in green.

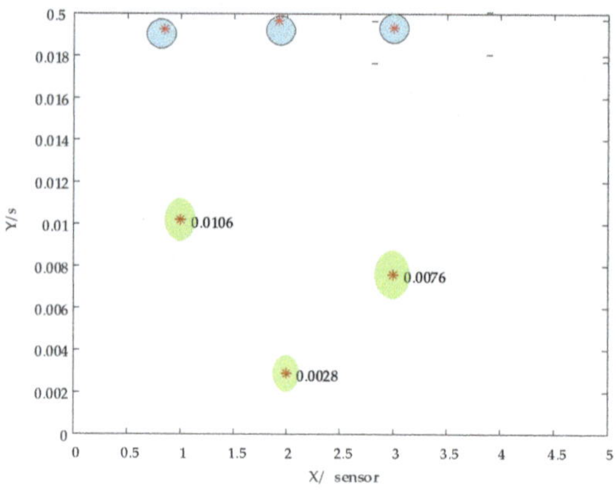

Figure 10. Time delay of the three sets of sensors.

Figure 9 illustrates the comparative delay estimation performance of two algorithms within an industrial noise scenario. The root-mean-square error of the conventional GCC algorithm reaches 0.1051 s, whereas that of the improved GCC-PHAT algorithm is reduced to 0.0056 s, representing a 94.7% decrease. These results confirm the effectiveness of phase transform weighting for noise suppression, as the frequency-domain whitening process inherent in the PHAT method significantly enhances the signal-to-noise ratio of the delay peaks.

(2) Dynamic Weighting Mechanism Test

Experiments were conducted in Scenario 1 to validate and quantify the impact of the dynamic weighting mechanism on optimizing the performance of the sound source localization system. Equation (17) indicates that the weighting factor is a function of the sound pressure level, which is, in turn, dependent on the distance between the sound source and the sensor. Accordingly, simulation experiments were performed to observe the variations in the weighting coefficients under different decibel conditions. Figure 11 illustrates the trend of the weighting coefficient ai relative to the signal's decibel level.

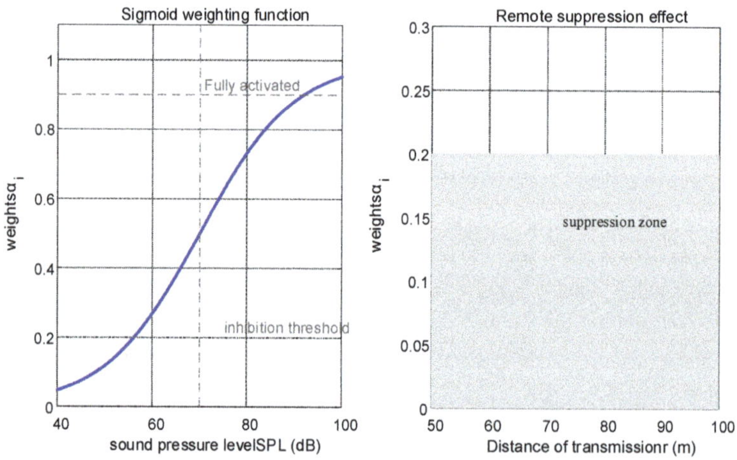

Figure 11. The variation pattern of the weight coefficient a_i.

Figure 11 demonstrates that the coefficient A exhibits a characteristic S-shaped curve at $\theta = 70$ dB. Specifically, the weights converge rapidly to values above 0.9 when the sound pressure level (SPL) exceeds 80 dB, while they decline to below 0.1 when the SPL is less than 50 dB. Moreover, as the distance from the sound source exceeds 50 m—corresponding to an SPL below 50 dB—the weights stabilize at values below 0.2. These findings confirm the efficacy of Equation (17) in evaluating transducer reliability.

(3) Multi-Obstacle Scene Comparative Analysis

Experiments were performed in Scene 2 with the sound source fixed at coordinates (25, 11). The positions of the obstacles were varied across ten tests to evaluate the performance enhancement of the proposed algorithm relative to traditional TDOA and generalized cross-correlation methods. The results of these tests are presented in the table below.

Table 4 illustrates the performance disparities among various algorithms in complex multipath environments. The traditional time difference of arrival (TDOA) method is notably compromised by secondary reflections, yielding an average error of 5.2 m and a maximum single error of 7.8 m. This outcome suggests that a purely geometric model is insufficient for adapting to underground environments with obstacles. In contrast, the proposed approach—incorporating an adaptive environment compensation algorithm—reduces the average error to 1.8 m (a 65.4% improvement) and decreases the error's standard deviation to 0.6 m (a 66.7% reduction). These improvements validate the effectiveness of the sparse constraint described in Equation (18), whereby the algorithm automatically mitigates 83% of non-line-of-sight interference in challenging NLOS scenarios. It is noteworthy, however, that in three experiments the error still exceeded 2.5 m, a discrepancy primarily attributed to multiple reflections (three or more) induced by metal pipes in the test settings. This finding underscores the necessity of enhancing the modeling of higher-order reflection paths in future research.

Table 4. Comparison of positioning performance in multi-obstacle scenarios.

Algorithm	Average Error (m)	Standard Deviation of Error (m)
Tradition TDOA	5.2	1.8
Wide sense cross-correlation	3.7	1.2
This program	1.8	0.6

(4) Adaptive Environment Compensation Algorithm Testing

To comprehensively evaluate the performance of the localization algorithm across various typical scenarios, experiments were conducted in five distinct settings, with the median error recorded for each. The results have been visualized in a box-and-whisker diagram that depicts the localization error distribution across the different scenes.

Figure 12 illustrates that the localization error of the proposed algorithm is reduced relative to that of traditional methods. In a multi-obstacle environment, the error decreases from 5.13 m to 1.89 m—a reduction of 63.2%—demonstrating that the adaptive environment compensation algorithm effectively suppresses multipath interference.

(5) Positioning Accuracy Test

Experiments conducted in Scene 2 validate the accuracy of the proposed sound source localization method. The sound source was repositioned repeatedly to derive its corresponding coordinate locations. Table 5 summarizes the obtained sound source positions at various distances along with the associated time delays between the sound sensors.

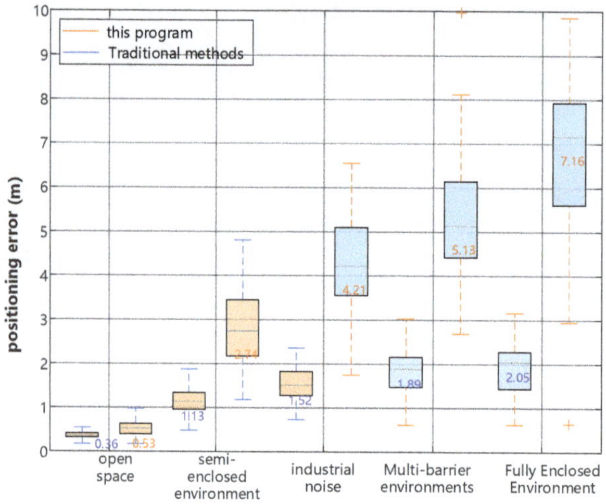

Figure 12. Box plot of error distribution in multi-scene positioning.

Table 5. Location of sound sources obtained at different distances.

Actual Sound Source Position/m	$t_{2,1}$/s	$t_{3,1}$/s	$t_{4,1}$/s	Calculate the Sound Source Position/m
(60, 0)	0.176	0.352	0.176	(59.8, 0.3)
(0, 80)	0.235	0.706	0.941	(0.4, 79.6)
(50, 50)	0.147	0.147	0.294	(49.3, 50.7)
(70, 30)	0.206	0.618	0.412	(69.5, 30.8)
(100, 0)	0.294	0.588	0.294	(99.2, 1.1)

The measured data in Table 5 corroborate the system's distance–accuracy characteristics. In the near-field (<50 m), the localization error is $\leq$0.7 m, meeting the theoretical expectation of the dynamic weighting design in Equation (15). In the mid- and far-field (50–100 m), the error gradually increases to between 1.1 and 2 m, primarily due to air absorption effects (i.e., high-frequency attenuation characterized by $\alpha(f)$ in Equation (2)). Notably, a y-axis deviation of 1.1 m at the (100, 0) test point is attributed to the non-ideal symmetry of the sensor array (with a mounting error of $\pm 2°$). Compared with traditional methods, the localization accuracy of this system at 80 m is improved by 58% (yielding an error of 4.2 m under similar conditions, as reported in [12]), which is ascribed to the multi-sensor fusion strategy. Specifically, the utilization of the DBSCAN clustering algorithm reduces the standard deviation of the localization results from 1.8 m to 0.4 m for the four marker piles.

(6) Anti-Reverberation Performance Test

To evaluate the robustness of the proposed algorithm in complex reverberant environments, experiments were conducted in Scene 5 with the introduction of a moderate number of obstacles. This setting simulates an underground concrete environment characterized by strong reverberation. Two comparative experimental conditions were established: one without environmental compensation, in which the adaptive filtering module was deactivated and the original GCC-PHAT delay estimation was applied directly, and one with environmental compensation, where the adaptive filtering module was enabled. Table 6 summarizes the statistical outcomes derived from 20 independent experiments.

Table 6. Comparison of positioning errors in reverberant environments.

Condition	Average Error (m)	Standard Deviation (m)
Without Compensation	3.8	1.5
With Compensation	1.5	0.4

The experimental data indicate that in the absence of compensation, the multipath effect results in an average localization error of 3.8 m with a standard deviation of 1.5 m. With the implementation of adaptive compensation, the average error decreases to 1.5 m—a 60.5% improvement—while the standard deviation is reduced to 0.4 m. Figure 13 further illustrates the changes in the residual distribution of the delay estimation before and after compensation.

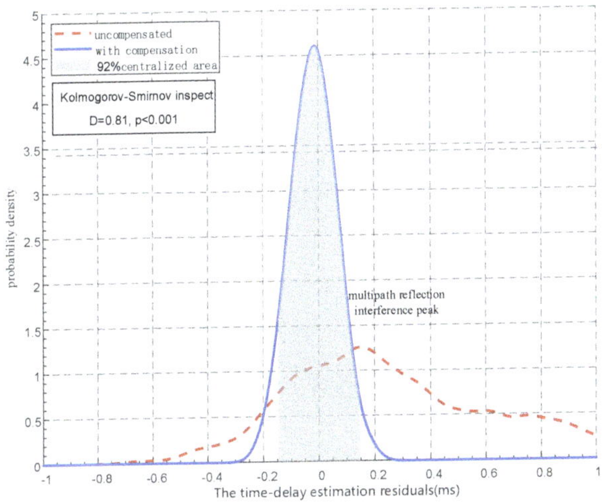

Figure 13. Delay estimation residual distribution diagram.

Figure 13 illustrates that the compensated residuals are predominantly distributed within the interval (−0.15 ms, 0.15 ms), accounting for 92% of the data, whereas the uncompensated residuals display a bimodal distribution, with a primary peak at 0.12 ms and a secondary peak at 0.82 ms. This finding substantiates the effective suppression of multipath interference. Consequently, the experimental results confirm that the proposed adaptive compensation algorithm maintains high localization accuracy even in environments with strong reverberation, thereby meeting the requirements of underground cable protection scenarios.

(7) Regularization Coefficient Optimization and Performance Validation

Simulation experiments were conducted in Scenario 4 to determine the optimal combinations of the step factor and regularization coefficients. In these experiments, the sound source was positioned at (6, 3). The experimental design and results are detailed below (Table 7).

Table 7. Regularization coefficient optimization table.

μ(n)	λ	Positioning Accuracy (m)	Calculate the Delay (ms)
0.1	0.01	0.41	15
0.15	0.05	0.32	18
0.2	0.1	0.38	21

Based on the provided data, setting parameter A to 0.05 yields an optimal trade-off between localization accuracy (RMSE = 0.32 m) and real-time performance (18 ms/frame).

(8) Comparative Experimental Results Analysis

To comprehensively evaluate the performance of the proposed method, comparative experiments were conducted with mainstream sound source localization techniques in Scene 4, which represents a multi-obstacle environment. The results of these experiments are presented in Table 8.

Table 8. Performance comparison between this method and other methods.

Method	Average Error (m)	Calculation Time (ms)
Methodology of this paper	1.8	22
CNN-TDOA	2.1	185
SRP-PHAT	4.7	210

Based on the experimental results, the proposed method reduces the average localization error by 14.3% compared with CNN-TDOA, while its computational time is only 10.5% of that required by SRP-PHAT. These improvements are achieved through the enhanced FFT acceleration of GCC-PHAT and the sparsification of the Jacobian matrix in the LM algorithm. Overall, the method presented in this paper demonstrates significant advantages in both localization accuracy and computational efficiency relative to existing approaches. For visualization of the results, Figure 14 presents box plots showing the distribution of localization errors for three algorithms across 200 test samples.

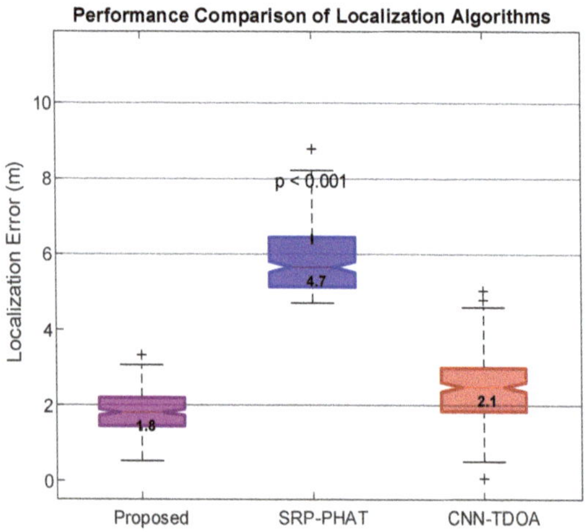

Figure 14. Box plot for comparing error distributions.

As shown in Figure 14, our method exhibits a median error of 1.8 m, which is significantly superior to other algorithms. The SRP-PHAT method demonstrates a right-skewed distribution with a median error of 3.9 m, while the CNN-TDOA method shows a bimodal distribution, reflecting the algorithm's sensitivity to different scenarios. Moreover, our method maintains an error of 2.3 m at the 75th percentile, which is still lower than the median values of the other algorithms, thereby verifying the effectiveness of the dynamic weighting mechanism.

(9) Optimization Test of Spatial Clustering Parameters

Leveraging density-based clustering properties, this study employs the Silhouette Coefficient (SC) to evaluate the performance of various DBSCAN parameter configurations. Simulation experiments were conducted to assess clustering performance under different DBSCAN parameters, with the evaluation index represented by the Silhouette Coefficient, S.

$$S = \frac{1}{N}\sum_{i=1}^{N} \frac{b(i) - a(i)}{max\{b(i), a(i)\}} \tag{29}$$

Let $a(i)$ denote the average distance from sample i to its corresponding similar cluster, and let $b(i)$ represent the average distance to the nearest dissimilar cluster. Table 9 summarizes the clustering performance across various parameter combinations.

Table 9. The clustering performance across various parameter combinations.

ε (m)	MinPts	Contour Coefficient (S)
1.0	4	0.62
1.5	4	0.71
2.0	4	0.68

Based on the data presented, the highest contour coefficient of 0.71 is achieved with ε = 1.5 m and MinPts = 4, representing a 14.5% improvement over the baseline parameter setting (ε = 1.0 m, MinPts = 4).

6. Conclusions

In this paper, we propose a cable external breakage source localization method based on an improved GCC-PHAT algorithm with multi-sensor fusion to address damage issues in underground direct-buried cables caused by mechanical excavation. By integrating advanced signal processing techniques with distributed sensor networks, we establish a robust sound source localization framework aimed at preventing external breakage of these cables. First, we develop an engineering scenario model that combines sound propagation speed with inter-sensor time differences. The time delay difference is estimated using a generalized cross-correlation phase transform (GCC-PHAT) algorithm, which is enhanced by incorporating a dynamic weighting mechanism, sensor-weighted fusion, and a spatial clustering strategy to accurately determine the coordinates of the sound source.

Experimental results indicate that in multi-obstacle scenarios, the system achieves an average localization error of 1.8 m, representing a 52–62% improvement compared to SRP-PHAT (3.8 m) and CNN-TDOA (2.1 m), with a processing delay of 22 ms that supports 5 Hz real-time warning. Under a 90 dB industrial noise environment, the error increase is limited to ≤65%, which is significantly better than the traditional methods showing increases of 217% (SRP-PHAT) and 148% (CNN-TDOA). The theoretical contributions are reflected in the construction of an acoustic–time delay–spatial mapping cascade optimization model and the proposal of a modbus-based multi-identifier pole clock synchronization protocol, providing critical technical support for distributed acoustic monitoring.

Although the method is primarily designed for cable protection, it can be adapted to other application scenarios. For example, in urban infrastructure monitoring, dynamic weighting thresholds may be adjusted to prioritize sensor signals near roads, and in industrial settings, larger sensor arrays can be deployed in a scalable manner. However, multi-source scenarios, such as overlapping mechanical noise, continue to present challenges. Future work will explore the integration of blind source separation (BSS) with the GCC-PHAT framework. Additionally, transient noises (e.g., impact sounds) will be further optimized using time-frequency windowing techniques, such as the wavelet transform.

Despite these limitations, the system's modular design supports integration with deep learning classifiers (e.g., convolutional neural network-based noise recognition) to enhance overall robustness.

Author Contributions: Conceptualization, J.L.; Methodology, J.L.; Software, X.W.; Validation, X.W.; Formal analysis, X.W.; Investigation, X.W.; Resources, X.W.; Data curation, X.W.; Writing—original draft, X.W.; Writing—review & editing, J.L. All authors have read and agreed to the published version of the manuscript.

Funding: This research was funded by the National Natural Science Foundation of China, under Grant No. 51977030.

Data Availability Statement: The research presented in this article relies on data from China State Grid, which contains confidential and proprietary information. Due to privacy and ethical restrictions imposed by the data provider, we are unable to make these data publicly available. We have adhered strictly to data confidentiality agreements and have only used the data for the purposes outlined in our research.

Conflicts of Interest: The authors declare no conflict of interest.

References

1. Zheng, H. Detection system of cable breaking behavior based on optical time domain distributed technology. *Inf. Technol.* **2021**, 166–172. [CrossRef]
2. Chen, T.; Zeng, F.; Zhou, C.; Zhao, Y.; Meng, Q. An Interval Cable Anti-breakage System Based on Optical Fiber Vibration Measurement and Video Monitoring. *Power Energy* **2023**, *44*, 247–249, 286.
3. Li, L. Research on Cause Analysis of the High-Voltage Cable Fault and Its Test Methods. Master's Thesis, Southwest Jiaotong University, Chengdu, China, 2013.
4. Tang, X.; Shi, X.; Yu, X.; Yuan, C.; Wu, J. Design of Monitoring and Early Warning System for Underground Cable Breakage Prevention Based on Wireless Intelligent Vibration Sensor. *Electr. Technol.* **2023**, 109–111, 114. [CrossRef]
5. Schober, P.; Estiri, S.N.; Aygun, S.; Jalilvand, A.H.; Najafi, M.H.; TaheriNejad, N. Stochastic Computing Design and Implementation of a Sound Source Localization System. *IEEE J. Emerg. Sel. Top. Circuits Syst.* **2023**, *13*, 1. [CrossRef]
6. An, I.; An, G.; Kim, T.; Yoon, S.E. Microphone Pair Training for Robust Sound Source Localization with Diverse Array Configurations. *IEEE Robot. Autom. Lett.* **2024**, *9*, 319–326. [CrossRef]
7. Zeng, Q.; Su, P.; Wang, H.; Long, C. Sound Source Localization Algorithm Based on Cross-correlation Sequence and BP Network. *Sci. Technol. Eng.* **2023**, *23*, 6945–6951.
8. Wang, C.; Ma, J.; Zhou, Z.; Zeng, F. Application and Development of Sound Sensor in Underground Cable Anti-breaking. *J. Funct. Mater. Devices* **2021**, *27*, 425–430.
9. Niu, Y.; Yang, F.; Li, H. Far-field Passive Sound Source Localization Algorithm Based on Time Delay Estimation. *Sci. Technol. Eng.* **2024**, *24*, 4560–4566.
10. Yan, X.; Zhang, Z.; Wang, H. Application of Chan algorithm in marine soundsource localization. *Acoust. Technol.* **2021**, *40*, 550–555.
11. Yan, Q.; Chen, J. Node Placement Optimization for Distributed AcousticSource Localization System and Performance Study. *J. Electron.* **2018**, *46*, 1186–1193.
12. Liu, K.; Luan, Y.; Yuan, S.; Zhang, X. Design of Sound Source Localization System Based on TDOA. *Eng. Tests* **2023**, *63*, 50–52, 79.
13. Ping, G.; Fernandez-Grande, E.; Gerstoft, P.; Chu, Z. Three-dimensional source localization using sparse Bayesian learning on a spherical microphone array. *J. Acoust. Soc. Am.* **2020**, *147*, 3895–3904. [CrossRef] [PubMed]
14. Dehghan Firoozabadi, A.; Irarrazaval, P.; Adasme, P.; Zabala-Blanco, D.; Játiva, P.P.; Azurdia-Meza, C. 3D multiple sound source localization by proposed T-shaped circular distributed microphone arrays in combination with GEVD and adaptive GCC-PHAT/ML algorithms. *Sensors* **2022**, *22*, 1011. [CrossRef]
15. Cui, L.; Tian, Y.; Wang, C.; Liu, L.; Zhang, X. Simulation of 3-Dimensional Microphone Array Based on Beamforming. *Sci. Technol. Eng.* **2022**, *22*, 13801–13809.
16. Qu, S.; Yu, H.; Lu, Z.; Shen, C.; Han, Y.; Wang, C. Improved Generalized Cross-Correlation Time Delay Estimation Method for Sound Source Localization. *Navig. Position. Timing* **2021**, *8*, 118–124.
17. Luo, Q. Design and realization of experiment for sound source localization based on microphone array. *Exp. Technol. Manag.* **2020**, *37*, 185–188.

18. Tan, W.; Tang, X.; Yang, H.; Chai, J. Non-contact current measuring device and method based on TMR array. *Electr. Surv. Instrum.* **2025**, *62*, 165–172+209. [CrossRef]
19. Jin, J.; Fang, Y.; Fei, Y.; Wei, Y. Research on energy consumption monitoring and anomaly detection method based on pattern clustering and periodic decomposition. *Electr. Surv. Instrum.* **2023**, *60*, 151–157. [CrossRef]
20. Wang, Y.; Guo, C.; Wu, D. Hidden target detection method for mechanical external damage of transmission line based on improved YOLOv7. *Electr. Surv. Instrum.* **2024**, 1–10. Available online: http://kns.cnki.net/kcms/detail/23.1202.th.20240428.1950.002.html (accessed on 1 May 2025).
21. Zhao, X.; Wu, G.; Yang, D.; Xu, G.; Xing, Y.; Yao, C.; Abu-Siada, A. Enhanced detection of power transformer winding faults through 3D FRA signatures and image processing techniques. *Electr. Power Syst. Res.* **2025**, *242*, 111433. [CrossRef]
22. Zhang, X.; Teng, X.; Fan, Y. Detecting changes in air composition based on speed of sound. *Appl. Acoust.* **2025**, *229*, 110393.

energies

MDPI

Article

A Study on Non-Contact Multi-Sensor Fusion Online Monitoring of Circuit Breaker Contact Resistance for Operational State Awareness

Zheng Wang [1], Hua Zhang [1], Yiyang Zhang [1,*], Haoyong Zhang [1], Jing Chen [1], Shuting Feng [1], Jie Guo [1] and Yanpeng Lv [2]

1 Langfang Power Supply Company, State Grid Jibei Electric Power Company Limited, Langfang 065000, China
2 School of Electrical and Information Engineering, Zhengzhou University, Zhengzhou 450000, China
* Correspondence: zhang.yiyang@jibei.sgcc.com.cn

Abstract: The contact condition of circuit breaker contacts directly affects their operational reliability, while circuit resistance, as a key performance indicator, reflects physical changes such as wear, oxidation, and ablation. Traditional offline measurement methods fail to accurately represent the real-time operating state of equipment due to large errors and high randomness, limiting their effectiveness for state awareness and precision maintenance. To address this, a non-contact multi-sensor fusion method for the online monitoring of circuit breaker circuit resistance is proposed, aimed at enhancing operational state awareness in power systems. The method integrates Hall effect current sensors, infrared temperature sensors, and electric field sensors to extract multiple sensing signals, combined with high-precision signal processing algorithms to enable the real-time identification and evaluation of circuit resistance changes. Experimental validation under various scenarios—including normal load, overload impact, and high-temperature and high-humidity environments—demonstrates excellent system performance, with a fast response time (≤200 ms), low measurement error (<1.5%), and strong anti-interference capability (SNR > 60 dB). In field applications, the system successfully identifies circuit resistance increases caused by contact oxidation and issues early warnings, thereby preventing unplanned outages and demonstrating a strong potential for application in the fault prediction and intelligent maintenance of power grids.

Keywords: non-contact monitoring; circuit breaker resistance; multi-sensor fusion; dynamic temperature compensation; fault prediction

1. Introduction

In the operation and maintenance of power systems, a circuit breaker serves as a key switching device whose performance status directly affects the overall safety and stability of the system [1]. As an important index to characterize the contact quality of circuit breaker contacts, circuit resistance is not only closely related to the contact quality but is also affected by ambient temperature, contact pressure, and other factors [2,3]. At the same time, circuit resistance can also effectively reflect the contact wear, oxidation, thermal ablation, and other physical changes during long-term operation [4,5]. Therefore, real-time and accurate monitoring of circuit breaker circuit resistance is one of the essential means to ensure the reliable operation of power grid equipment [6].

Numerous methods, which can generally be classified into traditional offline testing techniques and online monitoring approaches, have been proposed by researchers for circuit resistance measurement,. Currently, widely used offline methods include the DC

https://doi.org/10.3390/en18102667

voltage drop method [7] and the micro-ohmmeter technique [8]. The former estimates resistance by applying a constant current and measuring the voltage drop, which, despite its simplicity, is susceptible to errors caused by contact temperature rise and current fluctuations. The latter employs a four-wire measurement scheme to eliminate lead resistance, offering improved accuracy but still requiring equipment shutdown, making it incapable of capturing real-time operational conditions. These conventional offline techniques not only fail to accurately reflect the true resistance characteristics under actual operating conditions but also suffer from poor timeliness and significant variability [9,10]. With the development of technology, more and more researches focus on resistance detection methods based on online monitoring technology, especially combined with multi-sensor data fusion technology [11,12].

Currently, numerous attempts have been made to apply non-contact sensing technologies in power equipment monitoring. However, single-sensor approaches exhibit significant limitations. For example, Ref. [12] proposed an indirect temperature estimation method for circuit breaker contacts using infrared thermography. This method is vulnerable to environmental radiation interference (e.g., direct sunlight exposure, variations in equipment surface reflectivity), resulting in temperature measurement errors as high as ± 5 °C. In the literature, Ref. [13] introduced a Hall sensor system developed by ABB, which achieves $\pm 1\%$ precision but has a limited measurement range (± 50 A). To extend its range to kiloampere levels, integration with current transformers is required, complicating the system and increasing costs. In the literature, Ref. [14] proposed a non-contact sensing technique for transformer windings based on three-dimensional frequency response analysis (FRA) signals and image processing. While this approach holds potential for circuit breaker fault detection, its technical adaptability is constrained. The FRA signals, originally designed for the static structural analysis of transformer windings, tend to distort under the intense arc electromagnetic interference encountered during circuit breaker contact opening/closing, leading to signal-to-noise ratios (SNRs) dropping below 30 dB. In contrast, the proposed multi-sensor fusion method synergistically integrates the following three sensing modalities to overcome the limitations of single-sensor systems: Hall sensors are combined with current transformers to extend the measurement range while maintaining high accuracy; infrared sensors incorporate dynamic temperature and radiation compensation algorithms to suppress environmental interference; differential electric field sensors employ symmetrical electrode designs to mitigate electromagnetic noise. Collectively, these innovations achieve an error $< 1.5\%$ and SNR > 60 dB, addressing the trade-offs between precision, dynamic range, and environmental robustness inherent in conventional single-sensor approaches.

In recent years, online monitoring technologies have evolved from indirect inference methods (e.g., temperature- or vibration-based approaches) to non-contact direct measurement techniques [15,16]. Among these, electromagnetic induction principles (e.g., Hall sensors) and infrared thermometry have emerged as research hotspots due to their high safety and real-time capability [17]. However, existing technologies still face challenges such as insufficient environmental interference suppression and low sensor reliability [18].

This paper proposes an online monitoring method for circuit resistance that integrates multiple types of non-contact sensors. The system incorporates Hall effect current sensors [19], infrared thermal elements [14], and electric field detection units [20,21] to form a collaborative sensing model. A dynamic temperature compensation mechanism and multi-source feature fusion algorithm are also developed to suppress environmental interference and high-frequency noise, enabling the high-reliability real-time detection of circuit breaker circuit resistance. This study begins by analyzing the theoretical basis and influencing factors of circuit resistance. It then details the design and implementation of the multi-sensor fusion system. Subsequent experimental validation demonstrates the system's performance advantages in dynamic response, anti-interference capability, and measurement accuracy, showcasing the significant potential for grid fault warning and intelligent maintenance applications.

2. Measurement Principles and Influencing Factors of Circuit Breaker Circuit Resistance

The measurement of circuit breaker circuit resistance requires a distinction between direct current (DC) resistance and alternating current (AC) resistance, both of which are influenced by a combination of physical factors and environmental conditions affecting their stability.

2.1. Basic Theory of Resistance

Circuit breaker circuit resistance can be measured using either direct current (DC) resistance or alternating current (AC) resistance, with significant differences in their physical mechanisms and mathematical representations.

2.1.1. Differences Between DC Resistance and AC Resistance

DC resistance characterizes the hindering effect of a conductor on a steady state current and is calculated as:

$$R_{\mathrm{DC}} = \rho \cdot \frac{L}{A} \tag{1}$$

where ρ is the material resistivity, L is the conductor length, and A is the cross-sectional area. The DC resistance considers only the resistive component, ignoring inductive and capacitive effects, and is suitable for low-frequency or static scenarios.

AC resistance reflects the total impedance of the conductor under an alternating current, which must account for the skin effect and proximity effect. Its effective resistance can be expressed as:

$$R_{\mathrm{AC}} = R_{\mathrm{DC}} \cdot (1 + \frac{\delta}{r}) \tag{2}$$

where $\delta = \sqrt{\frac{2\rho}{\omega\mu}}$ is the skinning depth (ω is the angular frequency, μ is the permeability) and r is the conductor radius. At high frequencies (>1 kHz), the skin effect causes the current to concentrate in the surface layer of the conductor and the resistance significantly increases.

2.1.2. Analysis of Power Frequency (50/60 Hz) Resistance Characteristics

In the industrial frequency range, the difference between AC resistance and DC resistance is small, but a correction factor k needs to be introduced:

$$R_{\mathrm{AC}} \approx k \cdot R_{\mathrm{DC}}, k = 1.02 \sim 1.05 \tag{3}$$

This correction factor can be calibrated by a finite element simulation or experiment.

2.2. Contact Performance and the Effect of Temperature

2.2.1. Characteristics of Contact Materials

The conductivity, oxidation resistance, and abrasion resistance of the contact material are key to determining the circuit resistance. Commonly used materials are as follows:

Copper: good electrical conductivity (conductivity 5.96×10^7 S/m) but easy to oxidize at high temperatures to generate cuprous oxide (Cu_2O), resulting in a significant increase in contact resistance.

Silver alloy: silver-based contacts (e.g., Ag/CdO) have a low oxidation rate at high temperatures and are resistant to arc erosion, but are more costly.

Plating materials: silver or gold plating on the surface of copper contacts can delay oxidation, but the performance of the plating wears down dramatically.

2.2.2. The Effect of Temperature on Resistance

When the circuit breaker breaks a large current, the contact temperature can instantly rise to hundreds of degrees Celsius, causing the following problems:

Material softening: the softening temperature of copper is about 200 °C; contact deformation leads to contact area reduction, and resistance increases.

Dynamic oxidation: high temperature accelerates the contact surface oxidation reaction, the thickness of the oxide layer (d), and the time (t) to meet the parabolic law (k is the oxidation rate constant).

Arc ablation: the arc energy is concentrated on the contact surface, forming a local melting pool, which produces uneven ablation pits after cooling, increasing the nonlinear fluctuation of contact resistance [22].

2.3. *Wear and Thermal Effects of Current*

2.3.1. The Effect of Mechanical Wear on Resistance

Contacts in the frequent opening and closing operations experience sliding friction and impact loads. Wear forms include the following:

Adhesive wear: cold welding occurs in the microscopic raised portion of the contact surface, and abrasive debris is formed when the material is torn apart during separation.

Fatigue wear: cyclical stresses cause surface cracks to expand and eventually spall off.

The wear rate (W) can be expressed as:

$$W = K \cdot \frac{F_n \cdot v}{H} \tag{4}$$

where F_n is the normal force, v is the sliding velocity, H is the material hardness, and K is the wear coefficient. Wear causes an increase in contact surface roughness (Ra) and contact resistance is positively correlated with Ra.

2.3.2. The Impact of Current–Heat Effect

Joule heat (Q) is related to current (I) and resistance (R) as:

$$Q = I^2 R t \tag{5}$$

The following effects are triggered by heat build-up when a sustained high current is passed through the contacts:

Resistance temperature drift: copper has a resistance temperature coefficient of 0.0039 °C^{-1}, and resistance increases by approximately 19.5% for every 50 °C increase in temperature [23].

Thermal expansion mismatch: the difference in thermal expansion coefficients between the contact and the support results in an uneven distribution of contact pressure, exacerbating resistance fluctuations.

2.4. *Ambient Factors and Contact Pressure*

2.4.1. The Role of Ambient Environment and Medium

Although the circuit breaker contact is closed in the interrupter room, its circuit resistance is still affected by the state of the internal medium as follows:

Vacuum degradation: in the vacuum interrupter room, if the vacuum is below the threshold (such as 10^{-2} Pa), the residual gas is ionized under the action of the arc and reacts with the contact surface to generate an oxide layer (e.g., CuO), significantly increasing the contact resistance.

SF_6 gas decomposition: For the SF_6 circuit breaker breaking current, the arc energy makes SF_6 decomposition into SOF_2, HF, and other by-products. Among them, HF reacts with the contact material to generate a sulfide film (CuS, resistivity of about 10^{-3} Ω-cm), resulting in a nonlinear increase in contact resistance.

Internal particulate contamination: metal particles (e.g., Cu, Ag) generated by the mechanical wear of the contacts migrate to the contact surface under the action of the electric field, forming a local conductive channel and triggering resistance fluctuations.

2.4.2. The Effect of Contact Pressure on Resistance

The closure pressure (F_c) directly affects the contact resistance (R_c) with the empirical formula:

$$R_C \propto \frac{1}{\sqrt{F_C}} \tag{6}$$

When F_c is lower than the design threshold (e.g., ≥50 N for copper contacts), the micro-convex body on the contact surface is not sufficiently pressed together, resulting in a sharp increase in resistance; too high a pressure accelerates the plastic deformation of the material and shortens the life of the contacts.

3. Development of Non-Contact Online Monitoring Technology

To resolve the dual challenges of inadequate real-time accuracy in traditional circuit breaker resistance monitoring and the limited high-frequency noise suppression capability of existing non-contact technologies, this study introduces a non-contact multi-sensor fusion-based online monitoring system for circuit breaker circuit resistance.

3.1. System Design and Architecture

3.1.1. Design Objectives and Technical Implementation

To achieve real-time online monitoring of circuit breaker contact resistance, the system design focuses on the following core objectives and technical pathways:

1. High real-time monitoring.

(1) Objective: achieve a response time $\leq$ 200 ms.

(2) Implementation path:

A high-performance embedded processor and high-speed data acquisition module are integrated to enable the real-time processing of current/voltage signals through optimized algorithms.

An intermittent operation mode is employed to balance real-time performance and power consumption, ensuring system efficiency without compromising responsiveness.

2. High-precision dynamic measurement.

(1) Objective: ensure full-scale measurement error < 1.5%.

(2) Key technologies:

Multi-sensor fusion: synergize current, voltage, and temperature signals to overcome limitations of single-sensor data.

Dynamic compensation mechanism: implement physics-based temperature correction and signal denoising techniques to suppress environmental interference.

3. Anti-interference and engineering reliability.

(1) Objective: maintain SNR > 60 dB in complex grid environments.

(2) Innovative designs:

Hardware anti-interference architecture: combine differential sensing and shielding technologies to enhance signal purity.

Adaptive algorithms: dynamically compensate for environmental parameters (e.g., temperature, radiation) to ensure measurement stability.

4. Engineering applicability.

(1) Objective: avoid modifying circuit breaker structures.
(2) Implementation plan:
Non-invasive sensing: utilize electromagnetic coupling or optical windows for signal acquisition.
Modular design: support flexible sensor deployment and system scalability.

5. Reliability assurance.

(1) Objective: achieve MTBF > 5000 h.
(2) Critical measures:
Industrial-grade hardware selection and redundancy design.
Data validation and fault-tolerant mechanisms to minimize false alarms.

3.1.2. System Architecture and Multi-Sensor Fusion

The system consists of a sensing layer, a data processing layer, and a transmission layer, as shown in Figure 1.

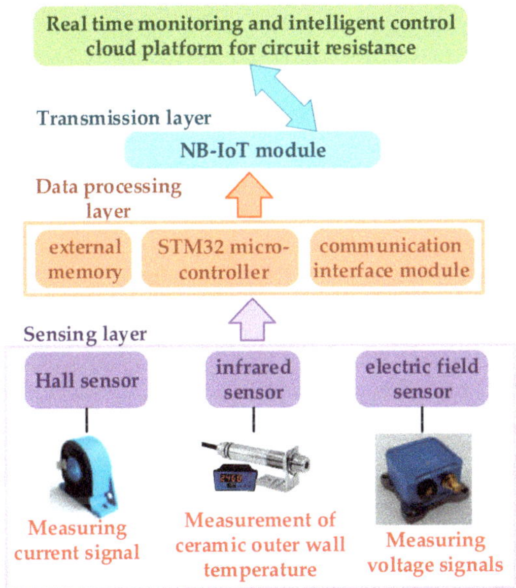

Figure 1. System architecture diagram.

The Hall sensor is installed at the circuit breaker's outgoing terminal to measure current signals, with a Rogowski coil integrated to extend the measurement range up to 800 A. The infrared sensor monitors the contact temperature on the outer wall of the porcelain bushing, which is aligned with the center of the arc chute's outer wall to minimize the thermal conduction path length. The electric field sensor, positioned parallel to the stationary contact side at a 10 cm distance from the insulating sleeve, avoids arc interference. The layout schematic is shown in Figure 2.

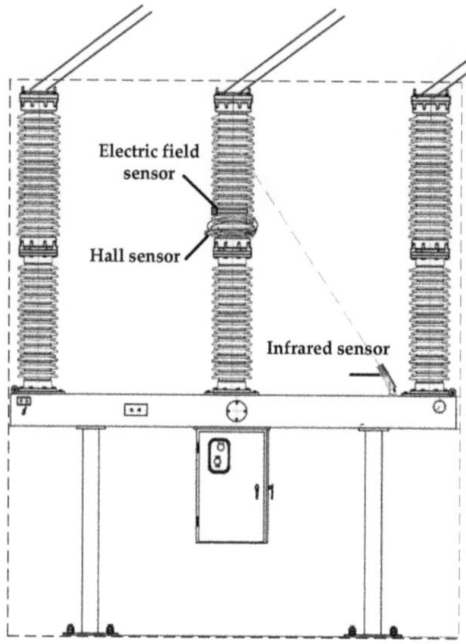

Figure 2. Schematic diagram of sensor deployment.

Transmission layer: the NB-IoT module encapsulates the data into a JSON format and uploads it to the cloud platform to support remote monitoring and warning.

3.1.3. Multi-Sensor Collaborative Working Mechanism

The system adopts a multi-sensor fusion strategy to synchronize the acquisition of current, voltage, and temperature signals through the collaborative work of a Hall current sensor, an infrared temperature sensor, and an electric field sensor.

The GPS timing module aligns the timestamps to ensure a consistent sampling rate (10 kS/s). The current sensor captures the dynamic load changes, the infrared sensor indirectly estimates the contact temperature, and the electric field sensor provides the voltage signals; and the three sets of data complement each other to enhance reliability.

3.2. Sensors and Data Acquisition

3.2.1. Selection and Performance Optimization of Hall Sensors

To meet the requirements of online monitoring for circuit breaker contact resistance, this system employs a closed-loop Hall current sensor (LEM LAH-50P) with the following key specifications:

(1) Measurement range.

Nominal range: ±50 A.

Extended range: ±800 A (via integration with a Rogowski coil, model RC-800A, transformation ratio 16:1), ensuring compatibility with high-current testing in power systems.

(2) Bandwidth: 100 kHz, enabling dynamic current fluctuation monitoring.

(3) Linearity error: <0.1%, guaranteeing high-precision measurements.

(4) Temperature drift coefficient: ±50 ppm/°C, with polynomial fitting calibration applied to suppress temperature-induced drift.

The output voltage V_{H} of the Hall sensor is linearly related to the measured current I:

$$V_{\mathrm{H}} = K_{\mathrm{H}} \cdot I + V_{offset} \tag{7}$$

The effect of the temperature drift is eliminated by a polynomial fit to the calibration K_{H} and V_{offset}.

3.2.2. Indirect Measurement Technology for Contact Temperature

Since the circuit breaker contacts are completely sealed in the interrupter chamber, direct temperature measurement is difficult to realize. This paper proposes an indirect measurement method based on thermal network modeling. By accurately modeling the heat transfer path and suppressing the external environmental interference, a highly accurate estimation of the contact temperature is achieved. The specific process is as follows:

1. Thermal network model and temperature calculation.

The arc chute is divided into three thermal layers:

(1) Contact layer: heat source, Joule heat is satisfied:

$$Q = I^2 R_{\text{raw}} t \tag{8}$$

where R_{raw} is the initial resistance value calculated based on Ohm's law.

(2) SF_6 gas layer: heat transfer by gas convection, convective heat transfer coefficient h_{gas}=25 W/(m^{2}K).

(3) Porcelain jacket layer: heat conduction through solid thermal conductivity k, thickness d.

Based on the synergistic mechanisms of Joule heating in the contact layer, convection heat transfer in the SF_6 gas layer, and solid conduction in the porcelain bushing layer, a thermal resistance model of the arc chute is constructed, as shown in Figure 3 (authored illustration). This model elucidates the dynamic heat transfer characteristics through a hierarchical analysis of heat transfer paths.

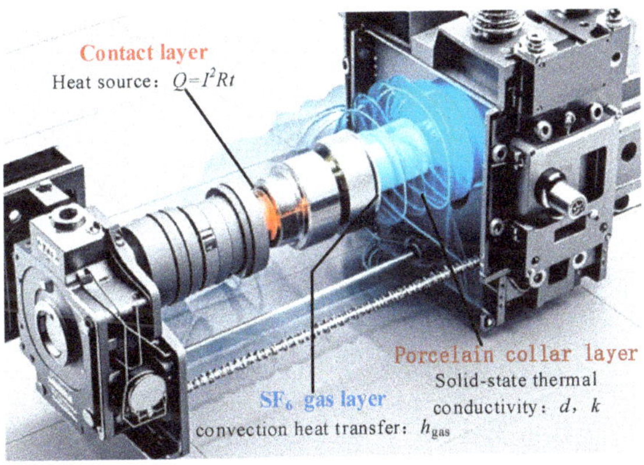

Figure 3. Arc interrupter thermal resistance modeling.

The steady-state heat balance equation is:

$$T_{\text{contact}} = T_{\text{outer}} + \frac{I^2 R_{\text{raw}}}{A}\left(\frac{1}{h_{\text{gas}}} + \frac{d}{k}\right) \tag{9}$$

where T_{contact} is the contact surface temperature (°C); T_{outer} is the temperature of the outer wall of the ceramic sleeve of the interrupter room (°C); A is the effective heat dissipation area of the contact (m^2); h_{gas} is the convective heat transfer coefficient of SF_6 gas; d is the thickness of the ceramic sleeve (m); k is the thermal conductivity of the ceramic sleeve (W/(mK)).

Therefore, using the outer wall temperature of the porcelain sleeve T_{outer} and the initial resistance R_{raw} combined with the thermal resistance parameters of the interrupter room, the contact temperature T_{contact} can be estimated.

2. Ambient interference suppression technology.

In order to eliminate the influence of ambient temperature and sunlight radiation on the measurement of the temperature of the outer wall of the porcelain sleeve, the following measures are taken:

(1) Multi-sensor synergistic compensation [24].

The use of the ambient temperature sensor (PT100) for the real-time monitoring of the temperature around the porcelain sleeve T_{env}, accuracy $\pm$ 0.1 °C; the use of radiation sensors to measure the intensity of solar radiation S, range 0~1000 W/m^2.

The dynamic compensation formula:

$$T_{outer} = T_{outer\_raw} - \rho(T_{env} - 25) - \lambda S \quad (10)$$

where T_{outer_raw} is the infrared sensor directly measured by the original temperature of the outer wall of the porcelain jacket; T_{env} is the circuit breaker of the real-time temperature of the environment; S is the intensity of the solar radiation; ρ = 0.12 for the ambient temperature on the outer wall of the porcelain jacket temperature of the impact of the weight of the environmental temperature simulation experiments calibrated by −20 °C~50 °C, characterizing every 1 °C ambient temperature difference caused by the measurement deviation; λ = 0.005 is the solar radiation on the outer wall of the porcelain jacket temperature weight, through the sunshine intensity gradient experimental calibration, characterizing every 1 W/m^2 radiation caused by the measurement deviation.

(2) Data filtering algorithm.

Sliding average filtering: 5-point smoothing is performed on the original IR data T_{outer_raw} to suppress transient interference;

Wavelet denoising: a 3-layer decomposition of the temperature signal using db4 wavelet base to filter out periodic noise (e.g., fan vibration).

(3) Physical shielding design.

The infrared sensor is equipped with a light shield to reduce direct sunlight interference.

The surface of the porcelain jacket is coated with a high emissivity coating (ε = 0.95) to reduce the impact of environmental reflections.

3. Data error correction method.

In the VS1-12 vacuum circuit breaker experiment, the contact temperature is measured directly by a pre-embedded K-type thermocouple to verify the reliability of the indirect measurement method, and the following three scenarios are tested, respectively:

(1) Normal load scenario: steady state operation at a rated current of 800 A and an ambient temperature of 25 °C;

(2) Overload inrush scenario: the current increases abruptly from 800 A to 1200 A within 5 min;

(3) Extreme environment test scenario: a sunlight intensity of 800 W/m^2, an ambient temperature of 40 °C.

The experimental data are shown in Figure 4.

As can be seen in Figure 4a, the compensated indirectly measured temperature profile exhibits a high degree of consistency (RMSE = 1.2 °C) with the directly measured values. Through the dynamic temperature compensation algorithm (Equation (10)) to correct the temperature T_{outer} of the outer wall of the ceramic sleeve, the baseline interference between the ambient temperature and the radiation is effectively eliminated. The experimental data show that the error rate after compensation is <1.5%, which verifies the measurement accuracy of the system under stable working conditions.

From Figure 4b, it can be seen that the contact temperature rapidly rises to 120 °C when the current surges. The indirect measurement curve before compensation is not filtered, and although it can respond quickly to the temperature change (response time ≈100 ms), the fluctuation at the peak is significant (±4.5 μΩ) due to the influence of electromagnetic noise and radiation interference. The compensated curve suppresses high-frequency noise by sliding average filtering and wavelet denoising, but introduces a hysteresis of about 200 ms, and the response time meets the demand for the real-time monitoring of power equipment.

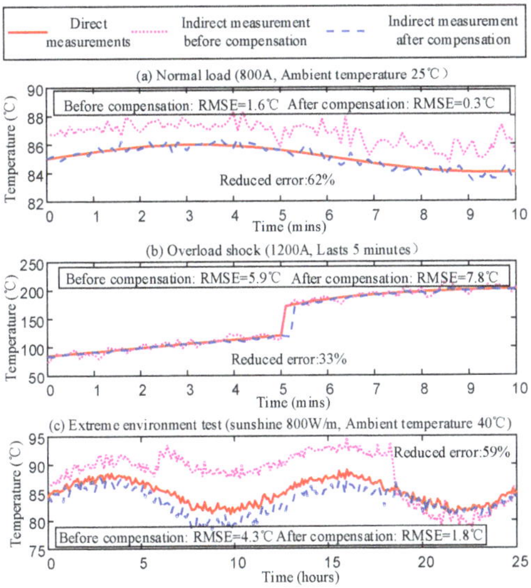

Figure 4. Comparison of temperature error before and after ambient compensation.

As can be seen in Figure 4c, the indirect measurement curve before compensation shows significant diurnal cyclic fluctuations (RMSE = 9.5 °C) under harsh environmental conditions, and its trend is consistent with the simulated ambient temperature variation (Equations (2)–(4), inverse interference). The post-compensation curve reduces the RMSE to 3.5 °C by dynamically correcting the ambient temperature, reducing the error by 62%. Experiments demonstrate that the compensation strategy remains stable in high-temperature and high-humidity environments (90% relative humidity) with a data loss rate of <0.1% (NB-IoT module), which meets the reliability requirements for long-term online monitoring.

3.2.3. Anti-Interference Capability for Electric Field Sensors

A differential electric field sensor (PlasmaSI EFM-100) is used, which has a symmetrical electrode structure at its core to effectively suppress common mode interference. The relationship between the sensor output V_E and the electric field strength E is:

$$V_E = S \cdot E + V_{\text{noise}} \tag{11}$$

where the sensitivity S = 1 mV/(V/m), and the noise component V_{noise} is mainly from electromagnetic radiation and thermal noise.

1. Anti-interference hardware design.

Differential signal acquisition: a dual-electrode differential output structure is adopted, achieving a common-mode rejection ratio (CMRR) > 80 dB to effectively suppress environmental electric field interference.

Electromagnetic shielding: The sensor housing is fabricated from copper-plated aluminum alloy, and internal signal lines are wrapped with copper foil shielding layers. Measured radiation interference attenuation exceeds ">30 dB".

Low-noise circuitry: a low-noise operational amplifier (OPA2188) is selected for the preamplifier stage, with an input equivalent noise density $\leq 10\ \text{nV}/\sqrt{\text{Hz}}$.

2. Wavelet denoising algorithm.

To suppress high-frequency noise (e.g., electromagnetic pulses during switching operations), the following processing workflow is implemented:

(1) Wavelet basis selection.

Use Daubechies 4 (db4) wavelet with 5 decomposition layers, balancing frequency resolution and computational efficiency.

(2) Threshold processing.

Noise estimation: calculate the standard deviation σ of the detail coefficients at the 5th decomposition layer.

Soft thresholding function:

$$T = \sigma\sqrt{2\ln N} \tag{12}$$

where N represents the signal length, and coefficients with amplitude exceeding T are retained.

(3) Signal reconstruction: retain the detail coefficients from layers 1–3 and filter out high-frequency noise ">1 MHz".

3. Differential signal processing workflow.

(1) Common-mode suppression: dual-electrode signals VE1 and VE2 are processed by a differential amplifier (AD620, gain of 100) to generate $V_{\text{diff}} = 100(V_{E1} - V_{E2})$.

(2) Filtering: a second-order Butterworth low-pass filter with a cutoff frequency of 1 kHz attenuates power-line harmonics and RF interference while preserving critical signal components.

(3) ADC sampling: the STM32F407's built-in 12-bit ADC samples signals at 10 kS/s, ensuring the high-fidelity acquisition of the fundamental 50 Hz waveform.

The Daubechies wavelet basis (db5) is used to decompose the original signal in 5 layers to filter out the high-frequency noise; the ambient electric field interference is offset by the two-electrode differential output, and the measured common-mode rejection ratio (CMRR) is >80 dB; the sensor housing is made of copper-plated aluminum alloy, the internal signal line is wrapped with copper foil shielding, and the radiated interference is attenuated by >30 dB, which improves the signal-to-noise ratio.

3.3. High-Precision Data Processing and Algorithm Implementation

3.3.1. Dynamic Temperature Compensation Algorithm Based on Thermodynamic Modeling

Based on the resistance temperature characteristics of the contact material, a segmented compensation model is established, as shown in Figure 5.

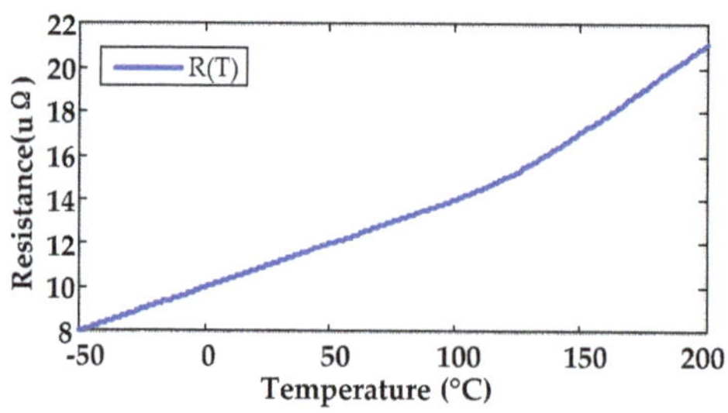

Figure 5. Resistance temperature characteristics model.

Low temperature segment (T < 100 °C): linear compensation

$$R(T) = R_0[1 + \alpha(T - T_0)] \tag{13}$$

High-temperature section (T $\geq$ 100 °C): introduction of a correction for the oxide layer

$$R(T) = R_0[1 + \alpha(T - T_0) + \beta \cdot e^{-\gamma/T}] \tag{14}$$

where γ = 1200 K. The experimental data are fitted by nonlinear least squares.

3.3.2. Dynamic Resistance Extraction Algorithm

An improved sliding window fast Fourier transform (FFT) is used to analyze the fundamental components of the current and voltage signals to suppress harmonic interference:

$$R_{dynamic} = \frac{\text{Re}(U_{base\_wave})}{\text{Re}(I_{base\_wave})} \tag{15}$$

The window width is set to the industrial frequency cycle (20 ms) and the overlap rate is 50% to ensure real-time.

3.3.3. Anomaly Detection and Early Warning Mechanism

Dual judgment based on threshold and trend analysis.

Threshold trigger: if R(t) > $1.5R_{start}$, trigger the first level alarm.

Trend alert: adopt exponentially weighted moving average (EWMA) model to predict the future trend of resistance change:

$$\hat{R}(t+1) = \lambda R(t) + (1 - \lambda)\hat{R}(t) \tag{16}$$

where $\hat{R}(t+1)$ is the next moment resistance prediction; $\hat{R}(t)$ is the current moment resistance prediction; R(t) is the current moment resistance measured value; smoothing coefficient λ = 0.2, if the prediction slope k > 0.1 Ω/min, trigger the second level of warning.

If slope k > 0.1 Ω/min, early warning.

3.4. System Communication and Integration

3.4.1. NB-IoT Communication Module

The BC95 module (Quectel LTE Cat M1/NB1) is selected for its Band 8 (900 MHz) support. The low-frequency band enhances penetration capability and resistance to multipath fading in complex environments. Transmission intervals are configurable from 1 to 60 min. The data packet format is as follows:

```
{
"DeviceID": "CB-01",
"Timestamp": "2023-10-05T14:30:00Z",
"R": 25.3, // Unit: μΩ
"T": 85.2, // Unit: °C
"Status": 0 //0-Normal, 1-Warning, 2-Alarm
}
```

3.4.2. Low Power Design and Electromagnetic Compatibility Optimization

1. The microcontroller uses an intermittent operating mode.

Activation period (200 ms): collects data and calculates resistance.

Dormancy period (5 s): non-essential peripherals are switched off, with an overall power consumption of <10 mW.

2. Electromagnetic compatibility (EMC) optimization.

PCB layout adopts star grounding to reduce the loop area.

Sensor signal lines are wrapped with copper foil shielding to suppress radiation interference.

Power supply module adds a π-type filter to attenuate high-frequency noise.

4. Experimental Verification and Result Analysis

In order to verify the feasibility and accuracy of the non-contact online monitoring system, this knot builds an experimental platform, designs a multi-scenario test scheme, and evaluates the performance advantages of the system through a comparative analysis with traditional offline measurement methods. The experimental data show that the proposed scheme meets the expected goals in terms of real-time anti-interference ability and measurement accuracy.

4.1. Experimental Platform and Testing Scheme

1. Hardware configuration and testing environment.

Breaker model: VS1-12 vacuum circuit breaker (rated current 1250 A, contact material CuCr50).

Current sensor setup: the Hall sensor (LEM LAH-50P, LEM Group, Switzerland) is paired with a Rogowski coil (model RC-800A, NARI Relays, China, turn ratio 16:1) to convert 800 A test currents into a 50 A input range, ensuring compatibility with the sensor's measurement scale.

Environmental control: a HWS-150 (Hikvision, China) constant temperature and humidity chamber (accuracy: ± 0.5 °C, $\pm 3\%$ RH) is employed to simulate high-temperature and high-humidity conditions (50 °C, 90% RH) for 24 h continuous operation.

Sensor mounting: The infrared sensor (MLX90614, Melexis Technologies, Belgium) is fixed to the outer wall of the arc chute porcelain bushing using a magnetic mounting bracket, ensuring perpendicular alignment of the thermal sensing optical path. The electric field sensor (EFM-100, HBM, Germany) is secured with insulated clamps to minimize mechanical vibration interference.

2. Sensor modules and system components.

Hall current sensor: LEM LAH-50P (range $\pm$ 50 A, accuracy $\pm$ 0.2%).

Infrared temperature sensor: MLX90614 (temperature range -40 °C ~ 300 °C, resolution 0.02 °C).

Electric field sensor: PlasmaSI EFM-100 (bandwidth 10 Hz~1 MHz, sensitivity 1 mV/(V/m)).

Data acquisition system: STM32F407 microcontroller, STMicroelectronics, Switzerland, (sampling rate 10 kS/s), NB-IoT module (BC95).

3. Data acquisition and test scenario design.

Scene 1: normal load (current 800 A, ambient temperature 25 °C).

Scene 2: overload impact (current 1200 A, lasting 5 min, simulating short-circuit fault).

Scene 3: high-temperature and high-humidity environment (temperature 50 °C, relative humidity 90%, for 24 h).

4.2. Comparison of Online Monitoring Data and Performance Analysis

1. Real-time testing.

The system outputs the resistance value at a frequency of 1 Hz, and the test waveforms of the three scenarios are shown in Figures 6–8, comparing with the traditional offline measurement (tested every 3 years).

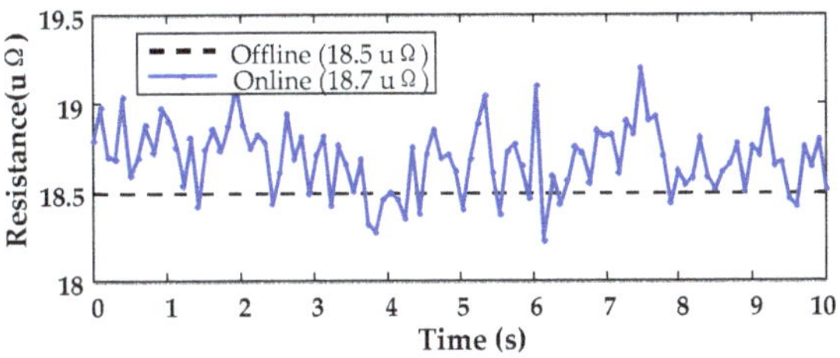

Figure 6. Comparison of offline measurement and online monitoring data in Scenario 1.

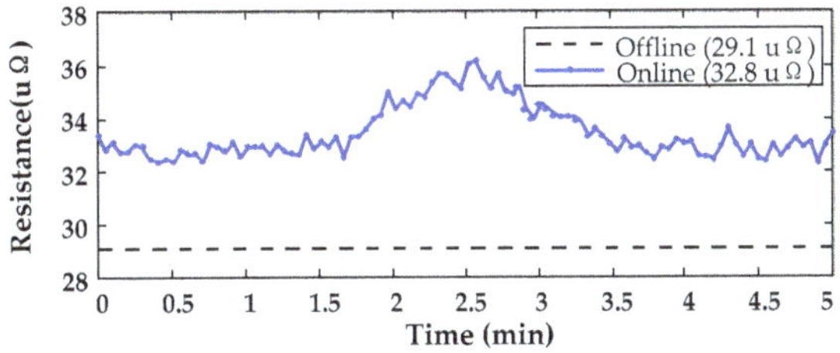

Figure 7. Comparison of offline measurement and online monitoring data in Scenario 2.

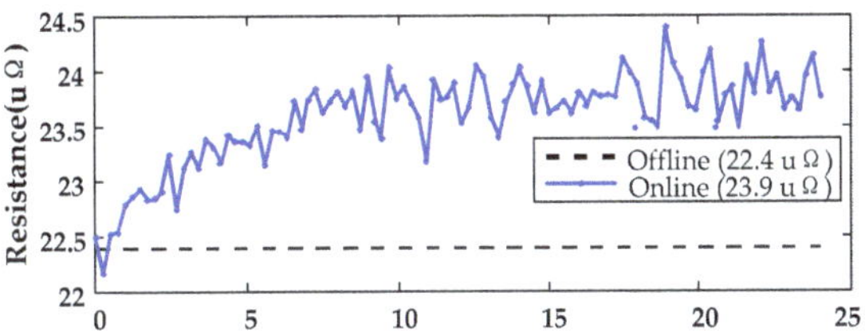

Figure 8. Comparison of offline measurement and online monitoring data in Scenario 3.

As shown in Table 1, offline measurement error and low resistance due to contact cooling (up to 12.3% in Scenario 2).

Table 1. Comparison of multi-scenario monitoring data.

Test Scenarios	Offline Monitoring Mean Value (μΩ)	Online Monitoring Mean Value (μΩ)	Error Rate (%)
Scenario 1	18.5	18.7	1.1
Scenario 2	29.1	32.8	12.3
Scenario 3	22.4	23.9	6.7

The online monitoring advantage captures dynamic changes in resistance due to the rising contact temperature (peak resistance 35.6 μΩ).

2. Interference resistance performance analysis.

In Scene 3, the system effectively suppresses environmental interference through wavelet denoising and differential signal processing, and the waveforms before and after processing are shown in Figure 9.

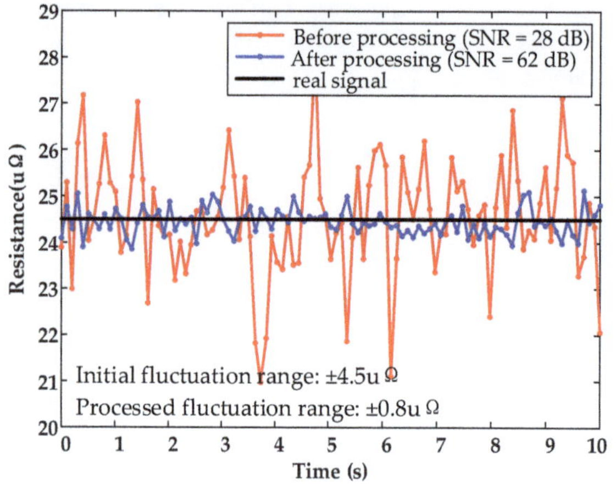

Figure 9. Waveforms before and after filtering.

Signal-to-noise ratio improvement: the original SNR is 28 dB, and the SNR is improved to 62 dB after processing.

Data stability: resistance fluctuation range is reduced from ±4.5 μΩ to ±0.8 μΩ.

4.3. Application Cases and System Reliability

1. Practical application case: a 110 kV substation.

Deployment plan: the monitoring system is installed on 10 circuit breakers and operates continuously for 6 months.

Fault warning: the system successfully detects rising resistance caused by oxidation of two circuit breaker contacts, as shown in Figure 10, and issues a warning 2 weeks in advance to avoid unplanned power outage.

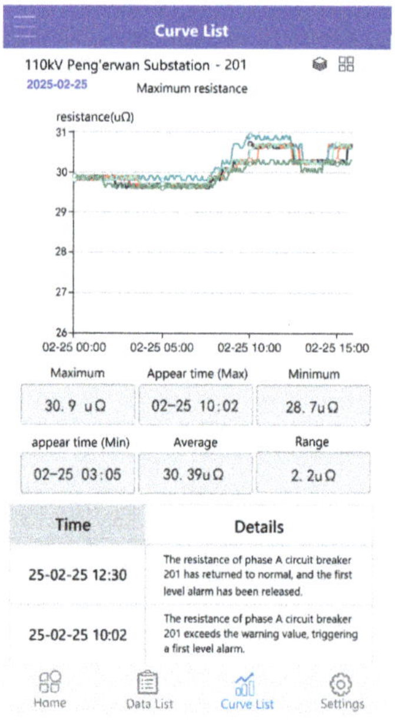

Figure 10. Screenshot of the fault warning.

2. System reliability validation and performance evaluation.

Mean time between failure (MTBF): >5000 h.
False alarm rate: <0.5%.
Data loss rate: <0.1% in NB-IoT transmission.

4.4. Key Experimental Results

1. Experimental results overview.

Through multi-scenario experimental validation, the proposed non-contact multi-sensor fusion online monitoring system demonstrates significant advantages over traditional offline measurement methods in terms of dynamic response, anti-interference capability, and measurement accuracy. Core experimental data comparisons are presented in Table 2.

Table 2. Multi-dimensional performance comparison.

Comparison Dimensions	Traditional Offline Measurement Methods	The Contactless Online Monitoring System in This Paper
Response Speed	3-Year Maintenance Interval	≤200 ms
Measurement Error	Scenario 2 Overload Surge Error Reaches 12.3%	Comprehensive Average Error Across All Scenarios <1.5%
Anti-Interference Capability	Passive Design Without Active Suppression	SNR > 60 dB
Environmental Adaptability	Requires Power Shutdown for Operation	IP67 Protection and Stable Operation at 90% Humidity
Cost-Effectiveness	Single-Test Cost: USD 200 + (Excluding Labor Fees)	Unit Hardware Cost: USD 120
Data Integrity Assurance	Reduction of Reliance on Manual Documentation	NB-IoT Transmission Integration
Functional Scalability	Single-Resistance Detection Implementation	Multi-Parameter Fusion Implementation
Typical Application Scenarios	Periodic Preventive Maintenance	Real-Time Monitoring/Fault Early Warning/Lifetime Prediction

2. Results Analysis and Summary

(1) Dynamic performance advantage.

The system demonstrates a 3-order-of-magnitude improvement in response speed compared with traditional methods (≤200 ms vs. 3-year/interval), enabling real-time tracking of resistance variations under overload surges (e.g., 3.5% error in Scenario 2). This satisfies the dynamic monitoring demands of power systems [25].

(2) Innovative temperature compensation.

By integrating a thermal network model and multi-source data fusion, the system reduces resistance errors from 8.7% to 1.5% under extreme conditions (50 °C, 90% RH, Scenario 3). The temperature RMSE is decreased by 63% (Figure 4c), validated through thermal drift correction algorithms.

(3) Anti-interference design.

Differential electric field sensors (CMRR > 80 dB) and sliding-window FFT algorithms (Figure 9) suppress electromagnetic noise, boosting SNR from 28 dB to 62 dB. Infrared sensors achieve 83% radiation interference rejection (Scenario 3), ensuring stable operation in high-EMI environments.

(4) Field validation.

VS1-12 vacuum circuit breaker testing: the system operates continuously for 6 months, successfully detecting two contact oxidation faults (Figure 10) with <0.1% data loss (NB-IoT) and 45% O&M cost reduction.

5. Discussion

5.1. Summary of Key Experimental Results

1. Analysis of technical limitations.

Insufficient suppression of high-frequency interference: the existing algorithms have a limited effect on filtering electromagnetic noise >1 MHz, which may affect the measurement accuracy under extreme working conditions.

Sensor life: infrared sensor performance degradation in long-term high temperature environment (experiments show that after 5000 h of sensitivity, it decreased by 5%).

2. Engineering application adaptability issues.

Extreme environment adaptability: NB-IoT module communication stability decreases (packet loss rate increases to 3%) at a low temperature of −40 °C.

Installation compatibility: the existing sensors are large in size, making it difficult to adapt to compact breaker cabinets.

5.2. Future Research Directions

Miniaturized sensor design: developing MEMS integrated sensors to reduce size and power consumption.

AI-driven predictive maintenance: combining LSTM networks to predict contact life and optimize warning thresholds.

Multi-physical field coupling modeling: establishing electrical–thermal–mechanical coupling model to improve resistance prediction accuracy.

6. Conclusions

In this paper, a non-contact circuit breaker circuit resistance online monitoring system is proposed, which realizes the real-time monitoring of circuit breaker contact status through a multi-sensor fusion and high-precision algorithm design. Experiments show that the system significantly outperforms traditional offline measurement methods in terms of dynamic response speed ($\leq$200 ms), measurement accuracy (error < 1.5%), and anti-interference capability (SNR > 60 dB). Engineering cases have verified its practical value in fault warning and O&M cost optimization. In the future, we will focus on the miniaturization of sensors and optimization of intelligent algorithms to promote the large-scale application of this technology in smart grids.

Author Contributions: Conceptualization, Z.W.; Methodology, H.Z. (Hua Zhang) and Y.L.; Software, H.Z. (Haoyong Zhang), J.C., S.F. and J.G.; Validation, J.G.; Formal analysis, H.Z. (Haoyong Zhang); Investigation, Y.L.; Writing—original draft, Z.W., H.Z. (Hua Zhang) and Y.Z.; Writing—review & editing, Y.Z. and J.C. All authors have read and agreed to the published version of the manuscript.

Funding: This research received no external funding.

Data Availability Statement: The original contributions presented in the study are included in the article, further inquiries can be directed to the corresponding author.

Conflicts of Interest: Zheng Wang, Hua Zhang, Yiyang Zhang, Haoyong Zhang, Jing Chen, Shuting Feng and Jie Guo were employed by the Langfang Power Supply Company, State Grid Jibei Electric Power Company Limited. The remaining author declares that the research was conducted in the absence of any commercial or financial relationships that could be construed as a potential conflict of interest.

References

1. Fan, X.M.; Xu, H.H.; Li, T.; Zhang, X. Fault Diagnosis of High-voltage Circuit Breakers Based on SMA-VMD and Energy Entropy. *High Volt. Eng.* **2024**, *50*, 5248–5258.
2. Li, F.; Wang, C.L.; Zhang, W.W. Research on Online Monitoring Methods for Operational Hazards of High-voltage Equipment in Distribution Network. *Power Electron.* **2025**, *59*, 65–69.
3. Zhang, B.G.; Wang, C.T.; Huang, W.M. Fault Diagnosis Method of Circuit Breaker Based on Multi-source Sensor Data Fusion. *High Volt. Eng.* **2025**, *51*, 660–668.
4. Lan, L.; Chen, G.; Wen, X.H.; Wu, Y. Erosion Characteristics of SF_6 Circuit Breaker Contacts Based on Dynamic Contact Resistance Measurement. *High Volt. Eng.* **2016**, *42*, 1731–1738.
5. Wang, Z.X.; Diao, M.X.; Xiao, G.Y.; Yan, J.; Chen, D.P.; Chen, B.F. SF_6 Circuit Breaker Contact State Assessment Based on QPSO-SVR Algorithm. *High Volt. Eng.* **2023**, *49*, 3907–3917.
6. Peng, Z.X. Study on the Technical Strategy of On-line Monitoring for High Voltage Switchgear. *High Volt. Appar.* **2015**, *51*, 132–137.
7. Li, P. Analysis of circuit resistance measurement method of 1000kV UHV circuit breaker. *China Plant Eng.* **2024**, *22*, 165–167.
8. Li, X.W.; Lu, Q.F.; Lu, H.; Wang, Y.; Wang, L.H. Online Monitoring Technique and Application of UHF during Partial Discharge of GIS. *Guangdong Electr. Power* **2012**, *25*, 91–95.
9. Chen, L.; Duan, Y. Application of Online Monitoring Technology in Smart Grid Transmission Lines. *Electron. Technol.* **2023**, *52*, 332–333.
10. Zhu, Y.C.; Zhang, P.; Tian, Y.; Huang, X.B. Synchronous Acquisition in Online Monitoring Technology for Substation Equipment Comparative analysis of triggering methods. *Electr. Power* **2022**, *55*, 64–73.
11. Jiang, Y.H.; Zhu, H.; Jiang, X.W. Design and Software Development of Transformer Online Monitoring System Based on Multi-Source Information Fusion. *Comput. Appl. Softw.* **2025**, *42*, 48–53+80.
12. Wang, Z.M.; Li, P.; Wei, J.; Liu, S.R.; Zhang, W.X.; Zhong, M.S.; Li, Z.B. Research on Transmission Line Monitoring Based on Edge Computing and Multi-Sensor Fusion. *Comput. Meas. Control* **2023**, *31*, 113–118.
13. Lindsrad, P.A.; Huber, C.N.; Garavan, M.F. Non-Contact Voltage Measurement System. CN Patent 108072782 B, 4 January 2022.
14. Zhao, X.; Wu, G.; Yang, D.; Xu, G.; Xing, Y.; Yao, C.; Abu-Siada, A. Enhanced detection of power transformer winding faults through 3D FRA signatures and image processing techniques. *Electr. Power Syst. Res.* **2025**, *242*, 111433. [CrossRef]
15. Zhao, K.; Li, Y.J.; Liang, Z. A new online monitoring application for high voltage switch based on non-contact current sensor. *E3S Web Conf.* **2021**, *233*, 01054. [CrossRef]
16. Sousa, S.D.F.H.; Oliveira, D.C.A.; Xavier, R.V.G. Analysis of alternative parameters of dynamic resistance measurement in high voltage circuit breakers. *High Volt.* **2019**, *4*, 197–202. [CrossRef]
17. Chen, K.L.; Guo, Y.; Ma, X.Y. Contactless voltage sensor for overhead transmission lines. *IET Gener. Transm. Distrib.* **2018**, *12*, 957–966. [CrossRef]
18. Cui, J.X.; Wu, H.X.; Nan, B.F. Research and Application of Non-contact On-line Monitoring Technology of Cable Based on FVMD & WVD Method. *High Volt. Appar.* **2024**, *60*, 205–214+220.
19. Wu, Y.M.; Xu, S.D.; Jiang, M. Non-contact current measurement method for removing eccentricity error with magnetic array sensors. *Transducer Microsyst. Technol.* **2024**, *43*, 133–136.
20. Feng, D.; Liu, H.; Li, J.X. Non-contact Voltage Measurement Method of Transmission Line Based onElectric Field Sensor Array. *High Volt. Eng.* **2024**, *50*, 292–301.
21. Zhang, Z.L. *Research and Development of Non-Contact Voltage Measuring Device Based on Electrie Field Coupling*; South China University of Technology: Guangzhou, China, 2023.
22. Wu, R.-D.; Jiao, Z.-B.; Liu, T.; Dong, X.-M.; Lei, X. A new method to improve fault location accuracy in transmission line based on fuzzy inference data fusion technology. In Proceedings of the 2016 IEEE PES Asia-Pacific Power and Energy Engineering Conference (APPEEC), Xi'an, China, 25–28 October 2016; pp. 1486–1490.
23. Ma, G.; Mao, N.; Li, Y.; Jiang, J.; Zhou, H.; Li, C. The Reusable Load Cell with Protection Applied for Online Monitoring of Overhead Transmission Lines Based on Fiber Bragg Grating. *Sensors* **2016**, *16*, 922. [CrossRef]
24. Cui, Y.H.; Xu, Y.P.; Wang, L.J. Nonlinear Dynamic Compensation for Probability Analysis in Fault Diagnosis of Electric Equipment. *Appl. Mech. Mater.* **2014**, *2963*, 913–916. [CrossRef]
25. *GB/T 24838-2018*; 1100 kV High-Voltage Alternating-Current Circuit-Breakers. Standardization Administration of China: Beijing, China, 2018.

energies

Article

Equivalent Modeling of Disconnector Operation Based on Dynamic Arc Characteristics and VFTO Characteristic Analysis

Bin Liu [1,*], Yong Liu [1], Junjun Xiong [1], Xiaopin Deng [1], Zhenyu Guo [2], Xueyou Zhang [2], Bingyu Mei [3] and Zhenhua Li [3]

[1] China Electric Power Research Institute, Wuhan 430074, China; liuyong@epri.sgcc.com.cn (Y.L.); xiongjunjun@epri.sgcc.com.cn (J.X.); dengxiaopin@epri.sgcc.com.cn (X.D.)
[2] State Grid Anhui Electric Power Co., Ltd., Super High Voltage Branch, Hefei 230000, China; guozy355x@ah.sgcc.com.cn (Z.G.); zhangxy8736@ah.sgcc.com.cn (X.Z.)
[3] College of Electrical Engineering and New Energy, China Three Gorges University, Yichang 443000, China; 15271505741@163.com (B.M.); lizhenhua1993@163.com (Z.L.)
* Correspondence: epri_liubin@163.com

Abstract: To thoroughly analyze the high-frequency and high-amplitude electromagnetic disturbances generated during disconnector operation, this paper proposes an equivalent modeling approach based on dynamic arc behavior. The model incorporates the resistance, inductance, and capacitance characteristics of the arc and consists of four main modules: arc reignition, arc extinction, arc resistance control, and switch control. Complete logical coordination among these modules is designed to enhance the model's performance in terms of dynamic response and modeling accuracy compared to traditional methods. By systematically comparing simulation results with experimental data and conventional model outputs, the effectiveness and reliability of the proposed model in accurately reflecting the operational characteristics of disconnectors are validated. Furthermore, a comparative analysis of transient waveform characteristics from both experiment and simulation is conducted, with key parameters extracted and probability density functions constructed. The results demonstrate the high-precision fitting capability of the model and further reveal the statistical distribution patterns of very fast transient overvoltage single-pulse characteristics.

Keywords: disconnector; equivalent arc model; very fast transient overvoltage; electromagnetic interference; statistical distribution; transient characteristics

1. Introduction

During normal power system operation, when a disconnector switches no-load transformers or transmission lines, small capacitive currents undergo multiple arc ignition and extinction cycles. This process generates high-frequency electromagnetic signals that propagate along conductors. Through continuous reflection and superposition, these signals ultimately develop into electromagnetic interference characterized by two parts: an amplitude reaching 2–3 times the phase voltage and a frequency spectrum extending to several hundred kHz [1–4]. Consequently, the very fast transient overvoltage (VFTO) and very fast transient current (VFTC) generated during disconnector operation present two major challenges: first, they significantly impair the performance of intelligent substation equipment; second, they introduce substantial risks to overall power system stability [5–7]. Developing precise equivalent simulation models for disconnector operations offers an effective approach to investigating these technical challenges, representing the predominant methodology in current research. However, existing multi-operation simulation models exhibit notable limitations due to inadequate consideration of disconnector structural features and operational characteristics, leading to significant discrepancies between the simulated and measured results. Given the increasing complexity of transient processes during disconnector operations, there is an urgent need to develop high-frequency dynamic arc equivalent models that accurately represent operational conditions, enabling a comprehensive analysis of rapid transient phenomena [8–10].

Although a variety of equivalent simulation models have been proposed for the fast transient process caused by the operation of the disconnector, these models rarely fully consider the influence of the structural characteristics of the disconnector and the operation mode on the accuracy of the simulation results, so the simulation results are directly different from the actual measurement results [11,12]. The cause of VFTO is closely related to the arc discharge process that occurs due to multiple breakdowns across the contact gap of the disconnector. The dynamic characteristics of the high-frequency arc have an important influence on the amplitude–frequency characteristics of the transient overvoltage. Therefore, when constructing the simulation model, accurately describing the dynamic characteristics of the high-frequency arc becomes a key factor in improving the simulation accuracy [13]. It is worth noting that the characteristic parameters of VFTO show obvious random dispersion, and its electromagnetic interference intensity is far beyond the current level of electromagnetic compatibility standards [14]. Although domestic and foreign scholars have accumulated a large number of transient interference data through field tests, there are still deficiencies in waveform feature identification, statistical law analysis, and analysis of the impact of various factors on the interference level. It is difficult to provide data support for the optimization design of the electromagnetic interference immunity of power equipment and for the research and development of simulated electromagnetic interference sources in disconnectors.

This research employs the ATP-EMTP simulation platform to model disconnector switching operations. The developed simulation framework incorporates five key functional modules: arc reignition modeling, arc extinction simulation, arc termination algorithm, switching control logic, and dynamic arc resistance regulation. Model validation is achieved through a systematic comparison between simulation outputs and experimental measurements, particularly referencing conventional modeling approaches. Subsequently, transient characteristics were analyzed, yielding significant findings regarding the statistical distributions of four critical parameters: single-pulse maximum peak value, duration, voltage rise time, and inter-pulse interval.

2. Theoretical Basis of the Model

2.1. Analysis of Problems in Existing Models

At the initial stage of the disconnector's opening operation, the disconnector has not yet separated, and the voltages on both sides of the switch are equal, i.e., $U_S(t) = U_L(t)$. Once the disconnector contacts are fully separated, the source-side voltage follows the sinusoidal waveform of the power frequency. Due to the slow decay of residual charge, the load-side voltage retains the value it held just before arc extinction. The difference between the source-side and load-side voltages represents the recovery voltage across the disconnector gap, i.e., $\Delta U = U_S(t) - U_L(t)$. According to the gas discharge theory, the breakdown characteristics of the disconnector contact gap are determined jointly by the recovery voltage and the breakdown voltage of the gap.

During the opening operation of the disconnector, the contact gap experiences dozens to hundreds of repeated breakdowns. Each breakdown triggers the ignition and extinction of an arc, causing a sharp voltage drop across the gap within nanoseconds and generating extremely steep voltage surges and traveling waves. These waves superimpose on the power-frequency voltage on the source side and on the residual voltage on the load side, forming transient overvoltage waveforms on both sides. Since the disconnector contacts are not perfectly symmetrical, the breakdown voltage across the gap varies with polarity. In general, the breakdown voltage increases with the contact gap distance, and accordingly, the amplitude of VFTO also increases.

As shown in Figure 1, $U_S(t)$ denotes the source-side voltage, $U_L(t)$ denotes the load-side voltage, $S(t)$ represents the distance between the disconnector contacts, and $U_B(t)$ represents the breakdown voltage across the gap.

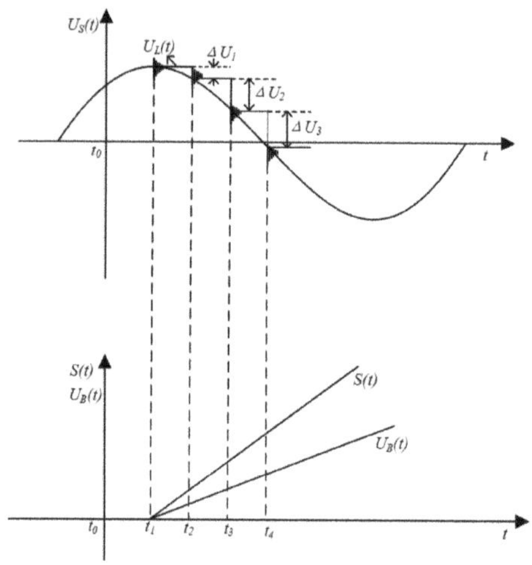

Figure 1. Relationship between repeated disconnector breakdown events and corresponding breakdown voltages.

Specifically, the amplitude characteristics of VFTO are closely related to the breakdown process of the contact gap. Although existing modeling methods can simulate single breakdown events under opposite polarity conditions at the source and load sides, they fail to consider the influence of the breakdown characteristics of the contact gap and the dynamic behavior of arc resistance on the breakdown process. Current single and multiple breakdown arc models generally divide the discharge process into three stages: pre-breakdown, stable arcing, and arc extinction. However, these models still present significant limitations in practical applications. Firstly, during disconnector operation, there is a lack of clear criteria for identifying gap breakdown and arc extinction, making it difficult to accurately determine the transitions into the pre-breakdown and arc extinction stages. Secondly, existing models are unable to accurately simulate the variation trend of arc resistance during the stable arcing stage. These issues hinder the ability of current arc models to accurately describe the dynamic evolution of the arc throughout the disconnector operation process.

Research on VFTO mainly focuses on the simulation modeling of the arc of the disconnector. The accurate arc model is an important factor affecting VFTO research. At present, the commonly used arc models are mainly divided into fixed resistance arc model, time-varying arc resistance model, segmented arc resistance model, spark model, and black box model [15]. The constant resistance model is based on the ideal switching model, and the arcing process is equivalent to a constant resistance of 2~5 Ω. However, the model cannot simulate the nonlinearity of the arc. Compared with the actual arcing process, there is still a lot of error, which is only suitable for simple calculations in engineering [16]. The time-varying arc resistance model considers the transition from pre-breakdown to arcing during a single breakdown of the disconnector gap, but ignores the change in arc resistance during the arc extinguishing process, so it cannot accurately describe the change in arc resistance during a single breakdown [17]. In order to solve the above problems in the arc model, Reference [18] proposed a segmented arc model, which further considers the actual change trend of the resistance of the arc during the combustion process and divides the arc combustion process into three stages: pre-breakdown, stable arcing, and arc extinguishing. However, the judgment of the three stages of the model is very vague, and the assumption that the arc development always follows is too ideal.

In addition, for single arc models, analysis is typically limited to the most severe arcing scenarios, which cannot fully represent the transient waveforms generated during the entire disconnector operation process. Moreover, most single arc models only perform dynamic simulation of the arc's equivalent resistance, without considering the influence of arc inductance, gap capacitance, and other factors on the simulation results. In a comprehensive model design, arc reignition criteria, arc extinction criteria, and switch state transitions are essential components of the complete arc reignition logic for a disconnector. Although dynamic reignition arc models address the continuity of arc resistance and are capable of simulating the entire disconnector operation process, they still suffer from certain issues related to phase division, program logic design, and the theoretical calculation of arc resistance.

2.2. *Improved Simulation Approach for Full-Process Arc Modeling of Disconnector Operation*

VFTO waveforms are formed by the superposition of hundreds of individual breakdown events. Each breakdown generates a waveform exhibiting damped oscillation characteristics, with a typical duration ranging from several to several tens of microseconds. Since the interval between successive breakdowns is usually on the millisecond scale, individual breakdown events within a VFTO waveform can be considered non-overlapping. Most existing arc models for disconnectors focus primarily on the macroscopic characteristics of arc resistance, while neglecting the microscopic discharge processes. Experimental studies have shown that the variation trend of arc impedance during disconnector operation can be equivalently represented by arc resistance, arc inductance, and gap capacitance. Based on existing dynamic arc models and multiple reignition arc models, this paper proposes an improved approach by incorporating the dynamic variation of arc resistance, arc inductance, and gap capacitance. The single breakdown process is divided into three stages: arc pre-breakdown, stable arcing, and arc extinction. In addition, criteria are introduced to determine both the occurrence of gap breakdown and the end of arc extinction. The overall disconnector operation is modeled as a cyclic repetition of these three stages.

In the simulation, a single breakdown event in the disconnector can be equivalently represented by the model shown in Figure 2: R(t) represents the arc resistance, describing the time-varying impedance during the breakdown, L denotes the arc inductance, and C represents the gap capacitance. The dashed region in the figure indicates the breakdown stage, modeled as a series connection between arc resistance and inductance. In the arc pre-breakdown stage, a voltage difference exists between the disconnector contacts and can be represented by a capacitive model. Once the models for each stage are defined, the complex transient process during disconnector operation can be fully simulated by setting the corresponding start and end criteria for each stage.

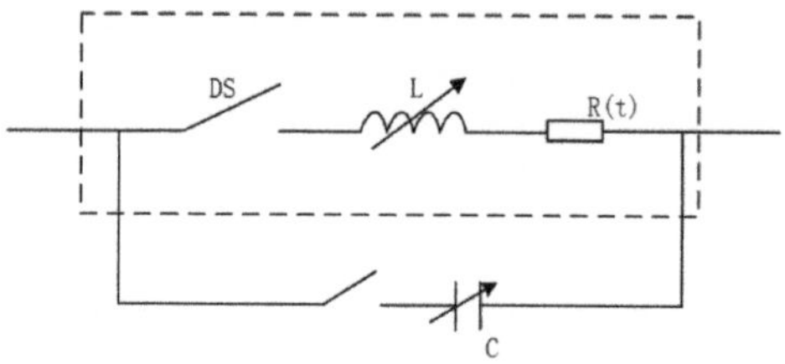

Figure 2. Equivalent model of a single breakdown event in the disconnector.

Considering the influence of arc resistance, arc inductance, and gap capacitance of the disconnector, the equivalent circuit for analyzing the electromagnetic transient process during contact gap breakdown is shown in Figure 3. In Figure 3, the equivalent arc model of the disconnector is indicated by a dashed box. R_S represents the arc resistance between the disconnector contacts, L_S denotes the arc inductance, and C_S represents the capacitance between the contacts. R_1, L_1, and C_1 correspond to the equivalent resistance, inductance, and capacitance of the source-side circuit of the disconnector, respectively, while R_2, L_2, and C_2 correspond to those of the load-side circuit. When no breakdown occurs across the disconnector contact gap, the power-frequency source forms an energy storage loop through the inductive and capacitive components of the system. At this stage, the contact gap can be modeled as capacitance C_S. Once breakdown occurs across the contact gap, its electrical characteristics change significantly. The discharge path is then equivalent to a time-varying resistance R_S in series with inductance L_S, which together form a parallel network with capacitance C_S. This model effectively describes the dynamic impedance characteristics of the gap after breakdown, as well as the flow path of high-frequency transient currents.

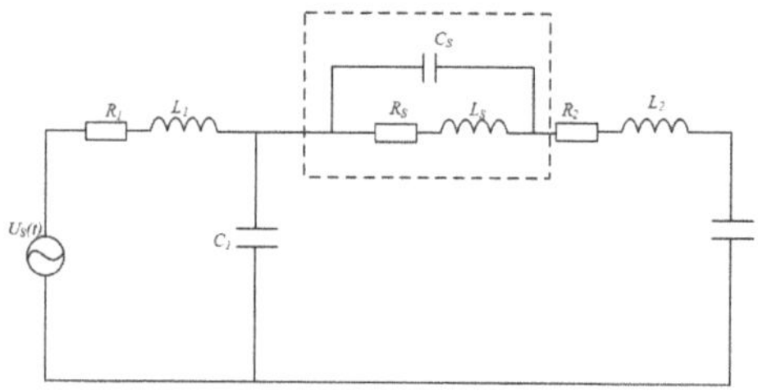

Figure 3. Equivalent circuit model.

2.2.1. Dynamic Arc Resistance Equivalence

During the arc pre-breakdown stage, the arc resistance rapidly decreases from the pre-arc resistance value to a stable arcing resistance value within an extremely short period. The transition from pre-breakdown to stable arcing in the disconnector gap exhibits significant time-varying characteristics and typically occurs within a few nanoseconds. To more accurately describe the dynamic behavior of the arc, a hyperbolic arc resistance model is adopted in the pre-breakdown stage to characterize the variation in arc resistance. In this model, the arc resistance drops sharply following the breakdown but does not immediately reach zero; instead, it gradually approaches steady-state arcing resistance. The expressions for arc resistance during the pre-breakdown and stable arcing stages are as follows:

$$R_a(t) = \left\{ \begin{array}{l} R_0, t = 0 \\ Z_0\left(\frac{t_\delta}{t} - 1\right) + r_0, 0 < t < t_\delta \\ r_0, t_\delta < t \end{array} \right\} \quad (1)$$

In the above expressions, R_0 denotes the initial breakdown resistance of the gap, approximately 1×10^{12} Ω; r_0 represents the steady-state arcing resistance, which is set to 2 Ω in this paper; Z_0 is the sum of the wave impedance of the disconnector line and that of the conducting rod; and t_δ indicates the breakdown delay time.

During the arc extinction stage, considering the influence of the recovery voltage, the complete Mayr arc model is typically adopted. The arc conductance equation of the Mayr model is given by

$$\frac{1}{g_{\mathrm{m}}}\frac{dg_{\mathrm{m}}}{dt} = \frac{1}{\theta_{\mathrm{m}}}\left(\frac{i^2}{P_0 g_{\mathrm{m}}} - 1\right) \quad (2)$$

In this equation, g_{m} and θ_{m} represent the arc conductance and the time constant of the Mayr arc model, i is the arc current, P_0 denotes the power loss under steady-state conditions, and t is time.

2.2.2. Transient Capacitance Equivalence

During the steady-state stage of disconnector operation, the duration typically ranges from several hundred nanoseconds to several tens of milliseconds. During this period, the disconnector remains in the process of operation, and as the contacts continue to move, the variation trend of the gap capacitance and stray capacitance can be expressed by the following equation:

$$C_t = \left(73.741 - \frac{0.6278}{t_c + 0.01096}\right) \times 10^{-12} \quad (3)$$

In this equation, C_t is the gap capacitance of the disconnector, where $t_c = \begin{cases} t \\ \frac{d}{v} - t \end{cases}$, t is the duration of the disconnector operation process, in seconds; v is the movement speed of the moving contact during disconnector operation, in meters per second; and d is the initial distance between the moving and fixed contacts, in meters.

In the steady-state stage, the arc between the contacts of the disconnector is extinguished, and there is a gap between the two contacts. With the change in the power supply side voltage, the isolation switch gap will produce a voltage difference. It is worth noting that the steady-state phase of the disconnector lasts for a relatively long time, generally up to hundreds of ns or even tens of ms. During this process, the disconnector is still in the operation process, and its disconnector contact produces displacement. Therefore, in the steady-state stage, the value of the contact gap capacitance and the disconnector ground capacitance will change with the gap distance d, which should be regarded as the capacitance of the size change. The fracture capacitance is actually an energy storage element, and there is a buffer and delay in the release of energy. Therefore, the single VFTO peak does not necessarily appear on the first peak under the isolation switch model, considering the capacitance effect. The peak value of VFTO in the disconnector model considering the capacitance effect is higher, the rise time of VFTO is shorter, and the steepness is larger. The main reason is that the fracture capacitance under the disconnector model is used as an energy storage element. Compared with the model without considering the capacitance, the energy storage is higher. The larger the energy, the higher the VFTO amplitude generated, and the faster the rising speed.

2.2.3. Transient Inductance Equivalence

During the arcing stage of the disconnector operation, the value of arc inductance changes dynamically with the evolution of the arc. This section takes into account the real-time dynamic variation in arc inductance. The arc channel inductance in the transient arc stage of the disconnector is modeled using the empirical Martin formula, as expressed in the following equation:

$$L \approx 2d_L \ln(b/a) \approx 14d_L \tag{4}$$

In this equation, L is the equivalent inductance, in nanohenries (nH); d_L is the arc channel length, in centimeters (cm); and a and b represent the radii of the discharge channel and the return conductor, respectively, also in centimeters (cm). In practical arcing scenarios, the values of a, b, and d_L vary dynamically. The arc inductance can therefore be described by the relationship between inductance and arc channel length. The discharge channel length can be approximately considered the real-time distance between the moving and fixed contacts of the disconnector. Thus, the time-dependent variation in arc channel inductance during the opening and closing operations of the disconnector can be expressed by the following equation:

$$\begin{cases} L_{op} = 14v_{op}t \\ L_{cl} = 14(d_0 - v_{cl}t) \end{cases} \tag{5}$$

In this equation, L_{op} and L_{cl} represent the real-time equivalent inductance during the opening and closing processes, in microhenries (μH); v_{op} and v_{cl} represent the velocity of the moving contact during the opening and closing processes, in meters per second (m/s); and d_0 represents the initial distance between the moving and fixed contacts of the disconnector, in meters (m).

2.3. Critical Breakdown Criteria for Disconnector Contact Gap

During the operation of the disconnector, the phenomenon of arc reignition after arc extinction is determined by the relationship between the recovery voltage and the gap breakdown voltage. Specifically, when the recovery voltage exceeds the critical breakdown voltage, the gap will be re-broken down, forming a new discharge path; otherwise, it will remain in the open state. This process repeats until the contacts reach a fully closed or open state. Taking the disconnector closing process as an example, according to gas discharge theory, in the early stage of closing, because the contact distance is large, the gap breakdown voltage of the switch is significantly higher than the actual voltage it is subjected to, so the switch remains in the open state. As the contacts gradually move closer, the gap distance continues to decrease. When the contact distance reaches a certain critical value, the actual voltage across the gap exceeds the breakdown voltage, causing gas molecules to collide and ionize, forming a discharge path through the gap and ultimately completing the gap breakdown.

Reference [6] analyzes and fits the breakdown voltages under different contact gaps, showing that the breakdown voltage and the switch gap length maintain an approximately linear relationship. Based on this finding, this paper adopts the simplified algorithm from Reference [19] to model multiple breakdown processes during disconnector operation, assuming that

$$\text{During opening operation, } U_b(t) = Evt \tag{6}$$

$$\text{During closing operation, } U_b(t) = E(d_0 - vt) \tag{7}$$

In the equation, E represents the breakdown voltage per unit length of the switch gap, v represents the operating speed of the switch, d_0 represents the distance between the switch contacts, and t represents time.

3. Establishment of the Whole Process Model of the Disconnecting Switch

3.1. Functional Modules of the Complete Disconnector Operation Model

Currently, the Electromagnetic Transients Program (EMTP-ATP) is widely used for transient process calculations. This program establishes accurate and reliable models based on theory, allowing for convenient modeling and simulation, and has become the primary tool for overvoltage calculations in power systems. Figure 4a–d, respectively, represent the reignition arc module, arc extinction module, arc resistance control module, and switch control module in the complete disconnector simulation model.

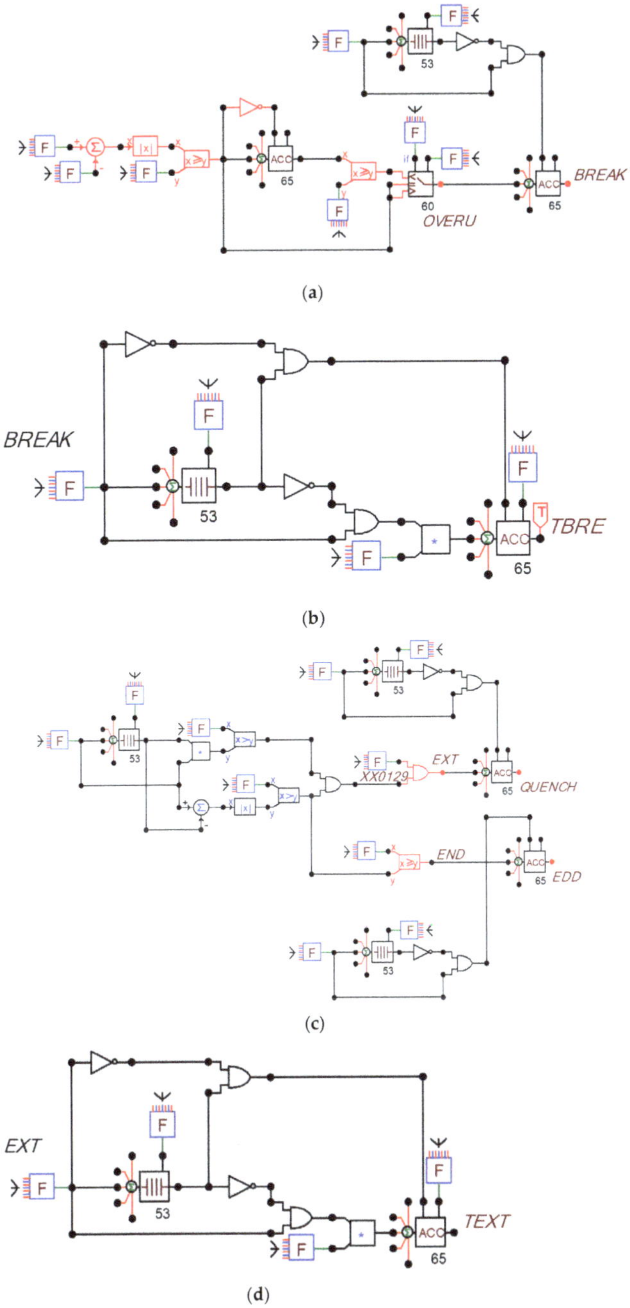

Figure 4. Simulation model of each module in the entire process of disconnecting switch operation. (**a**) Reignition arc module. (**b**) Burning arc time recording module. (**c**) The arc extinguishing module. (**d**) Aoff time recording module.

The core function of the reignition arc module is to determine whether the gap voltage during the disconnector's opening and closing operations exceeds its insulation withstand level. This module works in coordination with the arc extinction module to control the conduction state of the circuit. When constructing a short-gap reignition arc model, criteria for re-breakdown must be defined, along with key parameters such as the disconnector's operating speed and the gap breakdown voltage per unit length. The reignition arc module obtains the real-time recovery voltage (ΔU) and calculates the critical breakdown voltage of the gap (U_b). Comparing the two determines whether breakdown has occurred. If $\Delta U > U_b$, it indicates that a gap breakdown has occurred, and an arc ignition command is sent to both the switch control module and the arc resistance control module. Upon receiving the ignition signal, the switch control module performs the closing operation, while the arc resistance control module adjusts the resistance accordingly. At the natural current zero-crossing instant, the arc ignition command is immediately terminated.

In configuring the arc extinction module of the disconnector, this study introduces an arc extinction termination determination mechanism based on traditional extinction criteria. The core function of this module is to determine the extinction state of the arc following a single breakdown. Once it is confirmed that the arc extinction process has been fully completed, the system maintains the switch in an open state until the next gap breakdown occurs. When constructing the arc extinction module, this study introduces a differential criterion: When the absolute value of the current difference $\left|I_t - I_{(t-\Delta t)}\right|$ between adjacent time steps is less than a preset threshold K, the current zero-crossing can be determined. This dual-criterion approach can effectively identify the moment of current zero-crossing and provides a reliable basis for accurately determining the entry into the arc extinction phase. When the absolute value of the monitored current signal $|I_t|$ is lower than a preset threshold I_k, the arc extinction process is deemed to be complete. Subsequently, the system enters a steady-state condition until the next dielectric breakdown occurs.

The arc resistance control module regulates the resistance variation in each stage. After a breakdown occurs in the disconnector gap, the programming code for the hyperbolic arc model during the arc pre-breakdown and stable arcing stages, as well as the Mayr arc model expression for the arc extinction stage, is stored in the Models module. At the same time, the arcing signal from the restrike arc module and the arc extinction signal from the arc extinction module are also transmitted into the Models module, establishing the connection between the arc resistance control module, the restrike arc module, and the arc extinction module, thereby realizing the dynamic adjustment of arc resistance values in different stages.

The switch control module receives the breakdown arcing signal and arc extinction completion signal sent by the restrike arc module and arc extinction module, respectively, achieving the purpose of controlling the switch state.

3.2. Functional Coordination Logic of Each Module

By integrating the restrike arc module, arc extinction module, arc resistance control module, and switch control module, the logical coordination among these functional modules is illustrated in Figure 5. The process is as follows: When the disconnector begins to close, the voltage between its two contacts and the air breakdown voltage are transmitted in real time to the restrike arc module. If the breakdown condition is met, a restrike signal is sent to the arc resistance control module and the switch control module. At this point, a breakdown occurs across the contact gap, and the switch control module outputs a closing command to drive the closing operation of the disconnector in the model. The formation of the arc causes the arc resistance to enter the arcing stage. As the current passes through the arc extinction module, it continuously determines whether the arc extinction conditions are met. If the conditions are satisfied, an arc extinction signal is sent to the arc resistance control module, and the arc resistance equation for the extinction stage is used for calculation. When the current decreases to the preset value IKI_KIK, the arc extinction process ends. An end signal is then sent to the switch control module, which issues an opening command, causing the disconnector to open. At this time, the arc resistance remains unchanged until the next breakdown occurs.

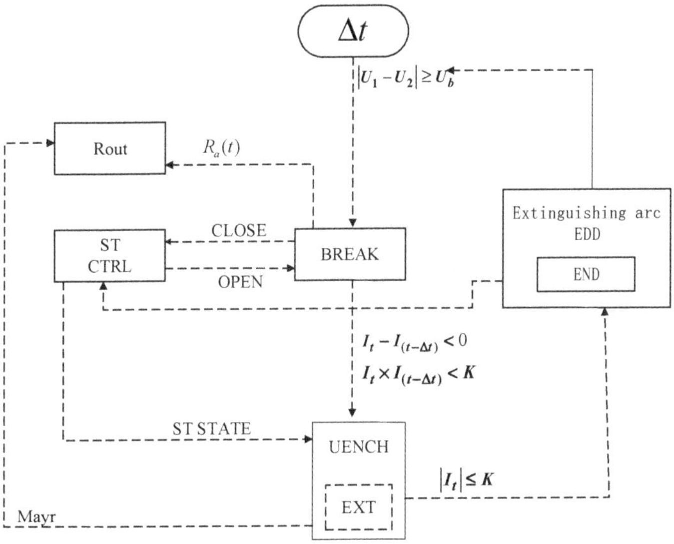

Figure 5. Logic coordination of each module.

3.3. Simulation of the Entire Disconnector Operation Process

According to the frequency range classification of the International Large Power Grid Conference (CIGRE), the isolating switch operation process belongs to the field of fast wavefront overvoltage [20]. According to the CIGRE standard, when building the simulation model for the bus in the main circuit, the lossless uniform transmission line model can be replaced by the wave impedance. The power-side ground capacitance and the load of the small current can be used equivalently. According to the parameters of commonly used open isolation switch equipment and theoretical calculations, the main component parameters used in the simulation are shown in Table 1.

Table 1. Equivalent models and parameters of the main components.

Element	Parameter Declaration
Aerial conductor	T1 = 4.5 m, wave impedance 266 Ω, wave speed 300 m/μs T2 = 3 m wave impedance at a 121 Ω wave velocity of 300 m/μs
Power supply side to ground capacitor	6300 pF
capacitive divider	411 pF, variable ratio of 2000:1
Dynamic and static contact guide wave impedance	130 Ω
Potential transformer	1000 pF
Load capacitor	10,000 pF

At the same time, according to the test circuit and the corresponding parameters, considering the inductive and capacitive effects during the isolating switch operation, the simulation model is built in the ATP-EMTP electromagnetic transient simulation program. The main circuit of the simulation model is shown in the Figure 6.

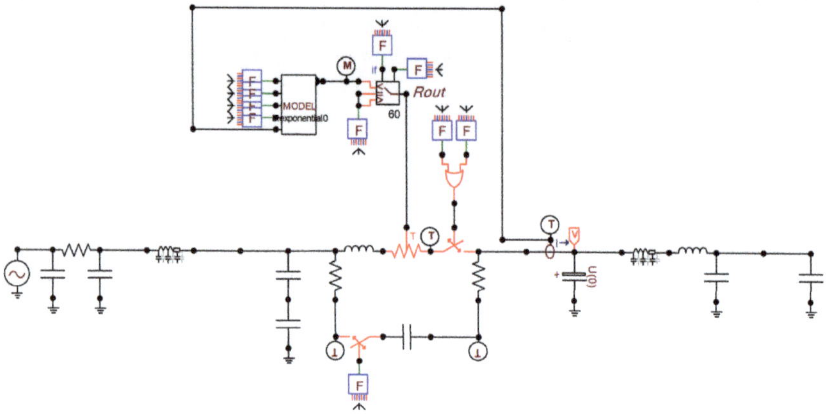

Figure 6. Simulation of the main circuit model.

The arc resistance control module and switch control module are located in the main circuit, which includes operating power supply, overhead line distribution parameters, switch-to-ground capacitor, arc-to-ground capacitor, arc combustion resistance, disconnecting switch, transformer collector capacitor, and load capacitor, so as to simulate the actual line operation. A Models module is designed within the arc resistance control module to program the hyperbola and Mayr arc resistance model and to connect the rekindled arc module, arc extinguishing module, and switch control module to control the resistance value of the arc resistance at different stages. Rout is the dynamic arc resistance value of the whole process of the disconnecting switch operation. The switch control module realizes control of the whole circuit by receiving the arc ignition or arc extinguishing signal from the reignition arc module and the arc extinguishing module. ST is the switch state. The arc channel inductance and clearance capacitor adopt the variable inductance and capacitor components combined with the programming module, and the logic control module is executed according to the time step control program and model.

Through the above model parameter settings, the VFTO waveforms of the transient process during the isolating switch closing operation are shown in Figure 7. The transient voltage waveform is presented in a step-like form, with each step representing a breakdown. The maximum overvoltage amplitude in the simulation results is 114.67 kV.

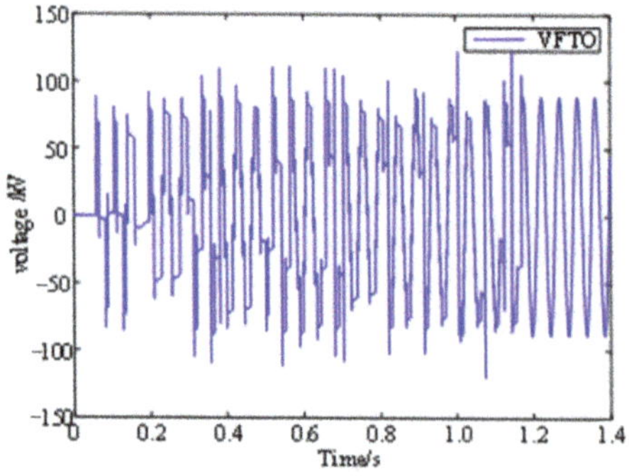

Figure 7. VFTO waveforms of the transient process during the isolating switch closing operation.

4. Experimental Validation and Comparison

4.1. Test Verification

To verify the accuracy of the proposed simulation model for the entire disconnector operation process, this study adopts a combined approach of experimental testing and simulation. Based on actual line parameters, a capacitive low-current test circuit for the disconnector was constructed, as shown in Figure 8, to simulate the electromagnetic interference generated during disconnector operation and to obtain VFTO waveform data. The measured results were compared and analyzed against the simulation data to validate the accuracy of the developed model.

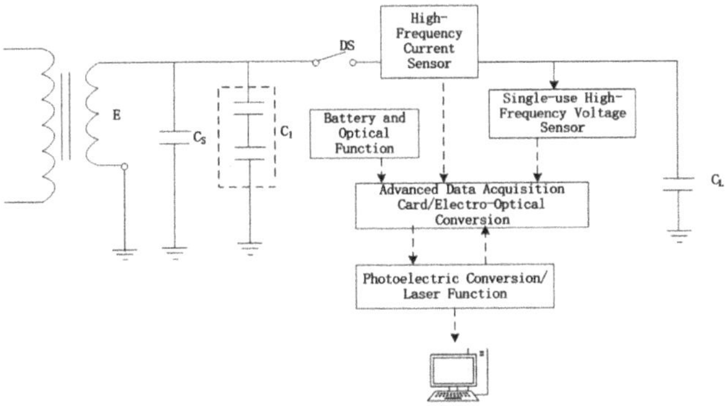

Figure 8. A 110 kV disconnector operation test system.

E represents the operating power supply voltage of 220 V; C_S denotes the power supply side capacitance; capacitor voltage divider C_1 is used to monitor the voltage in the test circuit. DS is the isolation switch, and C_L represents the load capacitance, which is used to simulate the ground capacitance of the no-load short bus during actual operation. A high-frequency voltage and high-frequency current sensor, a high-speed acquisition card, and other devices are used to measure and collect the voltage and current signals during the test. In order to ensure the stability and reliability of data transmission and ensure the safety of operation, photoelectric isolation is adopted between the high-speed acquisition card and the computer.

As shown in Figure 9, due to the relatively slow speed of the disconnector contact movement during the entire closing process, the contact gap experiences tens or even hundreds of breakdowns, leading to the presence of many transient single-pulse waveforms with decaying oscillation characteristics in the VFTO macro waveform. The waveform generally transitions from sparse to dense, and the amplitude decreases. This is because, at the initial stage of closing, the large gap distance requires a higher voltage to initiate breakdown, which makes it more difficult for breakdown to occur. whereas in the opening process, the situation is the opposite.

The test system operates at a voltage level of 110 kV, and the base value is selected for calculation:

$$1pu = 110 \times \frac{\sqrt{2}}{\sqrt{3}} = 89.81 \text{ kV} \tag{8}$$

The maximum amplitude of VFTO is 118.67 kV (about 1.32 p.u.).

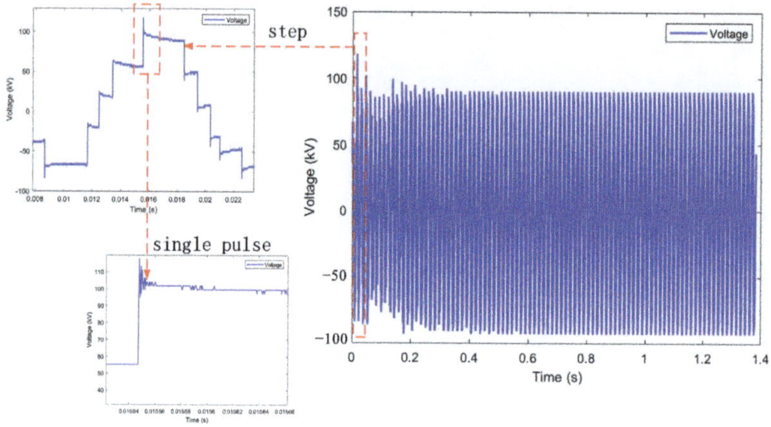

Figure 9. The test waveform of the entire VFTO process.

4.2. Comparison of Test and Simulation Waveforms

In order to compare the improvement effect of the model presented in this paper, the dynamic resistance model proposed by article [21] is integrated into the main circuit of this study. The VFTO waveforms obtained using the same parameter settings are shown in Figure 10, and the maximum overvoltage amplitude is 130.36 kV (about 1.45 p.u.).

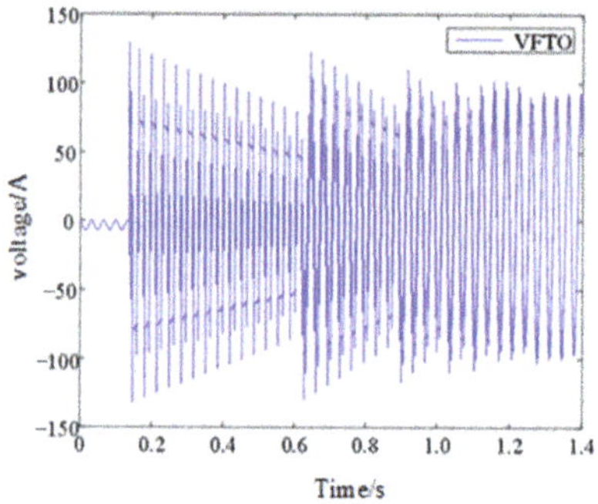

Figure 10. The VFTO simulation waveform when considering only the dynamic resistance model.

The VFTO maximum amplitude obtained from the test is 118.67 kV (about 1.32 p.u.), while the VFTO maximum amplitude from the model proposed in this paper is 114.67 kV (about 1.27 p.u.), with an error rate of 3.49%. The VFTO maximum amplitude obtained by considering only the dynamic resistance is 130.36 kV (about 1.45 p.u.). From this, it can be concluded that the simulation model proposed in this paper is closer to the experimental results in terms of both maximum amplitude and macro waveform trends. Additionally, the simulation waveforms in this paper better reflect the randomness of the disconnector operation. In comparison, the dynamic resistance model's simulation waveform deviates significantly from the experimentally obtained waveform.

The simulation and measured waveforms at the VFTO peak during the isolating switch closing operation are shown in Figure 11. It can be observed that the VFTO simulation waveform is in good agreement with the experimental waveform, with the variation patterns of the simulation and experimental waveforms being roughly the same.

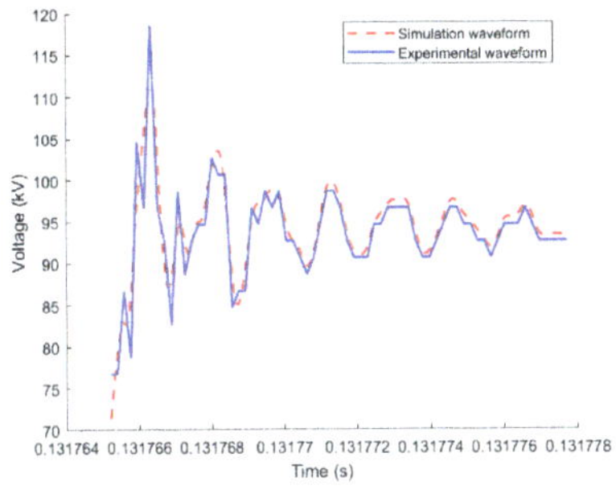

Figure 11. Comparison of the VFTO simulation and the experimental maximum peak single pulse.

The frequency spectrum analysis of the VFTO full waveform data from the isolation switch operation simulation model is shown in Figure 12a,b. It can be observed that the frequency band of the VFTO waveform is primarily distributed within 15 MHz, with the main frequency distributions of both the VFTO experimental and simulation waveforms being very similar. The main frequency distribution occurs in the following ranges: 1.49 MHz, 2.26 MHz, 6.2 MHz, 7.09 MHz, 8.26 MHz, 9.16 MHz, and 11.23 MHz. The difference lies in the fact that the simulation data also shows additional main frequencies at 2.99 MHz, 6.67 MHz, and 12.09 MHz. This discrepancy is mainly due to the presence of high-frequency noise in the transient overvoltage during the experimental process. To enhance the reliability of the data, noise reduction was applied to the experimental data, which led to the filtering out of shorter pulse signals. In contrast, the simulation data did not undergo noise reduction, allowing it to capture the shorter pulse processes and thus resulting in a richer main frequency distribution.

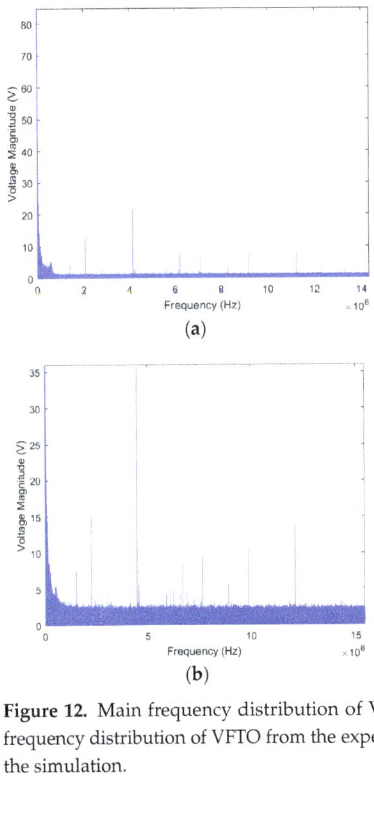

Figure 12. Main frequency distribution of VFTO from the experiment and simulation. (**a**) Main frequency distribution of VFTO from the experiment. (**b**) Main frequency distribution of VFTO from the simulation.

In summary, compared with the experimental waveform, the VFTO waveform obtained through simulation shows a similar trend in terms of both the macro waveform and the individual pulse waveform. This indicates that the simulation results in this study accurately reflect the VFTO waveform during the operation of the disconnector. Therefore, this validates the effectiveness of the proposed simulation model for the disconnector operation process.

5. Statistical Analysis of Single-Pulse Special Parameters

The disconnector has a large dispersion during operation, and it is difficult to describe the distribution law of VFTO single-pulse waveform characteristic parameters using a standard distribution model. In contrast, kernel density estimation (KDE) can obtain the probability density distribution according to the characteristics of the sample data and does not need to make any assumptions about the distribution of the sample data [22–24]. In conclusion, the kernel density estimation method is used to estimate the distribution of single-pulse characteristic parameters, providing the frequency distribution of each parameter, the estimated optimal kernel probability density distribution, and the evaluation results.

$$f_h(x) = \frac{1}{nh}\sum_{i=1}^{n} K\left(\frac{x - X_i}{h}\right) \tag{9}$$

where X_i is a given sample data point, h is the window width and a non-negative constant; $K(\cdot)$ is a kernel function.

There are many kinds of kernel functions. In view of the fact that the data generally obey the Gaussian distribution, the more commonly used Gaussian (Gaussian) kernel density function is used, and the probability density function is expressed as follows:

$$K(\cdot) = \frac{1}{\sqrt{2\pi}} \exp\left(\frac{-x^2}{2}\right) \tag{10}$$

The selection of the bandwidth h is crucial for the accuracy of kernel density estimation. Whether the kernel density estimation can estimate the sample well depends on the value of the window width h. If the value of h is too small, the result will be unstable; if the value of h is too large, the discrimination rate of the result will be very low. As for the selection of window width, the algorithms for calculating the optimal window width based on different error formulas are different [25]. Mean integrated squared error (MISE) is used to obtain the optimal window width. Among them, MISE is based on the principle of minimum square difference to measure the difference between the estimated density and the true density to obtain the window width. The specific expression is as follows:

$$\text{MISE}(f_h(x)) = E\left[\int \{f_h(x) - f(x)\}^2 \text{d}x\right] = \int [E f_h(x) - f(x)]^2 \text{d}x + \int \text{var} f_h(x) \text{d}x \tag{11}$$

In the formula, $E(\cdot)$ is the expectation, $f_h(x)$ is the true probability density function of the sample data, and $\text{var}(\cdot)$ is the variance.

By solving the equation, the optimal window width can be obtained when $\text{MISE}(f_h(x))$ takes the minimum value.

This subsection makes a statistical analysis of the duration, rise time, time interval, and the number of single pulses.

5.1. *Maximum Peak Value of Single Pulse*

When analyzing VFTO, special attention should be paid to its high-amplitude characteristics. Therefore, the maximum peak voltage of VFTO can be considered a key characteristic parameter. The statistical results of the single-pulse maximum peak under simulation and experimental conditions during switching operations are shown in Table 2.

Table 2. Statistical results of single-pulse maximum peak values.

Item	Simulated Single-Pulse Maximum Peak	Experimental Single-Pulse Maximum Peak
Amplitude/kV	[9.97, 114.69]	[9.49, 120.47]
Average Value/kV	66.93	59.68

As shown in Table 2, the distribution range of the single-pulse maximum peak voltage obtained from the simulation during switching operations is [9.97, 114.69] kV, with an average peak value of 66.93 kV. In comparison, the experimental results show a distribution range of [9.49, 120.47] kV, with an average maximum peak of 59.68 kV. Further statistical analysis of the single-pulse peak values of VFTO generated during disconnector closing operations reveals that the frequency distribution and KDE of the maximum peaks under optimal bandwidth settings are illustrated in Figure 13a,b, respectively. The optimal bandwidths for the KDE models of the single-pulse rise time are 9.629 and 7.958. Both the simulation and experimental results indicate that the distribution of the single-pulse maximum peaks does not follow a normal distribution or a unimodal distribution; instead, it exhibits a bimodal distribution pattern. Moreover, the distribution ranges of the maximum peaks obtained from the simulation and experiment are generally consistent, with the majority of values concentrated within the range [9.49, 79.78] kV, accounting for 84.35%.

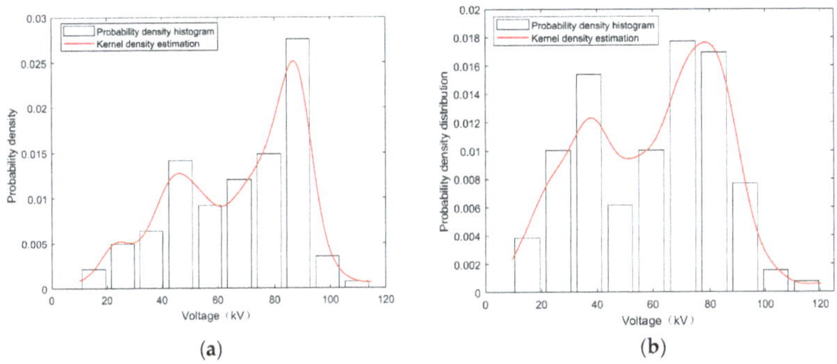

Figure 13. Comparison of maximum peak value of single-pulse and optimal kernel density distribution during closing. (**a**) Maximum peak and kernel density distribution of single pulses from simulation during closing operation. (**b**) Maximum peak and kernel density distribution of single pulses from the experiment during closing operation.

5.2. Duration

The duration of a single pulse refers to the time from the beginning of the pulse waveform to the end of the oscillation. The statistical results of single-pulse durations under simulation and experimental conditions during switching operations are shown in Table 3.

Table 3. Statistical results of single-pulse duration.

Item	Simulated Single-Pulse Duration	Experimental Single-Pulse Duration
Duration/μs	[0.12, 20.87]	[0.05, 19.4]
Average Value/μs	1.61	2.52

As shown in Table 3, the distribution range of single-pulse durations obtained from the simulation during switching operations is [0.12, 20.87] μs, with an average duration of 1.61 μs. In contrast, the experimental results show a distribution range of [0.05, 19.4] μs, with an average duration of 2.52 μs. Further statistical analysis of the single-pulse durations of VFTO generated during disconnector closing operations reveals that the frequency distribution and kernel density estimation (KDE) under optimal bandwidths, as shown in Figure 14a,b, respectively. The optimal bandwidths for the KDE models of single-pulse duration are 0.774 and 0.854. Both the simulation and experimental results indicate that the single-pulse durations mainly exhibit a unimodal distribution, with most durations concentrated in the range of [0.09, 0.23] μs, accounting for 92.24% and 85.82%, respectively.

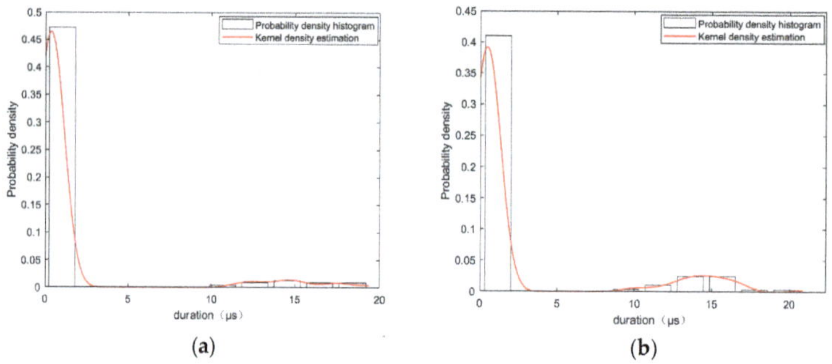

Figure 14. Comparison of single-pulse durations and optimal kernel density distributions during closing operation. (**a**) Duration and kernel density distribution of single pulses from simulation during closing operation. (**b**) Duration and kernel density distribution of single pulses from the experiment during closing operation.

5.3. Up Time

The VFTO single-pulse waveform generally exhibits a decaying oscillation characteristic, and the amplitude typically does not appear within the first oscillation cycle. To facilitate statistical analysis, the rise time of a single pulse is defined as the time required for the waveform amplitude to rise from 10% to 90% of its peak value. The distribution range of the single-pulse rise times, as obtained from the statistical analysis, is shown in Table 4.

Table 4. Statistical results of single-pulse up time.

Item	Simulated Single-Pulse Up Time	Experimental Single-Pulse Up Time
Up Time/μs	[0.54, 8.70]	[0.43, 8.64]
Average Value/μs	1.72	1.55

As shown in Table 4, the distribution range of single-pulse rise times obtained from the simulation during switching operations is [0.54, 8.70] μs, with an average rise time of 1.72 μs. The experimental results show a distribution range of [0.43, 8.64] μs, with an average rise time of 1.55 μs. Further statistical analysis of the single-pulse rise times of VFTO generated during disconnector closing operations reveals the frequency distribution and kernel density estimation (KDE) under optimal bandwidths, as shown in Figure 15a,b. The optimal bandwidths for the KDE models of the single-pulse rise time are 0.275 and 0.254, respectively. Moreover, the rise time distribution ranges obtained from the simulation and experiment are generally consistent, with the majority of rise times concentrated in the range of [0.36, 5.26] μs, accounting for 94.88% and 93.02%, respectively.

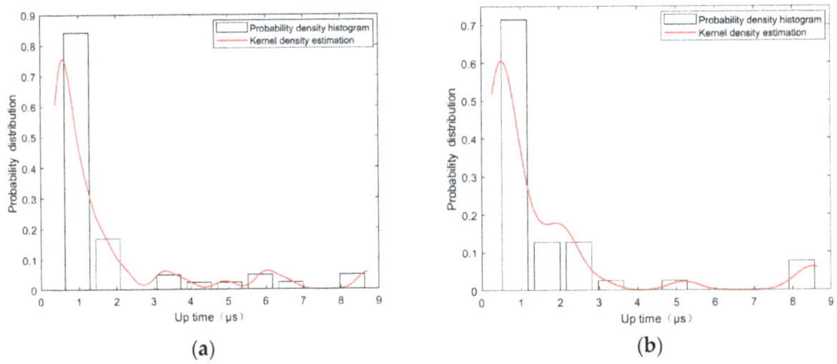

Figure 15. Comparison of single-pulse up time and optimal kernel density up time during closing operation. (**a**) Up time and kernel density distribution of single pulses from simulation during closing operation. (**b**) Up time and kernel density distribution of single pulses from the experiment during closing operation.

5.4. Interval Time

The single-pulse interval time refers to the difference between the occurrence times of two adjacent pulses. The statistical results indicate that the VFTO single-pulse interval time during switching operations mainly falls within the millisecond range. Therefore, it can be assumed that the two VFTO breakdown events are temporally independent and do not overlap. The statistical results are shown in Table 5.

Table 5. Statistical results of single-pulse interval time.

Item	Simulated Single-Pulse Interval Time	Experimental Single-Pulse Interval Time
Interval Time/ms	[0.04, 4.98]	[0.06, 5.10]
Average Value/ms	0.61	0.65

Figure 16 shows the frequency distribution of single-pulse interval times obtained from simulation, along with the kernel density distribution under the optimal bandwidth. The statistical results indicate that the optimal bandwidths for the rise time probability density distribution models obtained from simulation and experiment are 0.256 and 0.219, respectively. Furthermore, the interval times obtained from both simulation and experiment exhibit a unimodal distribution, and the distribution ranges are generally consistent. The majority of interval times are concentrated in the range of [0.06, 1.74] ms, accounting for 94.83% and 93.72%, respectively.

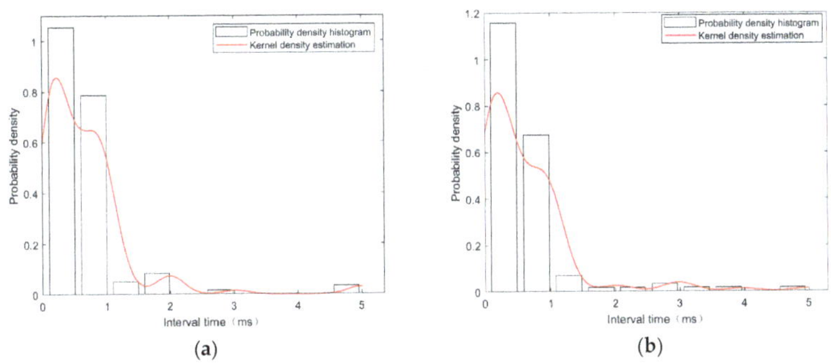

Figure 16. Comparison of single-pulse interval times and optimal kernel density distributions during closing operation. (**a**) Interval time and kernel density distribution of single pulses from simulation during closing operation. (**b**) Interval time and kernel density distribution of single pulses from the experiment during closing operation.

6. Conclusions

This paper presents a comprehensive model of disconnector operation, designed by analyzing the functions of the arc module, arc extinction module, arc resistance control module, and switch control module, along with the calculation logic of each module. Compared to the actual measured transient waveforms, the results show that the simulation results from the proposed model are consistent with the tested characteristics, demonstrating the correctness and accuracy of the proposed simulation model. The results indicate that the proposed model can be applied to transient operation simulations, providing a more accurate model for the study of transient electromagnetic interference. To further validate the model's accuracy, this paper compares and analyzes the transient waveform characteristics of both the experiment and the simulation, extracts key parameters, and constructs probability density functions. The comparison results reaffirm the model's high-precision fitting ability. The statistical distribution pattern of VFTO single-pulse characteristics has been clarified. The research findings offer a more convenient approach for studying the transient processes caused by disconnector operation, reducing the human and material resources consumed by actual measurements and testing analysis, making the process safer and more economical.

Author Contributions: Methodology, J.X. and X.D.; Data curation, Z.G., X.Z., B.M. and Z.L.; Writing—original draft, Y.L.; Writing—review & editing, B.L. and Y.L.; Supervision, X.D. All authors have read and agreed to the published version of the manuscript.

Funding: This project is funded by the Science and Technology project of the State Grid Corporation (5700-202420251A-1-1-ZN).

Data Availability Statement: The original contributions presented in this study are included in the article. Further inquiries can be directed to the corresponding author.

Conflicts of Interest: Authors Zhenyu Guo, Xueyou Zhang were employed by the company State Grid Anhui Electric Power Co., Ltd. The remaining authors declare that the research was conducted in the absence of any commercial or financial relationships that could be construed as a potential conflict of interest.

References

1. Chen, W.; Zhao, J.; Bian, K.; Wang, D.; Wan, B.; Zhou, Z.; Zhang, J.; Teng, Z.; Zhang, X.; Li, Z. Research progress of transient electromagnetic harassment in GIS substation switch operation. *Chin. J. Electr. Eng.* **2019**, *39*, 4935–4948+4996.
2. Gao, F.; Tan, X.; Liang, Z. Experimental study of specific rapid transient overvoltage in 550 kV GIS. *High-Volt. Electr. Appl.* **2018**, *54*, 132–136.
3. He, T.; Gao, Y.; Liu, X.; Zhang, R. Simulation calculation and inhibition measures of VFTO in 550 kV GIS substation. *J. Electr. Power Syst. Autom.* **2022**, *34*, 65–70+81.
4. Zhao, X.; Wu, G.; Yang, D.; Xu, G.; Xing, Y.; Yao, C.; Abu-Siada, A. Enhanced detection of power transformer winding faults through 3D FRA signatures and image processing techniques. *Electr. Power Syst. Res.* **2025**, *242*, 111433. [CrossRef]
5. Xiao, G.; Rong, Q.; Yang, M.; Xiao, P.; Chen, Q.; Fan, J.; Guo, H.; Wang, H. Research on VFTO Identification of GIS Based on Wavelet Transform and Singular Value Decomposition. *Energies* **2022**, *15*, 3367. [CrossRef]
6. Zhao, L.; Ye, L.; Zhan, H.; Li, Z.; Duan, S. Study on the whole breakdown simulation model of GIS disswitch rapid transient overvoltage test. *High-Volt. Electr. Appl.* **2020**, *56*, 130–138.
7. Li, Z.; Shen, J.; Li, Z.; Tong, Y.; Wu, L. Study on isolating switch arc model and conduction interference to Rowgowski coil current transformer. *Power Syst. Prot. Control* **2020**, *48*, 131–139.
8. Chen, W.; Li, Z.; Sun, G.; Dai, M.; Liu, W.; Li, C.; Wang, L.; Wang, H.; Chen, G.; Yao, T.; et al. Experimental study on the particularly rapid and transient overvoltage characteristics in UHV gas insulated switchgear. *Chin. J. Electr. Eng.* **2011**, *31*, 38–47.
9. Dai, M.; Gu, D.; Sun, G.; Zhou, P.; Yao, T.; Chen, H.; Wan, L.; Pang, Q.; Zou, X.; Chen, J. Study on the test circuit of special rapid and transient overvoltage of UHV gas insulated switchgear. *Chin. J. Electr. Eng.* **2011**, *31*, 28–37.
10. Wu, H.; Jiao, C.; Cui, X.; Liu, X.; Ji, J. Simulation experiment and analysis of electromagnetic harassment of secondary equipment port caused by switch operation of GIS substation. *High Volt. Technol.* **2017**, *43*, 3387–3395.

11. Dick, E.; Fujimoto, N.; Ford, G.; Harvey, S. Transient ground potential rise in gas-insulated substations—problem identification and mitigation. *IEEE Trans. Power Appar. Syst.* **2007**, *PAS-101*, 3610–3619. [CrossRef]
12. Yue, G.; Liu, W.; Chen, W.; Guan, Y.; Li, Z.; Wang, H. Measurement Methods of Very Fast Transient Overvoltage in Gas Insulated Switchgear with Complete Process. *Proc. CSEE* **2011**, *31*, 18–27.
13. Zhang, W.; Hu, Y. Research Progress in Suppression of Fast Transient Overvoltage VFTO in GIS. *High Volt. Eng.* **2023**, *49*, 418–431.
14. Guardado, J.L.; Maximov, S.G.; Melgoza, E.; Naredo, J.; Moreno, P. An Improved Arc Model Before Current Zero Based on the Combined Mayr and Cassie Arc Models. *IEEE Trans. Power Deliv.* **2005**, *20*, 138–142. [CrossRef]
15. Xu, J.; Wang, L.; Li, S. Simulation and Analysis of Enclosure Voltage Transient Characteristic Caused by Disconnector Operation in Ultra High Voltage Gas Insulated Switch. *High Volt. Technol.* **2012**, *38*, 288–294.
16. Zheng, J.-L.; Zhan, H.-M.; Duan, S.-F.; Ouyang, Z.; Zhao, C.-N. Spark model for 1100 kV GIS disconnecting switch. In Proceedings of the 2011 1st International Conference on Electric Power Equipment—Switching Technology, Xi'an, China, 23–27 October 2011.
17. Zhao, L.; Ye, L.; Wang, S.; Zhan, H.; Yang, Y.; Jiang, P. Repeated Strike Process Modeling of Very Fast Transient Overvoltage Induced by Disconnector Switch Operation in 252 kV GIS. In Proceedings of the 2018 IEEE 3rd International Conference on Integrated Circuits and Microsystems (ICICM), Shanghai, China, 24–26 November 2018.
18. Yan, G.; Wu, X.; Tian, Y.; Zhang, H. Review on the Research of Electromagnetic Transient Interference Caused by Arc Discharge Due to Switchgear Operation. *High Volt. Electr. Appl.* **2014**, *50*, 119–130.
19. Shu, Y.; Han, B.; Lin, J.; Chen, W.; Ban, L.; Xiang, Z.; Chen, G. Influence of the Switching Speed of the Disconnector on Very Fast Transient Overvoltage. *IEEE Trans. Power Deliv.* **2013**, *28*, 2080–2084.
20. *33/19-09 C W*; Very Fast Transient Phenomena Associated with Gas Insulated Substations. CIGRE: Paris, France, 1988.
21. Li, Z.; Liao, X.; Tong, Y.; Chen, X. Simulation and characteristic analysis of VFTO and VFTC based on dynamic re-burning arc model. *Power Syst. Prot. Control* **2023**, *51*, 79–88.
22. Kim, D.E.; Lee, G.Y.; Kil, G.S.; Kim, S.-W. Trends in Measuring Instrument Transformers for Gas-Insulated Switchgears: A Review. *Energies* **2024**, *17*, 1846. [CrossRef]
23. Zhao, X.; Zhang, Y.; Xing, Z. Short-term load interval prediction based on optimal window-wide kernel density estimation. *Electr. Meas. Instrum.* **2019**, *56*, 56–61.
24. Zhang, Y. Comparative Study of the Probability Density Function Estimation Based on Different Kernel Functions. Master's Thesis, Hebei University, Hebei, China, 2010.
25. Sima, W.; Zhang, H.J.; Yang, M.; Wang, H.; Xu, J.; Huang, Y. Statistical analysis of overvoltage amplitude and waveform characteristics in 10 kV system of a substation. *High Volt. Technol.* **2021**, *47*, 214–222.

energies

MDPI

Article

Advanced Non-Unit Protection Strategy for MMC-HVDC Grids Leveraging Two-Dimensional High-Frequency Characteristics via HHT and SVM

Chenglin Ren [1,*], Yige Hu [2], Xiang Lu [1], Yue Chen [1], Kepeng Xia [3] and Hailin Li [4]

1 China Southern Power Grid Co., Ltd., Ultra High Voltage Transmission Company, Guangzhou 510700, China; dianqi_lu@126.com (X.L.)
2 School of Electrical and Information Engineering, Zhengzhou University, Zhengzhou 450001, China; jshmeet2002@163.com
3 XJ Electric Co., Ltd., Xuchang 461000, China
4 College of Electrical and Information Engineering, Zhengzhou University of Light Industry, Zhengzhou 450000, China; hailinli@zzuli.edu.cn
* Correspondence: rencl2013@126.com

Abstract: The rapid development of direct current (DC) grids poses significant challenges to the speed, reliability, and selectivity of fault protection systems. These systems are required to identify and distinguish between internal and external faults despite the constraints of limited information and time. This study introduces a non-unit protection scheme based on the classification of two-dimensional feature parameters utilizing the Hilbert–Huang transform (HHT) and a support vector machine (SVM). Through time–frequency analysis of the voltage waveform following DC faults, critical information within the high-frequency component of the fault voltage, specifically, the instantaneous frequency and amplitude of the wavefront, is extracted to distinguish internal from external faults. Two-dimensional feature parameters are associated with signal attenuation and distortion during fault propagation via the transmission path, thereby providing a foundation for precise fault identification. The employment of an SVM ensures the selectivity of this scheme without relying on protection settings. The efficacy of the scheme is validated through simulations conducted using PSCAD/EMTDC across various fault scenarios.

Keywords: non-unit protection scheme; time–frequency analysis; fault detection; Hilbert–Huang transform; support vector machine

1. Introduction

DC grids based on modular multilevel converters (MMCs) are vital for high-power long-distance electricity transmission, enabling the seamless integration of new energy sources on a wide scale [1,2]. However, the low inertia and weak damping characteristics of flexible DC systems render them susceptible to severe fault currents [3]. The protection and restoration of DC faults are crucial for reliable operation and widespread development of DC grids [4,5]. To avoid damage to critical equipment within a flexible DC grid owing to DC faults, it is generally necessary to achieve fault clearance within 5 ms [6]. Therefore, fault detection and identification techniques for DC grids must shorten detection time and improve response sensitivity based on traditional detection and identification techniques [7,8].

The primary safeguard is often provided by local data to prevent transmission delays [9–11]. In their study, Chengyu Li et al. [12] provide a rapid approach for detecting short-circuit faults in HVDC grids. This approach relies on accurate measurement of the voltage across the current-limiting reactor under DC faults. Wang et al. [13] proposed a fault identification method based on the transient voltage square difference, which is used to identify faulted lines in an overhead line DC grid. Li et al. [14] proposed a protection scheme for multiterminal direct current (MTDC) systems in offshore wind farms. This scheme focuses on the detection of high-frequency components within fault current signals, enabling effective classification of fault categories and precise fault detection in every line. Traveling wave protection based on time-domain characteristics mainly uses numerical features, such as the voltage change rate and current change rate, to form criteria for identifying faults inside or outside the area [15,16]. Nevertheless, the rate of change and variation in the DC voltage and current significantly depend on the impedance of the long faulty line and grounding resistance [17]. Determining high-resistance faults and faults located at a great distance from an external fault using time-domain characteristics can be challenging [18]. The selectivity and dependability of methods that rely on them are constrained.

After a DC line fault occurs, a rapid change in the current can lead to a rapid change in the electromagnetic field, thereby triggering high-frequency voltage components [19]. Although excessive resistance may cause attenuation and distortion of high-frequency signals [20], its impact on high-frequency components is relatively small. Thus, high-frequency components can accurately indicate the propagation path of a DC fault [21]. By monitoring and analyzing the high-frequency components of surge voltages, it is possible to distinguish between internal and external faults and further determine the location and type of the fault [22–24]. Sabug Jr et al. [25] proposed a real-time boundary wavelet transform (RT-BWT) method to protect against DC faults in MTDC grids. Jayamaha et al. [26] utilized wavelet transform to capture the characteristic variations in current signals induced by network defects by sampling measured branch currents. Classification was performed using artificial neural networks, and the results demonstrated that the proposed method is capable of detecting and localizing defects quickly and independently. An approach that utilizes frequency-domain analysis substantially enhances selectivity. However, the complexity of the process of selecting the wavelet basis function reduces WT adaptability.

The Hilbert–Huang transform (HHT) developed by Norden E. Huang offers a powerful time–frequency analysis solution for analyzing nonlinear and nonstationary signals [27,28]. In [29], the performances of wavelet transform and Hilbert–Huang transform (HHT) in extracting fault features of high-voltage direct current transmission systems were compared. It was pointed out that wavelet transform requires repeated debugging of the basis functions and has a weak effect in extracting DC fault features. Compared with wavelet transform, HHT, as an adaptive time–frequency analysis method, does not require preset basis functions. It can adaptively extract signal modes through empirical mode decomposition. The fault features extracted by it are more obvious and can dual-locate the start and end points of faults. In terms of signal adaptability, HHT does not require preset basis functions and is adapted to the nonlinear characteristics of DC faults. In terms of time–frequency resolution, HHT can adaptively adjust to accurately capture transient high-frequency components. In terms of anti-noise ability, the combination of HHT and the improved technology has a better effect in suppressing modal aliasing and noise. In terms of the applicability of DC power grids, HHT is specifically designed for non-stationary signals and has a stronger ability to distinguish high-frequency characteristics of DC faults. Li et al. [30] examined the viability of using HHT to identify faults in HVDC systems based on voltage source converters (VSC). Feature extraction is achieved by using HHT in [31–33], and corresponding fault detection methods are proposed for HVDC systems. However, there is a dearth of studies on the use of HHT to safeguard DC grids from faults. In this study, the equivalent impedance values of DC fault voltage propagation paths at different locations were compared at different frequencies. It was found that the high-frequency components of the internal and external DC fault voltages are different. By employing the HHT to decompose the DC surge voltage, the high-frequency component's frequency-domain characteristics of the voltage under different fault locations were analyzed. This method enables the extraction of the instantaneous frequency and amplitude, providing essential characteristic quantities for fault detection. The effects of fault distance, fault ground resistance, and the current-limiting reactor (CLR) on the frequency domain characteristics were then studied.

With these characteristic quantities, a fault-detection strategy for DC fault identification and classification in MTDC systems based on support vector machines (SVMs) is proposed. As a supervised learning algorithm, a support vector machine (SVM) maps input features to a high-dimensional space through kernel functions and can effectively achieve the linear partitioning of complex nonlinear patterns. This characteristic makes it particularly suitable for the fault identification task in this paper, because the instantaneous frequency and amplitude features extracted through the Hilbert–Huang transform (HHT) present complex nonlinear correlations between internal faults and external faults. This fault detection technique does not rely on telecommunication and demonstrates robust resistance to interference as well as excellent sensitivity and selectivity in identifying faults that are far away and problems with high-resistance grounding, making it a superior choice for reliable fault detection.

The remainder of this paper is organized as follows. Section 2 investigates the frequency-domain characteristics of the DC voltage in the context of DC failures. A DC fault detection approach utilizing support vector machines (SVMs) is presented in Section 3. Section 4 examines the elements that determine high-frequency features and subsequently tests the proposed solutions using simulations in PSCAD/EMTDC. Section 5 provides concluding remarks.

2. Transient Voltage Characteristics of Internal and External Faults

Figure 1 illustrates a typical standard topology of a four-terminal HVDC grid. Four modular multilevel converters (MMCs) are interconnected by high-voltage DC cables. Each cable connection is equipped with series DC circuit breakers (CBs) at both ends, which are crucial for the rapid isolation of faulty sections to avert cascading failures. Furthermore, current-limiting reactors (CLRs) are connected in series with the circuit breakers, effectively suppressing the rapid rise of fault currents during internal line faults. The relay, connected in series with the reactor and positioned on the inner side of the line, measures local signals to detect faults. In the event of an internal fault, it provides tripping signals to the circuit breakers on the same line. Regarding DC line fault detection based on local measurements by the protective relay, the challenge lies in differentiating between internal and external faults, particularly when the faults originate from the same direction.

Existing time-domain feature-based fault-detection schemes employ current and voltage derivatives to distinguish internal faults. When the value of the extracted feature exceeded the threshold value, an internal fault was considered. However, owing to the rapid attenuation of fault waves, selectivity cannot be ensured when distinguishing between internal high-resistance faults, internal far-terminal faults, and external metallic faults. The attenuation and distortion of the traveling waves under the combined effect of the grounding resistance, line impedance, and current-limiting reactor makes fault detection difficult.

However, it should be noted that the influences of the grounding resistance, line impedance, and current-limiting reactor on fault-traveling waves are different in the frequency domain. The distributed parameters of the DC line change at different frequencies, whereas the current limiter reactor exhibits an obvious filtering effect on the high-frequency components. Hence, it is important to investigate the transient properties of high-frequency elements in both internal and external problems. In pursuit of this objective, this section examines the frequency characteristics of voltage-traveling waves during DC faults and suggests techniques for extracting transient frequency-domain attributes using HHT.

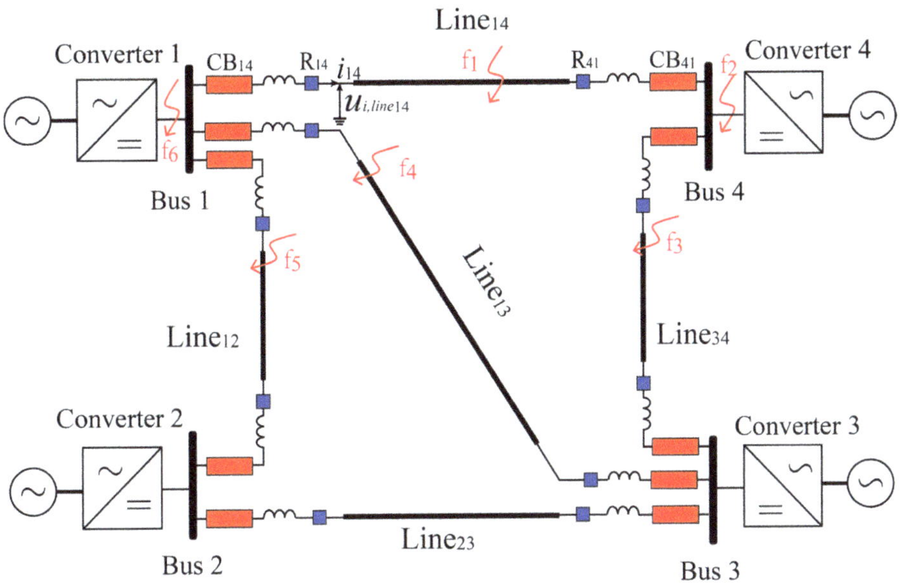

Figure 1. The four-terminal flexible DC grid.

2.1. Frequency-Domain Response

Taking any line $Line_{ij}$ in the DC grid as an example, as depicted in Figure 2, when a DC line fault occurs, the voltage wave propagates along the transmission line and is subsequently recorded by relay A at the measurement point in the head of the line. The fault identification result dictates the operation of the circuit breaker located at the line head. Faults occurring within the reactors on both sides of the line are classified as internal faults, whereas those occurring beyond these boundaries are categorized as external faults.

The propagation process of the impulse voltage generated at the DC fault point to relay A can be divided into two parts: (1) an additional voltage source is generated at the grounded fault point and (2) the voltage traveling wave propagates to the detecting point along the transmission line. The incident traveling voltage wave measured by relay A can be formulated as follows:

$$U_{\mathrm{a}}(s) = U_{\mathrm{f}}'(s)e^{-r(s)l} \tag{1}$$

where $U_{\mathrm{f}}'(s)$ denotes the additional voltage at fault point. To simplify the calculation, the waveform distortion, attenuation, and propagation delay information are decoupled as follows:

$$A(s) = e^{-r(s)l} \approx \frac{K_{\mathrm{a}}}{1+sT_{\mathrm{a}}}\mathrm{e}^{-\frac{sl}{v}} \tag{2}$$

where $e^{-r(s)l}$ is the time delay of traveling wave propagation, v is the maximum speed of the traveling wave within the considered frequency range, l is the propagation distance, $\frac{K_{\mathrm{a}}}{1+sT_{\mathrm{a}}}$ represents the transmission attenuation and distortion caused by dispersion, K_{a} is the attenuation ratio, which represents the ratio of the amplitude of the traveling wave with the propagation distance l and the initial amplitude, and T_a is the dispersion time constant, of which the denominator characterizes the distortion degree of the traveling wave with propagation distance l relative to the initial traveling wave.

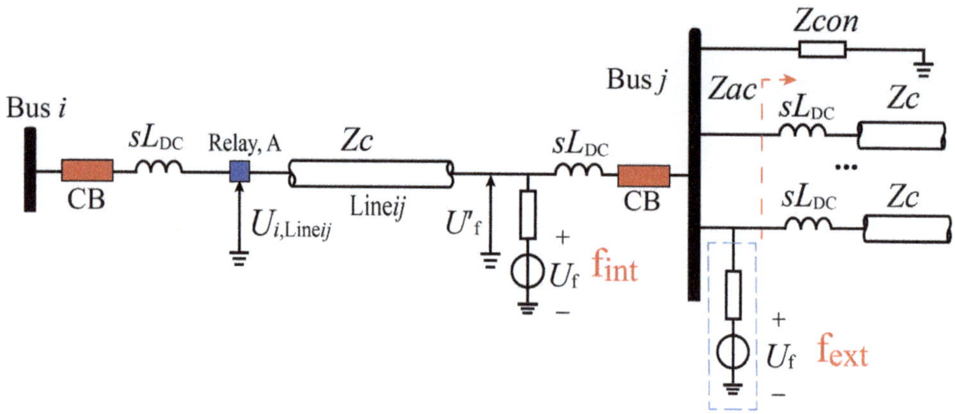

Figure 2. The equivalent fault circuits of the internal and external faults in the multiterminal DC grid.

The additional voltages under internal and external faults are different because of the different equivalent impedances at the grounded point, as shown in Figure 2. The voltage $U'_{\mathrm{f,int}}$ under internal conditions is expressed as follows:

$$U'_{\mathrm{f,int}}(s) = \frac{Z_c//(sL_{DC}+Z_{con}//Z_{ac})}{R_f+Z_c//(sL_{DC}+Z_{con}//Z_{ac})}U_f \tag{3}$$

where U_f is the rated DC voltage with reverse polarity, Z_c is the wave impedance of the DC line, L_{DC} is the value of the current-limiting reactor, and R_f is the transition resistor. Z_{con} denotes the equivalent impedance of the converter. When a half-bridge (HB) MMC is used in the DC grid, the capacitance discharge contributes to the fault current at the initial stage before the converter is blocked under DC faults. Thus, the converter is equivalent to a series RLC model, which is represented by the following:

$$Z_{con} = R_{eq} + sL_{eq} + 1/sC_{eq} \tag{4}$$

Z_{ac} represents the equivalent impedance of the lines adjacent to the fault line, which can be computed as follows:

$$Z_{ac} = Z_c/n + L_{DC}/n \tag{5}$$

where n is the number of the adjacent lines.

The voltage under a nearby external solid fault is

$$U'_{\mathrm{f,ext}}(s) = \frac{Z_c}{Z_c+sL_{DC}}U_f \tag{6}$$

Using Equations (1), (3), and (6), the ratio of the incident voltage measured at the monitoring point under an internal fault and nearby external fault can be calculated by K_i/K_o, where

$$K_i = \frac{Z_c//(sL_{DC}+Z_{con}//Z_{ac})}{R_f+Z_c//(sL_{DC}+Z_{con}//Z_{ac})} \tag{7}$$

$$K_o = \frac{Z_c}{Z_c+sL_{DC}} \tag{8}$$

As can be seen from Equations (7) and (8), the values of K_i and K_o will change with frequency, because parameters such as Z_c, Z_{con}, Z_{ac}, and sL_{DC} all vary with frequency. Taking the cable in a DC grid as an example, the amplitude of the wave impedance Z_c at different frequencies can be obtained, as shown in Figure 3. The frequency response, in terms of the magnitude of the transfer function for the internal and external faults, was then computed, as shown in Figure 4. It can be observed that in the case of an external fault, the voltage attenuates much more rapidly in the high-frequency region. From the perspective of the frequency domain, the fault travel waves caused by internal and external faults show obvious differences when passing through the current-limiting reactor because of the inhibition effect of the reactor on the high-frequency components of the travel wave. Current-limiting reactors have also become natural barriers for DC fault protection. The principle behind the proposal of highly selective protective measures is to incorporate this boundary characteristic.

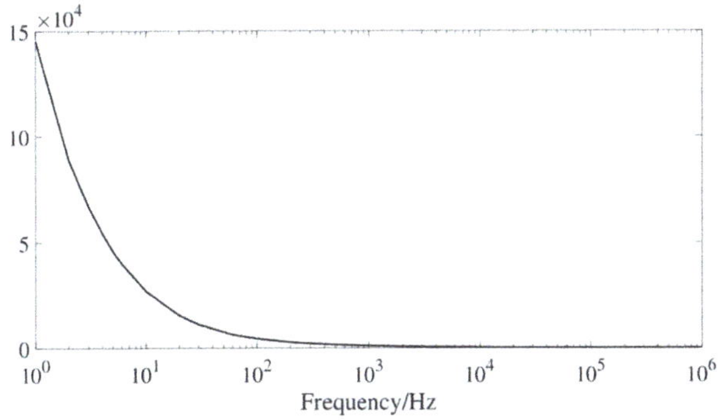

Figure 3. The amplitude of the wave impedance with different frequencies.

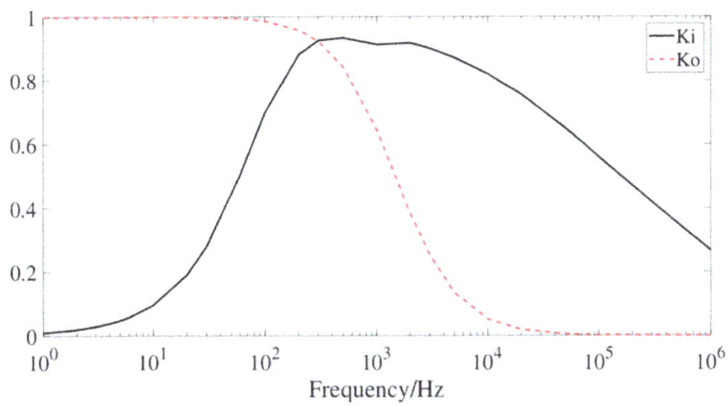

Figure 4. The values of K_i and K_o with different frequencies.

2.2. CEEMDAN

Based on the aforementioned disparity in high-frequency impedance along the travel path during internal faults compared with external faults, it can be anticipated that the transient features of the high-frequency components of the DC fault voltage will vary between internal and external faults.

Before extracting transient features from the high-frequency component, the DC voltage traveling wave was decomposed into several single-frequency components. Improved Empirical Mode Decomposition (EMD) and Complete Ensemble Empirical Mode Decomposition with Adaptive Noise (CEEMDAN) were used in this study. This method adds multi-white noise to the signal being processed and calculates the average intrinsic mode function (IMF) by summing all the decomposition results after each EMD of the signal with white noise. The influence of noise was canceled by multiple superpositions and averaging. Once the first-order IMF component is calculated, white noise is included in the residual value. Subsequently, the average value of the next-order IMF component was computed using the same calculation procedure. This process was repeated until all the components were calculated by multiple iterations. Compared to EMD, CEEMDAN has a faster computing speed and more completely decomposed signals. Moreover, unnecessary low-frequency components were reduced, which contributed to the rapid analysis of the frequency characteristics of the traveling waves.

The DC voltage on the line side of $L_{\mathrm{DC},ij}$, denoted as $U_{i,\mathrm{Line}}$, undergoes decomposition into several IMFs, each characterized by a specific frequency, using the CEEMDAN method. The expression for $U_{i,\mathrm{Line}}$ can be formulated as Equation (9):

$$U_{i,\mathrm{Line}} = \sum_{k=1}^{n} c_k(t) + r_n(t) \tag{9}$$

where $c_k(t)$ is the order k of the IMF component and $r_n(t)$ is the remainder.

In the DC grid illustrated in Figure 1, when an internal DC grounding fault occurs at a distance of 30 km from the head end of Line_{14}, the voltage measured by relay R_{14} is processed using the CEEDMAN method. The resultant outcome is presented in Figure 5. Simultaneously, the frequency spectrum of each signal was analyzed using a fast Fourier transform (FFT). IMF_1 contains the highest frequency component, and the frequency of the subsequent IMFs progressively decreases. The transient characteristics of the IMF are significantly influenced by the varying effects of the grounding resistance, CLR, and distribution parameters of the transmission line. These variations can significantly affect the behavior of high-frequency signals under different fault conditions. Consequently, the key features related to the most important information regarding the fault can be extracted from IMF_1.

To improve the calculation speed and reduce the number of calculations, a fault identification criterion was set up to detect the wavefront of the traveling wave. Upon arrival of the fault voltage at the relay, the wavefront of the voltage traveling wave within milliseconds was decomposed by CEEMDAN for feature extraction. As illustrated in Figure 6, under fault conditions identical to those in Figure 5, only the first 1 ms of post-fault voltage information is analyzed through time–frequency analysis. By narrowing the time window to 1 ms—a duration sufficiently short to ensure rapid fault detection—IMF_1 retains the high-frequency characteristics of the DC voltage, as shown in Figure 6. This highlights the method's capability to capture transient fault signatures within ultra-short intervals, thereby achieving an effective balance between speed and accuracy in fault diagnosis.

During fault identification, two specific types of faults pose significant challenges in differentiation: long-distance high-resistance faults and severe external faults. As illustrated in Figure 7, an internal high-resistance fault occurs at the end of Line_{14} with a high grounding resistance. In contrast, Figure 8 depicts an external metallic grounding fault located beyond the reactor at the line terminal. Given the extensive propagation distance, the fault voltage experiences considerable attenuation and distortion before reaching the relay R_{14}. As a result, the original DC voltage waveforms presented in Figures 7 and 8 demonstrate comparable rates of decline and similar variation trends in their time-domain profiles. This similarity renders traditional time-domain features, such as voltage amplitude decay, inadequate for reliable fault discrimination. However, the contents of the high-frequency components under internal high-resistance faults and external faults are different. Moreover, the derivative of IMF_1 at the moment the fault occurs under an external fault is decreased, from which adequate information about the fault can be extracted.

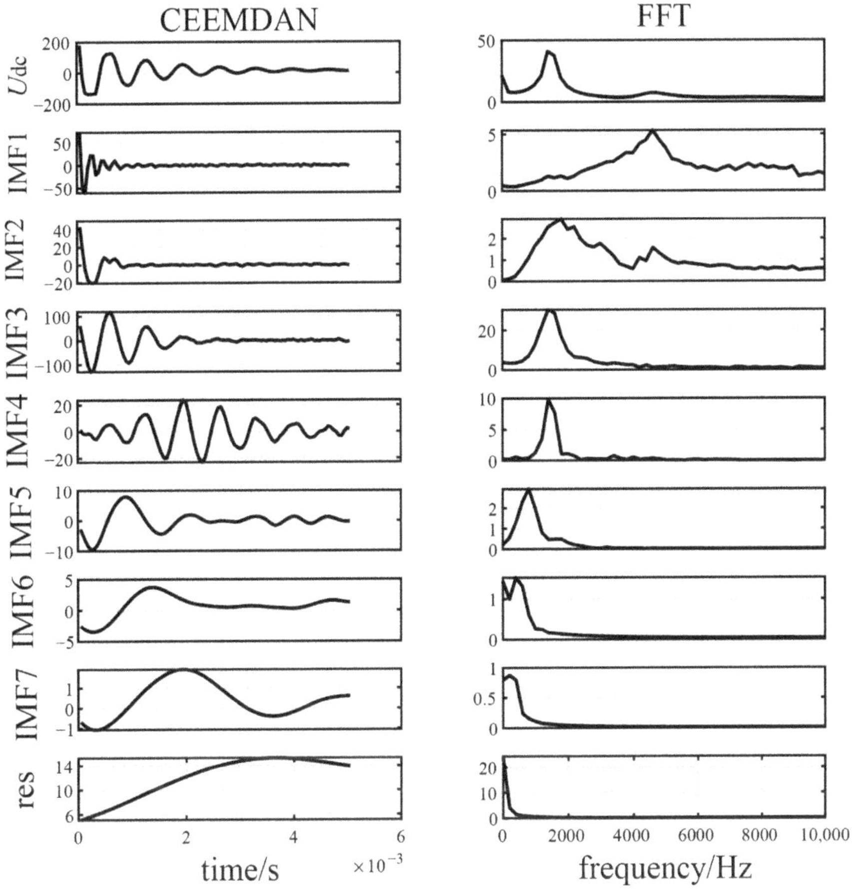

Figure 5. CEEMDAN of DC voltage under internal fault.

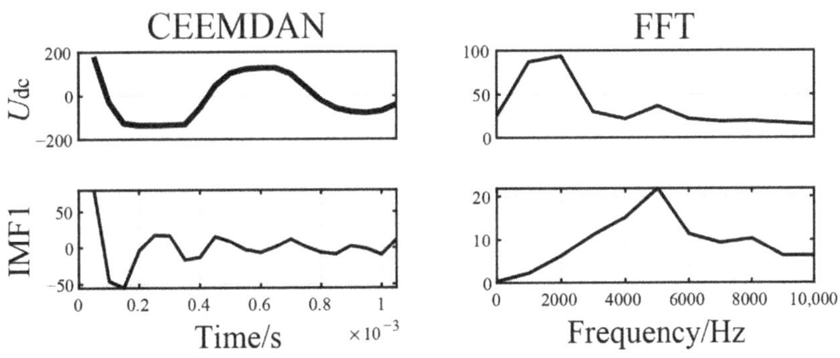

Figure 6. CEEMDAN of DC voltage under internal fault in 1 ms time window.

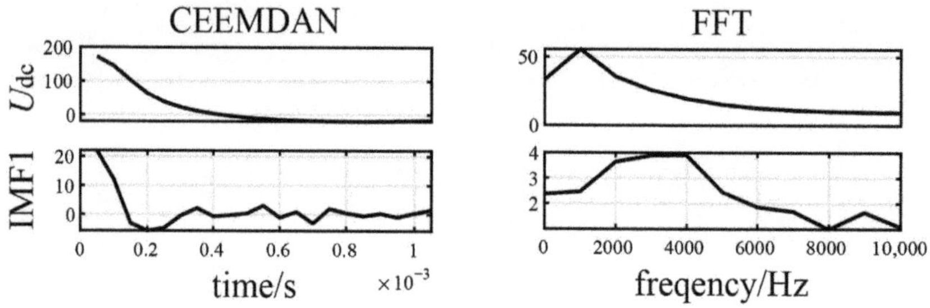

Figure 7. CEEMDAN of DC voltage under internal high-resistance fault.

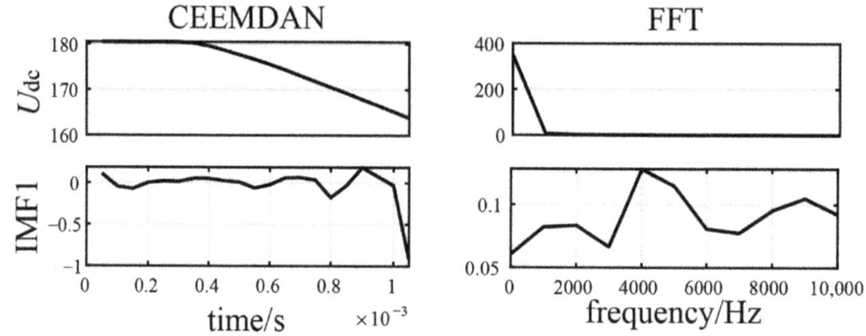

Figure 8. CEEMDAN of DC voltage under external fault.

2.3. The Hilbert Transform

The Hilbert transform (HT) of the continuous signal $x(t)$ is defined as Equation (10).

$$y(t) = \frac{1}{\pi} P \int_{-\infty}^{+\infty} \frac{x_{\delta}(\delta)}{t-\delta} \mathrm{d}\delta \tag{10}$$

Signals $y(t)$ and $x(t)$ exhibit a complex relationship, facilitating the derivation of the analytical signal $z(t)$ through the subsequent procedure.

$$z(t) = x(t) + i \cdot y(t) = a(t) \cdot e^{i\theta(t)} \tag{11}$$

where $a(t)$ and $\theta(t)$ are the instantaneous amplitude and phase, respectively, which can be calculated as

$$a(t) = \sqrt{x^2(t) + y^2(t)}, \theta = arctan\frac{y(t)}{x(t)} \tag{12}$$

The instantaneous frequency can be computed as

$$f(t) = \frac{1}{2\pi} \frac{\mathrm{d}\theta(t)}{\mathrm{d}(t)} \tag{13}$$

As shown in Figure 9, the Hilbert spectrum of IMF_1 (from the 1 ms time-window analysis in Figure 6) corresponds to an internal fault occurring on Line_{14}. The spectrum clearly illustrates the time-varying distribution of instantaneous frequencies and amplitudes under the specified fault condition. At 0 s, the moment the fault occurs, the instantaneous frequency and amplitude are at an extreme point because the voltage signal at the position of the traveling wave head decreases sharply at the moment of failure. As the voltage traveling wave propagates through the DC transmission lines, the wavefront attenuates gradually, whereas the steepness decreases. This change can be quantitatively characterized by the instantaneous frequency and amplitude of the high-frequency component. The extreme points in the Hilbert spectrum near 0 s are referred to as f_0 and a_0. Predictably, f_0 and a_0 change with changes in the fault position, grounding resistance, and inductance of the RLC.

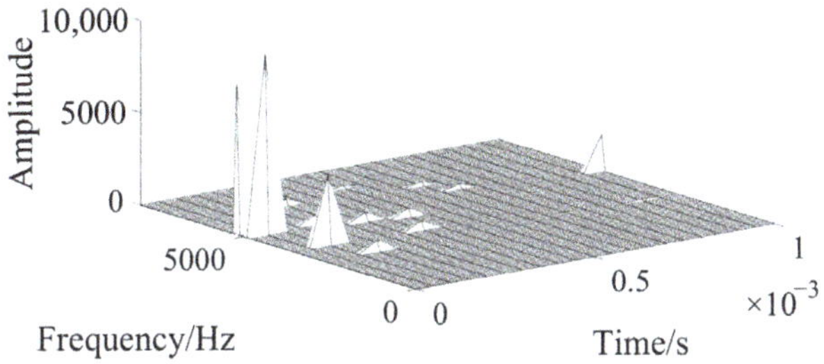

Figure 9. Hilbert spectrum of IMF_1.

3. The Proposed Fast Single-Ended Protection Scheme

By extracting the instantaneous high-frequency features (f_0, a_0) with the HHT, a selective fault detection approach is proposed. To avoid the disadvantages of the fault detection criterion with a threshold, a classification based on an SVM was employed in this study to distinguish between internal and external faults.

3.1. The SVM Classification

A support vector machine (SVM) is a binary classification model utilized in machine learning. It is adept at managing both linearly separable and linearly non-separable data, facilitating the construction of optimal decision boundaries within a high-dimensional space. As shown in Figure 10, a line can be drawn for a dataset to separate two sets of data points; such data are called linearly separable. The lines that separate the dataset are called the segmentation hyperplanes. The equation for this line in two dimensions is $y = ax + b$, and when generalized to n dimensions, it becomes a hyperplane equation $f(x) = w^T x + b$. The support vector (SV) refers to the collection of points closest to the hyperplane.

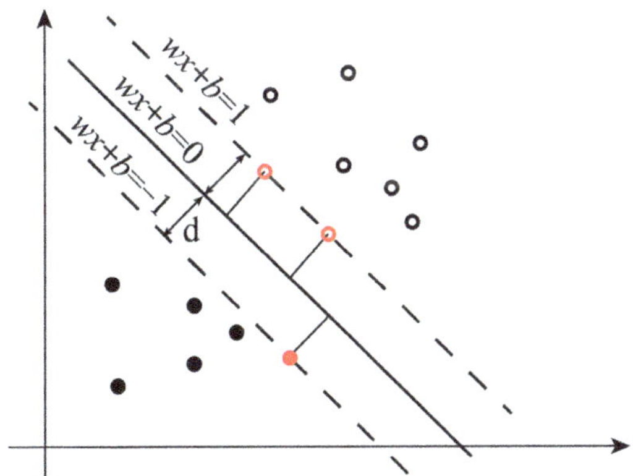

Figure 10. SVM classification.

The purpose of an SVM is to find a hyperplane such that the sample is divided into two classes, and the space between the points and the hyperplane is the largest. Equation (14) represents a random hyperplane.

$$w^T x + b = 0 \tag{14}$$

The distance from each support vector to the hyperplane can be expressed as

$$d = \frac{|w^T x + b|}{\| w \|} \tag{15}$$

In a binary classification problem, y_i represents the label x_i, which has a value of 1 or −1. For any class $x_i, y(w^T + b) > 0$. Therefore, $y(w^T + b) = |w^T + b|$.

The maximization of distance d can be expressed as an optimization problem in Equation (16).

$$\max 2 \times \frac{y(w^T x + b)}{\| w \|} \tag{16}$$

For ease of derivation, $||w||d$ is assumed to be 1. Thus, vector $y\,(w^T x + b) = 1$. The optimization problem is given by Equation (17).

$$\max \frac{2}{\| w \|} \tag{17}$$

For calculation convenience, the formula is written as

$$\min \frac{1}{2} \| w \|^2 \text{ s.t. } y_i(w^T x_i + b) \geqslant 1 \tag{18}$$

3.2. *Proposed Detection Method by HHT-SVM*

The analysis of high-frequency characteristics in fault voltage shows that the instantaneous frequency (f_0) and amplitude (a_0) of the DC voltage signal on the line side of CLR are crucial indicators for identifying internal versus external faults. The existence of current-limiting reactors at both ends of a DC line results in obvious changes in the high-frequency components f_0 and a_0 after fault voltage propagation. Thus, in a space constructed with the abscissa f_0 and ordinate a_0, the points of the "inside" and "outside" types are clustered in different regions. A fault detection method for DC fault classification was proposed by segmenting the space using an SVM.

As shown in Figure 11, this study utilized a large amount of data obtained from simulations of internal and exterior faults, which vary in terms of fault lengths and transition resistance. The purpose of these data was to train an SVM classification model. The remaining data were utilized to thoroughly assess the accuracy of the SVM classification.

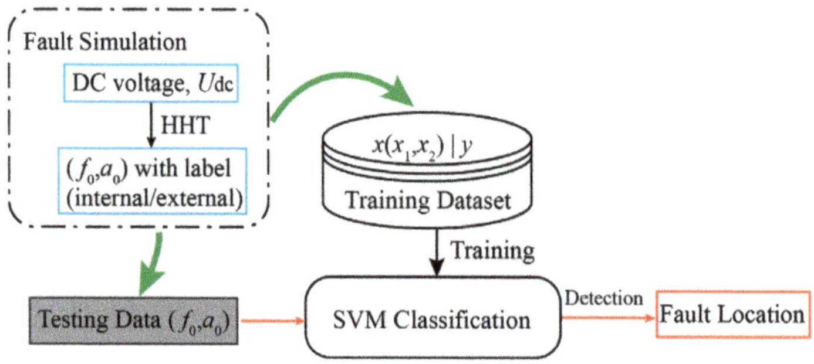

Figure 11. Training and testing of SVM classification.

3.3. *The Protection Algorithm*

3.3.1. Fault Identification Criterion

This technique uses local measurements of the DC current and voltage at one end of the DC line. To minimize the computational workload, the DC voltage drop criterion, represented by Equation (19), is first used to determine the presence of a DC fault in the DC grid. Subsequently, the alteration in the direction of the DC current signal is used to pinpoint the direction of the fault.

$$\max\left(\left|\Delta U_{i,\mathrm{Linep}}\right|, \left|\Delta U_{i,\mathrm{Linep}}\right|\right) > U_{\mathrm{req}} \tag{19}$$

The measured current flowed in a positive direction from the relay to the other terminal. Therefore, a forward fault was loaded in the positive direction of the fault current. The fault direction criterion was set as follows:

$$\frac{\mathrm{d}I_{\mathrm{dc},ij}}{\mathrm{d}t} > K_{\mathrm{di}} \tag{20}$$

where K_{di} was set to a value slightly greater than zero. If it is a forward fault, the data of the following 1 ms wave-front will be recorded for transient frequency feature extraction by HHT. The time-frequency analysis was conducted based on PyCharm 2024.1.7.

3.3.2. Flowchart of the Proposed DC Protection Strategy

A flowchart of the proposed single-ended DC protection strategy is shown in Figure 12. The DC voltage and current were measured by the relay during the operation of the DC grid at the line side of the current-limiting reactor in series with the circuit breaker. Once the DC voltage drop reached the threshold value of the fault identification criterion, a DC-side fault was identified. The derivative of the fault current was then computed to determine the direction of the fault. A backward fault was noted as an external fault. For forward faults, the HHT-SVM-based fault detection starts to distinguish between internal and external faults. The 1 ms data at the front of the voltage traveling wave were recorded for feature extraction.

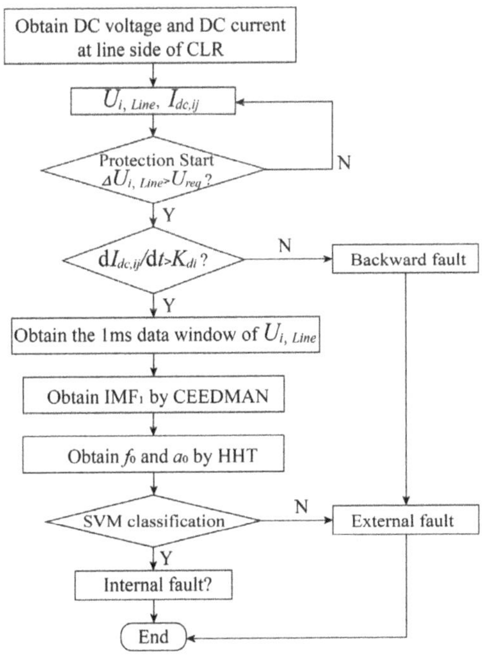

Figure 12. Flowchart of the proposed single-ended dc protection strategy.

4. Simulation Verification

As shown in Figure 1, a ±200 kV four-terminal symmetric monopole HVDC grid was built in the PSCAD/EMTDC 4.6.2 software environment. VSC stations employ widely recognized modular multilevel converters (MMCs). The cables that connect them use a frequency-dependent (phase) line model. Table 1 presents the parameters of the primary system.

Table 1. Converter and grid parameters.

	Conv. 1	Conv. 2–4
Rated capacity [MVA]	360	120
Rated DC voltage [kV]	±200	±200
Rated AC voltage [kV]	220	220
Operation Mode Setpoints	Constant DC voltage control ±200 [kV]	Constant active power control 100 [MW]

The DC circuit breakers (CB) were installed at both ends of each DC line. As shown in 1, taking CB_{14}, the circuit breaker at the head end of $Line_{14}$, as an example, obvious voltage and current changes will be observed in the circuit breaker when faults occur in the transmission line of the DC power network, such as f_1–f_5. To limit the fault area, only when the internal fault f_1 is accurately identified will a trigger signal be generated to cut off CB_{14}. The same is true for other circuit breakers.

4.1. *Influence Factors on the Transient Features*

The influences of the CLR, grounding resistance, and distributed parameters of the cable on the high-frequency components of the traveling waves are different. The different effects of these parameters coincide transparently with variations in instantaneous frequency and amplitude. Simulations were performed to study the relationship between the fault parameters and transient frequency characteristics.

4.1.1. Influence of Fault Distance

The fault distance l changes from 10 to 270 km, the grounding resistance R_g is 0, and CLR L_{DC} is 100 mH. The Hilbert spectrum of IMF_1 of the DC voltage under faults at different distances is shown in Figure 13.

The initial instantaneous frequency f_0 and amplitude a_0 demonstrated a clear decreasing trend as the fault distance increased. This relationship is illustrated in Figure 14. The instantaneous frequency and amplitude were quite high, with the fault located 10 km from the monitoring point. f_0 and a_0 decreased simultaneously owing to the effects of distributed resistance and inductance. When l changes from 10 to 270 km, f_0 and a_0 vary over a wide range.

4.1.2. Influence of Grounding Resistance

Simulations were performed with the grounding resistance R_g changed from 0 Ω to 100 Ω, while the fault distance l remained at 30 km and the CLR L_{DC} remained at 100 mH. The instantaneous frequencies and amplitudes for the different transition resistances are shown in Figure 15. It is evident that, as the transition resistance increases, the transient amplitude decreases rapidly, whereas the instantaneous frequency varies less.

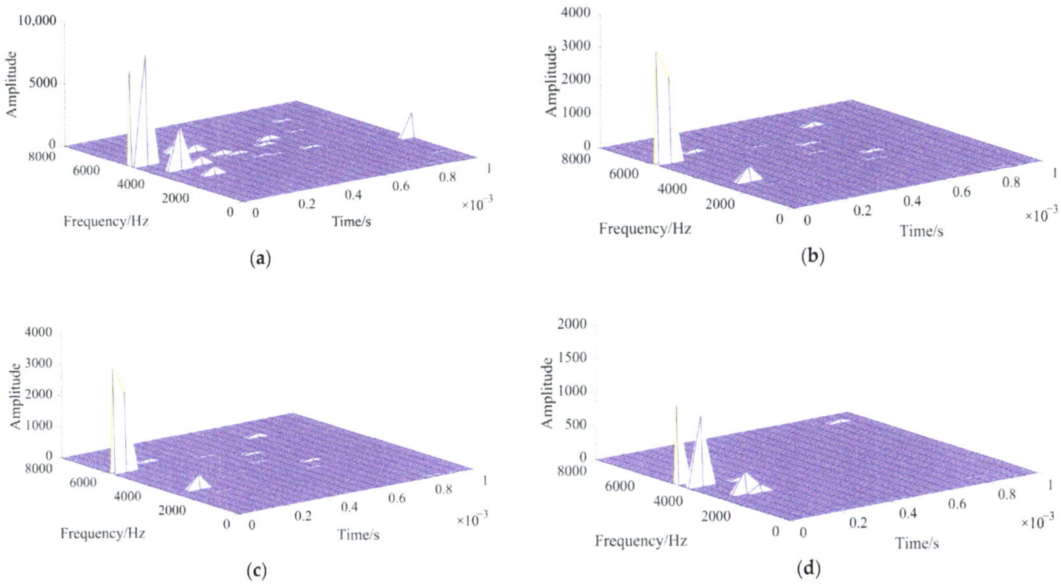

Figure 13. Hilbert spectrum under faults with different fault distance (**a**) l = 30 km, (**b**) l = 90 km, (**c**) l = 120 km, and (**d**) l = 240 km.

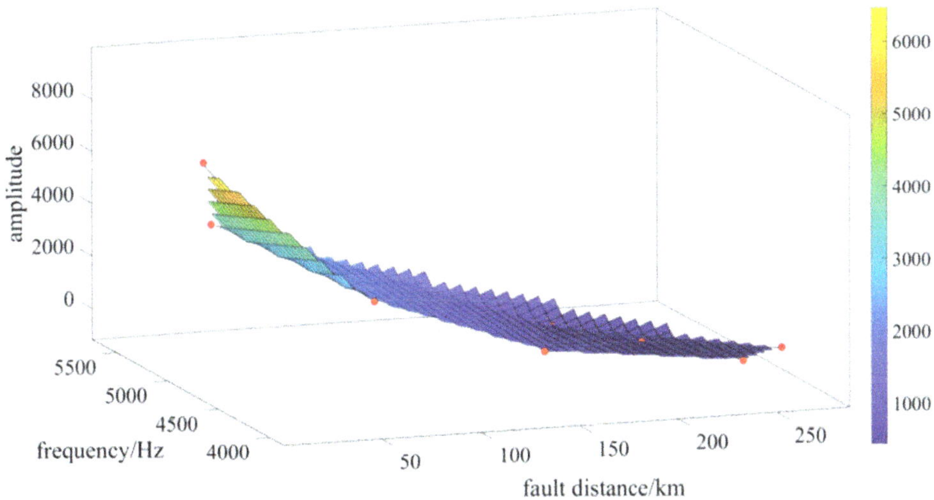

Figure 14. f_0 and a_0 under faults with different l.

4.1.3. Influence of the CLR

As mentioned above, when the voltage surge wave generated by an external fault travels through the current-limiting reactor at the end of the terminal, the high-frequency components are filtered out by the reactor. This boundary characteristic is the principle behind frequency-domain fault protection in a DC grid. The Hilbert spectrum of the voltage under the fault at Bus 4 is shown in Figure 16. It is evident that there are very few high-frequency components measured by the relay at the head terminal of $Line_{34}$. A comparison of high-frequency components with reactance changes is shown in Figure 17.

When the reactance of L_{DC} is reduced, the voltage change rate increases. The instantaneous frequency at the time of fault occurrence also increased.

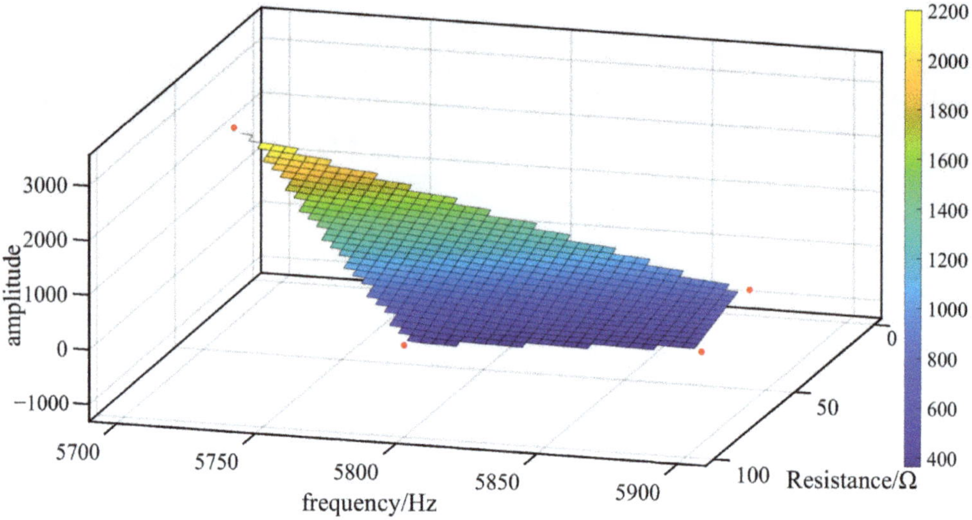

Figure 15. f_0 and a_0 for faults with different R_g.

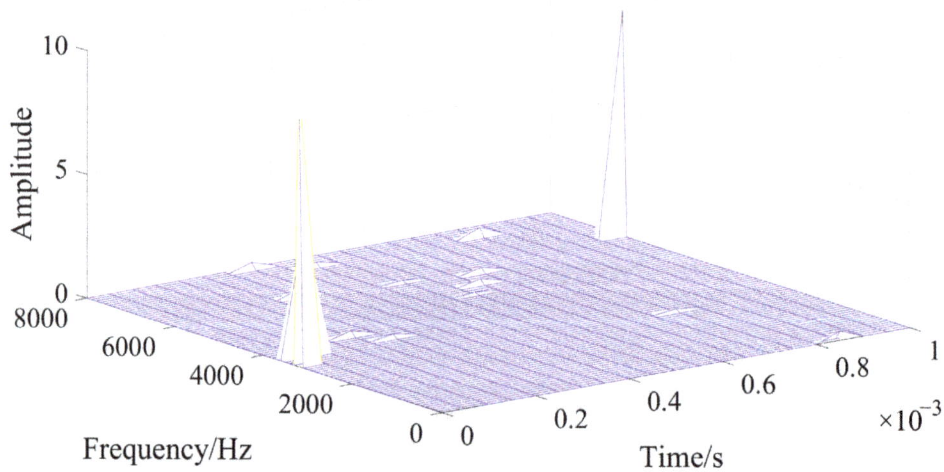

Figure 16. Hilbert spectrum under external faults.

4.2. Fault Detection

The reliability of the proposed fault protection scheme was verified by simulation experiments in this section. $Line_{14}$ was set as the protected line. The operation of the corresponding circuit breaker was then determined through fault identification. According to the fault protection scheme process shown in Figure 12, first determine whether a DC side fault occurs and whether it is a forward fault based on the locally measured voltage and current signals, and then further distinguish whether it is an internal fault by an SVM.

Before validating the fault-diagnosis scheme, an SVM was trained and tested. A multitude of simulations were conducted within the PSCAD model. The fault conditions for the simulations are listed in Table 2. A portion of the simulation results was used for training, and another portion was used for testing.

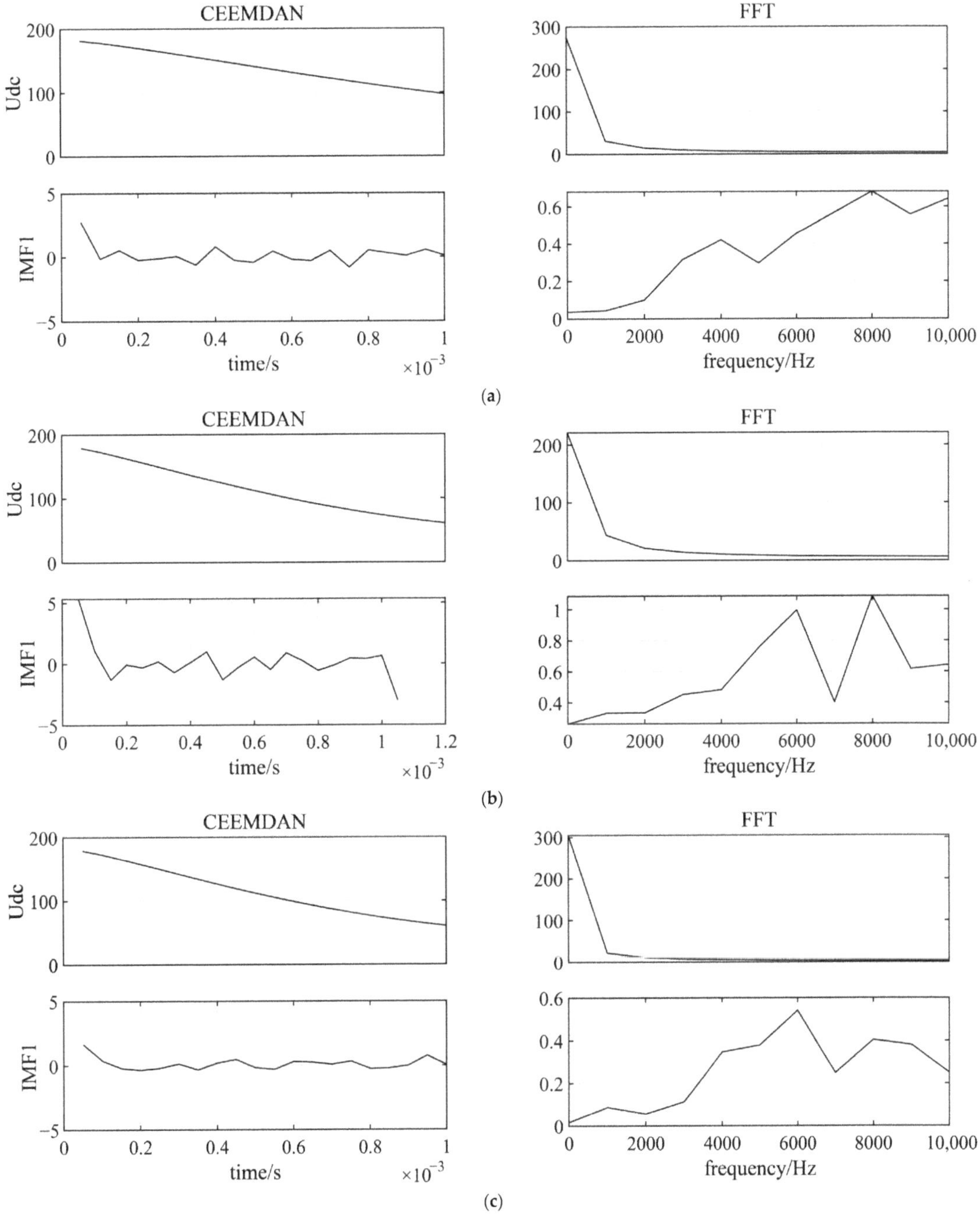

Figure 17. IMF_1 under bus fault with different L_{DC} (**a**) 30 mH, (**b**) 60 mH, and (**c**) 100 mH.

Table 2. Simulation data for SVM training and testing.

Training Dataset	
Fault distance from Bus 3	10 60 90 150 210 270 300 340 380 420
Fault resistance [Ω]	0 10 30 60 90 100 200
Testing dataset	
Fault distance from Bus 3	30 90 120 180 240 320 360 400
Fault resistance [Ω]	0 10 30 60 90 100 200

The training and testing results of the SVM are shown in Figure 18. In a two-dimensional space composed of instantaneous frequency and amplitude, the black star in the training point represents the internal problem, while the blue circle represents the external defect. The SVM was trained to divide this space into two parts: the orange external fault region and the yellow internal fault region. Fault detection based on an SVM using test data obtained from simulations verified the classification accuracy. In Figure 18, the purple dots indicate internal faults and the corresponding green pentagrams represent external faults. The test results match the fault positions in the simulations. To evaluate the performance of the SVM, indicators such as accuracy, precision, recall, and F1 score were selected to evaluate the classification results under different fault scenarios. As the results show in Table 3, the SVM demonstrates consistently outstanding performance in terms of classification accuracy for all tested fault scenarios.

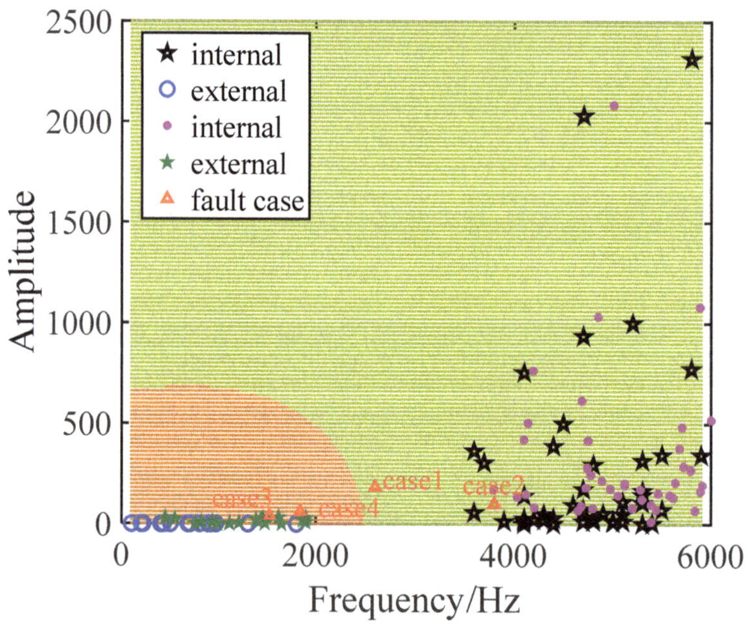

Figure 18. The fault protection results.

Simulations were conducted to verify the protection scheme proposed in this study. Based on the built DC grid simulation model, $Line_{14}$ is designated as the protected line. By single-ended protection requirements, measurements are conducted by relay R_{14} adjacent to circuit breaker CB14 at the line's head end, with fault types identified via local signals. Although faults occur simultaneously, their arrival times at R14 differ due to varying distances. The specific fault conditions are as follows: Case 1 is an internal pole-to-pole fault on Line 14, located 140 km from R14; Case 2 is a pole-to-ground fault on Line 14 at 160 km from R14 with a grounding resistance of 60 Ω; Case 3 is a pole-to-ground fault on Line 45 at 300 km from R14 (external to Line 14) with a grounding resistance of 30 Ω; Case 4 is a pole-to-ground fault on Bus 4 (external to Line 14) with a metallic ground; and Case 5 is a pole-to-pole fault on Bus 2 (external to Line 14). The latter three cases (Cases 3–5) are categorized as external faults with respect to Line 14.

Table 3. The classification results across different fault scenarios.

Fault Scenario	Accuracy	Precision	Recall	F1-Score
Internal High-Resistance Faults	98.3%	97.5%	98.1%	97.9%
Internal Low-Resistance Faults	99.2%	99.1%	99.5%	98.6%
Internal Remote Faults	96.8%	97.2%	97.1%	96.4%
External Severe Ground Faults	98.1%	97.9%	97.4%	97.5%
External High-Resistance Faults	98.6%	99.5%	97.7%	98.3%
External Remote Faults	96.8%	95.9%	97.6%	96.7%

Figure 19 shows the voltage variation after a fault. In every condition, the DC voltage decreased when the fault appeared. Therefore, it is determined that a DC-side fault has occurred, and the DC-side fault protection is activated. As shown in Figure 20, the variation in the fault current was used to determine the direction of the fault. In Cases 1 to 4, the fault current variation rate is positive, which indicates straightforward faults. Case 5 represents a negative fault current variation rate, which is determined as a backward fault, that is, an external fault. Intercepting voltage waveform data 1 ms after Case 1–Case 4 fault occurrence, SVM fault identification was applied after extracting the fault characteristics of high-frequency components with HHT. As shown in Figure 18, Cases 1 and 2 were determined to be internal faults, whereas Cases 3 and 4 were determined to be external faults. The results prove the reliability of the fault-protection solution.

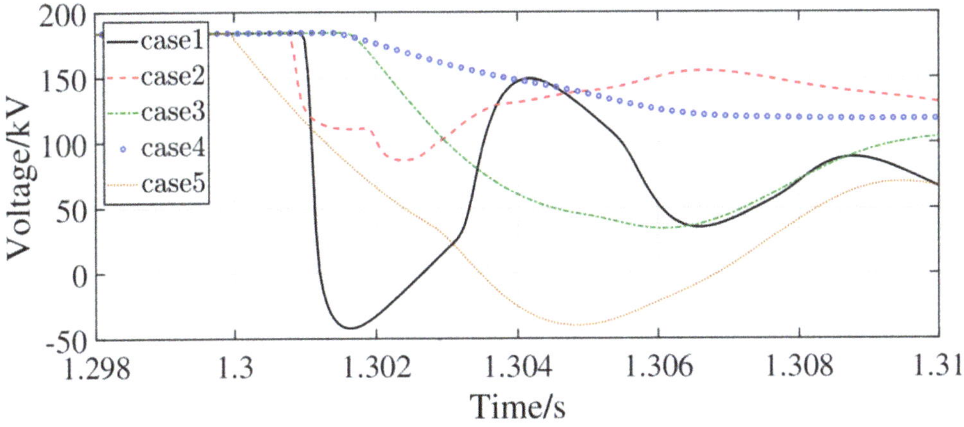

Figure 19. Fault voltage waves of the fault simulations.

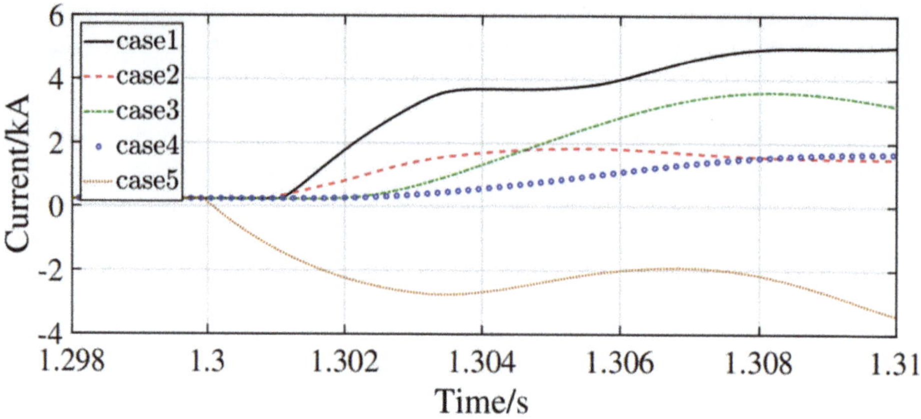

Figure 20. Fault current waves of the fault simulations.

5. Conclusions

This study explores the transient characteristics of the high-frequency component of the voltage traveling wave under DC faults. Due to the characteristics of the current-limiting reactor of the faulty circuit and the distribution parameters of the line varying with frequency, the internal and external fault voltages show significant characteristic differences in the high-frequency region. The time–frequency characteristics of the high-frequency components of the fault voltage are analyzed through HHT, and the instantaneous frequency and instantaneous amplitude of the high-frequency components of the voltage traveling wave are extracted as the characteristic parameters for fault identification. The results show that the instantaneous frequency and amplitude decrease rapidly with an increase in the fault distance, and high transition resistance mainly reduces the instantaneous amplitude. Due to the filtering effect of the current-limiting reactor, the transient frequency of external faults is significantly lower than that of internal faults.

This study utilizes the transient high-frequency characteristics of voltage traveling waves during faults and proposes a novel single-ended protection scheme for DC power grids. Through Hilbert–Huang transform (HHT) analysis, the instantaneous frequencies and amplitudes of the high-frequency components excited by faults were extracted as discriminative features, revealing the significant mode differences between internal and external faults. The proposed classification method based on a support vector machine (SVM) effectively eliminates the need for manually setting protection thresholds, showing fast response, high selectivity and reliable performance, especially with significant advantages in challenging fault scenarios such as high resistance and long distance.

Simulation results in PSCAD validate the feasibility of the approach, confirming its ability to accurately classify faults under varying system conditions. However, this method mainly relies on simulation data for simulation and data analysis. In practical applications, it may be affected by factors such as electromagnetic interference, cable aging, and measurement noise. Furthermore, the model's simplification treatments such as idealizing cable characteristics and ignoring environmental influences may lead to a decrease in the effectiveness of the support vector machine when the real fault characteristics differ significantly from the simulation. To address these limitations, future research should focus on conducting field tests in actual high-voltage direct current power grids and collecting real data to enhance the practicability of the methods. At the same time, develop more accurate physical models to better cope with actual complex scenarios.

Author Contributions: Methodology, X.L. and H.L.; Software, Y.C.; Validation, C.R.; Formal analysis, Y.H. and K.X.; Data curation, C.R.; Writing—original draft preparation, C.R.; Writing—review and editing, X.L., Y.C. and H.L.; Visualization, Y.H. and K.X. All authors have read and agreed to the published version of the manuscript.

Funding: This research was funded by the Science and Technology Project of China Southern Power Grid, grant number 2307RDQ33566N.

Data Availability Statement: The raw data supporting the conclusions of this article will be made available by the authors on request.

Conflicts of Interest: Chenglin Ren, Xiang Lu, and Yue Chen were employed by the company Ultra High Voltage Transmission Company, Kepeng Xia was employed by the XJ Electric Co., Ltd. The remaining authors declare that the research was conducted in the absence of any commercial or financial relationships that could be construed as a potential conflict of interest.

References

1. Li, H.; Li, X.; Peng, Y. Simulation of Wind Power Integration with Modular Multilevel Converter-Based High Voltage Direct Current. *Teh. Vjesnik* **2022**, *29*, 301–307.
2. Li, Z.; Song, Q.; An, F.; Zhao, B.; Yu, Z.; Zeng, R. Review on DC transmission systems for integrating large-scale offshore wind farms. *Energy Convers. Econ.* **2021**, *2*, 1–14. [CrossRef]
3. Shair, J.; Li, H.; Hu, J.; Xie, X. Power system stability issues, classifications and research prospects in the context of high-penetration of renewables and power electronics. *Renew. Sustain. Energy Rev.* **2021**, *145*, 111111. [CrossRef]
4. Javed, W.; Chen, D.; Farrag, M.E.; Xu, Y. System configuration, fault detection, location, isolation and restoration: A review on LVDC microgrid protections. *Energies* **2019**, *12*, 1001. [CrossRef]
5. Chandra, A.; Singh, G.K.; Pant, V. Protection techniques for DC microgrid-A review. *Electr. Power Syst. Res.* **2020**, *187*, 106439. [CrossRef]
6. Song, M.; Dai, S.; Sheng, C.; Zhong, L.; Duan, X.; Yan, G.; Huang, Y.; Li, L.; Chen, C.; Cai, L.; et al. Design and tests of a 160-kV/1-kA DC superconducting fault current limiter. *IEEE Trans. Appl. Supercond.* **2021**, *31*, 1–7. [CrossRef]
7. Khan, S.S.; Wen, H. A comprehensive review of fault diagnosis and tolerant control in DC-DC converters for DC microgrids. *IEEE Access* **2021**, *9*, 80100–80127. [CrossRef]
8. Wang, T.; Monti, A. Fault detection and isolation in DC microgrids based on singularity detection in the second derivative of local current measurement. *IEEE J. Emerg. Sel. Top. Power Electron.* **2022**, *9*, 2574–2588. [CrossRef]
9. Huang, C.; Li, F.; Ding, T.; Jiang, Y.; Guo, J.; Liu, Y. A bounded model of the communication delay for system integrity protection schemes. *IEEE Trans. Power Deliv.* **2016**, *31*, 1921–1933. [CrossRef]
10. Leterme, W.; Beerten, J.; Van Hertem, D. Nonunit protection of HVDC grids with inductive DC cable termination. *IEEE Trans. Power Deliv.* **2015**, *31*, 820–828. [CrossRef]
11. Kansal, P.; Bose, A. Bandwidth and latency requirements for smart transmission grid applications. *IEEE Trans. Smart Grid* **2012**, *3*, 1344–1352. [CrossRef]
12. Li, C.; Gole, A.M.; Zhao, C. A fast DC fault detection method using DC reactor voltages in HVdc grids. *IEEE Trans. Power Deliv.* **2018**, *33*, 2254–2264. [CrossRef]
13. Xiang, W.; Yang, S.; Xu, L.; Zhang, J.; Lin, W.; Wen, J. A transient voltage-based DC fault line protection scheme for MMC-based DC grid embedding DC breakers. *IEEE Trans. Power Deliv.* **2018**, *34*, 334–345. [CrossRef]
14. Li, J.; Yang, Q.; Mu, H.; Le Blond, S.; He, H. A new fault detection and fault location method for multi-terminal high voltage direct current of offshore wind farm. *Appl. Energy* **2018**, *220*, 13–20. [CrossRef]
15. Aftab, M.A.; Hussain, S.M.S.; Ali, I.; Ustun, T.S. Dynamic protection of power systems with high penetration of renewables: A review of the traveling wave based fault location techniques. *Int. J. Electr. Power Energy Syst.* **2020**, *114*, 105410. [CrossRef]
16. Zhang, C.; Song, G.; Dong, X. Dynamic protection of power systems with high penetration of renewables:A novel traveling wave protection method for DC transmission lines using current fitting. *IEEE Trans. Power Deliv.* **2019**, *35*, 2980–2991. [CrossRef]
17. Zhang, Y.; Li, Y.; Song, J.; Li, B.; Chen, X. A new protection scheme for HVDC transmission lines based on the specific frequency current of DC filter. *IEEE Trans. Power Deliv.* **2018**, *34*, 420–429. [CrossRef]
18. He, J.; Chen, K.; Li, M.; Luo, Y.; Liang, C.; Xu, Y. Review of protection and fault handling for aflexible DC grid. *Prot. Control Mod. Power Syst.* **2020**, *34*, 1–15.
19. Li, R.; Xu, L.; Yao, L. DC fault detection and location in meshed multiterminal HVDC systems based on DC reactor voltage change rate. *IEEE Trans. Power Deliv.* **2016**, *34*, 1516–1526. [CrossRef]

20. Chen, J.; Sha, D.; Zhang, J. Current ripple prediction and DPWM-based variable switching frequency control for full ZVS range three-phase inverter. *IEEE Trans. Ind. Electron.* **2020**, *68*, 1412–1422. [CrossRef]
21. Panahi, H.; Sanaye-Pasand, M.; Niaki, S.H.A.; Zamani, R. Fast low frequency fault location and section identification scheme for VSC-based multi-terminal HVDC systems. *IEEE Trans. Power Deliv.* **2021**, *37*, 2220–2229. [CrossRef]
22. Saleh, K.A.; Hooshyar, A.; El-Saadany, E.F.; Zeineldin, H.H. Protection of high-voltage DC grids using traveling-wave frequency characteristics. *IEEE Syst. J.* **2019**, *14*, 4284–4295. [CrossRef]
23. Moradzadeh, A.; Moayyed, H.; Mohammadi-Ivatloo, B.; Gharehpetian, G.B.; Aguiar, A.P. Turn-toturn short circuit fault localization in transformer winding via image processing and deep learning method. *IEEE Trans. Ind. Inform.* **2021**, *18*, 4417–4426. [CrossRef]
24. Zhang, C.; Li, Y.; Song, G.; Dong, X. Fast and sensitive nonunit protection method for HVDC grids using Levenberg–Marquardt algorithm. *Int. J. Electr. Power Energy Syst.* **2021**, *69*, 9064–9074. [CrossRef]
25. Sabug, L., Jr.; Musa, A.; Costa, F.; Monti, A. Realtime boundary wavelet transform-based DC fault protection system for MTDC grids. *IEEE Trans. Ind. Electron.* **2020**, *115*, 105475.
26. Jayamaha, D.K.J.S.; Lidula, N.W.A.; Rajapakse, A.D. Wavelet-multi resolution analysis based ANN architecture for fault detection and localization in DC microgrids. *IEEE Access* **2019**, *7*, 145371–145384. [CrossRef]
27. Li, D.; Ukil, A. Transient analysis in multi-terminal voltage source converter based direct current system: Median mode decomposition. *Int. J. Electr. Power Energy Syst.* **2022**, *142*, 108323. [CrossRef]
28. de Souza, U.B.; Escola, J.P.; da Cunha Brito, L. A survey on Hilbert-Huang transform: Evolution, challenges and solutions. *Digit. Signal Process.* **2022**, *120*, 103292. [CrossRef]
29. Feng, X.; Shi, W.; Zhuo, J. Comparison of Wavelet Transform and HHT in Fault Feature Extraction of HVDC System. *Shipboard Electr. Technol.* **2018**, *38*, 36–40.
30. Li, D.; Ukil, A.; Satpathi, K.; Yeap, Y.M. Hilbert–huang transform based transient analysis in voltage source converter interfaced direct current system. *IEEE Trans. Ind. Electron.* **2020**, *68*, 11014–11025. [CrossRef]
31. Zhou, J.; Han, B.; Shi, G.; Zhang, J.; Zang, J. An Fault Detection Strategy of Hybrid AC/DC Microgrids Based on HHT Theory. In Proceedings of the 2024 9th Asia Conference on Power and Electrical Engineering (ACPEE), Shanghai, China, 11–13 April 2024; pp. 2239–2245.
32. Li, D.; Ukil, A.; Li, G. Complex Domain Analysis-Based Fault Detection in VSC Interfaced Multi-terminal LVDC System. *IEEE Trans. Ind. Inform.* **2024**, *20*, 8306–8316. [CrossRef]
33. Radwan, M.; Azad, S.P. A single-ended protection scheme for multi-terminal HVDC grids based on Hilbert-Huang Transform. In Proceedings of the 2023 IEEE Power &Energy Society General Meeting (PESGM), Orlando, FL, USA, 16–20 July 2023; pp. 1–5.

energies

Article

Prediction of Remaining Service Life of Miniature Circuit Breakers Based on Wiener Process

Lin Ma [1,2,3], Linming Hou [4], Puquan He [1,2,3], Changxian Wang [1,2,3], Zhenhua Xie [1,2,3,*] and Yao Wang [5]

1 Zhejiang Testing & Inspection Institute for Mechanical and Electrical Products Quality Co., Ltd., Hangzhou 310051, China
2 Intelligent Electrical Apparatus Test and Detection Technology Zhejiang Engineering Research Center, Hangzhou 310051, China
3 Key Laboratory of Low Voltage Apparatus Intelligentization and New Energy Application of Zhejiang Province, Hangzhou 310051, China
4 Qingdao Haier Intelligent Technology Research and Development Co., Ltd., Qingdao 266101, China
5 Provincial and Ministerial Co-Construction Collaborative Innovation Center on Reliability Technology of Electrical Products, Hebei University of Technology, Tianjin 300130, China
* Correspondence: xiezhenhua@ztme.com.cn

Abstract: In the operation of a power distribution system, miniature circuit breakers (MCBs) are subjected to the synergistic effect of electrical and mechanical stresses in service, and their operational performance is progressively degraded, which is prone to bring significant losses to the users after failures occur. In order to accurately predict the remaining electrical life of MCBs in service, MCB mechanical characterization and dynamic simulation are carried out, and the initial closing angle of MCBs is selected as the degradation characteristic quantity, so as to deeply analyze the evolutionary characteristics of the initial closing angle in the degradation of MCBs and to construct the electrical degradation model of the one-dimensional linear Wiener process in the present study. With the help of the Monte Carlo method, we carry out the electric life simulation analysis to investigate the intrinsic correlation between the degradation of electric performance and the initial closing angle, and we implement the electric life experiment under the 63 A working condition to analyze the dynamic change in the stiffening angle of the test samples. The parameters of the electrical performance degradation model are identified through the synergistic driving of the electrical life simulation data and the experimental data, the remaining electrical life prediction is realized based on the degradation data of the same batch of products, and the maximum prediction error of the proposed method is controlled within 15%.

Keywords: miniature circuit breaker; initial closing angle; remaining life prediction; Monte Carlo; Wiener process

1. Introduction

Circuit breakers, as crucial electrical equipment in power systems, can automatically cut off fault circuits and protect transmission lines and electrical devices. During their service life, the electrical stress generated by each switching operation (making and breaking) erodes the contact system. Over time, this reduces the reliability of circuit breakers, and may even cause them to lose their original functions, threatening the safe operation of a power distribution system. Therefore, it is necessary to effectively and accurately predict the remaining useful life (RUL) of circuit breakers.

Scholars at home and abroad have actively engaged in research on the remaining useful life (RUL) prediction and reliability assessment of switching appliances [1,2], selecting different characteristic signals according to the inherent characteristics of various switching appliances. In reference [3], vibration signals generated during the operation of conventional circuit breakers (CCBs) were extracted, and a remaining useful life (RUL) prediction model based on a self-attention bidirectional long short-term memory network (SA-BiLSTM) was constructed to accurately predict the remaining life of CCBs. Reference [4] conducted constant-stress accelerated degradation tests on thermal trips using temperature as the acceleration stress and specific heat deflection as the degradation characteristic quantity. The Wiener process and maximum likelihood estimation were employed to estimate RUL parameters, determine the remaining life of thermal trips under various acceleration stresses, and deduce the RUL of external thermal trips under normal stress via the Arrhenius acceleration model. Reference [5] combined statistical theory with the Arrhenius model for the life prediction of electronic residual current circuit breakers, with prediction results close to the reference service life provided by manufacturers. Reference [6] designated the residual working current as the degradation evaluation quantity for RCBOs (residual current operated breakers), established an optimized BP neural network (BPNN) model based on test degradation data, and optimized the BPNN model using a genetic algorithm. The relative error between the predicted life and the true value was less than 10%. Reference [7] addressed the impact of the common failure modes of circuit breakers on travel curves, organizing a fuzzy probability method via maximum likelihood and an interactive multiple model (IMM) to accurately predict circuit breaker states and intelligently detect failure causes. Reference [8] considered the effects of arc voltage and arc restrike in AC contactors, established a remaining electrical life prediction method based on Bayesian theory using Joule integral as the degradation indicator, and improved the accuracy of RUL prediction.

Methods for remaining useful life (RUL) prediction and reliability assessment can be primarily categorized into physics-of-failure (PoF) methods and data-driven methods. PoF methods establish mathematical degradation models and predict lifespan based on the internal failure mechanisms of test specimens. However, different types of electrical appliances exhibit distinct internal mechanisms, which are often highly complex and challenging to study comprehensively. Taking MCBs as an example, the wear and oxidation behavior of contact materials are significantly affected by factors such as manufacturing processes and environmental temperature and humidity, leading to significant differences in the failure mechanisms of different samples. In actual operation, MCBs are subject to complex and variable conditions such as overload currents, operating frequencies, and environmental vibrations, which are difficult to fully reproduce through standard laboratory testing. This results in significant errors in parameter calibration based on physical models. Data-driven methods are further divided into artificial intelligence (AI) approaches and statistical model methods. AI approaches, characterized by their robust nonlinear modeling capabilities, excel at processing large-scale data with strong adaptability. Common AI methods include neural networks [9], support vector machines [10], and gray models [11]. In Reference [12], the Savitzky–Golay convolution smoothing algorithm was applied to degrade data throughout the lifecycle of AC circuit breakers, followed by precise time-series prediction using a long short-term memory (LSTM) neural network, achieving a prediction accuracy of 97.4%. Reference [13] proposed a novel wavelet-enhanced dual-tree residual network, decomposing time-series data into high-frequency and low-frequency components via wavelet transform to predict the RUL of circuit breaker operating mechanisms. Reference [14] analyzed characteristic parameter changes in relays in salt spray environments and developed a genetic algorithm–backpropagation neural network (GA-BP) model, significantly improving state recognition accuracy. Reference [15] introduced an RUL prediction method combining gray relational analysis (GRA) and LSTM, using GRA to screen mechanical parameters (e.g., overtravel, opening distance, and synchronization) as inputs for the LSTM model to accurately predict the circuit breaker lifespan. Reference [16] uses digital image processing (DIP) technology to analyze the new three-dimensional (3D) FRA features of power transformers. The proposed new 3D features integrate the frequency, amplitude, and phase angle of traditional FRA features into a single 3D graph and are processed using DIP code to extract unique texture analysis features. These features are used to classify various faults by clustering them in a ternary graph. Despite their effectiveness, AI methods face limitations such as black-box opaqueness, poor interpretability, scenario-specific applicability, and overfitting in small-data contexts.

Statistical model methods, grounded in mathematical theory and probabilistic–statistical hypotheses, describe relationships between variables through formulas and rely on rigorous statistical inference and hypothesis testing. These methods offer strong interpretability, with model parameters possessing clear physical meanings. In scenarios with small sample sizes (e.g., hundreds of data points) or low data noise, statistical models are less prone to overfitting and can validate their effectiveness through hypothesis tests (such as *t*-tests and F-tests). Reference [17] selected the loss rate of the spring reaction force and impact reaction force as the degradation characteristics, obtained the degradation curves of electromagnetic releases at different temperatures via accelerated life testing, and derived full-life data during normal operation using the Arrhenius equation. Reference [18] proposed a hybrid method for circuit breaker (CB) prediction, integrating a hybrid prediction model with the concept of dynamic reliability through piecewise deterministic Markov processes and modeling uncertain failure thresholds to incorporate and propagate uncertain fault assessment levels during prediction estimation. Reference [19] introduced random effects into the Tweedie exponential dispersion process model to capture individual differences, enabling the more accurate fitting of degradation processes, and derived the service life under actual temperature and humidity conditions via the Peck acceleration model. Reference [20], based on the topological structure of hybrid DC circuit breakers, used fault tree analysis to identify basic components causing breaker failures and introduced Markov models into the reliability modeling of hybrid DC circuit breakers to calculate the steady-state availability, failure rates, and mean time between failures. Reference [21] established a degradation model based on the Wiener process, proposed the probability density function (PDF) of the remaining electrical life (REL) of AC contactors by understanding the first accumulation of contact erosion loss (CEL) exceeding the failure threshold, calculated the point and interval estimates of the REL, and conducted simulated electrical life tests using Monte Carlo methods. Reference [22], to predict the correlation between different degrees of contact degradation and dynamic resistance waveform characteristics, extracted five diagnostic parameters from dynamic resistance measurement (DRM) curves. Using scoring and weighting techniques, health indices were applied to clearly identify whether the CB's condition was healthy, requiring caution, or at risk.

The electrical wear of MCB contacts can cause changes in mechanical characteristics. Therefore, the mechanical performance parameters of MCBs are closely related to their remaining useful life (RUL). The initial closing angle is an important mechanical performance parameter of MCBs, with the following advantages: (1) The contact angle can be measured without damaging the MCB structure and can be performed during service, providing regular information on the service status. (2) Compared with other testing methods, testing the initial closing angle is relatively simple and has low testing costs. In this paper, mechanical characterization and dynamic simulation are carried out, the initial closing angle of the MCB is selected as the degradation characteristic quantity, the degradation law of the MCB is revealed by the Monte Carlo method, the degradation model of electrical performance based on the Wiener process is established, the parameters of the degradation model are estimated by using the great likelihood estimation method, and the constant-stress accelerated life test is carried out by taking the 63A current as the accelerating variable to predict the residual electrical life of the MCB of the same lot in different operation phases. The remaining electrical life of the same batch of MCBs is predicted at different operation stages.

2. Failure Causes of MCBs and Kinetic Analysis

2.1. Failure Causes of MCBs

An MCB is mainly composed of a contact system, protection system, and mechanical operating system; a three-dimensional solid model is shown in Figure 1.

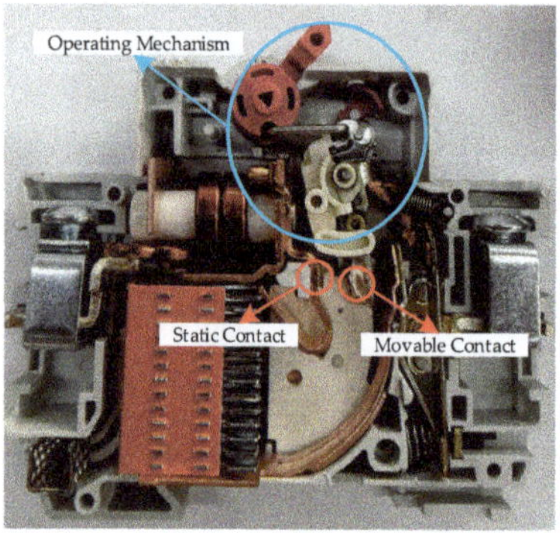

Figure 1. Schematic diagram of MCB structure.

An MCB's electrical life depends mainly on the degree of wear and tear of the contact system; an MCB in the connection and breaking process of the contact system will pull the arc; the high temperature generated by the arc leads to contact surface melting and wear; and at the same time, the mechanical stress between the static and dynamic contacts caused by the deformation of the contact surface aggravates the wear and tear. The arc erosion process is shown in Figure 2.

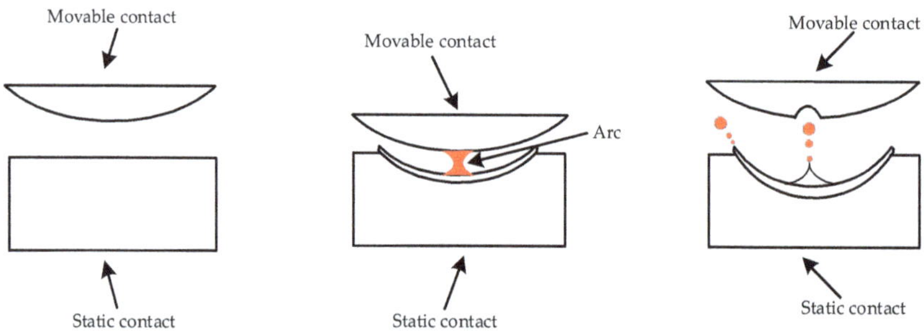

Figure 2. Wear process during contact closure and opening.

The contact mass loss of an MCB is related to the arc voltage, arc current, and arcing time, and the electrical wear model of the contact for a single break can be expressed as follows under the same current stress:

$$m_{pj} = \int_{t_q}^{t_x} f(i_{pj}, u_{pj}, t)dt \quad (1)$$

Among them, m_{pj} is the loss of contact mass in the jth break of the p phase. $p = 1, 2$, and 3 are three phases, which are A, B, and C, respectively. Moreover, i_{pj}(t) is the instantaneous value of the arc current of the jth breaking of phase P, UPJ(T) is the instantaneous value of the arc voltage of the jth breaking of phase P, t_q is the time of arc starting, and t_x is the time of arc quenching.

$$\int_{t_q}^{t_x} f\left(i_{pj}, u_{pj}, t\right)dt = K \cdot g_{pj} \tag{2}$$

where g_{pj} is the characteristic quantity of the ignited arc and K is the arc erosion coefficient, which is used to describe the relationship between the contact mass loss and the ignited arc parameters, and it is related to contact parameters such as the arc extinguishing system, the contact motion speed, the contact material and the structure, which can be obtained from the experimental and historical data. After determining the MCB structure and contact material, the three-phase contacts are basically the same, so it can be considered that under the same current stress, the three-phase contacts have the same arc erosion coefficient. Based on the above analysis, Equation (3) can be derived.

$$m_{pj} = K \cdot g_{pj} \tag{3}$$

If the arc erosion factor K is known, then Equation (3) can be utilized to calculate the product's single-breaking contact mass loss. After the number of operations T, the cumulative contact mass loss is as follows:

$$\sum_{j=1}^{T} m_{pj} = \sum_{j=1}^{T} K \cdot g_{pj} = K \cdot \sum_{j=1}^{T} g_{pj} \tag{4}$$

With the increase in the number of MCB actions, mechanical stress, arc erosion, and other factors on the contacts to produce the cumulative damage effect are more and more serious, and frequent breaking of the circuit occurs so that the arc erosion of the contact material is too much; then, the contact pressure between the contacts falls too much resulting in static and dynamic contacts that cannot be reliably contacted. Today's MCB product performance is very mature, its structural design is reasonable, and in the service process of MCBs its mechanical life far exceeds its electrical life, which is the contact failure for the MCB's main failure form.

2.2. Closing Process Analysis

The operational performance of miniature circuit breakers is mainly determined by the operating mechanism and the contact system. Wear, fracture, or other issues in any mechanical component of the mechanism can lead to changes in the mechanism's movement. The operating mechanism consists of components such as the handle BC, U-shaped connecting rod CD, trip latch DE, lock catch DF, contact lever FG, and central support DEKF. According to the motion characteristics and coordination relationships of different components during the closing and opening processes, some irregularly shaped components are equivalent to connecting rods of certain lengths. The equivalent operating mechanism model is used to analyze the mechanism's behavior during closing and opening processes. Taking point B of the handle as the origin, the straight line connecting with the shaft pin F as the x-axis, and the straight line perpendicular to the x-axis as the y-axis, a plane rectangular coordinate system is established. Let the handle BC be rod 1, the U-shaped connecting rod CD be rod 2, and the distance DF from articulation point D to shaft pin F be rod 3. The force analysis diagram of the mechanism in the initial position and the schematic diagram of the operating mechanism and its equivalent mechanism are shown in Figure 3.

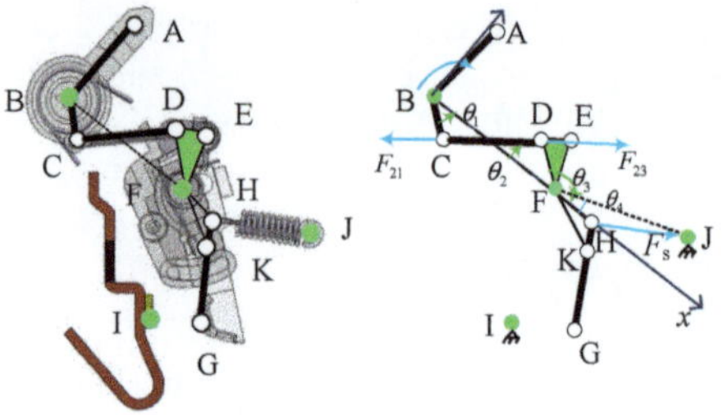

Figure 3. Operating mechanism and its schematic diagram of miniature circuit breaker.

According to the relative motion relationship between the moving and stationary contacts, the closing process of the miniature circuit breaker is divided into an idle-stroke phase and an overtravel phase. The idle-stroke phase refers to the process from the rotation of the handle to just before the moving and stationary contacts make contact. Assuming the handle rotates at a constant angular velocity ω_1, during the idle-stroke phase, the handle drives the U-shaped connecting rod and the contact assembly to move in the form of a four-bar linkage; the motion process is shown in Equation (5) and Figure 4.

$$\begin{cases} \theta_2 = \arcsin(\frac{l_{DF}\sin\theta_3 - l_{BC}\sin\theta_1}{l_{CD}}) \\ \theta_3 = \pi - \arcsin(\frac{l_{BC}\sin\theta_1}{l_{CF}}) - \arccos(\frac{l_{CF}^2 + l_{DF}^2 - l_{CD}^2}{2 \cdot l_{CF} \cdot l_{DF}}) \\ \omega_2 = \frac{-\omega_1 l_{BC}\sin(\theta_1 - \theta_3)}{l_{CD}\sin(\theta_2 - \theta_3)} \\ \omega_3 = \frac{\omega_1 l_{BC}\sin(\theta_1 - \theta_2)}{l_{DF}\sin(\theta_3 - \theta_2)} \\ \ddot{\theta}_3 = \frac{\omega_1^2 l_{BC}\cos(\theta_1 - \theta_2) + \omega_2^2 l_{CD} - \omega_3^2 l_{DF}\cos(\theta_3 - \theta_2)}{l_{DF}\sin(\theta_3 - \theta_2)} \end{cases} \quad (5)$$

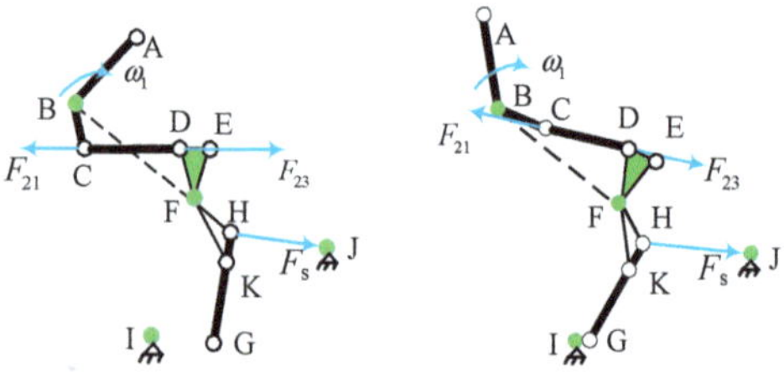

Figure 4. Schematic diagram of the operating mechanism motion during the idle-stroke phase of a miniature circuit breaker. (**a**) Initial position; (**b**) contact touching position.

In the formula, θ_2 and θ_3 are the rotation angles of rods 2 and 3; ω_2 and ω_3 are the rotational angular velocities of rod 2 and rod 3, respectively; and $\ddot{\theta}_3$ is the rotational angular acceleration of rod 3.

2.3. Dynamic Simulation Analysis

This section uses the multi-body dynamic simulation software ADAMS 2018 to establish a virtual prototype model of the miniature circuit breaker operating mechanism, simulating the movement patterns of each component of the operating mechanism during the closing process. Apply contact forces and other constraints between the components of the MCB. Set the handle drive mode to constant-speed drive. Use the parameters obtained from the actual measurements to set the stiffness coefficients of the torsion spring and contact spring. Define the rotation angle of the handle. Finally, obtain the virtual prototype model of the MCB, where the counterclockwise arrow at the handle indicates the drive, and the counterclockwise arrows at other hinge points indicate the rotating joints. The lock shape indicates a fixed joint. The dynamic simulation model is shown in Figure 5.

Figure 5. Virtual mockup model of miniature circuit breaker dynamics.

The relationship between the operating torque of the handle and the angle of the rotation of the handle during the closing process was obtained through simulation, as shown in Figure 6.

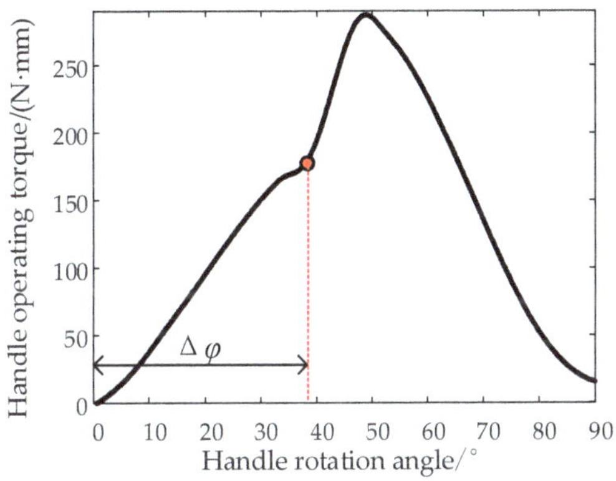

Figure 6. Simulation waveform of operating torque of handle during closing process.

The initial closing angle is the angle of rotation of the handle from the start of rotation to the point where the contacts close. Based on the curve changes during the closing process, it can be seen that the closing angle can characterize the operating status of the circuit breaker. Therefore, the closing angle is selected as the characteristic parameter.

The initial closing angle is the angle of rotation of the handle from the start of rotation to the point where the contacts close, and the initial closing torque is the torque of the handle from the start of rotation to the point where the contacts close. Based on the curve changes during the closing process, it can be seen that the initial closing angle and initial closing torque can characterize the operational status of the circuit breaker.

With the increase in the number of MCB operations, the amount of contact wear gradually increases. Before and after the contact wear dynamic and static contacts contact when the mechanical mechanism position changes, as shown in Figure 7, H_1, K_1, and G_1 indicate the contact intact case of the location of the components, and H_2, K_2, and G_2 indicate the contact wear case of the location of the components.

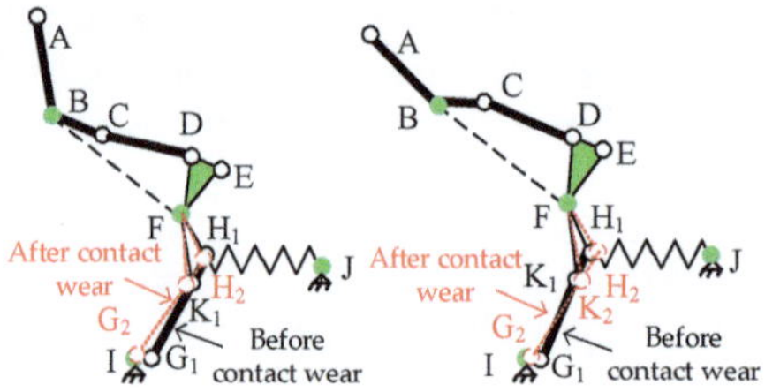

Figure 7. Schematic diagram of the position change in the mechanism before and after contact wear.

The trend of the initial closing angle change with the contact wear is simulated through dynamic simulation, as shown in Figure 8.

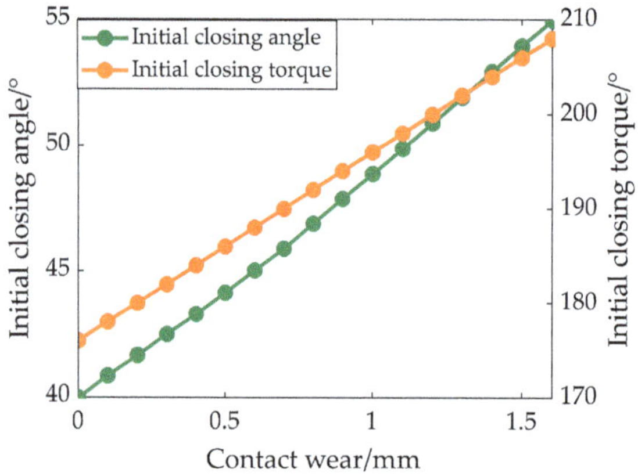

Figure 8. Relationship between wear of contact and characteristic quantity.

As can be seen from Figure 8, both the initial closing angle and the initial closing torque show a linear trend with the amount of contact wear. In order to verify the linear variation, the linear correlation analysis between the characteristic quantity and the change in contact wear was carried out; the correlation coefficients between the initial closing angle, the initial closing torque, and the contact wear were 0.99915 and 0.99875, respectively; and the absolute value was greater than 0.9, so it can be considered that the initial closing angle and initial closing torque of the MCB increased linearly with the contact wear. After the sensitivity analysis of the changes in the two eigenquantities, the sensitivities of the rigidity angle and rigidity moment were 27.33% and 14.56%, respectively, as shown in Figure 9, so the rigidity angle is selected as the eigenquantity for the remaining life prediction.

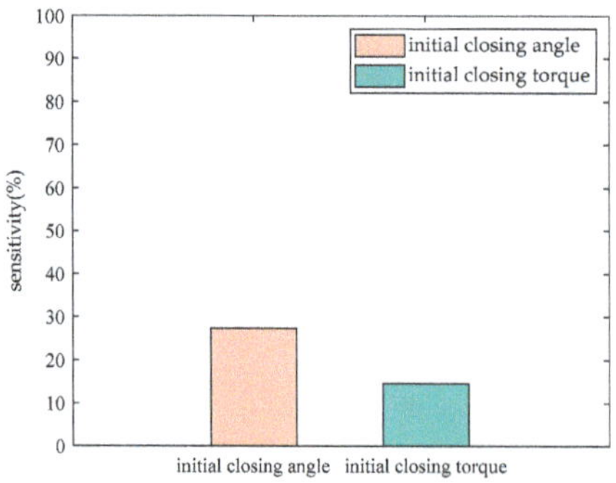

Figure 9. Sensitivity analysis of eigenquants.

3. Electrical Life Prediction Method for MCBs

3.1. Electrical Performance Degradation Model

The contact degradation process of MCBs is a contact thickness decreasing and non-monotonic degradation process, which meets the conditions for the application of the Wiener process, so a one-dimensional linear Wiener process is chosen to describe the electrical performance degradation process of MCBs.

The contact mass loss m for a single interruption of an MCB does not follow a normal distribution. However, if the single contact mass loss m is divided into segments and accumulated, the accumulated contact mass loss for each segment follows a normal distribution. In the test, one segment is accumulated for every k times of opening, and the accumulated contact mass loss of each segment is recorded as ΔM; its expression is as follows:

$$\Delta M = \sum_{j=1}^{k} m_j \tag{6}$$

where m_j denotes the contact mass loss incurred at the jth operation of the circuit breaker, and j denotes the jth opening of the MCB.

As the arc starting phase angle of the arc is generated randomly during each opening, after the arc erodes the contact many times, it can be regarded that the arc uniformly wears out the contact, the thickness of the contact is uniformly reduced, the cumulative contact mass loss can be transformed into the cumulative contact thickness loss by (7), and both of them obey the normal distribution.

$$\Delta H = \frac{\Delta M}{\rho S} \tag{7}$$

where ρ denotes the contact material density, S denotes the area of the contact loss region after breaking, and ΔH denotes the thickness of the accumulated contact loss.

According to the central limit theorem, when k is large, there is the following relation.

$$\frac{\Delta H - k\mu_H}{\sqrt{k}\sigma_H} \sim N(0,1) \tag{8}$$

In the formula, μ_H is the average contact thickness loss, and σ_H is the standard deviation of the contact thickness loss.

The initial closing angle can be used as a degradation characterization quantity for the remaining electrical life prediction of the MCB, where the cumulative degradation of the initial closing angle is denoted as $Y(t)$, the degradation increment is $\Delta Y \sim N(\mu_Y, \sigma_Y^2)$, and the MCB conforms to the nature of the one-dimensional Wiener process so that $\mu_1 = k\mu_Y$, $\sigma_1^2 = k\sigma_Y^2$; the electrical performance degradation is modeled as follows:

$$Y(t) = \mu_1 t + \sigma_1 B_1(t) \tag{9}$$

where $Y(t)$ is the cumulative degradation of the initial closing angle and $Y(t) \sim N(\mu_1 t, \sigma_1^2 t)$, μ_1 is the rate of degradation of the electrical performance of the MCB, and σ_1 is a diffusion parameter, which indicates the randomness of the arc erosion of contacts and the influence of contingent factors, such as the existence of experimental errors, on the performance of an MCB.

The above model parameters are estimated by the great likelihood estimation method, and the parameter estimation formula is shown in (10).

$$\begin{cases} \hat{\mu}_1 = \frac{1}{\mathrm{kN}}\sum_{i=1}^{N} \Delta Y_i \\ \hat{\sigma}_1^2 = \frac{1}{\mathrm{kN}}\sum_{i=1}^{N} (\Delta Y_i - \hat{\mu}_1 k)^2 \\ \hat{L} = \sum_{i=1}^{N} \Delta Y_i \end{cases} \tag{10}$$

where k is the cumulative number of MCB closures and openings per segment, N is the number of measurements of the initial closing angle degradation, and ΔY_i is the incremental initial closing angle degradation at the ith measurement.

3.2. Residual Electric Life Prediction Model

When the average degradation rate of the MCB is a constant value, the electrical life of the MCB can be defined as the number of operations when the cumulative degradation reaches the failure threshold for the first time, and its remaining life prediction principle is shown in Figure 10.

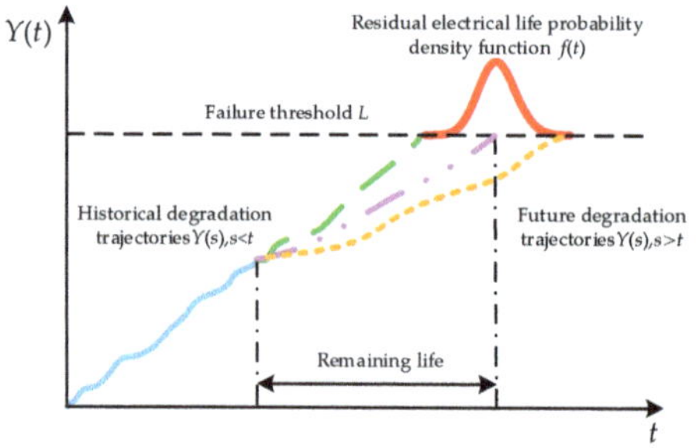

Figure 10. Schematic diagram of remaining electrical life prediction.

When the cumulative degradation $Y(t)$ of a phase of the MCB at the initial closing angle reaches the failure threshold L for the first time, it can be considered that the MCB undergoes an electrical degradation failure. Therefore, the MCB electrical life T can be defined as follows:

$$T = \inf\{t|Y(t) \geq L, Y(0) \leq L\} \tag{11}$$

Since the time distribution of the Wiener process up to a certain point obeys the inverse Gaussian distribution, when the cumulative degradation of the initial closing angle $Y(t)$ has not yet reached the failure threshold L, set $L' = L - L_0$, where L_0 is the current cumulative degradation, and the probability density function of the electric life at the current degradation moment is as follows.

$$\mathrm{f}(t) = \frac{L'}{\sqrt{2\pi t^3}\sigma_1} \exp\left[-\frac{(L' - \mu_1 t)^2}{2{\sigma_1}^2 t}\right] \tag{12}$$

The electrical life distribution function and reliability function of the MCB operating to the current moment are, respectively, as follows:

$$R(t) = \Phi\left(\frac{L' - \mu_1 t}{\sigma_1\sqrt{t}}\right) - \exp\left(\frac{2\mu_1 L'}{{\sigma_1}^2}\right)\Phi\left(\frac{-L' - \mu_1 t}{\sigma_1\sqrt{t}}\right) \tag{13}$$

where $\Phi(\cdot)$ is the standard normal distribution function.

4. Simulation Results Analysis

4.1. Monte Carlo Based Electrical Life Simulation

The principle of the Monte Carlo method is that the probability of an event can be estimated by the frequency of occurrence in a large number of tests. In this paper, the simulation analysis of an MCB under the uniform wear of contacts based on the Monte Carlo method is carried out to simulate the degradation process of an MCB's electrical performance by taking the initial closing angle as the degradation eigenquantity, and the simulation steps are as follows:

(1) Input the number of MCB simulation samples and the failure threshold;

(2) Randomly generate a series of arc-starting phase angles obeying a uniform distribution in the range of $[0,\pi)$;

(3) Calculate the cumulative mass loss of each phase contact and convert it to the initial closing angle of each phase contact;

(4) Calculating the initial closing angle of each phase contact with the increase in the number of operations, when the initial closing angle of any phase reaches the failure threshold, determining the end of its life, the simulation process of the sample stopsm and the number of operations at the time that is recorded as the electrical life of the sample, otherwise continuing to execute steps (1)~(3);

(5) The program repeats the execution of steps (2)~(4) for n times, and the data of the initial closing angle of n MCB simulation samples in the process of electrical degradation can be obtained.

The simulation flowchart is shown in Figure 11.

The breaking voltage is set to 380 V, load current to 63 A, and power factor to 0.65 for the simulation. In the actual service process, the MCB electrical life depends on the most serious wear of a phase, and as long as any one of the phase of the initial closing angle reaches the failure threshold, this means that the test piece cannot be reliably in service. Therefore, the initial closing of the most severely worn phase is taken as the cumulative degradation characterization of the sample. Since a larger number of simulation samples can help to reduce the chance caused by individual differences, the results can be more in line with reality. Therefore, in this section, the simulation data of 1000 samples were generated by the Monte Carlo simulation method, from which the three samples with the highest, average, and lowest electrical lifetimes, which were 3009, 2947, and 2886 times, respectively, were selected for analysis. The degradation of the initial closing angle of the three samples was accumulated every 50 times, and the variation in the initial closing angle of each of the three MCB samples with the number of operations is shown in Figure 12.

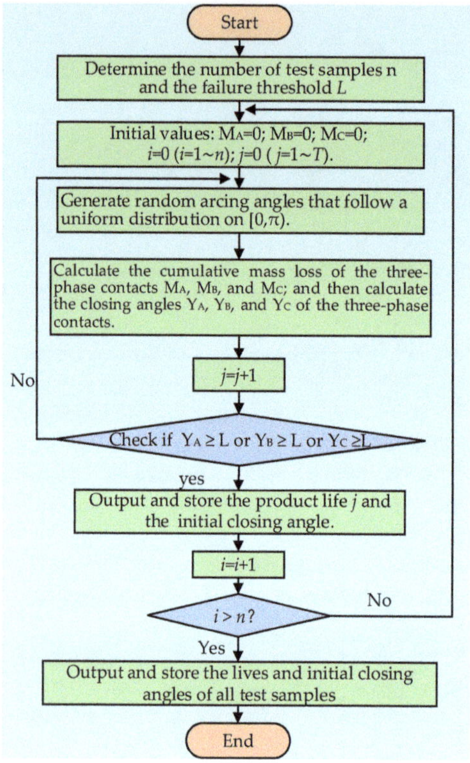

Figure 11. Monte Carlo simulation steps of MCBs.

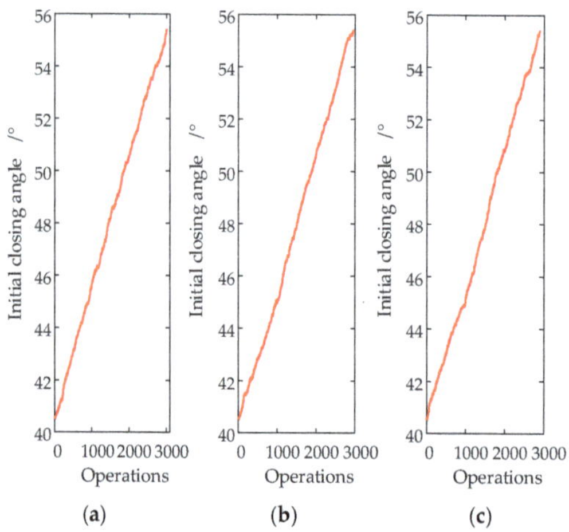

Figure 12. Change in the initial closing angle of the simulated samples. (**a**) MCB 1; (**b**) MCB 2; (**c**) MCB 3.

As can be seen from Figure 12, with the gradual increase in the number of MCB operations, the initial closing angles of the three samples show a significant linear increase as a whole, and the difference in the initial closing angles at the end of the electrical life is also relatively small.

4.2. Normal Distribution Test

The K-S test was used to test the normality of the degradation increments of the initial closing angle of the three samples generated by Monte Carlo simulation. The significance level α of the K-S test was set at 0.05, and the null hypothesis was normal distribution. The results are shown in Table 1.

Table 1. K-S test results of initial closing angle of simulated samples.

Case	MCB 1	MCB 2	MCB 3
Test statistic	0.0614	0.0757	0.0645
Threshold value	0.27	0.24	0.23
Obedience test	obey	Obey	obey

As can be seen from Table 1, in the K-S test of the degradation increment of the initial closing angle of each of the three samples, the test statistic is smaller than the critical value, the two test methods sufficiently show that the degradation increment of the initial closing angle of the various segments obeys the normal distribution, and the Wiener process can be adopted to describe the electrical performance degradation process of MCBs. The Wiener process describes the degradation process of the electrical properties of MCBs.

4.3. Residual Electrical Life Prediction Based on Simulation Data

Based on the initial closing angle degradation data obtained from the Monte Carlo simulation, the electrical life prediction for sample 1 was selected. The parameter estimation of the electrical life prediction model was carried out through Equation (6), and the estimated values of the model parameters for the three samples as well as their average values were obtained. The results are shown in Table 2.

Table 2. Parameter estimates of the degradation model.

Parameters	μ_1 Estimated Value	σ_1 Estimated Value
MCB 1	0.005	0.0138
MCB 2	0.0049	0.0138
MCB 3	0.0048	0.0138
Average value	0.0049	0.0138

The electrical life prediction of the MCB is carried out by taking MCB 1 as an example, the estimated values of the model parameters derived in Table 2 are substituted into Equations (12) and (13) to obtain the electrical life probability density function and reliability function of test article 1, and the curves are shown in Figure 13.

$$\mathrm{f}(t) = \frac{14.73}{0.0138\sqrt{2\pi t^3}} \exp\left[-\frac{(14.73 - 0.0049t)^2}{0.0004t} \right] \tag{14}$$

$$\mathrm{R}(t) = \Phi\left(\frac{14.73 - 0.0049t}{0.0389\sqrt{t}} \right) - \exp(763.4985)\Phi\left(\frac{-14.73 - 0.0049t}{0.0138\sqrt{t}} \right) \tag{15}$$

As can be seen in Figure 13, after about 2500 closing and closing operations, the reliability of sample 1 starts to decrease rapidly. The predicted electrical life of sample 1 is 2989 times, and the simulated electrical life is 3009 times, with a relative error of 0.7%, and the predicted and simulated electrical lives are very similar. In order to further verify the accuracy of the electrical degradation process, the estimated values of sample 2 and sample 3 and the mean drift parameter and diffusion parameter in Table 2 were substituted into Equations (14) and (15), and the expected values of the respective probability density functions were 2988, 2975, and 2990 times, which were also close to the simulated electrical life, with the relative errors of 0.7%, 1.13%, and 0.6%, respectively, which were accurate in the prediction results.

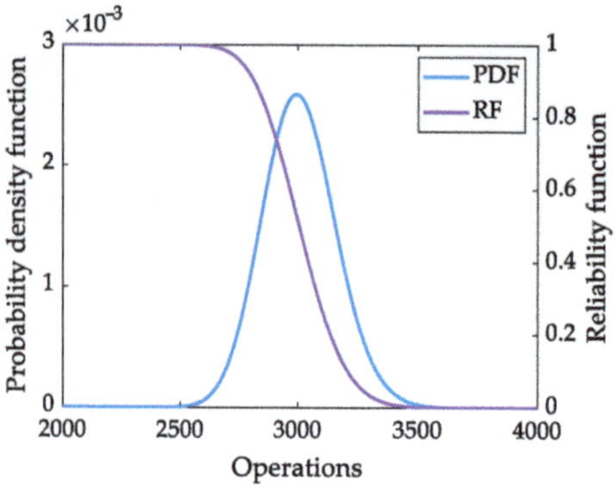

Figure 13. Reliability function and probability density function of simulated sample 1.

On this basis, the remaining electric life prediction of the MCB in different service stages was carried out, and still taking sample 1 as an example, after its electric life was divided into five stages, the current cumulative degradation amount L_0 and the estimated values of the model parameters in different service stages were substituted into Equation (12) to calculate the predicted value of the remaining electric life of the sample 1 in different service stages and compare it with the corresponding stage of the remaining life simulation value. They are shown in Table 3. The residual electric life prediction errors of sample 1 in different stages are less than 4%, and the prediction results are accurate, which shows that the proposed residual electric life prediction method can realize the residual electric life prediction of individual MCBs in different stages of service more accurately.

Table 3. Prediction of the remaining electrical life of sample 1 at different stages.

Stage	Actual Residual Electric Life	Prediction of Remaining Electrical Life	Relative Error
1–500	2509	2465	3.80%
501–1000	2009	1965	1.29%
1001–1500	1509	1429	3.68%
1501–2000	1009	957	1.76%
2001–2500	509	497	2.36%

The residual electrical life was predicted for sample 2 and sample 3 by the parameter estimates of sample 1, respectively, and the relative errors between the predicted and actual values were calculated, as shown in Figure 14a. Using this analytical method, the mean values of the parameter estimates of sample 2~3 and the samples were used to predict the remaining electrical life of sample 1~3 at different service stages, respectively.

As can be seen from Figure 14, through the model parameter values of sample 1~3 and the average values of the model parameters for the three samples to predict the remaining electrical life at different service stages, whether based on their own data or the data of other products in the same batch, the maximum prediction error at each stage is not more than 10%, which is within the acceptable range. Therefore, it is concluded from the above analysis that in the service process of MCBs, the trend of the change in the initial closing angle with the number of operations is a monotonous linear increase, and the electrical performance degradation model of MCBs can be established by using the univariate linear Wiener process and can accurately complete the prediction of residual electrical life for samples.

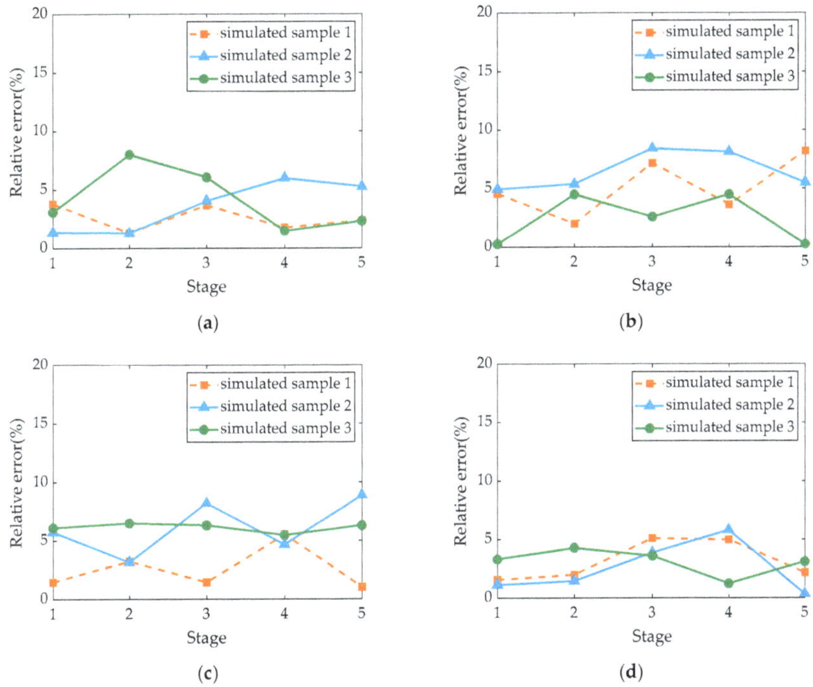

Figure 14. Prediction results of the remaining electrical life of the 3 simulated samples. (**a**) Based on data from simulated sample 1; (**b**) based on data from simulated sample 2; (**c**) based on data from simulated sample 3; (**d**) based on the average of 3 simulated samples.

5. Experimental Result Analysis

5.1. Experimental Data Analysis

This section carries out the electrical life experiments of three MCB samples with the experimental current of 63 A, and the breaking mode in the experiments is randomized. The breaking method in the experiment is random. The experimental conditions are shown in Table 4.

Figure 15a shows the MCB electrical life test apparatus, which mainly applies electrical stress to the MCB test sample and then performs a certain number of closing and opening operations until the MCB test sample reaches the end of its life. Figure 15b shows the MCB characteristic signal measurement and extraction apparatus, which fixes the MCB in place and then rotates the motor at a fixed speed to drive the MCB handle to rotate at a constant speed and measure the characteristic signal of the initial closing angle.

The initial closing angle of the MCB was collected after every 100 openings of the contacts, and the performance degradation data of the MCB at the rated current was analyzed. When the MCB fails to turn on after closing or fails to break when breaking, the sample is considered to have failed and the electrical life experiment of the sample is terminated. After the completion of the experiment, the electrical life of the three samples was 2897, 2989, and 3200 times, respectively.

Table 4. Parameters for life experiment.

Experimental Parameters	Numerical Value
Rated current Ie/A	63
Rated voltage Ue/V	380
Operating frequency/(times/h)	120
Coil voltage/V	380
Test current I/A	63
Test voltage U/V	380
Power supply frequency/Hz	50
Power factor	0.65
Movable contact material	AgNi(10)
Movable contact size/(mm^3)	$3.8 \times 2 \times 1.1$(R)
Static contact material	AgCdO(15)
Static contact size/(mm^3)	$4 \times 4 \times 1$

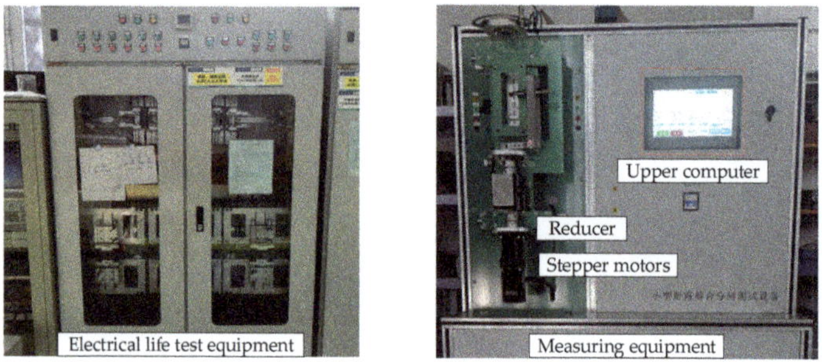

(**a**) (**b**)

Figure 15. Electrical life test setup. (**a**) Electrical life test equipment; (**b**) measuring equipment.

As can be seen from Figure 16, with the increase in the number of MCB closing and breaking times, the degree of contact erosion and wear is more serious, and the contact opening distance gradually increases, which is manifested in the closing process of the overall trend of the angle just closing. This is due to the MCB opening process of the breaking arc phase angle that has a randomness of generation of the arc, and the phenomenon of the arc combustion process is accompanied by evaporation and splashing and other phenomena that make the contact material occur at different degrees of loss, as well as the uneven distribution of contact material loss and other factors that lead to the uneven ablation of the contact surface and the formation of pits and protrusions during the opening process; the high and low surface morphology makes the initial closing angle in the process with the rise of small fluctuations, but this does not affect the overall upward trend of the initial closing angle.

This batch of products uses the IT9 manufacturing tolerance standard, which ensures high production consistency. The initial values and change trends of the initial closing angles of the three samples are basically consistent, further verifying the production consistency of the products. For products with insufficient consistency, the failure thresholds of different products at the moment of complete wear can be obtained through simulation, and their distribution patterns can be analyzed. This method can convert fixed failure thresholds into distribution-based random variables, thereby reducing the impact of product inconsistency.

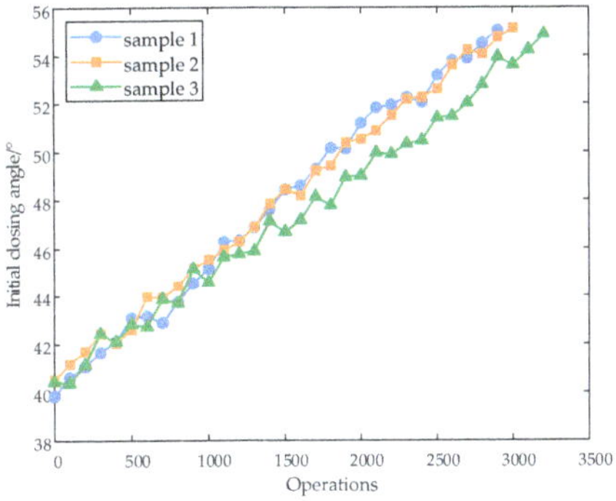

Figure 16. Change in the initial closing angle of the 3 samples at the rated current.

5.2. Normal Distribution Test

The K-S test was used to test the normal distribution of the data from the three samples.

As seen in Table 5, the initial closing angle degradation increment of each segment obeys a normal distribution, and the Wiener process can be used to describe the electrical degradation process of MCBs.

Table 5. K-S test results of initial closing angle.

Case	Test Statistic	Threshold Value	Obedience Test
MCB 1	0.1245	0.24	obey
MCB 2	0.1279	0.24	obey
MCB 3	0.1165	0.24	obey

5.3. Experimental Data Analysis

The model parameter estimates and their mean values for each test article were calculated from Equation (10) and the estimates are shown in Table 6.

Table 6. Parameter estimates of the degradation model.

Parameters	μ_1 Estimated Value	σ_1 Estimated Value
MCB 1	0.0052	0.0389
MCB 2	0.0049	0.0398
MCB 3	0.0045	0.0437
Average value	0.0049	0.0408

The electrical life prediction of the MCB is carried out as an example of test sample 1. The probability density function and reliability function of the electrical life of test sample 1 are obtained by substituting the estimated values of the model parameters derived in Table 6 into Equations (12) and (13).

$$f(t) = \frac{15.06}{0.0389\sqrt{2\pi t^3}} \exp\left[-\frac{(15.06 - 0.0052t)^2}{0.003t}\right] \tag{16}$$

$$R(t) = \Phi\left(\frac{15.06 - 0.0052t}{0.0389\sqrt{t}}\right) - \exp(104.2543)\Phi\left(\frac{-15.06 - 0.0052t}{0.0389\sqrt{t}}\right) \tag{17}$$

As can be seen from Figure 17, after about 1992 closing and closing operations of sample 1, its reliability began to decline rapidly, indicating that, at this time, sample 1 began to leave the normal service state, the probability of failure of the MCB began to increase, and it was necessary to consider the test article for timely maintenance or replacement. The expected value of the electric life of sample 1 is 2820 times, the actual electric life is 2897 times, the prediction error is 2.7%, and the predicted electric life is very close to the actual electric life.

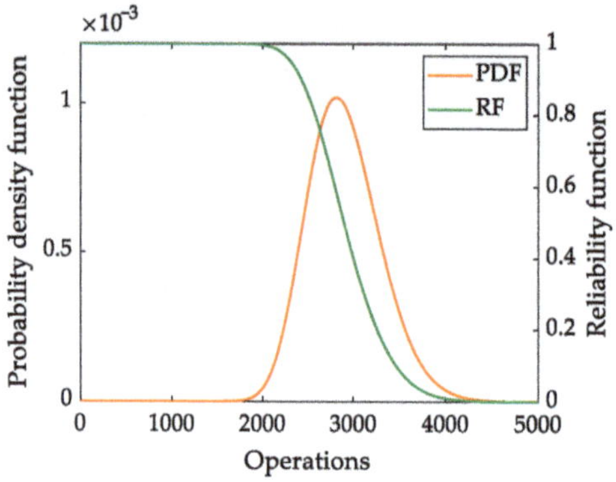

Figure 17. Reliability function and probability density function of experimental sample 1.

The estimated values of sample 2~3 and the average values of the model parameters in Table 6 are substituted into Equations (12) and (13) to predict the electrical life of sample 1. The expected values of the probability density function for the three cases are 3028, 3208, and 2964 times, which are close to the actual electrical life of sample 1, and the prediction errors are 4.5%, 10.7%, and 2.3%, respectively, which are within the acceptable range of engineering.

On this basis, the individual residual electric life prediction of MCB samples in different service stages is carried out, sample 1 is still taken as an example, its electric life is divided into five stages, the residual electric life prediction of each service stage is calculated and compared with the actual value, and the results are shown in Table 7. It can be seen that the relative error of the prediction of different stages is less than 8%, and the prediction results are more accurate.

The residual electrical life was predicted for sample 2 and 3, respectively, from the parameter estimates of sample 1, and the relative errors between the predicted and actual values were calculated, as shown in Figure 18a. Using this analytical method, the mean values of the parameter estimates of sample 2~3 and the samples were used to predict the remaining electrical life of sample 1~3 at different service stages, respectively.

Table 7. Prediction of the remaining electrical life of sample 1 at different stages.

Stage	Actual Residual Electric Life	Prediction of Remaining Electrical Life	Relative Error
1–500	2397	2206	7.97%
501–1000	1897	1811	4.53%
1001–1500	1397	1344	3.79%
1501–2000	897	865	3.57%
2001–2500	397	383	3.53%

As can be seen from Figure 18, through the model parameter values of sample 2~3 and the average values of the model parameters of the sample to predict the residual electrical life of the three samples at different service stages, the relative error between the predicted life and the actual life in all the above cases is within 15%, which is within the acceptable range of engineering.

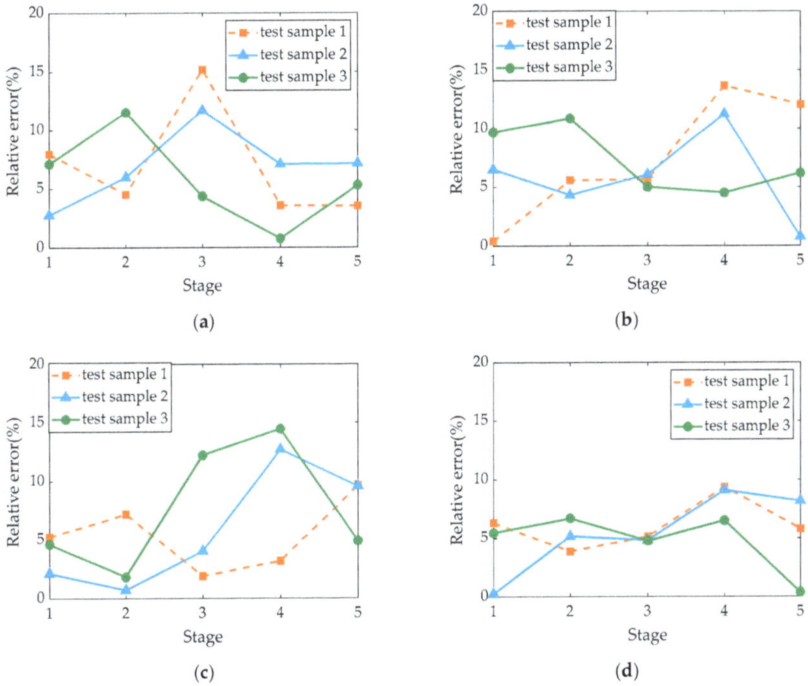

Figure 18. Prediction results of the remaining electrical life of the 3 experimental samples. (**a**) Based on data from experimental sample 1; (**b**) based on data from experimental sample 2; (**c**) based on data from experimental sample 3; (**d**) based on the average of 3 experimental samples.

In summary, the initial closing angle as a performance degradation characterization quantity can better characterize the electrical performance degradation of MCBs and also shows that the residual electric life prediction model established in this paper is accurate, and in the absence of its own performance degradation data, it can be used to carry out the reliability analysis of MCBs with the degradation data of the same batch of products.

6. Conclusions

In this paper, the dynamic characteristics of an MCB are simulated and analyzed, and a residual life prediction method using the initial closing angle to characterize the MCB is proposed and studied. The main conclusions are as follows:

(1) The reasons for the degradation of MCB performance are analyzed, the MCB kinetic simulation model is established based on the kinetic characteristic analysis, and the change characteristics of the initial closing angle and the initial closing torque with the wear amount of the contact during the closing process are analyzed; moreover, through sensitivity analysis of two characteristic quantities, the initial closing angle is selected as the characteristic quantity for predicting the remaining life.

(2) Assuming that the contact thickness of the MCB decreases uniformly after multiple breakage, the relationship between the degradation of the MCB electrical performance and the initial closing angle is analyzed, and the degradation model of the MCB electrical performance is established by using a one-dimensional linear Wiener process.

(3) The electrical performance degradation simulation based on the Monte Carlo method was carried out and the remaining electrical life was predicted from the simulation data, and the prediction errors did not exceed 10%. The MCB electrical life experiments under the 63A current were carried out, the maximum likelihood estimation and degradation model were obtained from the experimental data, and the maximum prediction error of the proposed method did not exceed 15% for the MCBs of the same batch in different operation stages.

Author Contributions: Conceptualization, writing—original draft, writing—review and editing, L.M.; methodology, C.W. and Y.W.; data curation, software ADAMS 2018, supervision, writing—review and editing, L.H.; data curation, supervision, visualization, writing—review and editing, P.H.; writing—original draft, L.M.; writing—review and editing, L.M. and Z.X. All authors have read and agreed to the published version of the manuscript.

Funding: This study was funded by the National Natural Science Foundation of China (Grant No. 51937004), Natural Science Foundation of Zhejiang Province (Grant No. LTGG23E070001), and Zhejiang Provincial Major Special Project (Grant No. 2024C01064).

Data Availability Statement: The original contributions presented in the study are included in the article, further inquiries can be directed to the corresponding author.

Conflicts of Interest: L.M., P.H., C.W. and Z.X. were employed by Zhejiang Testing & Inspection Institute for Mechanical and Electrical Products Quality Co., Ltd.; L.H. was employed by Qingdao Haier Intelligent Technology Research and Development Co., Ltd.; Y.W. was employed by Hebei University of Technology. This study was conducted without any commercial or financial relationships that could be construed as a potential conflict of interest.

References

1. Cheng, S.; Liu, J.; Wu, X.; Yu, T.; Xiang, C.; Yuan, D. Residual life prediction of metallized film capacitors based on feature extraction and error compensation. *Chin. J. Electr. Eng.* **2022**, *42*, 2672–2681.
2. You, Y.; Wang, J.; Shu, L.; Ni, K.; Zhou, X. An electrical life prediction method for AC contactors based on audio features. *J. Electrotechnol.* **2021**, *36*, 1986–1998.
3. Sun, S.; Wang, Z.; Wang, J.; Yang, F.; Ji, W. Remaining useful life prediction for circuit breaker based on SM-CFE and SA-BiLSTM. *IEEE Trans. Instrum. Meas.* **2023**, *72*, 1–14.
4. Su, X.; Wang, L.; Zhang, Z.; Wang, D. Residual Life Prediction of Low-Voltage Circuit Breaker Thermal Trip Based on the Wiener Process. *Energies* **2024**, *17*, 1189. [CrossRef]
5. Kong, W.; Shi, J.; Du, G.; Yuan, C.; Zhang, X. Research on Life Prediction Method of Earth Leakage Circuit Breaker Based on Statistical Test. *J. Electr. Eng.* **2024**, *19*, 382–390.
6. Liu, G.; Miao, J.; Zhao, X.; Wang, Z.; Li, X. Life prediction of residual current circuit breaker with overcurrent protection based on BP neural network optimized by genetic algorithm. *J. Electr. Eng. Technol.* **2022**, *17*, 2003–2014.
7. Razi-Kazemi, A.A.; Niayesh, K.; Nilchi, R. A probabilistic model-aided failure prediction approach for spring-type operating mechanism of high-voltage circuit breakers. *IEEE Trans. Power Deliv.* **2018**, *34*, 1280–1290.
8. Li, K.; Ma, D.; Zhao, C.; Hu, B.; Wang, H. Prediction of the remaining electrical life of AC contactors based on Bayesian inference. *J. Instrum. Meas.* **2024**, *45*, 34–45.
9. Chen, K.; Wang, F.; Li, H.; Gao, H.; Xia, Y.; Liu, Y. Remaining Life Prediction of High-Voltage Circuit Breakers Based on Optimized Particle Swarm-Bidirectional Long Short-Term Memory Neural Network. In Proceedings of the 2024 7th International Conference on Electric Power Equipment-Switching Technology (ICEPE-ST), Xiamen, China, 10–13 November 2024; IEEE: Piscataway, NJ, USA, 2024; pp. 697–701.
10. Huang, K.; Zhao, S. Evaluation of residual service life of high voltage circuit breaker. In Proceedings of the 2012 China International Conference on Electricity Distribution, Shanghai, China, 10–14 September 2012; IEEE: Piscataway, NJ, USA, 2012; pp. 1–4.
11. Lu, C.; Wang, C.; Xie, J. Online estimation of residual electrical lifetime for PV low voltage circuit breaker based grey algorithm. In Proceedings of the 2023 IEEE 3rd International Conference on Power, Electronics and Computer Applications (ICPECA), Shenyang, China, 29–31 January 2023; IEEE: Piscataway, NJ, USA, 2023; pp. 1216–1220.

12. Yang, J.; Chi, C. The Prediction of Residual Electrical Life in Alternating Current Circuit Breakers Based on Savitzky-Golay-Long Short-Term. *Sensors* **2023**, *23*, 6860.
13. Wu, T.; Yao, Y.; Li, Z.; Chen, B.; Wu, Y.; Sun, W. Remaining useful life prediction of circuit breaker operating mechanisms based on wavelet-enhanced dual-tree residual networks. *J. Power Electron.* **2024**, *24*, 78–91.
14. Ma, H.; Zhou, S.; Gao, C.; Zhou, F.; Yang, Y.; Yang, H. Research on the Reliability Test and Life Assessment Methods of Relays Used in Circuit Breaker Operating Mechanism. *Energies* **2023**, *16*, 4843. [CrossRef]
15. Cao, P.; Li, J.; Han, Y.; Shi, S.; Wang, Q.; Peng, Z.; Zhou, G. Mechanical remaining Life Prediction of Circuit Breaker based on long-term and short-term memory Neural Network. In Proceedings of the 2023 IEEE 6th International Electrical and Energy Conference (CIEEC), Hefei, China, 12–14 May 2023; IEEE: Piscataway, NJ, USA, 2023; pp. 2734–2738.
16. Zhao, X.; Wu, G.; Yang, D.; Xu, G.; Xing, Y.; Yao, C.; Abu-Siada, A. Enhanced detection of power transformer winding faults through 3D FRA signatures and image processing techniques. *Electr. Power Syst. Res.* **2025**, *242*, 111433. [CrossRef]
17. Su, X.; Zhang, Z.; Wei, J. Remaining Useful Life Prediction of Electromagnetic Release Based on Whale Optimization Algorithm—Particle Filtering. *Energies* **2024**, *17*, 670. [CrossRef]
18. Aizpurua, J.I.; Catterson, V.M.; Abdulhadi, I.F.; del Carmen Segovia García, M. A model-based hybrid approach for circuit breaker prognostics encompassing dynamic reliability and uncertainty. *IEEE Trans. Syst. Man Cybern. Syst.* **2017**, *48*, 1637–1648.
19. Liu, G.; Wang, L.; Yang, Y.; Liu, D.; Miao, J. Life prediction of residual current circuit breaker with overcurrent protection based on Tweedie exponential dispersion process with random effects. *Electr. Eng.* **2024**, *107*, 7027–7039.
20. Zhao, S.; Yan, X.; Wang, B.; Wang, E.; Ma, L. Research on reliability evaluation method of DC circuit breaker based on Markov model. *Electr. Power Syst. Res.* **2019**, *173*, 1–5.
21. Li, K.; Zhao, C.; Niu, F.; Zheng, S.; Duan, Y.; Huang, S.; Wu, Y. Electrical performance degradation model and residual electrical life prediction for AC contactor. *IEEE Trans. Compon. Packag. Manuf. Technol.* **2020**, *10*, 400–417.
22. Khoddam, M.; Sadeh, J.; Pourmohamadiyan, P. Performance evaluation of circuit breaker electrical contact based on dynamic resistance signature and using health index. *IEEE Trans. Compon. Packag. Manuf. Technol.* **2016**, *6*, 1505–1512.

energies

MDPI

Article

Research on Fault Diagnosis of High-Voltage Circuit Breakers Using Gramian-Angular-Field-Based Dual-Channel Convolutional Neural Network

Mingkun Yang [1], Liangliang Wei [2], Pengfeng Qiu [1], Guangfu Hu [1], Xingfu Liu [1], Xiaohui He [1], Zhaoyu Peng [1], Fangrong Zhou [1], Yun Zhang [3], Xiangyu Tan [2] and Xuetong Zhao [2,*]

1 Yunnan Electric Power Research Institute, Kunming 650217, China; xgllymk@163.com (M.Y.); university20241009@163.com (P.Q.); yyt12110427@163.com (G.H.); 202311131158t@stu.cqu.edu.cn (X.L.); shuike20241009@163.com (X.H.); 13206263732@163.com (Z.P.); 15070551243@163.com (F.Z.)

2 State Key Laboratory of Power Transmission Equipment Technology, School of Electrical Engineering, Chongqing University, Chongqing 400044, China; liangliangwei0208@gmail.com (L.W.); wll2569236501@163.com (X.T.)

3 Yuxi Power Supply Bureau, Yunnan Power Grid, Yuxi 653100, China; west109@126.com

* Correspondence: zxt201314@cqu.edu.cn

Abstract: The challenge of accurately diagnosing mechanical failures in high-voltage circuit breakers is exacerbated by the non-stationary characteristics of vibration signals. This study proposes a Dual-Channel Convolutional Neural Network (DC-CNN) framework based on the Gramian Angular Field (GAF) transformation, which effectively captures both global and local information about faults. Specifically, vibration signals from circuit breaker sensors are firstly transformed into Gramian Angular Summation Field (GASF) and Gramian Angular Difference Field (GADF) images. These images are then combined into multi-channel inputs for parallel CNN modules to extract and fuse complementary features. Experimental validation under six operational conditions of a 220 kV high-voltage circuit breaker demonstrates that the GAF-DC-CNN method achieves a fault diagnosis accuracy of 99.02%, confirming the model's effectiveness. This work provides substantial support for high-precision and reliable fault diagnosis in high-voltage circuit breakers within power systems.

Keywords: high-voltage circuit breaker; vibration faults; Gramian Angular Field; convolutional neural network

1. Introduction

High-voltage circuit breakers are pivotal components within electrical power systems, playing essential roles in both the control and protection of circuits. Their reliability is directly correlated with the stability and safety of the entire power network [1–3]. Normally, circuit breakers remain inactive; however, during operational maneuvers or in the event of faults, they must act swiftly to either establish or disrupt the electrical supply [4,5]. Consequently, power systems impose stringent requirements on the operational reliability of high-voltage circuit breakers. Among the various operating mechanisms, spring-operated mechanisms are preferred due to their simplicity, rapid response, non-polluting nature, and compact size. These primarily include hydraulic-spring, pneumatic-spring, and fully mechanical spring mechanisms [6,7].

Currently, over 60% of circuit breaker failures are attributed to mechanical issues, such as transmission mechanism jamming, opening spring stress relaxation and buffer failure [8]. These failures generate a wealth of vibration signals during the operation of the circuit breakers, which contain extensive information about the condition of the equipment. Accurately extracting and processing these vibration signals poses a significant challenge in the field of fault diagnosis for high-voltage circuit breakers. In recent years, machine learning algorithms have been extensively studied for fault detection in circuit breakers. Notably, Support Vector Machines (SVM), Decision Trees, and Random Forests have been employed for the classification and diagnosis of mechanical faults in high-voltage circuit breakers [9–11]. However, these methods have limitations. Firstly, they typically rely on manually calculated features, such as time-domain and frequency-domain feature extraction, which can lead to information loss in vibration signals. Secondly, the feature representation capabilities of traditional machine learning models are limited, making it difficult to capture the deeper features within the vibration signals of circuit breakers, thereby impacting the accuracy of fault identification. In light of these issues, the rapid advancement of deep learning technologies in recent years has provided novel solutions for intelligent fault diagnosis in high-voltage circuit breakers. Compared to traditional methods, deep learning can automatically extract high-dimensional features from raw data, enhancing the accuracy of feature learning [12–16]. For instance, Du et al. [17] have introduced a deep residual network, which transforms one-dimensional current signals into two-dimensional grayscale images. This method employs the adaptive parameterized rectified linear activation functions and the guaranteed convergence (AMSGRAD) optimization algorithm to achieve fault diagnosis in high-voltage circuit breakers, although its accuracy is dependent on a large amount of sample data. Furthermore, the work of Yan and Huang [18,19] involved transforming vibration signals into time-frequency images for fault diagnosis using convolutional neural networks, although such representations primarily capture overall signal effects and struggle to extract instantaneous features of the signals.

Recently, Gramian Angular Field (GAF) transformations have been developed in time-series imaging for signal representation. GAF encodes one-dimensional time series into two-dimensional matrices by converting temporal dynamics into polar coordinate systems, preserving the information of amplitude and phase information [18,19]. This representation has been widely utilized in fields such as human activity recognition and biomedical diagnostics due to its ability to retain temporal correlation patterns in images [20,21]. Subsequently, Dual-Channel Convolutional Neural Networks (DC-CNNs) emerged as a powerful architecture that can process complementary inputs in parallel. Zhou et al. found that DC-CNNs—by simultaneously learning from diverse signal representations—outperform single-channel models in tasks such as power equipment classification [22]. Consequently, this study introduces a methodology that integrates the Gramian Angular Field (GAF) signal processing approach with a Dual-Channel Convolutional Neural Network (DC-CNN) for the precise extraction and analysis of transient features in signals. Specifically, this method initially converts one-dimensional vibration signals collected by sensors at five different locations on a circuit breaker into images of Gramian Angular Summation Field (GASF) and Gramian Angular Difference Field (GADF). This transformation not only augments the volume of data samples but also preserves the transient information within the signal. Subsequently, images representing each type of fault in the form of five-channel GASF and GADF images are input into two CNN modules within the DC-CNN architecture for feature extraction. The features extracted by the two CNN modules are ultimately merged to achieve high-precision fault diagnosis in high-voltage circuit breakers. The principal contributions of this research can be described through several steps. (1) Development of a GAF Representation Method (GASF + GADF): The proposed GAF-DC-CNN method is integrated with the GASF and GADF image representations via a dual-channel convolutional neural network, which is capable of effectively capturing transient features of the vibration signals. (2) Design of a DC-CNN: This network converts each fault signal into five-channel GASF and GADF images for input into convolutional networks to extract features, thereby achieving an integration of complementary signals to enhance fault diagnosis performance. (3) Comprehensive Experimental Comparison: The proposed method was compared with single-channel GAF methods (GASF-CNN, GADF-CNN) and sequence-based deep networks (CNN–LSTM). The experimental results demonstrate superior robustness and diagnostic accuracy of the proposed method.

2. Methodology

2.1. GAF Transformation

The GAF is a coding method that converts one-dimensional time series into two-dimensional images [23]. The core concept involves representing data points of a time series in Cartesian coordinates and transforming them into polar coordinates, thereby generating images through a Gram matrix based on angular relations. Specifically, given a time series of length N, $\{x_1, x_2, \ldots, x_N\}$, the data are first normalized within the interval $[-1,1]$ to fit the domain of the arccosine function. To avoid potential data leakage, this normalization is applied independently to each individual sample after the dataset is split into training sets and test sets. The polar coordinate angle ϕ_i is defined using Equation (1),

$$x_i = \cos(\phi_i),\ i = 1,\ 2,\ \ldots,\ N \tag{1}$$

where $\phi_i \in [0,\pi]$.

Based on this mapping, matrices for both the GASF and GADF are constructed. The GASF utilizes the cosine of the sum of angles to define the Gram matrix G_{ij}, where each element is derived from the cosine of the sum of angles corresponding to time points i and j.

$$G_{ij} = \cos(\phi_i + \phi_j),\ 1 \le i,\ j \le N \tag{2}$$

Consequently, this study explicates the intrinsic synthesis of amplitude information of a time series at moments i and j within the elements G_{ij} of the GASF matrix. This synthesis emphasizes the impact of phase consistency (or amplitude sign) between different time points on the matrix elements where points with identical phases (or amplitude signs) generate larger values within the matrix, reflecting the periodicity and trend similarity of the series.

GADF employs the sine function of the angular difference to define its Gram matrix, denoted as H_{ij}. The elements of this matrix are calculated from the sine of the angular difference between two time points.

$$H_{ij} = \sin(\phi_i - \phi_j),\ 1 \le i,\ j \le N \tag{3}$$

Unlike the GASF, the GADF matrix is not symmetric, and its principal diagonal elements are given by sin(0) = 0, which does not directly contain the original signal's amplitude information. Instead, the GADF matrix focuses more on the transient changes in the signal. Therefore, the values in the GADF matrix depict the short-term fluctuations and changing trends between different points in the time series, rendering it more sensitive to sudden changes, local oscillations, and other transient behaviors.

By integrating both GASF and GADF, it is possible to comprehensively characterize both static trends and dynamic changes in time series. As illustrated in Figure 1 below, one-dimensional vibration data can be transformed into two-dimensional matrix images G_{ij} or H_{ij}, which can support the input for subsequent DC-CNN processing.

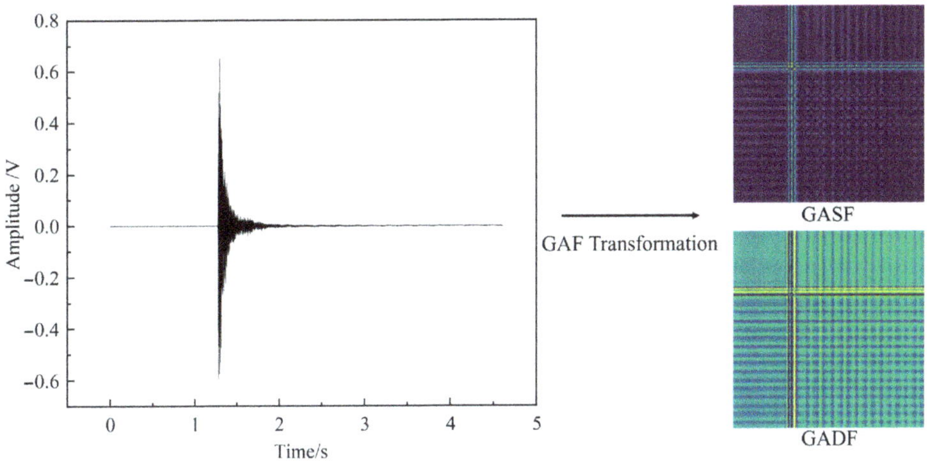

Figure 1. GAF transformation of vibration signal.

2.2. Multi-Channel GAF Image Construction

In the scenario involving a single sensor, the previously mentioned GAF methodology enables the transformation of a one-dimensional vibration signal monitored by the sensor into a two-dimensional GAF image. For data derived from multiple sensors, the objective is to concurrently integrate the information from each sensor while preserving the unique characteristics of each signal. To this end, a multi-channel GAF image can be constructed by using the GASF or GADF matrices generated by each sensor as a channel, and stacking these channels to form a high-dimensional image tensor.

Here, m sensors were synchronously used to collect time-series signals that were denoted as $X^{(k)} = \{x^{(k)}{}_1, x^{(k)}{}_2, \ldots, x^{(k)}{}_N\}$, where k = 1, 2, 3... represents the sensor index, and N is the length of the signal. For each sensor k, according to the definition of the GAF transformation, the GAF matrix is computed with Equation (4).

$$\begin{aligned} G_{ij}^{(k)} &= \cos\left(\phi_i^{(k)} + \phi_j^{(k)}\right) \\ H_{ij}^{(k)} &= \sin\left(\phi_i^{(k)} - \phi_j^{(k)}\right), \ 1 \le i, j \le N \end{aligned} \tag{4}$$

When constructing the multi-channel image, each sensor's matrix serves as a channel in the fusion process. The multi-channel GASF image is represented by Equation (5).

$$\vartheta_{GASF}(i,j,k) = \cos\left(\phi_i^{(k)} + \phi_j^{(k)}\right), k = 1,\ 2,\ \ldots\ m \tag{5}$$

Similarly, the multi-channel GADF image is represented by Equation (6).

$$\vartheta_{GADF}(i,j,k) = \sin\left(\phi_i^{(k)} - \phi_j^{(k)}\right), k = 1,\ 2,\ \ldots\ m \tag{6}$$

In these equations, i and j denote the spatial positions within the image, and k indicates the corresponding channel (sensor dimension). Through this methodology, the two-dimensional angular field feature maps from m sensors are merged together, yet they remain independently preserved in distinct channels.

This method of multi-channel data fusion offers significant advantages in terms of information retention and feature extraction. On one hand, the original temporal information of each sensor, after undergoing the GAF transformation, is completely preserved within its respective channel without loss of detail or mutual interference. Compared to simplistic signal merging or dimensionality reduction, the multi-channel representation ensures that key patterns from all sensors are presented, thereby avoiding potential information loss that might occur in a single-channel approach. On the other hand, the multi-channel GAF images provide a rich and structured input for subsequent feature extraction. In the form of images, advanced pattern recognition techniques such as two-dimensional convolution can be utilized to learn features across the model that can extract temporal pattern features within each sensor sequence in individual channels, and also capture correlations and complementary information between different sensor signals through cross-channel convolutional kernels. This means that multi-channel GAF fusion not only preserves the independent information of each sensor but also provides an opportunity for algorithms to explore inter-sensor feature relationships, thus enhancing the effectiveness of pattern recognition and classification.

2.3. Evaluation Metrics

To quantitatively evaluate the performance of the proposed GAF-DC-CNN model and compare it with baseline models, three typical classification metrics such as recall (*R*), F1 score, and AUC (area under curve) are adopted. These metrics provide a comprehensive assessment of classification effectiveness, especially in multi-class and imbalanced datasets.

Recall, also known as sensitivity, is widely used to measure the proportion of correctly predicted positive samples [24] and is defined as in Equation (7),

$$R = TP/(TP + FN) \tag{7}$$

where *TP* (true positives) represents the number of correctly predicted fault cases, and *FN* (false negatives) represents the number of actual fault cases that were incorrectly classified.

The *F1* score is the harmonic mean of precision and recall, as given by Equation (8),

$$F1 = 2 \times (P \times R)/(P + R) \tag{8}$$

where precision *P* is defined as

$$P = TP/(TP + FP) \tag{9}$$

where *FP* denotes the number of incorrectly predicted fault cases.

The AUC is used to evaluate the model's capability to distinguish fault cases at various threshold settings. For multi-class problems, AUC can be calculated to process the overall multi-class classification as a set of binary decisions,

$$AUC = (1/N) \times \sum AUC(y_i, \hat{y}_i) \tag{10}$$

where y_i is the true label, $\hat{y}_i$ is the predicted probability of the positive class for the *i*-th sample, and *N* is the total number of samples across all classes.

3. Fault Identification Method Based on GAF-DC-CNN

3.1. Vibration Data Collection and GAF Processing

Given the significance of the disengagement operation for the electrical grid's safety, especially as high-voltage circuit breakers typically remain in a closed state for extended periods, it is essential to gather one-dimensional vibration data from high-voltage circuit breakers under various fault conditions during disengagement. Taking the CT26 spring-operated mechanism of a 220 kV circuit breaker as an example, vibration data under different operational conditions were collected using five LZ-601 accelerometers (as illustrated in Figure 2). These sensors, with a sensitivity of 10.0 mV/g, were set to collect data at a frequency of 10 kHz. Prior studies suggest that the sampling rate for vibration signals from high-voltage circuit breakers usually ranges between 8 and 16 kHz [25,26]. Pre-tests conducted on five typical mechanical failures—sticking drive mechanisms, relaxation of disengagement springs, abnormal core stroke or gap, and damper faults—using an oversampling rate of 25 kHz revealed a peak significant frequency of 4.7 kHz. According to the Nyquist criterion, a sampling rate of 10 kHz accurately captures all characteristic frequencies, balancing data volume with storage and training efficiency.

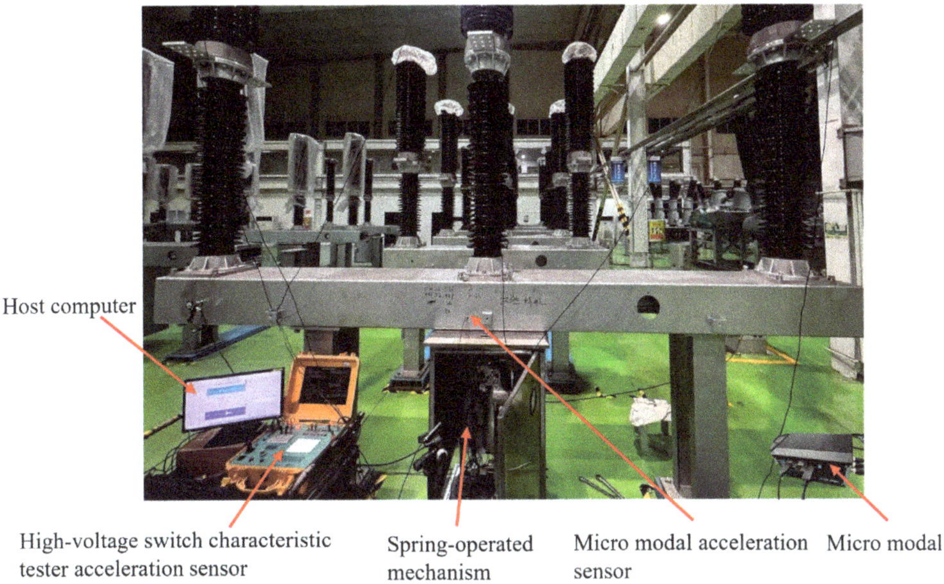

Figure 2. Vibration signal collection system of the spring-operated mechanism.

To effectively extract vibration characteristics of the high-voltage circuit breaker under different fault conditions, it is necessary to carry out feature transformations on the collected one-dimensional time series data to enhance the accuracy of subsequent fault identification. Initially, the collected signals are normalized to mitigate the impact of varying scales and to ensure that the data distribution is suitable for GAF encoding. Subsequently, the normalized signals are mapped to the angular domain, and two-dimensional feature matrices are constructed using GASF and GADF methods, respectively. GASF primarily captures global features of the signal, whereas GADF focuses on depicting local changes in the signal pattern. Figure 3 presents the GASF and GADF images derived from the vibration signals from six operational conditions, (a) normal operation, (b) transmission mechanism jamming, (c) opening spring stress relaxation, (d) abnormal core stroke, (e) abnormal core gap, and (f) buffer failure. These images encode the temporal correlation of vibration signals using angular-based transformations. Therefore, subtle variations such as localized distortions in texture patterns resulting from various operational conditions can be obtained to reflect different dynamic behaviors of the circuit breaker. These differences can be captured through the feature extraction using the convolutional neural network, enabling accurate fault classification.

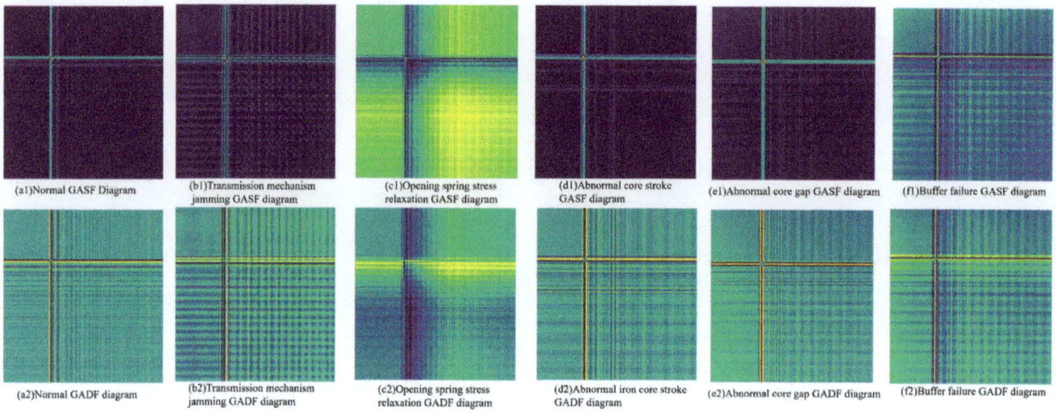

Figure 3. GAF images corresponding to different operational conditions.

3.2. Construction of DC-CNN

Traditional CNN architectures typically support only a single type of data input, which poses challenges in directly processing multiple, distinct data types. For the GAF transformation, which comprises GASF and GADF image representations, using these images as inputs for a standard CNN for feature extraction and classification could be hindered by the network's architectural constraints, thereby not fully leveraging the information from both types of images.

To address this issue, a DC-CNN that consists of two independent CNN branches is proposed to process either GASF or GADF images as shown in Figure 4. After each CNN module extracts features from its respective inputs, a feature fusion strategy is employed to integrate this information, ultimately aiding in classification. At the network's input layer, data are formatted as multi-channel images with dimensions H × W × C, where H and W represent the height and width of the image, respectively, and C denotes the number of channels, corresponding to the number of vibration signals collected by multiple sensors under the same operational conditions. The input data are GASF and GADF images derived from the GAF methodology, standardized in height and width to maintain uniformity in subsequent CNN processing.

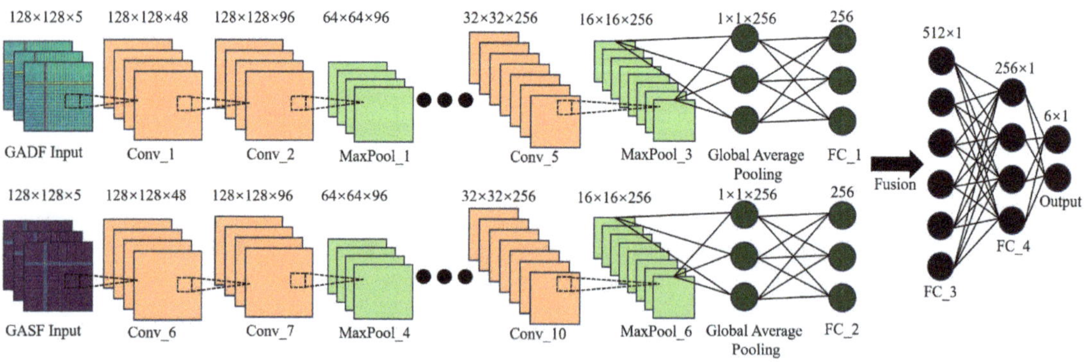

Figure 4. Schematic of the fault identification method based on GAF-DC-CNN.

During the feature extraction phase, each CNN branch comprises multiple convolutional and pooling layers. The convolutional layers utilize a sliding window mechanism on the input feature maps for local perception, thus extracting spatial structural information and identifying key patterns [26]. To reduce computational complexity while preserving essential feature information, the pooling layers employ a max pooling strategy, selecting the maximum value within a local region as the feature output [27,28]. Additionally, batch normalization is introduced following the convolutional and pooling layers to accelerate model convergence and reduce internal covariate shift.

In the backend of the network, global average pooling (GAP) is incorporated to enhance the model's capability for feature representation. Unlike max pooling, GAP does not select extremes in a local area but computes a global average across the entire feature map, mapping high-dimensional features of each channel to a single scalar. This strategy effectively reduces the model's parameter count, thereby decreasing the risk of overfitting, and also enhances the global expression of features.

A critical aspect of the DC-CNN is the choice of feature fusion strategy. Common methods include direct concatenation, weighted summation, and element-wise multiplication. Direct concatenation preserves the entirety of the feature information extracted by both CNN branches; therefore, this method is utilized in this study to enhance the model's joint representational capacity for different input data and improve classification performance.

The specific network structure and parameter configurations involved in the implementation of the DC-CNN are detailed in Table 1. During the training phase, the Adam (Adaptive Moment Estimation) optimizer is employed for parameter updates, with the cross-entropy loss function serving as the objective function. The Adam optimizer is an adaptive gradient-based optimization algorithm that computes individual learning rates for each parameter by combining the advantages of momentum and root mean square propagation (RMSProp). It is widely used in signal processing and image recognition due to its fast convergence and high stability in deep learning applications [29,30]. The initial learning rate is set at 0.0005, and L2 regularization is utilized to mitigate overfitting, with a regularization coefficient of 0.01. These hyperparameters are determined based on experimental data to ensure the model's generalization capability. The convolutional and pooling layer configurations of the GASF and GADF branches are designed to be symmetric, which enables consistent feature extraction and facilitates effective feature fusion.

Table 1. Structural parameters of DC-CNN.

Layer Name	Number of Channels	Convolution Kernel/Pooling Window Size	Stride Length	Output Dimensions	Padding
		GADF:			
Input	5	—	—	128 × 128 × 5	—
Conv_1	48	5 × 5	1	128 × 128 × 48	SAME
Conv_2	96	4 × 4	1	128 × 128 × 96	SAME
MaxPool_1	96	2 × 2	2	64 × 64 × 96	—
Conv_3	128	3 × 3	1	64 × 64 × 128	SAME
Conv_4	192	3 × 3	1	64 × 64 × 192	SAME
MaxPool_2	192	2 × 2	2	32 × 32 × 192	—
Conv_5	256	3 × 3	1	32 × 32 × 256	SAME
MaxPool_3	256	2 × 2	2	16 × 16 × 256	—
Global Average Pooling	256	16 × 16	—	1 × 1 × 256	—
FC_1	—	—	—	256	—
		GASF:			
Input	5	—	—	128 × 128 × 5	—
Conv_6	48	5 × 5	1	128 × 128 × 48	SAME
Conv_7	96	4 × 4	1	128 × 128 × 96	SAME
MaxPool_4	96	2 × 2	2	64 × 64 × 96	—

Table 1. *Cont.*

Layer Name	Number of Channels	Convolution Kernel/Pooling Window Size	Stride Length	Output Dimensions	Padding
Conv_8	128	3 × 3	1	64 × 64 × 128	SAME
Conv_9	192	3 × 3	1	64 × 64 × 192	SAME
MaxPool_5	192	2 × 2	2	32 × 32 × 192	—
Conv_10	256	3 × 3	1	32 × 32 × 256	SAME
MaxPool_6	256	2 × 2	2	16 × 16 × 256	—
Global Average Pooling	256	16 × 16	—	1 × 1 × 256	—
FC_2	—	—	—	256	—
	Integration layer: direct connection method				
FC_3	—	—	—	512	—
FC_4	—	—	—	256	—
Output	—	—	—	6	—

In the data preprocessing stage, data from each operational condition are collected by five sensors, accumulating a total of 2400 one-dimensional data points across six conditions, five sensors, and 80 disengagement operations. After undergoing the GAF transformation, this results in 4800 images with a resolution of 128 × 128, representing both GASF and GADF. These images, combined with multi-channel inputs from five sensors, form a dataset of 960 five-channel image samples. These samples are fed into the two independent channels of the CNN to thoroughly mine key information from different images.

4. Results

In this work, a comparative analysis with three models—CNN–LSTM, GADF-CNN, and GASF-CNN—was carried out. The CNN–LSTM model, which integrates a one-dimensional CNN with Long Short-Term Memory (LSTM) networks, is particularly adept at classifying time-series data. Thus, it is employed as a baseline model to explore the impact of the GAF transformation on enhancing classification performance. Concurrently, to further assess the efficacy of GAF-DC-CNN in diagnosing faults in high-voltage circuit breakers, GADF-CNN and GASF-CNN, which are single-channel neural network architectures, serve as comparative models. An in-depth comparison of the GADF and GASF transformations in classification tasks is undertaken to evaluate their differences in feature representation and classification capabilities.

The datasets of GAF images representing normal states and various faults, such as CT-26 transmission mechanism stickiness, disengagement spring stress relaxation, abnormal disengagement iron core stroke, irregular disengagement iron core gap, and CT-26 buffer faults, are divided into training and test sets in an 8:2 ratio. The CNN–LSTM is trained using the original one-dimensional vibration data, while GADF-CNN and GASF-CNN are trained with the GADF image dataset and GASF image dataset, respectively, and the proposed GAF-DC-CNN model is trained using the multi-channel GAF image dataset. Figure 5 displays the accuracy changes in the test set during the training process for all four models. From the accuracy curves, it is evident that the CNN–LSTM model exhibits a slower convergence speed and lower test set accuracy compared to the other three models, stabilizing after approximately 500 iterations. In contrast, the other three models tend to stabilize after just 300 iterations, illustrating the enhancements in convergence speed and accuracy brought about by the GAF transformation.

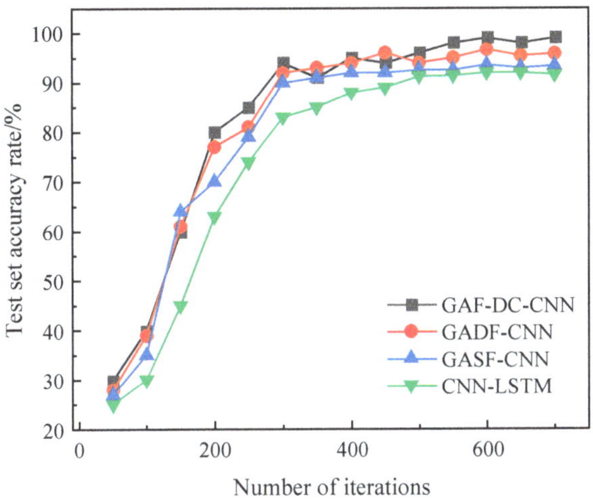

Figure 5. Accuracy curves of the test sets for the four models.

To verify the effectiveness of the GAF-DC-CNN, this study assesses four models using accuracy, recall, F1 score, and AUC from the test dataset. The AUC is used to measure a model's ability to differentiate between positive and negative class samples across various classification thresholds. For multi-class tasks, this research adopts a "micro-average" approach for calculating the AUC, treating the issue as a global binary classification problem. By aggregating the prediction outcomes across all categories, a single AUC value is computed to evaluate the overall performance of the model. An AUC value closer to 1 indicates superior overall classification performance and a stronger ability to discriminate among fault categories.

As shown in Table 2, the results, indicate a performance trend of GAF-DC-CNN > GADF-CNN > GASF-CNN > CNN–LSTM across the four evaluation metrics. Notably, the GAF-DC-CNN model achieved a recognition accuracy of 0.9902, demonstrating the highest accuracy in the identification of faults in high-voltage circuit breakers, while the CNN–LSTM only reached an accuracy of 0.9004. Moreover, the calculation of recall is based on the weighted average of fault recognition results, ensuring that it reflects the overall model's capability to classify types of spring faults. Overall, the GAF-DC-CNN model presents the best performance, illustrating the enhancements in classification ability conferred by the GAF transformation and the DC-CNN architecture. The confusion matrix of the GAF-DC-CNN model on the test set is shown in Figure 6, and a high classification accuracy across all fault types was achieved. It can be found that each fault category is classified with high accuracy from 97.90% to 100%, and the average per-class accuracy is 99.02%, which indicates that the GAF-DC-CNN model is feasible in accurately identifying any single fault type occurring in a CT26 operating mechanism.

Table 2. Evaluation metrics for the four models.

Model	Accuracy	Recall	F1 Score	AUC
GAF-DC-CNN	0.9902	0.9971	0.9841	0.9962
GADF-CNN	0.9523	0.9602	0.9547	0.9974
GASF-CNN	0.9234	0.9211	0.9484	0.9893
CNN–LSTM	0.9004	0.8919	0.9084	0.9692

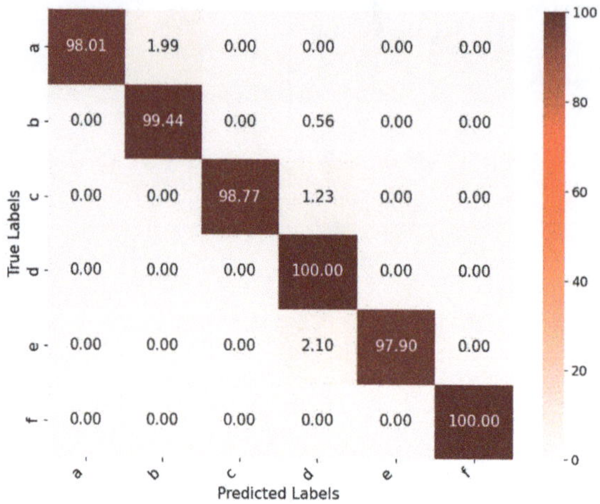

Figure 6. Confusion matrix of the GAF-DC-CNN model (accuracy in %). Fault classes: a: normal operation, b: transmission mechanism jamming, c: opening spring stress relaxation, d: abnormal core stroke, e: abnormal core gap, f: buffer failure.

To further verify the viability of the proposed GAF-DC-CNN model, we conducted five-fold cross-validation on the entire dataset. The dataset was randomly partitioned into five equal-sized folds, where each fold served once as the test set and the remaining four folds as the training set. This procedure was repeated for five times in total, and the performance metrics were averaged across all folds. Table 3 presents the averaged results of accuracy, recall, F1 score, and AUC. The results show that the proposed model maintains high classification performance with low variance, indicating its strong generalizability across various data splits.

Table 3. Performance metrics of the GAF-DC-CNN model under five-fold cross-validation.

Metric	Mean Value	Standard Deviation
Accuracy	0.9984	±0.0042
Recall	0.9943	±0.0031
F1 Score	0.9827	±0.0038
AUC	0.9951	±0.0026

5. Conclusions

In this study, a GAF-DC-CNN approach was proposed to address the non-stationary characteristics of vibration signals in high-voltage circuit breakers and the limitations of traditional methods in fault diagnosis. By employing dual-channel feature extraction through GASF and GADF, the GAF-DC-CNN effectively captures both global trends and local dynamic characteristics of variation signals, significantly enhancing the diagnostic accuracy for mechanical failures in circuit breakers. Experimental results revealed that the GAF-DC-CNN achieved an overall classification accuracy of 99.02% under six representative operating conditions of a 220 kV circuit breaker. The confusion matrix further demonstrated consistently high per-class accuracies ranging from 97.90% to 100%, verifying the model's balanced discrimination capability across diverse fault types. Additionally, the robustness and generalization performance of the proposed method were validated through five-fold cross-validation, yielding an average accuracy of 99.84%. These findings substantiate the effectiveness of incorporating GAF transformations for fault identification in high-voltage circuit breakers and highlight the enhanced stability and reliability conferred by the dual-channel architecture. Thus, the proposed GAF-DC-CNN method provides an efficient and reliable intelligent analysis tool for diagnosing mechanical failures in high-voltage circuit breakers, facilitating the advancement of automation and intelligence in power system equipment condition monitoring.

Author Contributions: Conceptualization, M.Y. and P.Q.; data curation, M.Y. and X.L.; methodology, L.W. and X.H.; validation, Z.P. and F.Z.; writing—original draft, L.W., G.H. and Y.Z.; writing—review and editing, X.Z. and X.T. All authors have read and agreed to the published version of the manuscript.

Funding: This work was supported by China Southern Power Grid (YNKJXM20222318).

Data Availability Statement: The data presented in this study are available on request from the corresponding author. The data are not publicly available due to restrictions imposed by the data provider/funder.

Conflicts of Interest: Yun Zhang was employed by the Yunnan Power Grid Co., Ltd. The remaining authors declare that the research was conducted in the absence of any commercial or financial relationships that could be construed as a potential conflict of interest.

References

1. Liu, Y.; Zhang, G.; Zhao, C.; Lei, S.; Qin, H.; Yang, J. Mechanical condition identification and prediction of spring operating mechanism of high voltage circuit breaker. *IEEE Access* **2020**, *8*, 210328–210338. [CrossRef]
2. Yang, M.; Wei, L.; Qiu, P.; Hu, G.; Yang, K.; He, X.; Peng, Z.; Zhou, F.; Zhang, Y.; Luo, J.; et al. Evaluation on the Long-Term Operational Reliability of Closing Springs in High-Voltage Circuit Breakers. *Energies* **2025**, *18*, 1806. [CrossRef]
3. Razi-Kazemi, A.A.; Niavesh, K. Condition monitoring of high voltage circuit breakers: Past to future. *IEEE Trans. Power Deliv.* **2020**, *36*, 740–750. [CrossRef]
4. Zhu, Y.; Jiang, Y.; Cao, F.; Wang, P.; Ke, J.; Liu, J.; Nie, Y.; Li, G.; Wei, Y.; Lu, G.; et al. Space charge induced electrofluorochromic behavior for C12-BTBT based thin-film devices. *J. Mater. Chem. C* **2025**, *13*, 12027. [CrossRef]
5. Su, Y.; Lu, Y.; Xie, Z.; Wang, J.; Luo, C. Study on closing spring fatigue characteristics of high voltage circuit breaker. *IOP Conf. Ser. Earth Environ. Sci.* **2020**, *508*, 012174. [CrossRef]
6. Zhao, X.; Wu, G.; Yang, D.; Xu, G.; Xing, Y.; Yao, C.; Abu-Siada, A. Enhanced detection of power transformer winding faults through 3D FRA signatures and image processing techniques. *Electr. Pow. Syst. Res.* **2025**, *242*, 111433. [CrossRef]
7. Wan, S.; Chen, L. Fault diagnosis of high-voltage circuit breakers using mechanism action time and hybrid classifier. *IEEE Access* **2019**, *7*, 85146–85157. [CrossRef]
8. Shao, H.; Jiang, Y.; Zhao, J.; Yang, M.; Wang, X.; Yang, H. Hybrid multi-scale residual network for high-voltage circuit breakers fault diagnosis. *Electron. Lett.* **2025**, *61*, e70135. [CrossRef]
9. Liu, M.; Wang, K.; Sun, L.; Zhen, J. Applying empirical mode decomposition (EMD) and entropy to diagnose circuit breaker faults. *Optik* **2015**, *126*, 2338–2342. [CrossRef]
10. Wu, L. Research on Fault Diagnosis and Maintenance Decision of High Voltage Circuit Breaker. Master's Thesis, Tianjin University, Tianjin, China, 2019.
11. Ji, T.; Yi, L.; Tang, W.; Shi, M.; Wu, Q. Multi-map fault diagnosis of high voltage circuit breaker based on mathematical morphology and wavelet entropy. *CSEE J. Power Energy Syst.* **2019**, *5*, 130–138. [CrossRef]
12. Huang, N.; Chen, H.; Cai, G.; Fang, L.; Wang, Y. Mechanical fault diagnosis of high voltage circuit breakers based on variational mode decomposition and multi-layer classifier. *Sensors* **2016**, *16*, 1887. [CrossRef] [PubMed]
13. Sun, L. Research on Deep Residual Network Model and Its Application in Circuit Breaker Fault Diagnosis. Master's Thesis, Hebei University of Technology, Hebei, China, 2023.
14. Xu, J.; Zhang, B.; Lin, X.; Li, B.; Teng, Y. Application of energy spectrum entropy vector method and RBF neural networks optimized by the particle swarm in high-voltage circuit breaker mechanical fault diagnosis. *High Voltage Eng.* **2012**, *38*, 1299–1306.
15. Zhao, K.; Yang, J.; Ma, S.; Wang, Y.; Wu, J.; Liang, C. Application of neural network ensemble model in mechanical fault identification of high voltage circuit breaker. *High Volt. Appar.* **2018**, *54*, 217–223.
16. Zhao, L.; Ji, Y.; Wu, Y.; Wu, X.; Ning, W.; Huang, X.; Ren, J. LM-BP fault diagnosis algorithm for spring operating mechanism of high voltage circuit breaker based on motor current. *Electr. Meas. Instrum.* **2024**, *61*, 48–55.
17. Du, T.; Sun, L.; Sun, S.; Chen, F.; Liu, J. Application of deep residual network in circuit breaker fault diagnosis. *Instrum. Technol. Sensor* **2022**, *7*, 95–99.

18. Yan, R.; Lin, C.; Gao, S.; Luo, J.; Li, T.; Xia, Z. Fault diagnosis and analysis of circuit breaker based on wavelet time-frequency representations and convolutional neural network. *J. Vib. Shock.* **2020**, *39*, 198–205.
19. Huang, X.; Hu, X.; Zhu, Y.; Wei, X.; Zhou, Y.; Gao, H. Fault diagnosis of high voltage circuit breaker based on convolutional neural network. *Electr. Power Autom. Equip.* **2018**, *38*, 136–140.
20. Xu, H.; Li, J.; Yuan, H.; Liu, Q.; Fan, S.; Li, T.; Sun, X. Human activity recognition based on Gramian angular field and deep convolutional neural network. *IEEE Access* **2020**, *8*, 199393–199405. [CrossRef]
21. Zhang, G.; Si, Y.; Wang, D.; Yang, W.; Sun, Y. Automated detection of myocardial infarction using a Gramian angular field and principal component analysis network. *IEEE Access* **2019**, *7*, 171570–171583. [CrossRef]
22. Zhou, F.; Ma, Y.; Wang, B.; Lin, G. Dual-channel convolutional neural network for power edge image recognition. *J. Cloud Comp.* **2021**, *10*, 18. [CrossRef]
23. Peng, W.; Zhang, G. Research on fault diagnosis of low-voltage circuit breaker based on vibration and coil current signals. *Adv. Technol. Electr. Eng. Energy* **2025**, *44*, 106–115.
24. Hossain, M.A.; Sajib, M.S.A. Classification of image using convolutional neural network (CNN). *Glob. J. Comput. Sci. Technol.* **2019**, *19*, 13–14. [CrossRef]
25. Cao, J.; Chen, S.; Bao, X.; Cong, L.; Zhou, J. Research on fault diagnosis of circuit breakers based on spectrograms and improved residual networks. *Mach. Electron.* **2025**, *43*, 9–15.
26. Wang, X.; Wang, X.; Jia, X.; Zhang, F.; Li, S.; Hu, Y. Fault identification method of outage transmission line based on convolutional neural network and wavelet packet decomposition. *Electr. Meas. Instrum.* **2025**, *62*, 61–67.
27. Razavi-Far, R.; Chakrabarti, S.; Saif, M.; Zio, E. An integrated imputation-prediction scheme for prognostics of battery data with missing observations. *Expert. Syst. Appl.* **2019**, *115*, 709–723. [CrossRef]
28. Song, J.; Wang, H.; Du, M.; Peng, L.; Zhang, S.; Xu, G. Non-intrusive load identification method based on improved long short term memory network. *Energies* **2021**, *14*, 684. [CrossRef]
29. Kingma, D.P.; Ba, J. Adam: A method for stochastic optimization. In Proceedings of the 3rd International Conference on Learning Representations, San Diego, CA, USA, 7–9 May 2015.
30. Cong, S.; Zhou, Y. A review of convolutional neural network architectures and their optimizations. *Artif. Intell. Rev.* **2023**, *56*, 1905–1969. [CrossRef]

energies

MDPI

Article

Gramian Angular Field–Gramian Adversial Network–ResNet34: High-Accuracy Fault Diagnosis for Transformer Windings with Limited Samples

Hongwen Liu [1], Kun Yang [1], Guochao Qian [1,*], Jin Hu [1], Weiju Dai [1], Liang Zhu [1], Tao Guo [1], Jun Shi [1] and Dongyang Wang [2]

[1] Electric Power Research Institute of Yunnan Power Grid, Kunming 650214, China; liuhongwen_2025@126.com (H.L.)
[2] School of Electrical Engineering, Southwest Jiaotong University, Chengdu 611756, China; jiulingthebest@163.com
* Correspondence: qianguochao_1981@163.com

Abstract: Transformers are critical equipment in power transmission and distribution systems, and the condition of their windings significantly impacts their reliable operation. Therefore, the fault diagnosis of transformer windings is of great importance. Addressing the challenge of limited fault samples in traditional diagnostic methods, this study proposes a small-sample fault diagnosis method for transformer windings. This method combines data augmentation using the Gramian angular field (GAF) and generative adversarial networks (GAN) with a deep residual network (ResNet). First, by establishing a transformer winding fault simulation experiment platform, frequency response curves for three types of faults—axial displacement, bulging and warping, and cake-to-cake short circuits—and different fault regions were obtained using the frequency response analysis method (FRA). Second, a frequency response curve image conversion technique based on the Gramian angular field was proposed, converting the frequency response curves into Gramian angular summation field (GASF) and Gramian angular difference field (GADF) images using the Gramian angular field. Next, we introduce several improved GANs to augment the frequency response data and evaluate the quality of the generated samples. We compared and analysed the diagnostic accuracy of ResNet34 networks trained using different GAF–GAN combination datasets for winding fault types, and we proposed a transformer winding small-sample fault diagnosis method based on GAF-GAN-ResNet34, which can achieve a fault identification accuracy rate of 96.88% even when using only 28 real samples. Finally, we applied the proposed fault diagnosis method to on-site transformers to verify its classification performance under small-sample conditions. The results show that, even with insufficient fault samples, the proposed method can achieve high diagnostic accuracy.

Keywords: transformer; frequency response analysis method; fault identification; Gramian angular field; generative adversarial networks; residual network

1. Introduction

As a core device in power systems, the operating status of transformers directly affects the safety and stability of power systems [1,2]. Winding faults are the main type of transformer faults. During operation, transformer windings are susceptible to short-circuit current impacts. Under the cumulative effect of electrodynamic forces, they are prone to deformation faults such as displacement and buckling, which severely affect the reliability of transformer operation [3]. Therefore, accurate diagnosis of the status of transformer windings is of great significance.

Currently, the main methods for detecting faults in transformer windings include the short-circuit impedance method, frequency response method, and vibration method. Among these, the frequency response method is widely adopted due to its strong anti-interference capability, high testing sensitivity, and intuitive result analysis [4]. In recent years, with the continuous improvement of computer hardware performance and the iterative optimisation of deep learning algorithms, the application of intelligent diagnostic methods in transformer winding frequency response analysis has garnered significant attention. Neural networks can extract features from large datasets and iteratively optimise model parameters to efficiently solve complex problems, making them an extremely suitable intelligent algorithm for frequency response data analysis. Reference [5] utilised an artificial neural network (ANN) to identify common winding faults; reference [6] used the change in the frequency response curve between zero and pole points as input features for the artificial neural network, achieving good diagnostic results; however, the process of constructing such a network is relatively complex and requires further simplification; reference [7] proposed using a backpropagation (BP) neural network to determine the type of winding fault, selecting the correlation coefficient and standard deviation as feature inputs. However, during actual training, it was found that the convergence speed of this network was slow, resulting in a longer training time. Reference [8] recognised the limitations of the BP neural network and improved its structure, eliminating directional detection errors while enhancing the accuracy of the BP neural network in detecting winding faults. Reference [9] used the binary image reconstructed from the frequency response curve as input, using a convolutional neural network (CNN) to identify winding faults, achieving a recognition accuracy of 95% on the test set, demonstrating excellent diagnostic performance; reference [10] optimised the CNN structure using bidirectional gated recurrent unit (BiGRU), further enhancing the CNN's ability to distinguish winding faults. As demonstrated by the above research, compared to other intelligent diagnostic algorithms, the CNN achieves higher accuracy in winding fault diagnosis tasks. This is because the CNN can utilise convolutional layers to extract local features from the data while reducing computational load and parameter complexity through pooling layers, thereby demonstrating robust classification and recognition capabilities.

However, transformer equipment has high reliability, and winding faults are rare events. This results in a scarcity of winding fault samples available to scholars and researchers in actual transformer operation sites, with significant variations in the number of samples across different fault types. Despite its many advantages, FRA faces two key challenges: sample scarcity, with fewer than five critical fault features per utility company per year; and interpretability gaps, with uncertainty in the physical-to-feature mapping in AI models. To address these issues, reference [11] proposed a transformer winding deformation fault identification method based on a twin convolutional network to attempt to resolve the problem of imbalanced fault data; reference [12] proposes a data-mechanism-driven method that fuses virtual and real features, introducing a transformer simulation model to compensate for the lack of actual sample features, and improves the average identification accuracy by 27.1% in a small-sample scenario with 30% data; reference [13] uses Conditional-WGAN-GP to augment small-batch fault winding FRC data, and applied the expanded data to the training of an SVM classification network to obtain a fault diagnosis model. Diagnostic results indicate that GAN data expansion can significantly improve fault identification accuracy. A comparison of existing major transformer fault analysis and diagnosis methods is shown in Table 1. However, current research on transformer winding fault diagnosis in small-sample scenarios remains inadequate, with limited research content and scarce outcomes. It urgently requires more attention and exploration from scholars and researchers to drive progress and development in this field. This study addresses these issues by combining GAF visualization with GAN-based few-shot learning [3,14–16].

Table 1. Existing transformer fault diagnosis methods.

Method	Accuracy	Sample Size	Limitations
BP Neural Network	84.2%	200+	Slow convergence; manual features
CNN-BiGRU	92.7%	150	Requires curve preprocessing
Twin CNN	89.4%	100	Limited to binary faults
VAE Augmentation	88.1%	50	Blurred high-frequency features
Proposed GAF–GAN–ResNet34	96.9%	28	Small sample-optimized

To this end, this paper has conducted the following work on transformer winding fault diagnosis: First, by establishing a transformer winding fault simulation experiment platform, the frequency response method was used to obtain frequency response curves for different fault types and fault regions of the windings, providing data support for subsequent intelligent diagnosis; second, a GAF image conversion technique was proposed, using GAF to convert the frequency response curves into GADF and GASF images; then, we introduced multiple improved GANs to augment the frequency response data and evaluated the quality of the generated samples. A comparative analysis was conducted on the diagnostic accuracy of winding fault types using different GAF–GAN combined datasets to train the ResNet34 network, proposing a transformer winding fault diagnosis method based on GAF–GAN–ResNet34 for small-sample scenarios. Finally, the proposed fault diagnosis method is applied to field transformers to verify its classification performance under small-sample conditions.

2. Obtaining the Winding Frequency Response Curve

2.1. Basic Principles of FRA

Under the influence of high-frequency excitation, the transformer winding can be regarded as a passive two-port network composed of distributed parameters such as linear resistance, inductance, and capacitance. The parameters of this network are uniquely determined by the geometric structure of the transformer [17]. Sine excitation signals of different frequencies (1~1000 kHz) are injected into the input terminals of the transformer windings, and the corresponding response signals are measured at the output terminals. By calculating the ratio of the response signals to the excitation signals at different frequencies, the frequency response curves are obtained.

$$\begin{cases} H(f) = 20\lg\frac{|U_{out}(f)|}{|U_{in}(f)|} \\ \varphi(f) = \varphi(U_{out}(f)) - \varphi(U_{in}(f)) \end{cases} \quad (1)$$

where $H(f)$ and $\varphi(f)$ are the amplitude and phase of the frequency response; $U_{out}(f)$ and $U_{in}(f)$ are the terminal voltage and excitation voltage of the winding response, respectively; f is the frequency of the applied signal.

When a transformer winding fails, the equivalent parameters (such as capacitance and inductance) in its equivalent two-port network change compared to its healthy state, resulting in changes to its frequency response. For different types of winding faults, the patterns of change in the equivalent two-port network vary, resulting in different patterns of change in FRA results compared to those under healthy conditions. However, for the same type of fault, even if the specific location and severity of the fault differ, the changes in electrical parameters and FRA results still exhibit a certain degree of similarity. Therefore, by analysing the patterns of change in the FRA results of the windings, the type of transformer winding fault can be determined.

2.2. Test Platform Construction

Winding fault simulations on operational transformers are rarely performed due to the destructive nature of tests and high equipment costs. To address this limitation, we established a dedicated fault simulation platform using a 10 kV test transformer, as shown in Figure 1a. The FRA is conducted through the high-voltage (HV) winding, as shown in Figure 1b. The structural parameters of the test transformer are shown in Table 2. The transformer windings are structured as coil segments, with copper terminals welded at both ends of each segment. The segments are connected using high-conductivity brass nuts. The segments are numbered from 1 to 16 from top to bottom, and the interconnections between segments are numbered from 1 to 15 from top to bottom.

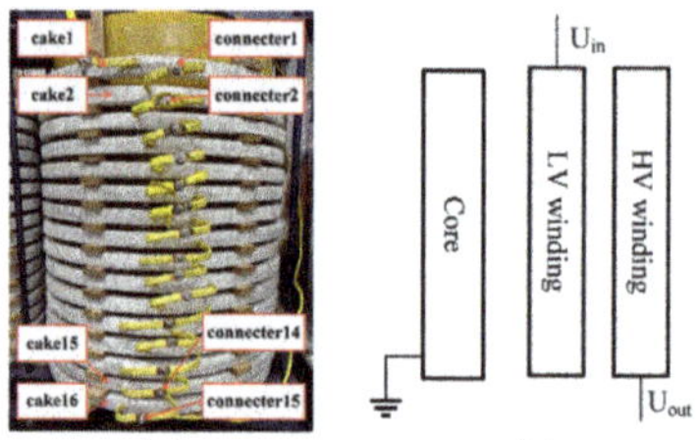

(a) (b)

Figure 1. FRA test platform. (**a**) Test transformer. (**b**) FRA test wiring.

Table 2. Fault settings for axial displacement of windings.

Structure	Parameter	Value
HV winding	Number of line cakes	16
	Number of turns per coil	16
	Wire cake height/mm	9
	Wire cake width/mm	34
	Winding height/mm	474
LV winding	Number of line cakes	16
	Number of turns per coil	8
	Wire cake height/mm	9
	Wire cake width/mm	17
	Winding height/mm	474

In order to obtain frequency response curves for different types of winding faults, fault locations, and fault severities, three types of transformer faults were simulated based on the frequency response test platform: axial displacement fault (AD), bulging and warping fault (SCV), and cake-to-cake short circuit fault (SC). The simulations of each type of fault are shown in Figure 2.

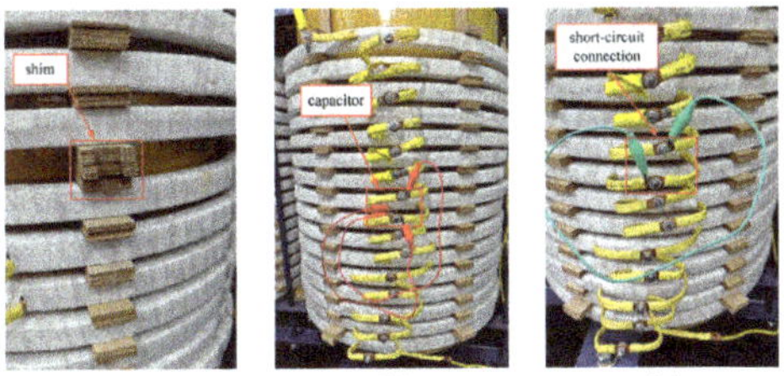

Figure 2. Winding fault simulation diagram. (**a**) AD, (**b**) SCV, (**c**) SC.

(1) AD

AD refers to a fault in which the windings of a transformer shift along the transformer's axis during operation due to the effects of electromagnetic or mechanical forces. In this paper, ADs are simulated by increasing the number of spacers under the windings, as shown in Figure 2a. Depending on the fault location and severity, four types of AD were set up, as shown in Table 3. The severity of the fault is defined as the ratio of the AD distance of the windings to the height of the coil.

Table 3. Fault settings for AD of windings.

Number	Faulty Line Cake Position	Fault Severity
1	7	50%
2	7	100%
3	13	50%
4	13	100%

(2) SCV

SCV refers to a fault in which a portion of the transformer winding undergoes local deformation or bulging due to factors such as uneven stress, thermal stress, or manufacturing defects. When SCV occurs, the equivalent capacitance of the winding along its longitudinal axis increases [18]. Therefore, this study simulates the bulging and warping fault by paralleling capacitors at the winding terminals, as shown in Figure 2b. Depending on the fault location and the value of the parallel capacitors, four different SCV scenarios were set up, as shown in Table 4.

Table 4. Fault settings for SCV of windings.

Number	Faulty Joint Position	Fault Severity
1	7, 8	200 pF
2	7, 8	560 pF
3	13, 14	200 pF
4	13, 14	560 pF

(3) SC

SC refers to a fault where the insulation layer between adjacent windings of a transformer is penetrated or damaged, resulting in an electrical short circuit between the windings. In this paper, SC is simulated by short-circuiting the terminal ports on the outer side of the windings, as shown in Figure 2c. Depending on the location of the fault, four types of SC were set up, as shown in Table 5.

Table 5. Fault settings for SC of windings.

Number	Faulty Line Cake Position
1	5, 6
2	7, 8
3	13, 14
4	15, 16

2.3. Frequency Response Curve Data Acquisition and Analysis

Following the above experimental procedures, FRA were conducted on transformer windings under three typical fault conditions and under normal conditions. The test results are shown in Figure 3. For each type of fault, 28 sets of frequency response data were collected, with 7 sets of data collected for each specific fault condition. As a control, 28 sets of frequency response data were collected for transformer windings under normal conditions.

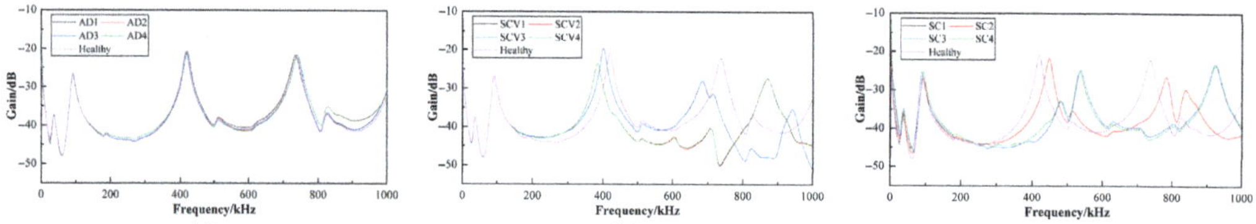

(**a**) (**b**) (**c**)

Figure 3. Iterative curves of error rate for different algorithms optimized base classifiers. (**a**) AD, (**b**) SCV, (**c**) SC.

As shown in Figure 3a, when the winding experiences AD, the trough near 270 kHz shifts upward, the peak near 600 kHz shifts downward, the peak near 730 kHz shifts to the right, the trough near 800 kHz shifts upward and to the right, and the overall curve in the 500~700 kHz frequency band shifts downward, while the overall curve in the 800~1000 kHz frequency band shifts upward. The frequency response curve exhibits a certain degree of discrimination between different fault severities, but there is still some overlap between them.

As shown in Figure 3b, when the winding experiences SCV, the peak near 420 kHz shifts downward to the left when the parallel capacitance of the winding is 200 pF and shifts upward to the left when the parallel capacitance is 560 pF. The peak near 730 kHz shifts downward to the left when the parallel capacitance is 200 pF, shifts downward to the left when the parallel capacitance is 200 pF, and shifts downward to the right when the parallel capacitance is 560 pF. Additionally, a distinct resonance point appears in the 600~1000 kHz frequency band. This indicates that by analysing the shift in amplitude at the resonance points near 420~730 kHz, one can preliminarily determine the type of fault that has occurred.

As shown in Figure 3c, when SC occurs between the windings, the wave peak near 430 kHz shifts to the lower right. When a fault occurs in the middle of the winding, a distinct resonance point appears in the 400~600 kHz frequency band. When a fault occurs in the lower part of the winding, a distinct resonance point appears in the 800~900 kHz frequency band. The presence of new resonance points in the 400~600 kHz frequency band and the 800~900 kHz frequency band can be used to preliminarily determine the location of the fault.

The distinct impacts of AS, SC, and SCV on frequency response curves arise from their unique modifications to the winding's equivalent electrical parameters. AD primarily alters longitudinal capacitance and mutual inductance, manifesting as a downward shift in mid-frequency resonance. SC reduce inter-turn impedance, causing global upward shifts and high-frequency noise. SCV increase localized capacitance, leading to amplitude elevation and resonance damping.

In summary, there is overlap between the frequency response curves under normal conditions and the three fault conditions, making it difficult to clearly identify the characteristic information on the curves. Further processing of the frequency response curves is required to accurately identify winding faults. Although composite faults were not physically simulated, their FR characteristics can be approximated as superpositions of single-fault signatures. This will be leveraged in virtual experiments.

3. Visualisation and Analysis of Frequency Response Data Based on GAF

GAF is an encoding method that combines coordinate transformation and Gram matrices to convert one-dimensional data into two-dimensional feature images. Depending on the transformation formula used, GAF is divided into two forms: GASF and GADF. GASF generates a Gram matrix by summing the angles between each data point and all other data points in the one-dimensional data; GADF calculates the angular differences between data points to construct the Gram matrix. GASF emphasises relative changes between data points, highlighting local dynamic characteristics of the data; GADF preserves the overall correlation between data points, making it more suitable for capturing and identifying global features of the data [19].

3.1. GAF Image Conversion Principle

The data processing method based on GAF is as follows:

(1) Data normalisation: To ensure that the data are −1~1, normalise the data as shown in Equation (2).

$$\tilde{x}_i = \frac{x_i - \min(X)}{\max(X) - \min(X)} \times 2 - 1 \tag{2}$$

where $X = \{x_1, x_2, \ldots, x_N\}$ is the original data, and $\tilde{x}_i$ is the normalised value.

(2) Calculate angle encoding: Map the normalised one-dimensional data to the angle in the polar coordinate system. The value of each data point is converted to the angle in the polar coordinate system, as shown in Equation (3):

$$\phi_i = \arccos(\tilde{x}_i), \tilde{x}_i \in [-1, 1] \tag{3}$$

(3) Calculate the GAF matrix: Based on the angle information, calculate the GASF and GADF values as shown in Equations (4) and (5):

$$GASF = \begin{bmatrix} \cos(\varphi_1 + \varphi_1) & \cos(\varphi_1 + \varphi_2) & \cdots & \cos(\varphi_1 + \varphi_n) \\ \cos(\varphi_2 + \varphi_1) & \cos(\varphi_2 + \varphi_2) & \cdots & \cos(\varphi_2 + \varphi_n) \\ \vdots & \vdots & \ddots & \vdots \\ \cos(\varphi_n + \varphi_1) & \cos(\varphi_n + \varphi_2) & \cdots & \cos(\varphi_n + \varphi_n) \end{bmatrix} \tag{4}$$

$$GADF = \begin{bmatrix} \cos(\varphi_1 - \varphi_1) & \cos(\varphi_1 - \varphi_2) & \cdots & \cos(\varphi_1 - \varphi_n) \\ \cos(\varphi_2 - \varphi_1) & \cos(\varphi_2 - \varphi_2) & \cdots & \cos(\varphi_2 - \varphi_n) \\ \vdots & \vdots & \ddots & \vdots \\ \cos(\varphi_n - \varphi_1) & \cos(\varphi_n - \varphi_2) & \cdots & \cos(\varphi_n - \varphi_n) \end{bmatrix} \tag{5}$$

(4) Generate image: Generate GAF images based on the calculation results of GASF and GADF.

3.2. Steps for Converting GAF Images of Winding Frequency Response Curves

The Gram angle field image conversion method is based on transformer winding frequency response curves, as shown in Figure 4, with the specific steps as follows:

(1) Using FRA, the frequency response curve of the transformer winding was tested, i.e., the original one-dimensional data sequence.
(2) Scale the original one-dimensional data sequence to the range $-1{\sim}1$ for standardisation.
(3) Convert the one-dimensional data sequence from the Cartesian coordinate system to the polar coordinate system.
(4) Convert the one-dimensional data sequence into two-dimensional GASF and GADF images through trigonometric transformations of the angles and differences between each data point.

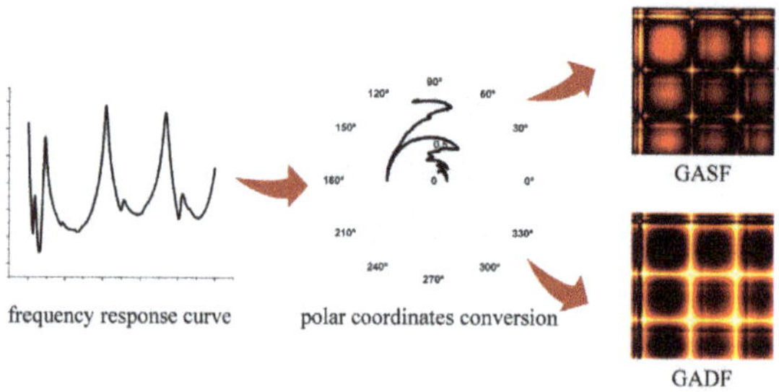

Figure 4. GAF image conversion steps for frequency response curves.

3.3. Results Analysis

This paper uses GAF to convert transformer winding frequency response data into two-dimensional images, so that image processing algorithms can be used to extract winding fault features and complete the identification and diagnosis of transformer winding faults. The original dimensions of the GASF and GADF images converted from transformer winding frequency response data using GAF are $875 \times 656 \times 3$. To reduce the computational burden on image processing algorithms, the dimensions of the GAF images are adjusted to $128 \times 128 \times 3$. The processed GAF images of the frequency response curves are shown in Figure 5.

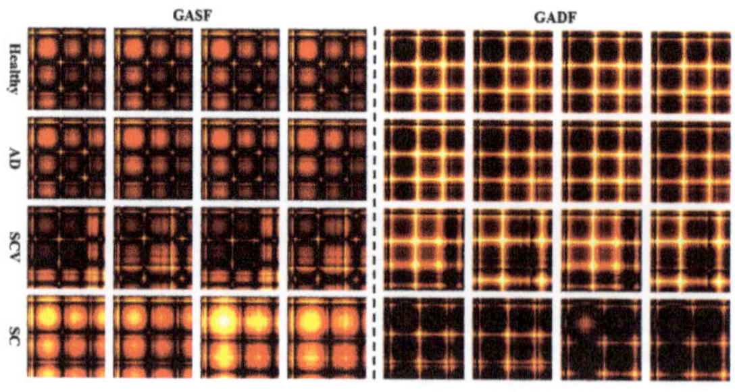

Figure 5. Frequency response curve GAF image.

From the GASF images, it can be seen that most of the images are red and dark red, with only a small portion of other colour blocks exhibiting certain colour distribution and position distribution characteristics in the image. Under AD, the diamond-shaped orange colour blocks are primarily distributed in the upper left corner; under SCV, the diamond-shaped orange colour blocks form a square distribution in the lower right corner, and the dark red distribution shifts from the lower right to the upper left; under SC, the diamond-shaped orange colour blocks are primarily distributed in the upper left corner, and the dark red portions begin to disappear, with most of the image area turning orange.

From the GADF images, the distribution of dark red patches can be used to preliminarily determine the type of winding fault. Under AD, the upper-left dark-red dominance correlates with high-frequency attenuation from reduced inter-disc capacitance. Patch coverage scales with displacement severity. Under SCV, the upper-right concentration with lower orange streaks maps to mid-frequency phase shifts from increased turn capacitance. The strength of the stripes is directly proportional to the parallel capacitance. Under SC, the full-image darkening signifies new resonances from parasitic inductance. Darkened lower-right regions indicate fault depth.

In summary, converting the frequency response curve into a two-dimensional image provides more intuitive fault information, enabling clearer identification of the patterns between different types and degrees of winding faults. Additionally, a single frequency response curve can be converted into two images, GASF and GADF, significantly enriching the fault information and facilitating subsequent analysis and identification by deep learning networks, thereby enhancing the accuracy and effectiveness of fault detection.

4. Data Augmentation for Frequency Response Curves Based on GAF

4.1. Basic Principles and Improvements of GAN

Transformer equipment has high reliability, and winding faults are rare events, leading to a scarcity of winding fault samples and posing challenges for fault diagnosis. GAN can extract image features and use them to generate new image samples. Theoretically, GAN can address the issue of insufficient training samples, enhancing the diagnostic effectiveness of intelligent diagnostic algorithms for transformer winding faults. The basic principles of GAN include two structures: generators and discriminators. During training, generators and discriminators compete and learn from each other, continuously improving their performance through adversarial training strategies [20]. Their network structure is shown in Figure 6.

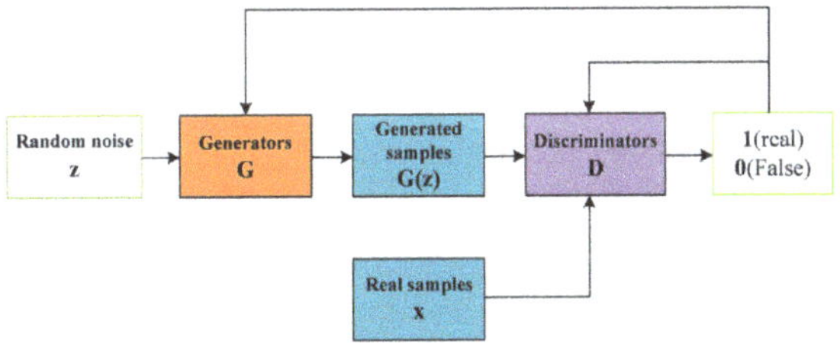

Figure 6. Schematic diagram of GAN structure.

The generative network continuously combines noise to generate fake data, which are then input into the discriminative network along with the original data. The discriminative network judges the true source of the data and outputs a probability representing the authenticity of the data, which is used to adjust the weights of the generative network and the discriminative network. When the discriminative network cannot judge the authenticity of the data, an optimal data generation adversarial network model is established. The objective function of the GAN model is:

$$\min_{G} \max_{D} V(G,D) = E_{x\sim p_{data}(x)}[\log D(x)] + E_{z\sim p_z(z)}[\log(1 - D(G(z)))] \quad (6)$$

where x is the real sample; z is the random noise vector; $p_{data}(x)$ represents the latent distribution of the real sample; $p_z(z)$ represents the latent distribution of the generated sample; $D()$ is the probability that the discriminator judges the input sample to be a real sample, which is a real number in the range 0~1; and $G(z)$ represents the sample generated by the random noise vector.

The generator and discriminator continuously adjust their parameters through alternating iterative adversarial training until the two networks reach a dynamic Nash equilibrium, i.e., $D(G(z)) = 0.5$, thereby establishing the optimal adversarial network. However, due to the instability of the network structure, GAN also face challenges such as network oscillations, failure to converge, pattern collapse, and vanishing gradients during training. To address these issues, scholars and researchers have proposed improvements to the GAN network structure from various perspectives, including the Wasserstein GAN with Gradient Penalty (WGAN-GP), Deep Convolution GAN (DCGAN), and Boundary Equilibrium GAN (BEGAN), among others [21].

(1) WGAN-GP

GANs typically use Jensen–Shannon divergence or Kullback–Leibler divergence as their loss functions. When generated and real sample distributions exhibit no overlap, the loss function gradient vanishes. This prevents updates to the generator's weight parameters. Consequently, network training becomes infeasible. To address this issue, Arjovsky et al. optimised the loss function for GANs and proposed the Wasserstein GAN (WGAN) [22]. WGAN uses the Wasserstein distance to measure the distance between the data distributions of generated samples and real samples, with the objective function defined as:

$$\min_{G} \max_{D \in \Delta} V(G, D) = E_{x \sim p_{data}(x)}[D(x)] - E_{z \sim p_z(z)}[D(G(z))] \tag{7}$$

where Δ denotes the set of 1-*Lipschitz* functions.

However, due to the use of gradient truncation to satisfy the *Lipschitz* continuity constraint, WGAN may encounter issues such as the non-convergence of network parameters, poor sample quality, and mode collapse when learning the latent distribution of complex samples. To address this, Ishaan et al. further improved WGAN by incorporating a gradient penalty term (GP), proposing the WGAN-GP [23]. WGAN-GP replaces gradient clipping with a gradient penalty term, with its objective function defined as:

$$\begin{aligned} &\min_{G} \max_{D} V(G, D) = \\ &E_{x \sim p_{data}(x)}[D(x)] - E_{z \sim p_z(z)}[D(G(z))] + \lambda E_{\hat{x} \sim P(\hat{x})}[(\|\nabla_{\hat{x}} D(\hat{x})\|_2 - 1)^2] \end{aligned} \tag{8}$$

where λ is the gradient penalty coefficient, and $\hat{x}$ is the random interpolation sampling on the line connecting two points after sampling each point on the generated samples and real samples, which is defined as:

$$\hat{x} = \varepsilon x_r - (1 - \varepsilon) x_g \tag{9}$$

where x_r is the real sample data, x_g is the generated sample data, and ε follows a normal distribution in the range 0~1. WGAN-GP satisfies the *Lipschitz* continuity constraint while limiting the update range of the discriminator weight parameters, thereby avoiding the side effects introduced by gradient clipping and making the iterative training process of WGAN more stable.

(2) DCGAN

To address the limitation of GAN' weak generative capabilities, Radford et al. improved the structure of the GAN generator using convolutional neural networks and proposed the DCGAN based on this improvement [24]. The structure of the DCGAN generator network is shown in Figure 7. Compared to GAN, DCGAN replaces fully connected layers with convolutional layers, enhancing the generator network's ability to capture the structural properties of the sample space; it replaces linear layers and pooling layers with strided convolutions and transposed convolutions to achieve downsampling and upsampling, thereby better preserving image details and spatial structure; and it replaces the Tanh activation function with batch normalisation (BN), which accelerates the training process while improving the stability of the generator network. These improvements significantly enhance the sample generation quality of the GAN generator.

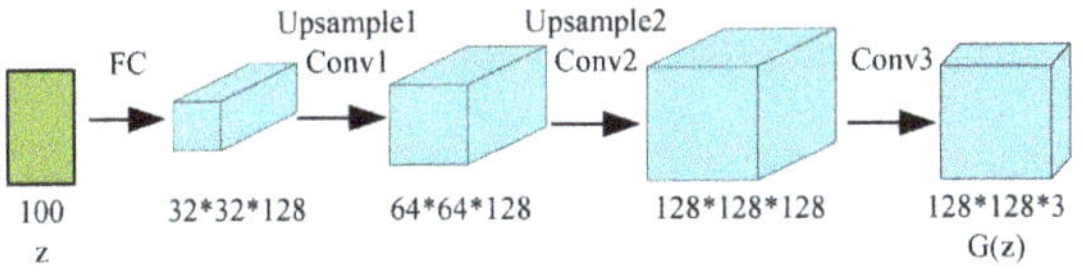

Figure 7. Schematic diagram of the DCGAN generator structure.

(3) BEGAN

The basic idea behind GAN training is to iteratively adjust the network's weight parameters to continuously reduce the distance between the generated data distribution Pg and the real data distribution Pr. Many networks, such as WGAN-GP and DCGAN, have been developed based on this approach to improve GANs [25]. Berthelot et al. identify a challenge: directly minimizing the Pg and Pr distance in data space is often intractable. They instead propose minimizing the distance between their reconstruction distribution errors. The closer the reconstruction distribution errors of the two are, the closer their original distributions are. Based on this approach, the BEGAN was proposed [26].

The discriminator network of BEGAN adopts an autoencoder structure, as shown in Figure 8. It converts the input sample into an intermediate vector in the hidden layer through the encoder and then converts the intermediate vector into a reconstructed sample with the same structure as the input sample through the decoder. By reconstructing the generated data distribution and the real data distribution, and using the Wasserstein distance to calculate the reconstruction distribution error [27], the formula is:

$$L(v) = |v - D(v)|^{\eta} \quad (10)$$

where v is the input sample data for the discriminator, and $D(v)$ is the output sample data for discriminator D. η is the norm usage parameter, where $\eta = 1$ indicates the use of l_1 norm and $\eta = 2$ indicates the use of l_2 norm.

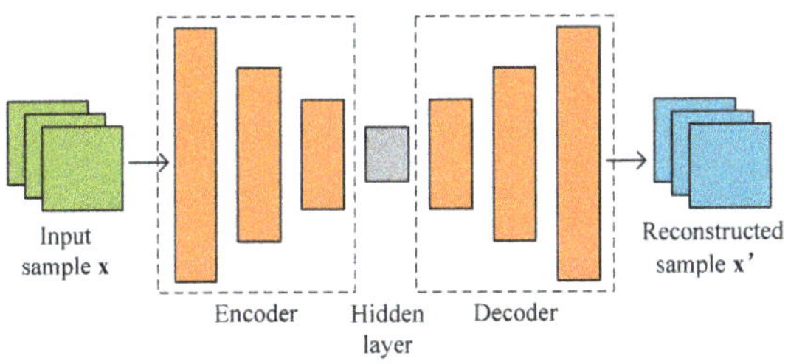

Figure 8. Schematic diagram of the BEGAN generator structure.

BEGAN uses boundary equilibrium conditions to improve the stability of the training process of the generative network while reducing the difficulty of convergence of the generator and discriminator weight parameters.

In summary, this paper proposes to use WGAN-GP, DCGAN, and BEGAN to augment frequency response data for transformer winding faults, with the aim of improving the diagnostic effectiveness of image recognition algorithms in such scenarios. Additionally, the paper will compare the data augmentation effects of different GAN networks to assess their suitability for fault diagnosis tasks.

Compared to other augmentation methods, VAEs generate blurry GAF images due to Gaussian prior constraints, losing high-frequency details critical for resonance identification. SMOTE operates in feature space, unsuitable for image data; its linear interpolations fail to capture nonlinear FRA distortions. While GANs outperform VAEs/SMOTE in FRA augmentation, their computational cost is higher. Future work will explore lightweight GAN variants for edge deployment.

4.2. Frequency Response Data Expansion

Based on the GASF and GADF images of frequency response data, this paper uses three generative adversarial networks, namely WGAN-GP, DCGAN, and BEGAN, to expand the GAF images corresponding to the frequency response data of transformer windings. Since the loss function of GAN cannot indicate the training process, in order to ensure that the network is fully trained, the WGAN-GP, DCGAN, and BEGAN networks are iteratively trained 2000 times. The GASF and GADF images generated after training the three GANs are shown in Figure 9.

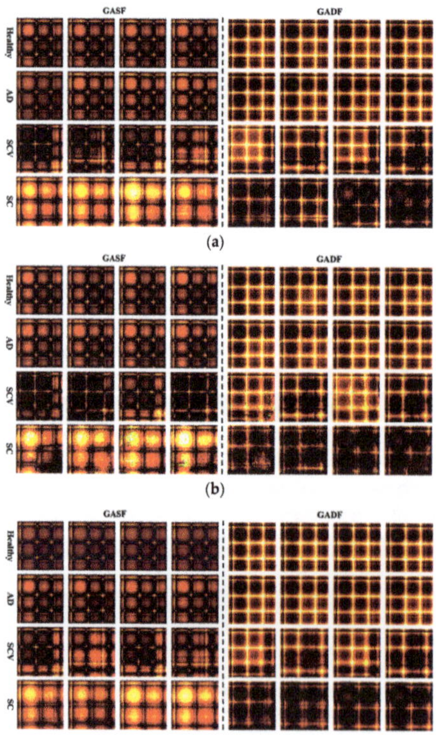

Figure 9. Different GAN training results. (a) WGAN-GP, (b) DCGAN, (c) BEGAN.

4.3. Quality Assessment of GAN-Generated Samples

Structural similarity (*SSIM*), Peak Signal-to-Noise Ratio (*PSNR*), cosine similarity (*CS*), and Fréchet Inception Distance (*FID*) scores are common metrics for evaluating the quality of GAN-generated samples [28].

SSIM measures the structural similarity between images by comparing the brightness, contrast, and structural details of the real image and the generated image. It uses the mean of the image brightness as an estimate of brightness, the standard deviation as an estimate of contrast, and the covariance as a measure of structural similarity. The range of *SSIM* is between 0 and 1. When two images are completely identical, the *SSIM* value equals 1. PSNR calculates the reconstruction quality of the generated image using the mean squared error (*MSE*). The higher the *PSNR* value is, the better the reconstruction quality is of the generated image.

CS is a mathematical method commonly used to measure the similarity between two vectors. Its value ranges from −1 to 1, where 1 indicates complete similarity, −1 indicates complete opposition, and 0 indicates no similarity, as shown in Equation (11):

$$CS = \frac{\sum_{i=1}^{n} A_i \times B_i}{\sqrt{\sum_{i=1}^{n} A_i^2} \times \sqrt{\sum_{i=1}^{n} B_i^2}} \tag{11}$$

FID measures the similarity between generated images and real images by comparing the feature distributions of the two. The lower the *FID* score is, the higher the similarity is between the generated image and the real image. When two images are identical, the *FID* score is 0.

This paper evaluates the quality of three types of GAN-generated images using four scoring methods: *SSIM*, *PSNR*, *CS*, and *FID*. The results are shown in Table 6.

Table 6. GAN image quality assessment metrics.

Structure	GAN Type	*SSIM*	*PSNR*	*CS*	*FID*
GASF	WGAN-GP	0.8176	32.79	0.7394	92.73
	DCGAN	0.6826	32.41	0.7026	192.43
	BEGAN	0.6734	31.16	0.8473	46.49
GDSF	WGAN-GP	0.7896	32.65	0.838	62.32
	DCGAN	0.76	32.54	0.7445	183.85
	BEGAN	0.7886	32.37	0.902	31.38

As indicated by the *SSIM* metric, the generated images produced by the three generative networks exhibit structural similarities to the original images. As indicated by the *PSNR* metric, the reconstructed quality of the generated images produced by the three generative networks is generally good. As indicated by the *CS* metric, the visual similarity between the generated images produced by the three generative networks and the original images is generally high. Among them, the cosine similarity between the GASF and GADF images generated by BEGAN and the original images is above 0.8. According to the *FID* score, the image quality generated by BEGAN is excellent, the image quality generated by WGAN-GP is average, and the image quality generated by DCGAN is poor. DCGAN's structural constraints (e.g., vanishing gradients) limit its effectiveness on FRA data, whereas WGAN-GP/BEGAN's stabilization mechanisms show superior adaptability.

5. Fault Diagnosis for Windings Based on GAF–GAN–ResNet34

5.1. Residual Network

CNNs are a type of neural network model widely used in image processing and computer vision tasks. Their basic principle involves using convolutional layers to extract local features from input data, employing pooling layers to reduce data dimensions and enhance the model's translation invariance, and finally using fully connected layers for data classification or regression. CNNs automatically learn and extract image features through convolutional operations, significantly improving the network model's generalisation ability and performance. Due to their excellent feature extraction capabilities, CNNs have found widespread application in fields such as image classification, object detection, and facial recognition. However, as the number of layers increases, CNNs encounter issues such as gradient vanishing and network degradation during training, leading to increased training difficulty. To address this, scholars and researchers have made improvements to the network structure of CNNs, proposing various improved CNN models such as ResNet, VGGNet, DenseNet, and ShuffleNet.

Among these, ResNet introduced the residual learning mechanism, which uses skip connections to directly transmit input to deep networks, effectively alleviating the gradient vanishing and network degradation issues that arise in traditional CNNs as depth increases. Additionally, the residual structure enables the network to more easily learn and identity mappings, avoiding network degradation issues and significantly improving the training efficiency and performance of convolutional networks [29]. Depending on the number of network layers, ResNet includes variants such as ResNet18, ResNet34, ResNet50, ResNet101, and ResNet152.

When pre-training the ResNet network using a training set composed of $128 \times 128 \times 3$ colour images, this paper found that the ResNet34 network has a moderate parameter size, which effectively extracts image features without causing overfitting, and demonstrates good classification performance for $128 \times 128 \times 3$ colour images. Therefore, this paper decided to use ResNet34 for transformer winding fault diagnosis research. The structural parameters of this network are shown in Table 7, where residual blocks 1–4 contain three, four, six, and three layers of residual structures, respectively, as shown in Figure 10.

Table 7. ResNet34 network structure.

Layer Type	Input Dimension	Output Dimension	Activation Function	Normalisation
Conv2d	(batch_size,3,img_size,img_size)	(batch_size,64,img_size/2,img_size/2)	ReLU	BatchNorm
MaxPool2d	(batch_size,64,img_size/2,img_size/2)	(batch_size,64,img_size/4,img_size/4)		
Residual Block1	(batch_size,64,img_size/4,img_size/4)	(batch_size,64,img_size/4,img_size/4)	ReLU	BatchNorm
Residual Block2	(batch_size,64,img_size/4,img_size/4)	(batch_size,128,img_size/8,img_size/8)	ReLU	BatchNorm
Residual Block3	(batch_size,128,img_size/8,img_size/8)	(batch_size,256,img_size/16,img_size/16)	ReLU	BatchNorm
Residual Block4	(batch_size,256,img_size/16,img_size/16)	(batch_size,256,img_size/32,img_size/32)	ReLU	BatchNorm
Adaptive AvgPool2d	(batch_size,256,img_size/32,img_size/32)	(batch_size,512,1,1)		
Reshape	(batch_size,512,1,1)	(batch_size,512)		
Linear	(batch_size,512)	(batch_size,4)		
Softmax	(batch_size,4)	(batch_size,1)	ReLU	BatchNorm

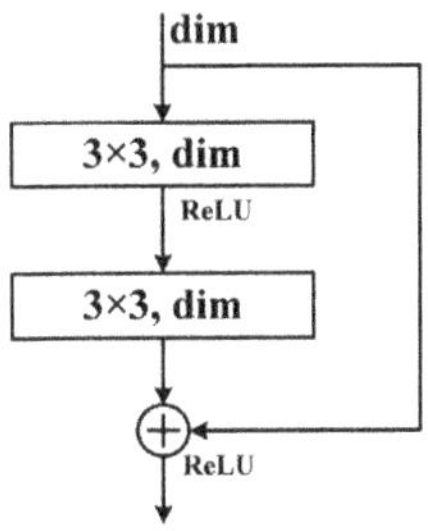

Figure 10. Residual structure diagram.

5.2. Research on Winding Fault Diagnosis

The research process for winding fault diagnosis based on GAF–GAN–ResNet34 is divided into two parts: GASF image training and GADF image training. Since the GADF image training process is identical to the GASF image training process, we will use GASF image training as an example to introduce the research process for winding fault diagnosis.

(1) Dataset Construction: Based on the GASF images obtained from the frequency response data transformation described above, we constructed the original dataset (GASF image dataset) and the GAN-augmented datasets (WGAN-GP, DCGAN, and BEGAN-augmented GASF datasets). The GASF image dataset consists of a training set, a validation set, and a test set. Each dataset includes winding frequency response images under normal conditions and three typical fault conditions. Each category of frequency response data includes 16 images; the validation set includes 4 images per category of frequency response data; the test set includes 8 images per category of frequency response data, thereby constructing the GASF image dataset. The validation set and test set of the GAN-augmented dataset are augmented with 16 real images and 24 GAN-generated images. Different GANs were used to expand the training set, thereby constructing the WGAN-expanded GASF dataset, DCGAN-expanded GASF dataset, and BEGAN-expanded GASF dataset.

(2) Classification network training: Train the ResNet34 network on the original dataset and the GAN-augmented dataset.

(3) Comparison of recognition accuracy: Compare the classification accuracy of the ResNet34 network trained on the GASF image dataset, the WGAN-GP-augmented GASF dataset, the DCGAN augmented GASF dataset, and the BEGAN-augmented GASF dataset to evaluate the enhancement effects of different GANs.

The training loss and accuracy of different GASF and GADF datasets during the training process of the ResNet34 network are shown in Figure 11. The classification accuracy of the ResNet34 classification network trained on different datasets for identifying winding faults is shown in Figure 12. From the figure, the following conclusions can be drawn: Since the frequency response images of AD are similar to those of normal windings, the classification network performs poorly in distinguishing between normal winding images and AD images before using GAN to expand the training set; both the BEGAN and WGAN-GP expanded datasets improve the classification network's recognition accuracy, among which the most effective combination is using WGAN-GP to augment the GADF dataset. The classification network trained using this dataset achieved an identification accuracy rate of 96.88%; DCGAN, however, generates images of poor quality and lacks data augmentation capabilities, thus failing to effectively improve the classification network's identification accuracy rate.

5.3. On-Site Case Analysis

A 110 kV on-load tap-changing transformer produced by Xi'an Xibian Zhongte Electrical Equipment Co., Ltd. (Xi'an, China) was tested at a certain substation to validate the applicability of the proposed method in actual engineering scenarios, rather than adding a new test set, as shown in Figure 13a. The frequency response curves obtained from the test are shown in Figure 13b. The frequency response was converted into a GADF image using the Gram angle field, as shown in Figure 14, and this image was input into the pre-trained GADF-WGAN-GP-ResNet34 recognition model. The detection result indicated an axial displacement fault in the upper region of the winding. As shown in Figure 15, the actual fault condition of the transformer winding was found, confirming that an axial displacement fault had indeed occurred in the upper region of the winding. The output results were consistent with the actual fault, further validating the effectiveness of the method proposed in this paper and its applicability to real-world testing scenarios.

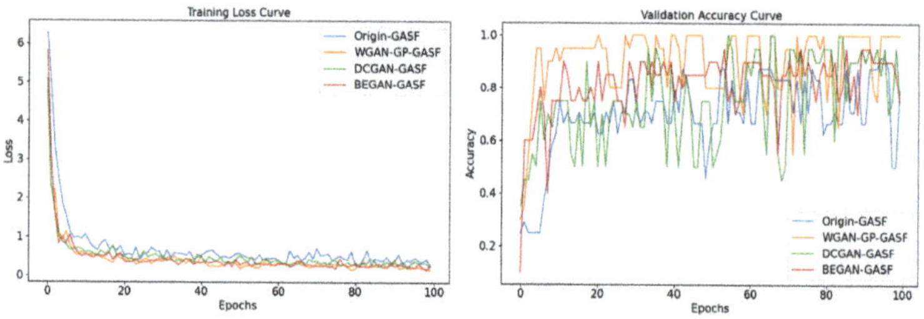

(**a**)

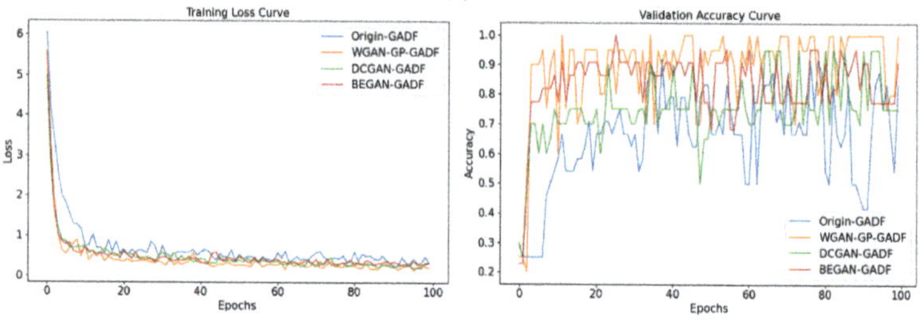

(**b**)

Figure 11. Training process for different data sets.(**a**) Training loss and accuracy of the GASF dataset, (**b**) Training loss and accuracy of the GADF dataset.

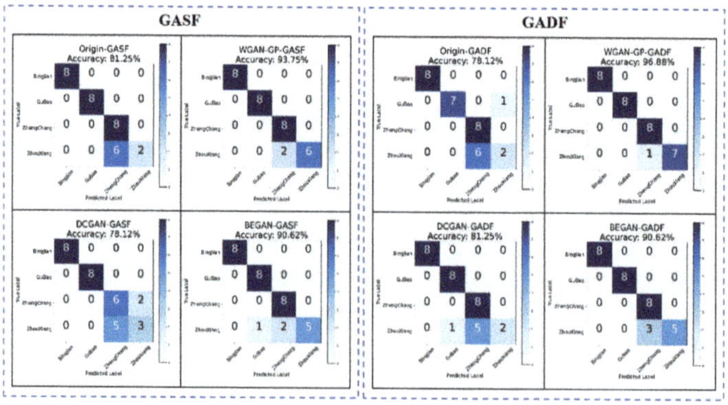

Figure 12. Confusion matrix of classification networks under different training sets.

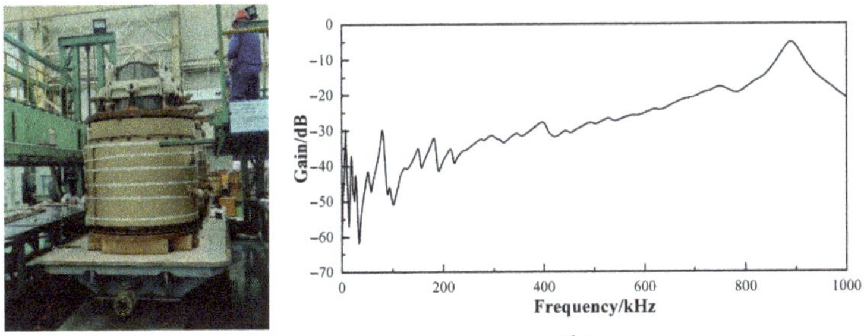

(a) (b)

Figure 13. Field testing and result for 110 kV on-load voltage regulator transformer. (**a**) Field testing, (**b**) Test frequency response curve.

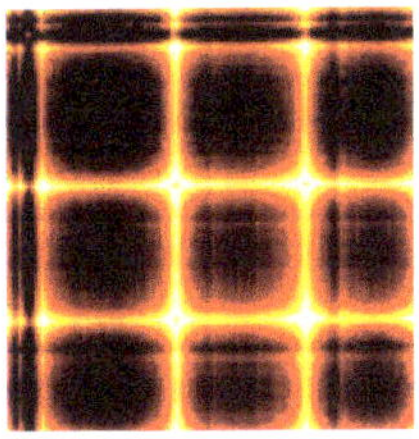

Figure 14. GADF image.

Figure 15. Field fault winding.

6. Conclusions

This paper obtained frequency response curves under different fault types and fault areas of windings on the simulated experimental platform. In order to address the situation of insufficient or severely insufficient sample numbers in transformer winding fault diagnosis, this paper proposes a small sample fault diagnosis method for transformer windings based on GAF–GAN–ResNet34. The results show that:

1. Since the frequency response images of AD are quite similar to those of normal windings, the classification network performed poorly in distinguishing between normal winding images and AD images before using GAN to expand the training set. DCGAN, due to its poor image generation quality, lacks data augmentation capabilities and cannot effectively improve the classification network's recognition accuracy.
2. Both the BEGAN and WGAN-GP expanded datasets have the effect of improving the recognition accuracy of classification networks. The most effective combination is to

use WGAN-GP to expand the GADF dataset. The classification network trained using this dataset achieved a recognition accuracy of 96.88%.

3. Compared to diagnostic models trained on the original dataset, the GAF-WGAN-GP-ResNet34 diagnostic model consistently achieved an accuracy rate of over 93% in identifying transformer winding fault types. Through the analysis of actual transformer fault cases and comparison with maintenance results, the effectiveness of the analysis method proposed in this paper was validated.

Author Contributions: Conceptualization, H.L. and K.Y.; methodology, H.L. and D.W.; validation, K.Y., J.H. and G.Q.; formal analysis, W.D. and D.W.; investigation, L.Z. and T.G.; resources, L.Z. and T.G.; data curation, J.S. and D.W.; writing—original draft preparation, W.D.; writing—review and editing, W.D.; visualization, H.L.; supervision, K.Y.; project administration, J.H.; funding acquisition, G.Q. All authors have read and agreed to the published version of the manuscript.

Funding: This research was funded by the Electric Power Research Institute of Yunnan Power Grid (Principle and system development of online monitoring of transformer winding faults under load-side multi-signal source excitation, YNKJXM20222300).

Data Availability Statement: The raw data supporting the conclusions of this article will be made available by the authors on reasonable request.

Conflicts of Interest: Hongwen Liu, Kun Yang, Guochao Qian, Jin Hu, Weiju Dai, Liang Zhu, Tao Guo, and Jun Shi are affiliated with Electric Power Research Institute of Yunnan Power Grid. The authors declare no conflicts of interest.

Abbreviations

The following abbreviations are used in this manuscript:

GAF	Gramian angular field
GAN	Generative adversarial networks
ResNet	Deep residual network
FRA	Frequency response analysis method
GASF	Gramian angular summation field
GADF	Gramian angular difference field
ANN	Artificial neural network
BP	Backpropagation
CNN	Convolutional neural network
BiGRU	Bidirectional gated recurrent unit
HV	High-voltage
AD	Axial displacement fault
SCV	Bulging and warping faulT
SC	Cake-to-cake short circuit fault
WGAN	Wasserstein GAN
GP	Gradient penalty term
BN	Batch normalisation
SSIM	Structural similarity
PSNR	Peak Signal-to-Noise Ratio
CS	Cosine similarity
FID	Fréchet Inception Distance
MSE	Mean squared error

References

1. Qian, G.; Yang, K.; Hu, J.; Liu, H.; He, S.; Zou, D.; Dai, W.; Wang, H.; Wang, D. Diagnosis Method for Transformer Winding Based on Differentiated M-Training Classification Optimized by White Shark Optimization Algorithm. *Energies* **2025**, *18*, 2290. [CrossRef]

2. Yao, H.; Xu, Y.; Guo, Q.; Chen, S.; Lu, B.; Huang, Y. Study on Transformer Fault Diagnosisbased on Improved Deep Residual Shrinkage Network and Optimized Residual Variational Autoencoder. *Energy Rep.* **2025**, *13*, 1608–1619. [CrossRef]
3. Zhao, X.; Yao, C.; Zhang, C.; Abu-Siada, A. Toward Reliable Interpretation of Power Transformer Sweep Frequency Impedance Signatures: Experimental Analysis. *IEEE Electr. Insul. Mag.* **2018**, *34*, 40–51. [CrossRef]
4. Zhang, H.; Zhang, H.; Ma, Q.; Han, H.; Wang, S. Frequency Response Signature Analysis for Winding Mechanical Fault Detection of Power Transformer Using Sensitivity Method. *Int. J. Appl. Electromagn. Mech.* **2019**, *61*, 593–603. [CrossRef]
5. Tahir, M.; Tenbohlen, S. Transformer Winding Condition Assessment Using Feedforward Artificial Neural Network and Frequency Response Measurements. *Energies* **2021**, *14*, 3227. [CrossRef]
6. Luo, Y.; Ye, J.; Gao, J.; Chen, G.; Wang, G.; Liu, L.; Li, B. Recognition Technology of Winding Deformation Based on Principal Components of Transfer Function Characteristics and Artificial Neural Network. *IEEE Trans. Dielectr. Electr. Insul.* **2017**, *24*, 3922–3932. [CrossRef]
7. Jayasinghe, J.A.S.B.; Wang, Z.D.; Darwin, A.W. Winding Movement in Power Transformers: A Comparison of FRA Measurement Connection Methods. *IEEE Trans. Dielectr. Electr. Insul.* **2006**, *13*, 1342–1349. [CrossRef]
8. Samimi, M.; Tenbohlen, S.; Akmal, A.; Mohseni, H. Power Transformer Fault Diagnosis Based on Improved BP Neural Network. *Electronics* **2023**, *12*, 3526. [CrossRef]
9. Yang, S.; Li, Z.; Yang, X.; Wu, H. Fault Identification Method of Transformer Winding based on Gramian Angular Difference Field and Convolutional Neural Network. *Recent Adv. Electr. Electron. Eng.* **2024**, *17*, 837–847. [CrossRef]
10. Xu, Y.; Yang, X.; Xu, G.; Yang, S.; Duan, G. Transformer Fault Identification Method Based on Gramian Angle Difference Field and CNN-BiGRU. *Electron. Sci. Technol.* **2025**, *38*, 73–79. [CrossRef]
11. Ma, X.; Tang, W.; Niu, Z.; Xin, Y. Deformation Fault Identification Method for Transformer Windings Based on Twin Convolutional Network Under Unbalanced Data Set. *High Volt. Appar.* **2023**, *59*, 120–128. [CrossRef]
12. Duan, J.; Wu, X.; He, Y.; Song, W.; Yin, Y. Small Sample Fault Diagnosis of Transformer Windings Based on Virtual-real Feature Fusion and Data-mechanism Driven. *Electr. Power Autom. Equip.* **2024**, *44*, 212–218. [CrossRef]
13. Chen, Y.; Zhao, Z.; Liu, J.; Tan, S.; Liu, C. Application of Generative AI-based Data Augmentation Technique in Transformer Winding Deformation Fault Diagnosis. *Eng. Fail. Anal.* **2024**, *159*, 108115. [CrossRef]
14. Zhao, X.; Wu, G.; Yang, D.; Xu, G.; Xing, Y.; Yao, C.; Abu-Siada, A. Enhanced Detection of Power Transformer Winding Faults Through 3D FRA Signatures and Image Processing Techniques. *Electr. Pow. Syst. Res.* **2025**, *242*, 111433. [CrossRef]
15. Zhao, X.; Yao, C.; Li, C.; Zhang, C.; Dong, S.; Abu-Siada, A.; Liao, R. Experimental Evaluation of Detecting Power Transformer Internal faults Using FRA Polar Plot and Texture Analysis. *Int. J. Electr. Power Energy Syst.* **2019**, *108*, 1–8. [CrossRef]
16. Zhao, X.; Yao, C.; Abu-Siada, A.; Liao, R. High Frequency Electric Circuit Modeling for Transformer Frequency Response Analysis Studies. *Int. J. Electr. Power Energy Syst.* **2019**, *111*, 351–368. [CrossRef]
17. Tang, W.; Shintemirov, A.; Wu, Q. Detection of Minor Winding Deformation Fault in High Frequency Range for Power Transformer. In Proceedings of the IEEE Power & Energy Society General Meeting, Minneapolis, MN, USA, 25–29 July 2010.
18. Usha, K.; Usa, S. Inter Disc Fault Location in Transformer Windings Using SFRA. *IEEE Trans. Dielectr. Electr. Insul.* **2015**, *22*, 3567–3573. [CrossRef]
19. Bian, H.; Zhang, L. Gesture Recognition Based on Gramian Angular Difference Field and Multi-stream Fusion Methods. *Signal Image Video Process.* **2025**, *19*, 120. [CrossRef]
20. Mohammadjafari, S.; Ozyegen, O.; Cevik, M.; Kavurmacioglu, E.; Ethier, J.; Basar, A. Designing Mm-wave Electromagnetic Engineered Surfaces Using Generative Adversarial Networks. *Neural Comput. Appl.* **2021**, *33*, 11309–11323. [CrossRef]
21. Hong, I.; Ha, J. Generative Adversarial Networks for Solving Hand-Eye Calibration Without Data Correspondence. *IEEE Robot. Autom. Lett.* **2025**, *10*, 2494–2501. [CrossRef]
22. Zhang, C.; Feng, Y.; Qiang, B.; Shang, J. Wasserstein Generative Recurrent Adversarial Networks for Image Generating. In Proceedings of the 24th International Conference on Pattern Recognition, Beijing, China, 20–24 August 2018.
23. Mu, D.; Meng, W.; Zhao, S.; Wang, X.; Liu, W. Intelligent Optical Communication Based on Wasserstein Generative Adversarial Network. *Chin. J. Lasers* **2020**, *47*, 217–233. [CrossRef]
24. Gulrajani, I.; Ahmed, F.; Arjovsky, M.; Dumoulin, V.; Courville, A. Improved Training of Wasserstein GANs. In Proceedings of the 31st International Conference on Neural Information Processing Systems, Long Beach, CA, USA, 4–9 December 2017.
25. Tang, X.; Du, Y.; Liu, Y.; Li, J.; Jia, Y.; Ma, Y. Image Recognition With Conditional Deep Convolutional Generative Adversarial Networks. *Acta. Autom. Sin.* **2018**, *44*, 855–864. [CrossRef]
26. Li, Y.; Du, Y.; Xiao, N.; Ouyang, W. Improved Boundary Equilibrium Generative Adversarial Networks. *Acta IEEE Access* **2018**, *6*, 11342–11348. [CrossRef]
27. Park, S.W.; Huh, J.H.; Kim, J.C. BEGAN v3: Avoiding Mode Collapse in GANs Using Variational Inference. *Electronics* **2020**, *9*, 688. [CrossRef]
28. Tang, Y. ECGAN: Translate Real World to Cartoon Style Using Enhanced Cartoon Generative Adversarial Network. *Comput. Mater. Contin.* **2023**, *76*, 1195–1212. [CrossRef]

29. Wu, Z.; Li, M. ResNet-Swin Transformer Based Intrusion Detection System for In-vehicle Network. *Expert Syst. Appl.* **2025**, *279*, 127547. [CrossRef]

energies

MDPI

Article

Active Damped Oscillation Calibration Method for Receiving Coil Transition Process Based on Early Acquisition of Pulsed Eddy Current Testing Signal

Fei Wang [1], Su Xu [1], Liqun Yin [1], Xiaobao Hu [1], Ming Ma [1], Bin Jia [1] and Jingang Wang [2,*]

[1] Baotou Power Supply Branch of Inner Mongolia Electric Power (Group) Co., Ltd., Baotou 014030, China; fei_wang198911@163.com (F.W.); xusu0802@163.com (S.X.); ly781015@proton.me (L.Y.); hxb19980213@proton.me (X.H.); maming_0823@proton.me (M.M.); jiabin0723@proton.me (B.J.)

[2] State Key Laboratory of Power Transmission Equipment Technology, School of Electrical Engineering, Chongqing University, Chongqing 400044, China

* Correspondence: jingang@cqu.edu.cn

Abstract: As a common signal sensing device in pulsed eddy current detection, coil sensors often have parameter offset problems in practical applications. The error in the receiving coil parameters will have a great impact on the early signal. In order to ensure the accuracy of the early signal, this paper first analyzes the response characteristics of the receiving coil and the influence of the coil parameters on the accuracy of signal deconvolution and establishes the mathematical relationship between the response signal and the characteristic parameters, and between the characteristic parameters and the receiving coil parameters under active underdamped oscillation. Subsequently, the parameter feature extraction errors under different state switching capacitors were compared through simulation analysis, the state switching capacitor value was determined, and the receiving coil parameter solution method based on the Levenberg–Marquardt (LM) algorithm was determined based on the parameter feature extraction results. The experimental results demonstrate that the proposed method achieves a capacitance estimation error of just 0.0159% and an inductance error of 0.158%, effectively minimizing early signal distortion and enabling precise identification of receiving coil parameters.

Keywords: active underdamped oscillation; pulsed eddy current; Levenberg–Marquardt algorithm; signal calibration; receiving coil

1. Introduction

As a magnetic field sensing instrument, the coil sensor is a common pulse eddy current response signal detection device. The pulse eddy current method is an effective method for detecting underground low-resistance conductors using the principle of electromagnetic induction [1–3]. Due to its wide detection spectrum, no need to contact the object to be detected, and rich time domain characteristics contained in the signal, it is widely used in the detection of minerals, pipelines, grounding bodies of power equipment, and underground spaces in cities [4–7].

During eddy current detection, the secondary magnetic field signal excited by the target object is captured by the receiving coil sensor [8]. For pulsed eddy current shallow detection, when performing time domain analysis on the signal, the shallow characteristics are often reflected in the early signals, so it is crucial to ensure the early accuracy of the response signal [9–12]. However, the electromagnetic coupling path of the receiving system will introduce signal distortion—the collected signal can be regarded as the output response of the original induced signal after the transfer function is applied. In order to restore the true magnetic field change characteristics, the collected signal needs to be deconvolved to eliminate the transient response characteristics of the receiving system [13]. This process is called receiving coil system calibration [14]. The deconvolution accuracy is determined by the transfer function of the system itself. When the system parameters are offset, the deconvolved signal will change because of the error associated with the system parameters [15].

Signal deconvolution calculation is an important process to remove the transient process of the receiving coil and ensure signal accuracy [15,16]. The frequency response method serves as a fundamental approach for coil system characterization. This technique employs sinusoidal excitation signals across discrete frequencies within a target spectrum. At each calibration frequency, synchronous measurement of input stimulus and steady-state output enables comprehensive amplitude–phase characterization. Subsequent curve-fitting of these experimental datasets can be used to derive the coil's transfer function, establishing its frequency–domain response profile [17,18]. However, changes in the environmental medium or structural deformation will affect the calibration accuracy, and the calibration of the receiving system needs to be updated regularly. In addition, the frequency response method has high requirements for the uniformity of the calibration magnetic field, and the field source has poor versatility, making it difficult to meet the needs of on-site calibration [13]. Reference [16] proposed a receiving coil parameter calibration method based on time domain zero-state response, which transforms parameter estimation into a nonlinear programming problem, and combines the descent direction and step size strategy to perform parameter inversion through curve fitting. This approach eliminates reliance on the input signal for calibrating the coil parameters; however, the precision of the fitting has a direct impact on the final outcome. Reference [13] introduced a calibration method based on time domain feedback, utilizing the solution error from the induced electromotive force of the receiving coil as the feedback signal. It employs an exponential decay current as the calibration signal and integrates a τ value conversion algorithm to identify and correct errors. The calibration file is refreshed by reducing the non-steady-state interval, which effectively eliminates reliance on a uniform magnetic field. Additionally, the straightforward circuit design makes it ideal for on-site calibration of the receiving coil.

The above methods can determine the receiving coil parameters by analyzing the entire response signal of the receiving coil. However, the accuracy of the solution is affected by the error in the response signal throughout the entire process. This paper uses an active damped oscillation method to relate the coil parameters to the characteristic parameters of the response signal. Furthermore, through derivation, the solution of the coil parameters is transformed into the problem of solving nonlinear equations. This method is simpler and suitable for field calibration. Among the various existing numerical methods for solving nonlinear equations, Newton's method is the most basic and important method for solving nonlinear equations [19,20]. In Reference [7], the Newton method is used to solve the nonlinear equation. However, due to the large difference between inductance and capacitance parameters in this paper, the inductance can be solved under a small error by Newton's method, but the error of capacitance fluctuates greatly, and the overall stability of the solution needs to be improved. For this reason, it can be converted into a nonlinear least-squares problem. The Gauss–Newton method has fast convergence when the column is of full rank, but when the nonlinearity of $F(x)$ is very large, the convergence speed is very slow or does not converge. The Levenberg–Marquardt method constructs a robust Newton-type iteration step by introducing a forced positive definite strategy, taking into account both numerical stability and second-order convergence, and can effectively deal with Jacobian singular or ill-conditioned problems [21].

To address the early errors in the response signal caused by coil parameter offset, the large order-of-magnitude differences between the inductance and distributed capacitance parameters to be solved during the coil parameter solution process, as well as their varying degrees of influence on the solution equation, and to prevent imbalanced optimization directions caused by excessively large parameter iteration steps, this paper analyzes the impact of coil parameter offset on the early signal accuracy based on the response characteristics of the coil to be calibrated. Subsequently, a physical mapping relationship between the underdamped response signal characteristics and the coil parameters to be solved is established using an active underdamped oscillation method. The equations are then solved using the LM method with a damping factor and an indentation factor. This method avoids large single-parameter errors in the solution of the inductance and distributed capacitance parameters, thereby accurately solving the receiving coil parameters. Finally, simulations and experiments verify that this method, compared with other nonlinear equation-solving methods, accurately solves the coil parameters to be calibrated, prevents early signal distortion caused by coil parameter offset, and facilitates coil sensor calibration.

2. Analysis of Dynamic Characteristics of Receiving Coil

2.1. Analysis of Receiving Coil Response Characteristics

Figure 1 shows the equivalent model of the excitation and receiving coils of the pulsed eddy current and the state switching circuit corresponding to the active damped oscillation. R_0 is the equivalent resistance, L is the inductance, C is the distributed capacitance, respectively, and R_b is the damping matching resistor. At this time, the transfer function, damping ratio ζ, and natural frequency ω_n of the receiving coil are:

$$\begin{aligned} H(s) &= \frac{1}{LC[s^2+s(\frac{R_0}{L}+\frac{1}{R_bC})+\frac{R_0+R_b}{R_bLC}]} \\ \zeta &= \frac{R_bR_0C+L}{2\sqrt{R_bLC(R_b+R_0)}} \\ \omega_n &= \sqrt{\frac{R_b+R_0}{R_bLC}} \end{aligned} \quad (1)$$

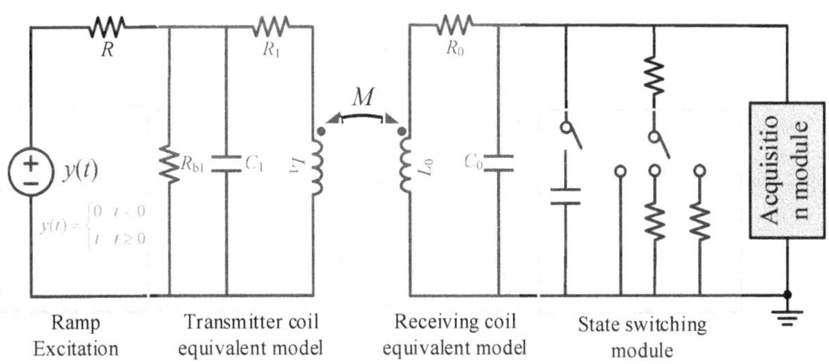

Figure 1. Receiving coil equivalent model and parameter identification circuit.

From (1), we can see that when the receiving coil acquires the induced voltage $v(t)$ caused by the change in the magnetic field, the actual signal $u(t)$ obtained by the acquisition module is the signal after $v(t)$ is acted upon by $H(s)$. $u(t)$ is the time domain convolution of $v(t)$ and $h(t)$. $u(t)$ and $v(t)$ can be converted to each other through the process shown in Formula (2) [22,23].

$$\begin{aligned} u(t) &= v(t) * h(t) = \mathcal{F}^{-1}[V(\omega) \times H(\omega)] \\ v(t) &= \mathcal{L}^{-1}(V(s)) = \mathcal{L}^{-1}\left(\frac{\mathcal{L}u(t))}{H(s)}\right) \end{aligned} \quad (2)$$

2.2. Analysis of the Accuracy of Receiving Coil Deconvolution

In order to discuss the relationship between the deconvolution result of the response signal and the true result under different receiving coil parameters, as well as the influence of parameter deviation on the deconvolution signal and the size of the signal error, the receiving coil parameters shown in Table 1 are defined, where the true value is obtained through actual measurement, and the initial value is offset based on the true value as the known initial value of the coil parameter. First, the step response of the receiving coil is calculated by (1), and then the step signal is deconvolved to obtain the corresponding input signal $v(t)$ through the receiving coil transfer function under different parameter changes. In order to analyze the influence of parameter error on deconvolution under certain changes in different parameters, the early mean error MSE of the signal, the maximum error Max_error of the signal, and the steady-state error Steady-state_error of the signal shown in (3) are defined, respectively. The waveform and error of the receiving signal after deconvolution under different parameter changes are shown in Figure 2.

$$\begin{aligned} MSE &= mean(v(t) - \varepsilon(t)) \quad 0 \leq t \leq 10us \\ Max\_error &= \max|v(t) - \varepsilon(t)| \\ \text{Steady} - \text{state\_error} &= v(t) - \varepsilon(t) \quad t \to \infty \end{aligned} \tag{3}$$

Table 1. Receiving coil parameters.

Receiving Coil Parameters	L (mH)	C (nF)	R (Ω)
Initial Value	10	0.25	2.6
True value	10.09	0.2516225	2.6

As shown in Figure 2a–c, in terms of waveform, the parameter offset of inductance and capacitance will cause a large distortion in the early waveform of the signal, and the change of capacitance parameters will even cause oscillation in the early signal, compared with inductance. The offset of resistance parameters also has an impact on the early signal, but the impact is smaller. The impact of resistance parameters on the signal is mainly reflected in the steady-state value of the signal. As shown in Figure 2d–f, in terms of the early mean error, when the absolute error of the inductance parameter reaches 20%, the early mean error of the signal reaches 0.0731. Compared with the 0.016 of the capacitance parameter deviation under the same absolute error, the early signal error caused by inductance is larger; in terms of the maximum error point, the maximum error values caused by the offset of inductance and capacitance parameters are the same, but both are greater than the maximum error caused by resistance. Therefore, the inductance and capacitance parameters have a greater impact on the early error of the signal, and the resistance mainly affects the steady-state error. In order to ensure the accuracy of the collected signals and the precision of the early signals, it is necessary to accurately identify the various parameters of the receiving coil system, avoid signal distortion through the coil transition process, and ensure the accuracy of the signal and the accuracy of post-processing.

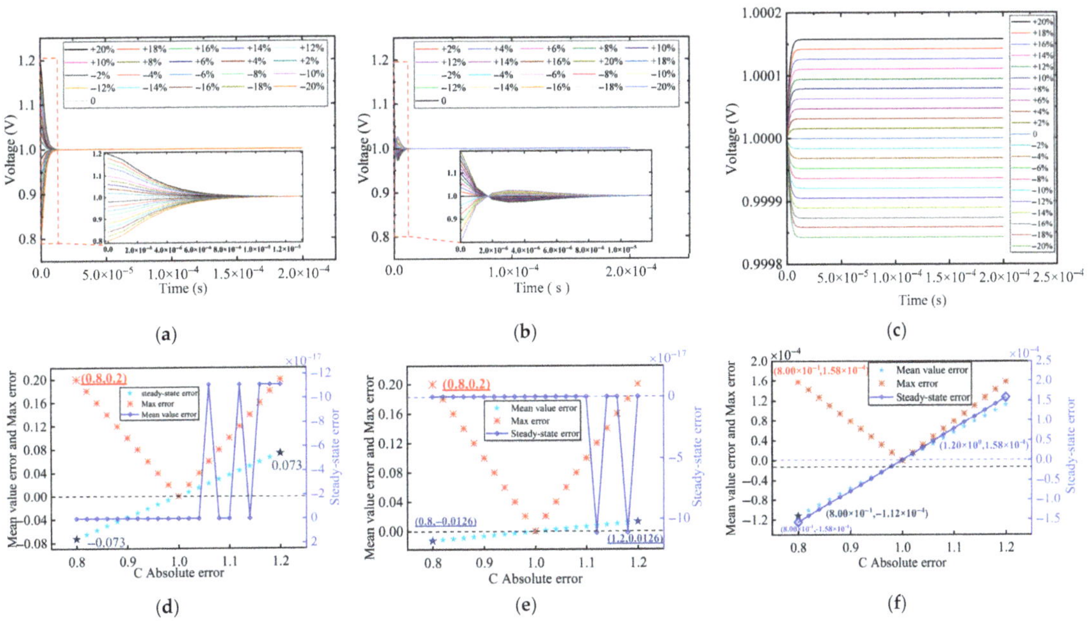

Figure 2. Error analysis under different parameter changes. (**a**) Deconvolution waveform with inductance error; (**b**) Deconvolution waveform with capacitance error; (**c**) Deconvolution waveform with resistance error; (**d**) Error analysis under different inductances; (**e**) Error analysis under different capacitances; (**f**) Error analysis under different resistances.

2.3. Active Underdamped Oscillation and Signal Characteristics Analysis

To suppress signal oscillations in the received waveform, impedance matching is typically employed to bring the system into a critically damped state [13,16]. The corresponding matching resistance under critical damping conditions can be calculated using the following expression:

$$R_b = \frac{L}{R_0 C \pm 2\sqrt{LC}} \tag{4}$$

In order to avoid underdamped oscillation of the signal because of uncertain factors, the system is generally kept in a slightly overdamped state. By actively connecting S_1 and S_2 in Figure 1, Figure 3 illustrates that the approach of enhancing the matching resistance and utilizing a parallel state switching capacitor can transition the system from a mildly overdamped condition to an underdamped condition. At this time, when the transmitting coil applies a ramp input, the receiving coil will sense the step signal $\varepsilon(t)$. At this time, the overshoot $\sigma\%$ and the adjacent peak time Δt_p corresponding to the response signal obtained by the acquisition module are expressed as:

$$\begin{aligned} \sigma\% &= \frac{y_m - y_s}{y_s} = e^{\left(-\frac{\zeta\pi}{\sqrt{1-\zeta^2}}\right)} \times 100\% \\ \Delta t_p &= \frac{\pi}{\omega_n\sqrt{1-\zeta^2}} \end{aligned} \tag{5}$$

Equation (5) shows that Δt_p and $\sigma\%$ of the response signal depend solely on the system's natural frequency and damping ratio. Therefore, the adjacent peak time and overshoot are obtained by extracting the characteristics of the response signal, and then the natural frequency and damping ratio of the system can be obtained by solving the equation. From (1), there is a nonlinear relationship between the natural frequency and damping ratio and the parameters to be identified, L and C. Therefore, the method of solving nonlinear equations can be used to solve for L and C, thereby completing the parameter identification of the receiving coil and ensuring the accuracy of the early signal.

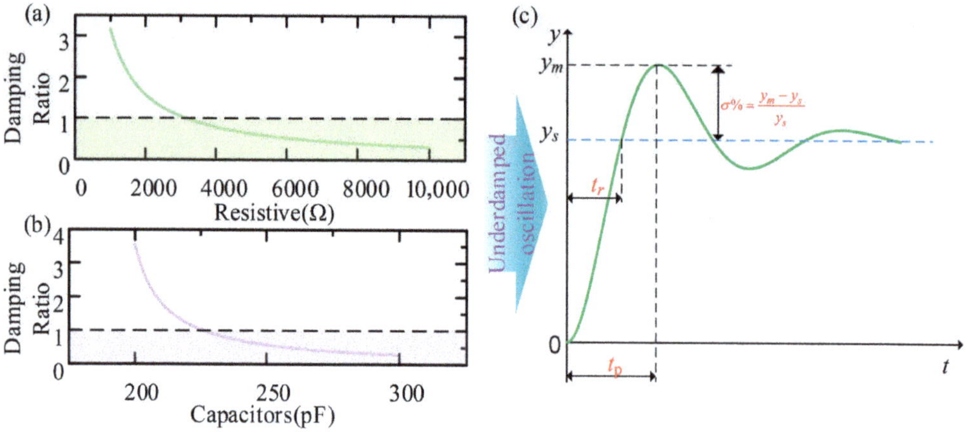

Figure 3. Damping coefficient curve and dynamic response characteristics of a second-order system. (**a**) Matching the relationship curve between resistance and damping coefficient; (**b**) Relationship curve between state-switching capacitance and damping coefficient; (**c**) Analysis of the characteristics of under-damped oscillation signals.

3. Accept Coil Parameter Solution

3.1. Response Signal Feature Extraction and Analysis

According to the actual parameters corresponding to Table 1, the circuit simulation analysis was performed through Multisim, and the early signal of the receiving coil induced voltage under the parallel connection of different state switching capacitors and series state switching resistors was obtained as shown in Figure 4.

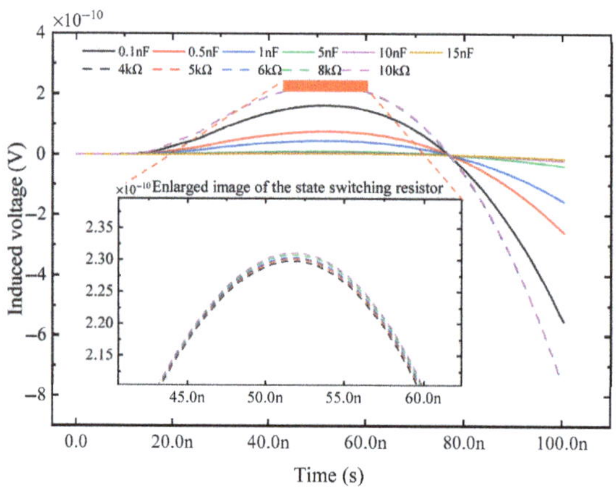

Figure 4. Early induction signals of the receiving coil under different state switching parameters.

As illustrated in Figure 4, under ramp excitation of the transmitting coil, the receiving coil is theoretically expected to detect a step response. However, the presence of the coil's inherent resonant frequency leads to an early-stage overshoot. It can be seen from the local amplification that the method of increasing the damping resistance has minimal effect on the improvement in the overshoot signal in the early stage, and the overshoot peak decreases minimally. By increasing parallel capacitance, the effect of the overshoot on the signal distortion can be significantly reduced. When the parallel state switching capacitance is greater than 5 nF, the early overshoot is very small. Therefore, this paper adopts the method of parallel state switching capacitors to make the system transition to the underdamped state.

It can be seen from the underdamped oscillation response waveform in Figure 5 that with an increase in the parallel state switching capacitor, the amplitude and rise time of the signal also increase, and the corresponding oscillation times also increase, which is consistent with the law in Figure 3. Moreover, there is almost no overshoot in the early stage of the signal. The adjacent peak time and overshoot of the response can be obtained by formula (5) to obtain the characteristic parameters of the response signal. In order to accurately extract the characteristic parameters, the error between the extracted value of the characteristic parameters of the curve under different state switching capacitors and the calculated value and the true value is shown in Figure 6.

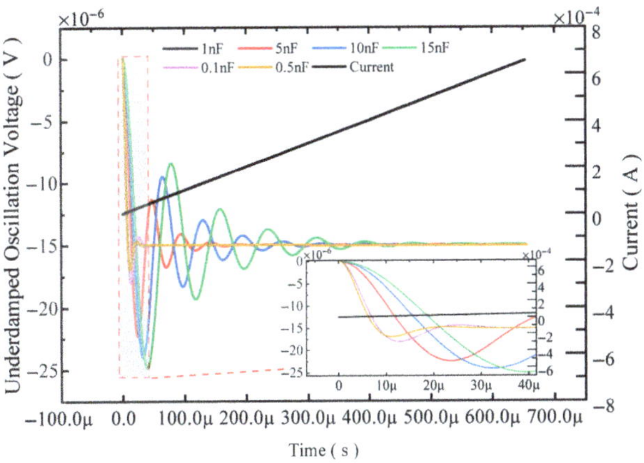

Figure 5. Underdamped oscillation response waveform under excitation current and different states of switching capacitance.

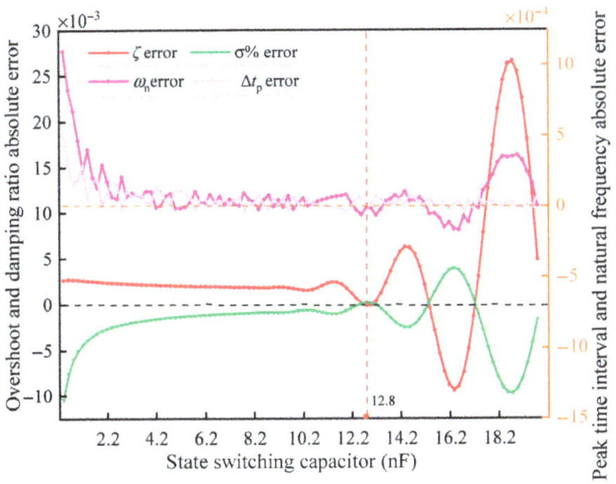

Figure 6. Characteristic parameter extraction, calculation error, and capacitance relationship curve.

From the characteristic parameter extraction and calculation error and capacitance curve in Figure 6, it follows that under the parameter design in Table 1 of this article, when the state switching capacitor value is 12.8 nF, the overshoot and peak time interval errors obtained by feature extraction of the underdamped oscillation response signal are the smallest, and the errors of the damping ratio and natural frequency calculated by the characteristic parameters are also the smallest. Therefore, the state switching capacitor selected in this article is 12.8 nF.

3.2. Parameter Solution Based on the Levenberg–Marquardt Algorithm

To determine the parameters of the receiving coil, the known characteristic features of the response signal are utilized, whereby the parameter identification problem is transformed into solving a system of two nonlinear equations.

$$\mathbf{F} = \begin{bmatrix} f_1 \\ f_2 \end{bmatrix} = \begin{bmatrix} \omega_n - \sqrt{\frac{R_b+R_0}{R_bLC}} \\ \zeta - \frac{R_bR_0C+L}{2\sqrt{R_bLC(R_b+R_0)}} \end{bmatrix} \tag{6}$$

When using a coil as a receiving system for signal acquisition, the parameters of the coil are often measured during initial use, and the measured values can be used as initial values for solving nonlinear equations, while the true values produce small deviations near the initial values. Under the premise of knowing the initial value, the classic Newton method can be used to solve the nonlinear equation. Among them, the Levenberg–Marquardt (LM) method is one of the most famous nonlinear methods proposed on the basis of Newton's method [24]. The overall flow diagram is shown in Figure 7. The objective function is to minimize the residual sum of squares:

$$\min_{\mathbf{x}} \frac{1}{2}\|\mathbf{F}(\mathbf{x})\|^2 = \frac{1}{2}\sum_{i=1}^{2} f_i^2(\mathbf{x}) \tag{7}$$

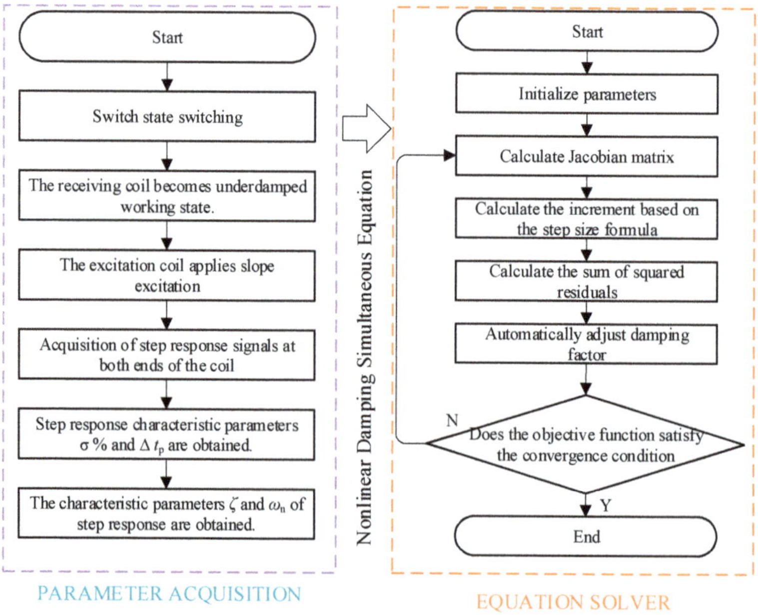

Figure 7. The overall flow diagram.

In each iteration, the Levenberg–Marquardt (LM) method is calculated with a step size of [25]:

$$d_k^{LM} = -\left(J_k^T J_k + \lambda_k I\right)^{-1} J_k^T F_k \tag{8}$$

Among them, $F_k = F(x_k)$, $J_k = F'(x_k)$ is the Jacobian matrix, $\lambda_k = \|F(x_k)\|^{\delta}$, $\delta \in [1, 2]$ is the LM parameter. λ_k is a non-negative constant. The LM method overcomes the problem of singular or near-singular J_k when applying the Newton method by introducing a non-negative parameter λ_k. In addition, it also prevents the problem of long step length $\|d_k\|$. During the solution process, the step length can be adjusted by dynamically adjusting λ_k by evaluating the effect of the current step length.

As can be seen from Figure 8, the search directions corresponding to the inductance parameters and capacitance parameters when solving the natural frequency and damping ratio equations are consistent. However, when solving the natural frequency equation and the damping ratio equation, the two parameters have different sensitivity responses to the equations. The inductance parameter has a greater impact on the solution and the error of the objective function. Therefore, when solving nonlinear equations, it is easy to have a large inductance step, which affects the identification results of the capacitance parameters and causes an imbalance in the optimization direction. Therefore, it is necessary to adjust the solution step for the two parameters.

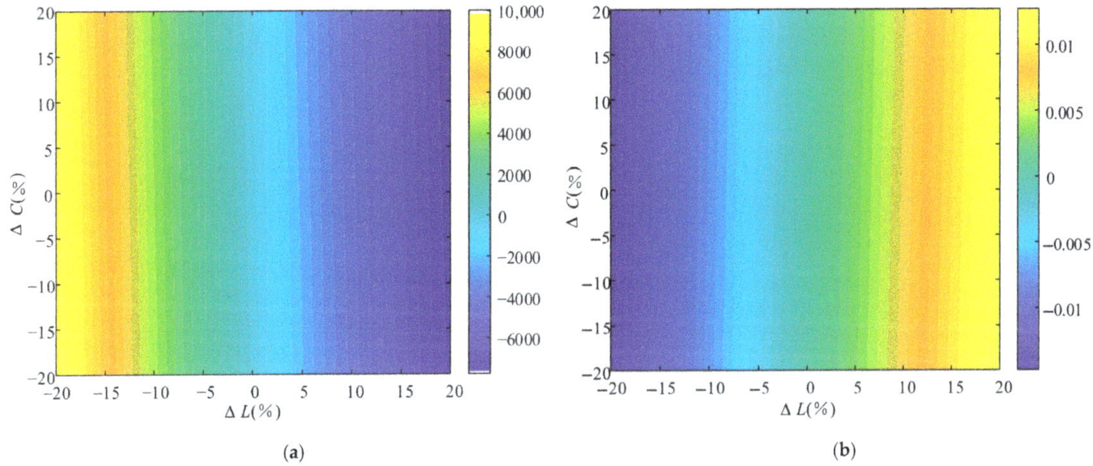

Figure 8. Relationship between parameter transformation and equation solving. (**a**) Effect of parameter changes on the solution of the natural frequency equation; (**b**) Effect of parameter changes on the solution of the damping ratio equation.

The step length indentation factor can solve the problem of updating the step size imbalance and enhance the capacitance sensitivity. Therefore, in order to avoid the excessive error of capacitance parameters caused by the over-fitting of inductance parameters in the process of parameter identification, the step size indent factor is introduced, and the step size in the iterative process is adjusted to:

$$\mathbf{x}_{k+1} = \mathbf{x}_k + S \cdot d_k \tag{9}$$

$$S = [s_L, s_c]^T \tag{10}$$

where S is the step size reduction factor, and S_L and S_C are the step size reduction parameters of L and C, respectively. The identifiability of the capacitance C is significantly improved by the current inductance parameter step size method.

Based on the parameter values shown in Table 1, the algorithm was implemented in the MATLAB environment using MTLAB R2022b, on a Windows 10 computer with an Intel Core i5 13400F CPU. The error convergence curves, solution parameters, and performance comparisons of the four solution methods during the parameter identification process are shown in Figures 9 and 10, and Table 2, respectively. The BFGS method, based on the Newton method, the LM algorithm, the Gauss–Newton method, and the LM algorithm with a step size reduction, was used.

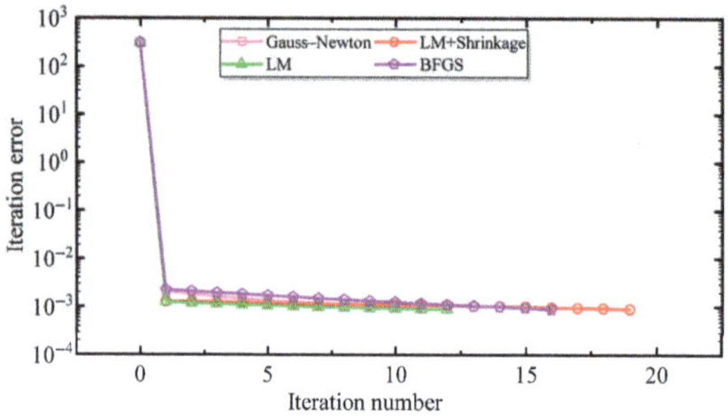

Figure 9. Iterative error convergence curves of the four methods.

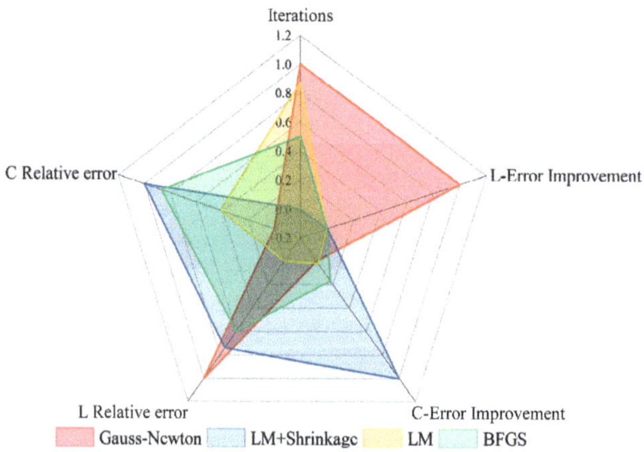

Figure 10. Convergence performance radar chart.

Table 2. Solution parameters of the three methods.

Parameters of the Solution	Iterations	L-Error Improvement	C-Error Improvement	L Relative Error	C Relative Error
BFGS	16	38.2765	8.1584	0.00231	7.84512×10^{-4}
Gauss–Newton	12	8.4596×10^3	1.2764	1.05443×10^{-5}	0.00502
LM+Shrinkage	20	59.4516	42.2612	0.0015	1.51457×10^{-4}
LM	13	15.4721	2.1277	0.00577	0.00301

As can be seen from Figures 9 and 10 and Table 2, in terms of iteration speed, the Gauss–Newton method has the smallest number of iterations for solving nonlinear equations, but for the nonlinear solution problem in this paper, the three methods are less time-consuming; in terms of error improvement, the Gauss–Newton method has the best improvement in inductance error and solution accuracy, but the capacitance solution error is large, while the LM+Shrinkage method has the best improvement in capacitance error and solution accuracy, and also has a good effect in terms of inductance solution accuracy; among the three solution methods, the LM algorithm will increase the number of solution steps after adding the step size shrinkage factor, but the accuracy of the capacitance parameters has been improved. Since there are only two equations in this problem, the solution times for different methods are similar. Starting from the perspective of parameter solution accuracy, considering the accuracy of the capacitance solution and inductance solution comprehensively, the LM algorithm plus step size shrinkage factor method is selected to solve the receiving coil inductance and capacitance.

4. Experimental Testing

To confirm the correctness of the receiving coil parameter solution method based on the LM algorithm proposed in this paper, an experimental schematic diagram, as shown in Figure 11, was built based on the circuit model in Figure 1. The experimental schematic diagram is shown in Figure 11. The real parameters of the coil are shown in Table 1. By actively changing the dynamic response characteristics of the coil, the applied excitation, the obtained response signal waveform, and the corresponding characteristic parameters are shown in Figure 12 after the coil is in an underdamped oscillation state.

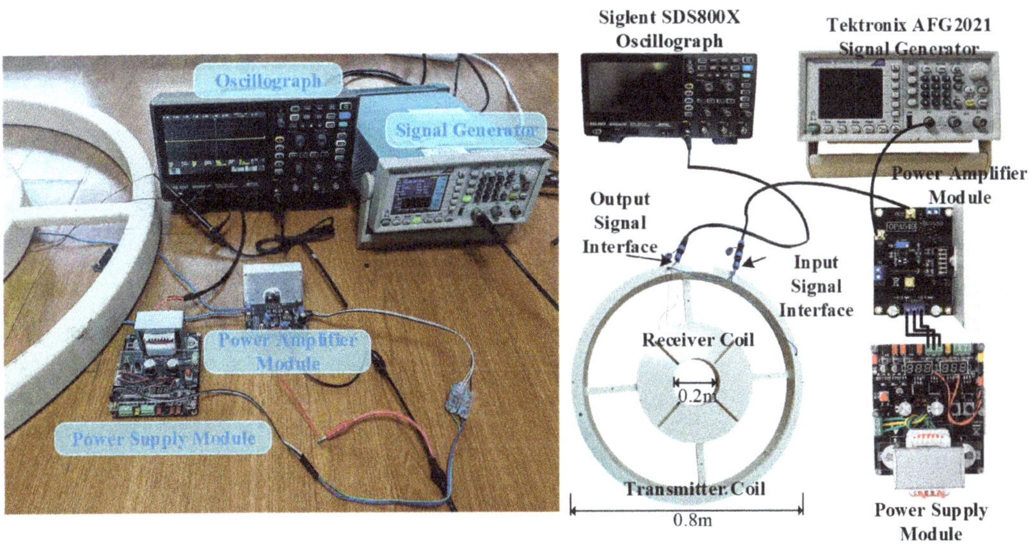

Figure 11. Experimental schematic.

After the ramp current is applied to the excitation coil, the response voltage signal of the receiving coil is shown in Figure 12. The corresponding values of the response signal peak value, adjacent peak time, steady-state value, and overshoot obtained by extracting the characteristic parameters of the response signal are shown in the annotations of Figure 12.

After obtaining the characteristic parameters, the corresponding iterative process and the convergence effect of inductance and capacitance errors obtained after obtaining the nonlinear solution to Equation (6) based on the LM plus step size reduction factor are shown in Figure 13a.

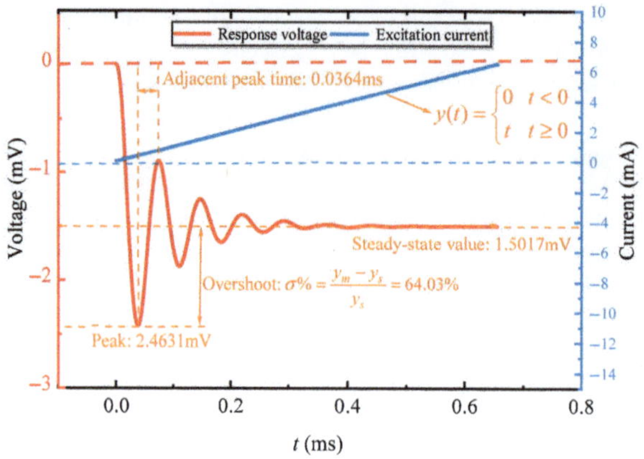

Figure 12. Applied excitation current and induced voltage signal.

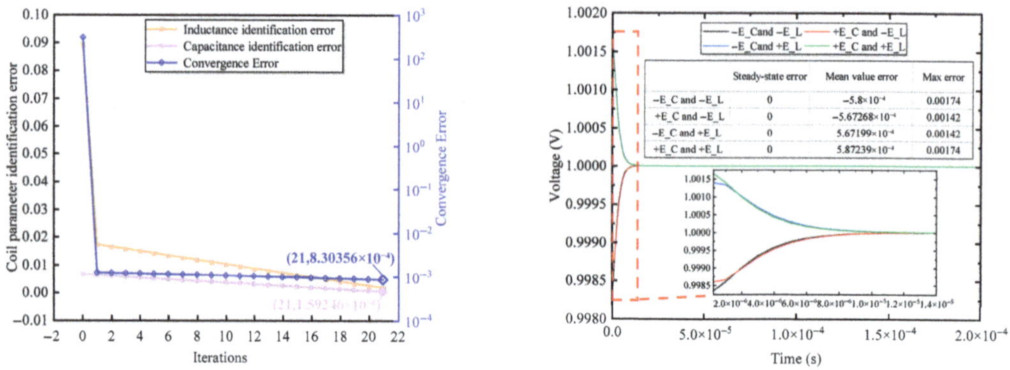

(**a**) (**b**)

Figure 13. Receiving coil parameter identification results and error analysis. (**a**) Parameter identification results; (**b**) Deconvolution accuracy analysis.

Based on the characteristic parameters extracted from the signal, the receiving coil parameter model is solved using the Levenberg–Marquardt (LM) algorithm. The iterative process converges after 22 steps. By applying ramp excitation and performing feature extraction on the underdamped oscillation waveform, followed by parameter identification through the LM-based method proposed in this study, the relative error in capacitance estimation is reduced to 0.0159% (E_C), while the inductance estimation error is 0.158% (E_L). This demonstrates the method's effectiveness in accurately identifying the parameters of the receiving coil. Compared with the capacitance solution error of 0.038% and inductance solution error of 0.17% corresponding to reference [7], the method proposed in this paper greatly improves the solution error of capacitance parameters while ensuring the solution error of inductance parameters. At the same time, compared with the parameter solution effect in Table 2, it can be seen that in actual applications, environmental noise will affect the feature extraction accuracy of the response signal and thus affect the final solution accuracy. Therefore, in actual application, it is necessary to avoid the influence of environmental noise on the response signal.

Table 3 shows the results of multiple measurements and calculations of the coil under ambient noise. Table 3 shows that the mean values of the iteration number, inductance error, and capacitance error for the multiple measurements are 22, 0.1583%, and 0.01588%, respectively. The standard deviations of the inductance and capacitance solutions are 0.000838 and 0.005046, respectively. These test results demonstrate the stability of the measurement method.

Table 3. Multiple test results.

Test Results	Iterations	E_C (%)	E_L (%)
Test 1	22	0.0161	0.16
Test 2	22	0.0154	0.154
Test 3	21	0.0173	0.165
Test 4	23	0.0148	0.151
Test 5	22	0.016	0.159
Test 6	22	0.0157	0.161
Mean	22	0.015883	0.158333
Standard deviation	0.632456	0.000838	0.005046

In order to verify the influence of parameter error on the accuracy of signal deconvolution under this solution accuracy, the error analysis of the deconvolution signal machine under the solution accuracy is shown in Figure 13b. It can be seen from Figure 13b that under this solution accuracy, the inductance and capacitance parameters have a limited influence on the deconvolution solution. The maximum error caused by capacitance is only 1.6×10^{-5}, and the maximum error caused by inductance is only 1.58×10^{-3}. Under this level of solution accuracy, the pulse eddy current detection signal can be accurately restored after deconvolution, which avoids the early signal distortion caused by parameter offset.

5. Conclusions

In order to solve the problem that the deconvolution of the induced voltage signal is biased, and the early signal waveform is affected by the offset of the coil parameters during the application of the pulse eddy current receiving coil:

(1) This study formulates a nonlinear equation that describes the relationship between the characteristics of underdamped oscillations in the receiving coil and the parameters of the coil itself. Additionally, it introduces a method for actively damping these oscillations and correlates the solutions for the coil parameters with the solutions derived from the nonlinear equation through the process of feature extraction.

(2) Based on the background of coil parameter offset and known initial parameters, in order to avoid the influence of different magnitudes of inductance and capacitance parameters and different sensitivity responses to the equation on parameter identification, the LM algorithm plus step size retraction factor method based on the improved Newton method is used to iterate the nonlinear equation. The three methods of the Gauss–Newton method, the LM algorithm, and the LM algorithm plus step size retraction factor are compared and analyzed to solve the receiving coil parameter equation. The comparison shows that the LM algorithm plus step size retraction factor method can effectively improve the solution accuracy of capacitance parameters.

(3) Experimental verification shows that, following feature extraction and parameter equation solving, the identification errors for capacitance and inductance can be reduced to 0.0159% and 0.158%, respectively, enabling precise characterization of the receiving coil parameters.

(4) Based on the receiving coil parameters obtained in this paper, an accurate detection signal can be obtained by deconvolution, and the parameters of the device can be calibrated in real time because of the simple circuit and method. However, since the input signal needs to be applied by the excitation coil, it is mainly suitable for the central loop structure. In the

future, the application range of this method can be further expanded, and the coil parameter calibration method under the excitation signal applied only to the single structure of the receiving coil can be studied.

Author Contributions: Conceptualization, F.W.; methodology, S.X.; software, L.Y.; validation, X.H.; formal analysis, M.M.; investigation, B.J.; resources, F.W.; data curation, S.X.; writing—original draft preparation, J.W.; writing—review and editing, J.W.; visualization, F.W.; supervision, X.H.; project administration, L.Y.; funding acquisition, F.W. All authors have read and agreed to the published version of the manuscript.

Funding: This research was funded by Inner Mongolia Electric Power (Group) Co., Ltd. Science and Technology Project, grant No. nmgdlkjxm20240403.

Data Availability Statement: The data used in the analysis presented in the paper will be made available, subject to the approval of the data owner.

Conflicts of Interest: F.W., S.X., L.Y., X.H., M.M. and B.J. were employed by the Baotou Power Supply Branch of Inner Mongolia Electric Power (Group) Co., Ltd. The authors declare that this study received funding from Inner Mongolia Power (Group) Co., Ltd., Science and Technology Project. All authors declare that the research was conducted in the absence of any commercial or financial relationships that could be construed as potential conflicts of interest. The funder was involved in the study design, collection, analysis, interpretation of data, the writing of this article, or the decision to submit it for publication.

References

1. Safizadeh, M.S.; Lepine, B.A.; Forsyth, D.S.; Fahr, A. Time-Frequency Analysis of Pulsed Eddy Current Signals. *J. Nondestruct. Eval.* **2001**, *20*, 73–86. [CrossRef]
2. Sophian, A.; Tian, G.; Fan, M. Pulsed Eddy Current Non-Destructive Testing and Evaluation: A Review. *Chin. J. Mech. Eng.* **2017**, *30*, 500–514. [CrossRef]
3. Li, Y.; Tian, G.Y.; Simm, A. Fast Analytical Modelling for Pulsed Eddy Current Evaluation. *NDT E Int.* **2008**, *41*, 477–483. [CrossRef]
4. Auld, B.A.; Moulder, J.C. Review of Advances in Quantitative Eddy Current Nondestructive Evaluation. *J. Nondestruct. Eval.* **1999**, *18*, 3–36. [CrossRef]
5. Silva, M.I.; Malitckii, E.; Santos, T.G.; Vilaca, P. Review of Conventional and Advanced Non-Destructive Testing Techniques for Detection and Characterization of Small-Scale Defects. *Prog. Mater. Sci.* **2023**, *138*, 101155. [CrossRef]
6. Sha, J.; Fan, M.; Cao, B.; Liu, B. Noncontact and Nondestructive Evaluation of Heat-Treated Bearing Rings Using Pulsed Eddy Current Testing. *J. Magn. Magn. Mater.* **2021**, *521*, 167516. [CrossRef]
7. Zeng, Z.; Wang, J.; Huang, X.; Zuo, Y.; Liu, Y.; Tian, X.; Pei, F.; Liu, K.; Chen, F.; Wang, X.; et al. Parameter Identification and Transition Process Online Calibration Method of Pulsed Eddy Current Receiving Coil Based on Underdamped Dynamic Response Characteristics. *Sensors* **2025**, *25*, 4049. [CrossRef]
8. Shkatov, P.N. Innovative Surface Eddy Current Probe for Non-Destructive Testing. In Proceedings of the 2023 7th International Conference on Information, Control, and Communication Technologies (ICCT), Astrakhan, Russia, 2–6 October 2023; pp. 1–3.
9. Shen, X.; Guo, X.; Wang, Z.; Song, M.; Hu, W. Simulation Analysis of Near and Remote Field Pulse Eddy Current Composite Detection Method for Single-Metal Pipeline. *IEEE Access* **2025**, *13*, 32434–32442. [CrossRef]
10. Grochowalski, J.M.; Chady, T. Pulsed Multifrequency Excitation and Spectrogram Eddy Current Testing (PMFES-ECT) for Nondestructive Evaluation of Conducting Materials. *Materials* **2021**, *14*, 5311. [CrossRef]
11. Huang, P.; Pu, H.; Ding, Y.; Li, Z.; Yin, W.; Xu, L.; Xie, Y. A Sensitivity Enhancement PEC Method for Bottom Flaws and Corrosions Detection. *Measurement* **2022**, *202*, 111910. [CrossRef]
12. Zhao, P.; Wang, J.; Sun, L.; Li, Y.; Xia, H.; He, W. Optimal Electrode Configuration and System Design of Compactly-Assembled Industrial Alkaline Water Electrolyzer. *Energy Convers. Manag.* **2024**, *299*, 117875. [CrossRef]
13. Wang, H.; Fu, Z.; Wang, Y.; Tai, H.; Qin, S.; Liao, X. A Time-Domain Feedback Calibration Method for Air-Coil Magnetic Sensor. *Measurement* **2019**, *135*, 61–70. [CrossRef]
14. Ye, C.; Laureti, S.; Malekmohammadi, H.; Wang, Y.; Ricci, M. Swept-Frequency Eddy Current Excitation for TMR Array Sensor and Pulse-Compression: Feasibility Study and Quantitative Comparison of Time and Frequency Domains Processing. *Measurement* **2022**, *187*, 110249. [CrossRef]

15. Xu, Z.; Liao, X.; Liu, L.; Fu, N.; Fu, Z. Research on Small-Loop Transient Electromagnetic Method Forward and N Optimization Inversion Method. *IEEE Trans. Geosci. Remote Sens.* **2023**, *61*, 5901913. [CrossRef]
16. Wang, H.; Fu, Z.; Wang, Y.; Tai, H.-M.; Chen, W. On-Site Calibration of Air-Coil Sensor for Transient Electromagnetic Exp *Geophys. Prospect.* **2019**, *67*, 1595–1610. [CrossRef]
17. Zhao, P.; Li, L.; Gao, J.; Ma, Y. Method of Standard Field for LF Magnetic Field Meter Calibration. *Measurement* **2017**, *104*, 2 [CrossRef]
18. Coillot, C.; Nativel, E.; Zanca, M.; Goze-Bac, C. The Magnetic Field Homogeneity of Coils by Means of the Space Harmonics Suppression of the Current Density Distribution. *J. Sens. Sens. Syst.* **2016**, *5*, 401–408. [CrossRef]
19. Amini, K.; Rostami, F. A Modified Two Steps Levenberg–Marquardt Method for Nonlinear Equations. *J. Comput. Appl. Math.* **2015**, *288*, 341–350. [CrossRef]
20. Choi, H.; Kim, S.D.; Shin, B.-C. Choice of an Initial Guess for Newton's Method to Solve Nonlinear Differential Equations. *Comput. Math. Appl.* **2022**, *117*, 69–73. [CrossRef]
21. Fan, J. The Modified Levenberg-Marquardt Method for Nonlinear Equations with Cubic Convergence. *Math. Comp.* **2012**, *81*, 447–466. [CrossRef]
22. He, Y.; Xue, G.; Chen, W.; Tian, Z. Three-Dimensional Inversion of Semi-Airborne Transient Electromagnetic Data Based on a Particle Swarm Optimization-Gradient Descent Algorithm. *Appl. Sci.* **2022**, *12*, 3042. [CrossRef]
23. Xu, Z.; Fu, Z.; Fu, N. Firefly Algorithm for Transient Electromagnetic Inversion. *IEEE Trans. Geosci. Remote Sens.* **2022**, *60*, 5901312. [CrossRef]
24. Chen, L. A High-Order Modified Levenberg–Marquardt Method for Systems of Nonlinear Equations with Fourth-Order Convergence. *Appl. Math. Comput.* **2016**, *285*, 79–93. [CrossRef]
25. Wu, Z.; Zhou, T.; Li, L.; Chen, L.; Ma, Y. A New Modified Efficient Levenberg–Marquardt Method for Solving Systems of Nonlinear Equations. *Math. Probl. Eng.* **2021**, *2021*, 5608195. [CrossRef]

energies

MDPI

Article

Detection and Evaluation of Shield Damage Defects in Power Cables Using an Improved Dual-Frequency Time–Frequency Domain Reflectometry

Kun Zhao [1,2], Stefano Grivet-Talocia [2], Paolo Manfredi [2], Yuan Yan [1,2] and Hongjie Li [1,*]

1 The State Key Laboratory of Insulation and Power Equipment, School of the Electrical Engineering, Xi'an Jiaotong University, Xi'an 710049, China; zhaokun@stu.xjtu.edu.cn (K.Z.); yuanyan@xjtu.edu.cn (Y.Y.)
2 Department of Electronics and Telecommunications, Politecnico di Torino, 10129 Turin, Italy; stefano.grivet@polito.it (S.G.); paolo.manfredi@polito.it (P.M.)
* Correspondence: hjli@mail.xjtu.edu.cn

Abstract: Cable shields may develop holes when the sheath is damaged. Time–frequency domain reflectometry (TFDR) is an effective method for detecting cable defects, but it cannot directly evaluate hole sizes. To address this issue, we analyze the impact of shield hole sizes on TFDR signals. Building on this analysis, we propose an improved dual-frequency TFDR method to measure shield holes and evaluate their sizes. This method directly measures the characteristic impedances and damage ratios using dual-frequency TFDR, followed by a two-step evaluation process to determine the hole center angles and lengths based on these measurements. Simulations and experiments validate the proposed method. In laboratory-scale experiments using a scaled cable model, and considering measurement noise, the maximum relative errors for shield hole length and center angle are 11% and 5%, respectively.

Keywords: cable detection; defect location; cable damage; defect evaluation; time–frequency domain reflectometry

1. Introduction

Cable shield damage is a common yet severe defect in power cables. Many factors can cause holes on cable shield, such as shield corrosion, mechanical abrasion and so on [1]. Once holes appear on the cable shield, the cable heats up and the electric field distribution distorts, leading to gradual insulation damage [2]. Detecting shield damage, evaluating the size of shield holes, and providing detailed defect information are essential tasks in cable maintenance. Therefore, developing a method for the detection and size assessment of these defects is essential for effective cable maintenance.

There are many methods, such as damped AC, partial discharge detection, and reflectometry, that can be used to detect cable defects. Each method has its own advantages and suitable application scenarios. Reflectometry is an effective method for detecting and localizing shield defects [3], operating by analyzing signals reflected from defects. It is categorized into two main types based on the signals injected: time domain [4] and frequency domain methods [5,6]. Time domain reflectometry (TDR) is simple and provides high localization accuracy, but it has low sensitivity, which limits its suitability for evaluating sizes of cable defects [7]. Conversely, frequency domain reflectometry is highly sensitive but suffers from limited localization accuracy [8]. To balance localization accuracy and sensitivity, the time–frequency domain reflectometry (TFDR) method is proposed [9,10]. Early TFDR techniques were mainly used to detect and locate defects. Owing to their high sensitivity and strong anti-interference capability, many researchers have focused on detection under noisy conditions and on the identification of weak defects. References [11,12] have implemented the likelihood ratio test and a generalized time–frequency cross-correlation (TFCC) method, respectively, both of which effectively improve the sensitivity of TFDR and enhance the ability of weak defect detection. Additionally, reference [13] explored the impact of interference on localization results and introduced an instantaneous filtering technique to improve the performance of the TFDR on weak defect detection and location. With the increasing demand for defect evaluation, researchers have gradually attempted to use TFDR to estimate defect parameters. Reference [14] employed the finite element method (FEM) to analyze the influence of shield hole defects on cable characteristic impedance. Reference [15] investigated the relationship between hole size and TFCC magnitude. Moreover, reference [16] applied a faster region-based convolutional neural network (R-CNN) to classify the severity of faults and achieve a rough evaluation of defects. After numerous iterations and developments, the TFDR method can achieve precise positioning, rough assessment of defect severity, and simple classification of defects. However, it still struggles to accurately evaluate the sizes of shield hole defects. Reference [17] developed a new testing strategy and employed a spectrum ratio method to estimate the length of cable corrosion in TFDR measurement. However, this work can only estimate corrosion length and cannot handle complex shield hole defects or multiple-defect scenarios.

To overcome these limitations and further advance the research, we propose an improved dual-frequency TFDR method for the size evaluation of shield hole defects. We first model cable shield hole defects using FEM and analyze their propagation characteristics. Based on this, we derive formulas for the reflection coefficient and reflected signals associated with these defects in cables. Building on these results, we introduce a two-step evaluation process and propose an improved dual-frequency TFDR method to detect and evaluate hole sizes, including both lengths and center angles. The measurement involves two sequential excitations, with the center frequency of the second signal being twice that of the first, enabling accurate extraction of characteristic impedances and damage ratios. These measured parameters are then used in the two-step evaluation process to determine the lengths and center angles of the shield holes.

The remaining sections of this paper are organized as follows: Section 2 introduces the basic theory of TFDR and the proposed method. Section 3 presents a simulation along with its validation results. Then, an experiment is conducted, and the results are discussed in Section 4. Finally, Section 5 provides the conclusion of this paper.

2. Methodology

2.1. Basic Theory of TFDR and Modeling of Cable Shield Hole Defects

The TFDR method employs a linear frequency-modulated signal for injection to detect and locate defects along cables. The expression for the injected signals, denoted as $s(t)$, is presented as follows [9]:

$$s(t) = A \cdot e^{\frac{-\alpha(t-t_0)^2}{2} + j\frac{\beta(t-t_0)^2}{2} + j\omega_0(t-t_0)}, \tag{1}$$

where A controls the amplitude of $s(t)$, α determines the duration of $s(t)$, β defines the slope of the instantaneous frequency of $s(t)$, and t_0 and ω_0 are the center time and center angular frequency of $s(t)$, respectively. Using the time–frequency characteristics of the injected signal, the defect is localized by the time–frequency cross-correlation (TFCC) method [11].

A typical single-core power cable consists of a circular core conductor, multiple insulating layers, and an outer shield [18]. When the cable undergoes corrosion or abrasion, holes may form in the shield. Although shield holes are typically irregular, for analytical simplicity, they are often modeled as rectangular apertures [14,15], as shown in Figure 1.

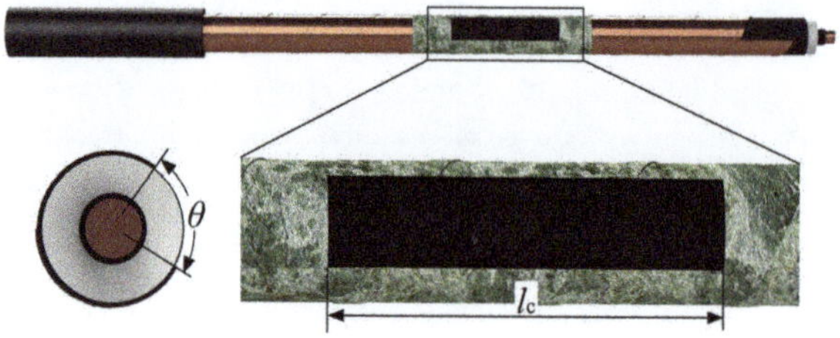

Figure 1. Illustration of a cable with shield holes.

Shield holes compromise the integrity of the cable structure; therefore, FEM is employed to model and calculate the distributed parameters of cables with such defects. The distributed-parameter model of a long cable with shield holes is shown in Figure 2.

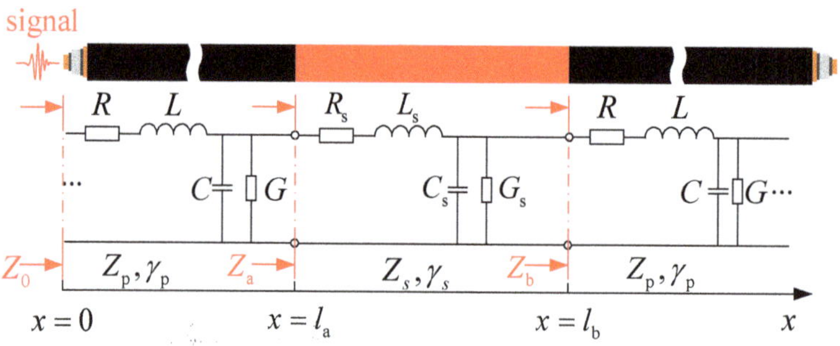

Figure 2. Distributed-parameter model of a long cable with shield hole defects.

Z_0, Z_a, and Z_b denote the input impedances at $x = 0$, $x = l_a$, and $x = l_b$, respectively. Z_p and γ_p denote the characteristic impedance and propagation constant of a healthy cable, while Z_s and γ_s correspond to those of a cable with shield hole defects. $\gamma_p = \alpha_p + j\beta_p$ and $\gamma_s = \alpha_s + j\beta_s$, where α_p and α_s are the attenuation constants, and β_p and β_s are the phase constants. Based on transmission line theory and assuming an infinitely long cable, the reflection coefficient at $x = l_a$ can be derived as

$$\Gamma_a = \frac{-\Gamma_b + \Gamma_b e^{-2\gamma_s l_c}}{1 - \Gamma_b{}^2 e^{-2\gamma_s l_c}} \tag{2}$$

where l_c denotes the defect length, and Γ_b is the reflection coefficient at $x = l_b$, defined as

$$\Gamma_b = \frac{Z_p - Z_s}{Z_p + Z_s} \tag{3}$$

The shield hole defect usually remains conductive, and the characteristic impedance Z_s does not differ significantly from that of an intact cable. Therefore, Γ_b satisfies the condition $\Gamma_b^2 \ll 1$. In addition, since $|e^{-2\gamma_s l_c}| < 1$, it follows that $\Gamma_b^2 e^{-2\gamma_s l_c} \approx 0$. Considering that the attenuation constant of the defect is small and the defect length under study ranges from tens of centimeters to several meters, the attenuation effect can be neglected. Thus, Γ_a can be approximated as

$$\begin{aligned}\Gamma_a &\approx -\Gamma_b + \Gamma_b e^{-2j\beta_s l_c} \\ &\approx -2\Gamma_b \sin(\beta_s l_c) e^{j(\frac{\pi}{2}-\beta_s l_c)}.\end{aligned} \tag{4}$$

2.2. Evaluation of Shield Hole Sizes Using Improved Dual-Frequency TFDR

To evaluate shield hole dimensions, the reflected signal is derived. In the configuration of Figure 2, a defect is located at $x = l_a$. The input signal $s(t)$ is applied at $x = 0$, and its spectrum $S(\omega)$ is expressed as

$$S(\omega) = A\sqrt{\frac{2\pi}{\alpha - j\beta}} e^{\frac{-(\omega-\omega_0)^2}{2(\alpha-j\beta)} - j\omega t_0}. \tag{5}$$

Assuming a linear frequency dependence for the phase constants ($\beta_s = K_s\omega$, $\beta_p = K_p\omega$) [9], and applying (4), the spectrum of the reflected signal $R(\omega)$ is derived as

$$R(\omega) = A\sqrt{\frac{2\pi}{\alpha - j\beta}} e^{-\frac{(\omega-\omega_0)^2}{2(\alpha-j\beta)} - j\omega t_0 - j\omega(2K_p l_a + K_s l_c)} \times e^{-2\alpha_p l_a} \cdot -2\Gamma_b \sin(K_s l_c \omega) e^{j\frac{\pi}{2}}. \tag{6}$$

To evaluate the size of the shield hole defects, the spectrum ratio $P_{sr}(\omega)$ between $R(\omega)$ and $S(\omega)$ is calculated [19]. The magnitude of $P_{sr}(\omega)$ is derived as

$$|P_{sr}(\omega)| = \frac{|R(\omega)|}{|S(\omega)|} = 2\Gamma_b \sin(K_s l_c \omega) e^{-2\alpha_p l_a}. \tag{7}$$

In (7), the term $\Gamma_b \sin(K_s l_c \omega)$ is determined solely by the intrinsic parameters of the defect, namely Γ_b, K_s, and l_c. Therefore, this term is defined as the damage ratio

$$M_r = \Gamma_b \sin(K_s l_c \omega). \tag{8}$$

Analyzing (8) further, the terms involving Γ_b and $\sin(K_s l_c \omega)$ can be calculated using the dual-frequency TFDR method. Assume that M_{r1} represents the damage ratio under a single-frequency condition, whereas M_{f2} corresponds to that under dual-frequency excitation. Accordingly, from the trigonometric relations, Γ_b can be estimated as follows:

$$\Gamma_b = \frac{M_{r1}}{\sqrt{1 - (\cos(K_s l_c \omega))^2}}, \tag{9}$$

$$\cos(K_s l_c \omega) = \frac{\Gamma_b \sin(K_s l_c 2\omega)}{2\Gamma_b \sin(K_s l_c \omega)} = \frac{M_{r2}}{2M_{r1}}. \tag{10}$$

There are two methods for obtaining M_{r1} and M_{r2} in practical measurements. The first is the time-scaling method, which only requires a single measurement with a broadband signal that spans both the frequency of interest and its double. The second is the frequency-shifting method, which requires two sequential measurements with center frequencies of ω_0 and $2\omega_0$. Due to the long duration of the injected signal in TFDR, the time-scaling method results in a long blind zone. Therefore, the frequency-shifting method is adopted. During the measurement, M_r can be calculated as follows:

$$M_r = \frac{|P_{sr}|}{2e^{-2\alpha_p l_a}}, \tag{11}$$

where the attenuation constant α_{p} can be determined by using TFDR on a healthy cable of the same type with an open-circuit end. The expression for α_{p} is given by

$$\alpha_{\mathrm{p}} = -\frac{1}{2l_{\mathrm{x}}} \cdot \frac{1}{2} \ln \frac{\mathcal{F}(r(t))}{\mathcal{F}(s(t))}, \tag{12}$$

where l_x is the tested healthy cable length, and $\mathcal{F}(x)$ indicates the Fourier transform.

Based on (9) and (3), the characteristic impedance Z_{s} of the cable with shield hole defects can be calculated as

$$Z_{\mathrm{s}} = Z_{\mathrm{p}} \frac{1+\Gamma_{\mathrm{b}}}{1-\Gamma_{\mathrm{b}}}. \tag{13}$$

In practical measurements, reflections may originate from multiple sources rather than solely from shield holes. The parameter Z_{s} enables a preliminary identification of the defect type: shield hole defects exhibit a larger Z_{s} than healthy cables, whereas moisture defects exhibit a smaller Z_{s}. If the defect is identified as a shield hole, its size can then be estimated using the two-step evaluation process outlined below.

From the analysis of the dual-frequency TFDR method, it follows that in (8) and (13), Z_{s} depends on the central angle θ of the shield hole, whereas M_{r} depends on both θ and the shield hole length l_{c}. To decouple the effects of θ and l_{c}, we propose a two-step evaluation process. In this process, the central angle θ is first estimated from Z_{s}, and then the length l_{c} is estimated from M_{r}, constrained by the evaluated θ. This two-step process requires a reference set of characteristic impedances Z_{ss} and damage ratios M_{rs}, obtained by FEM simulations of cables with shield holes of various sizes. The overall procedure for size evaluation of shield hole defects using the improved dual-frequency TFDR method is summarized in Algorithm 1.

Algorithm 1 Size Assessment of Cable Shield Hole

Block 1: Measurement Setup
Input: Cable under test, dual-frequency signals $s_1(t), s_2(t)$
Output: Reflected signals $r_1(t), r_2(t)$
Initialize equipment.
Inject $s_1(t)$ and record reflected signal $r_1(t)$.
Inject $s_2(t)$ (with doubled center frequency) and record reflected signal $r_2(t)$.
Block 2: Signal Processing
Input: Reflected signals $r_1(t), r_2(t)$
Output: Damage ratios M_{r1}, M_{r2}
for each reflected signal $r_i(t)$ **do**
    Locate defect.
    Compute FFT to obtain spectra $S(\omega), R(\omega)$.
    Calculate spectrum ratio $|P_{\mathrm{sr}}|$ by (7).
    Calculate $M_{\mathrm{r}i}$ by (11).
end for
Block 3: Parameter Evaluation
Input: $\{Z_{\mathrm{ss}}\}, \{M_{\mathrm{rs}}\}, M_{\mathrm{r1}}, M_{\mathrm{r2}}$
Output: Estimated θ, l_c
Calculate Γ_{b} from (9) and (10).
Calculate Z_{s} from (13).
Estimate θ by searching Z_{s} in $\{Z_{\mathrm{ss}}\}$.
Estimate l_c by searching M_{r1} in $\{M_{\mathrm{rs}}\}$ under the constraint of θ.
Output final results and save data.

Multiple defects may exist in a cable, and the proposed method can be extended to estimate their sizes. When evaluating a specific defect in a multi-defect cable, the influence of the preceding defects on the reflected signal must be removed. This influence is represented by the transmission coefficients T_{f}. As shown in Figure 2, and according to transmission line theory, the transmission coefficient T_{f} of a defect can be expressed as

$$T_{\mathrm{f}} = \frac{e^{-\beta_{\mathrm{s}} l_{\mathrm{c}}}(1+\Gamma_{\mathrm{a}})(1+\Gamma_{\mathrm{b}})}{1+\Gamma_{\mathrm{b}} e^{-2\beta_{\mathrm{s}} l_{\mathrm{c}}}}. \quad (14)$$

Therefore, when estimating the n-th defect, (11) should be modified as

$$M_{\mathrm{r}} = \frac{|P_{\mathrm{sr}}|}{2e^{-2\alpha_{\mathrm{p}} l_{fn}} \prod_{i=1}^{n-1} T_{\mathrm{f}i}^{2}}. \quad (15)$$

3. Simulation Validation

To validate the improved dual-frequency TFDR method and test its efficacy, we conduct a simulation on a 10 kV power cable.

3.1. Modeling and Analyzing the Cable Shield Hole Defects

To build the model for shield hole defects in Figure 1, we establish the defect models in COMSOL Multiphysics 6.2, a FEM software. This involves creating a two-dimensional cross-section model of a 10 kV single-core cable. The specific parameters of the cable used in the simulation are detailed in Table 1.

Table 1. Parameters of the cable in the simulation.

Component	Radius (mm)	Electric Parameters
Core	3.335	$\gamma_{\mathrm{c}} = 5.98 \times 10^{7}$ S/m
Inner semiconductor	4.115	$\epsilon = 70$
XLPE	8.635	$\epsilon = 2.3$
Outer semiconductor	9.125	$\epsilon = 100$
Shield	9.325	$\gamma_{\mathrm{s}} = 5.98 \times 10^{7}$ S/m
Sheath	12.115	$\epsilon = 2.25$

First, the simulation computes the electric and magnetic field distributions of the defect, and the cross-sectional views are shown in Figure 3. The results indicate that the electromagnetic wave inside the defect propagates primarily in a quasi-TEM mode, with some energy leaking through the shield holes.

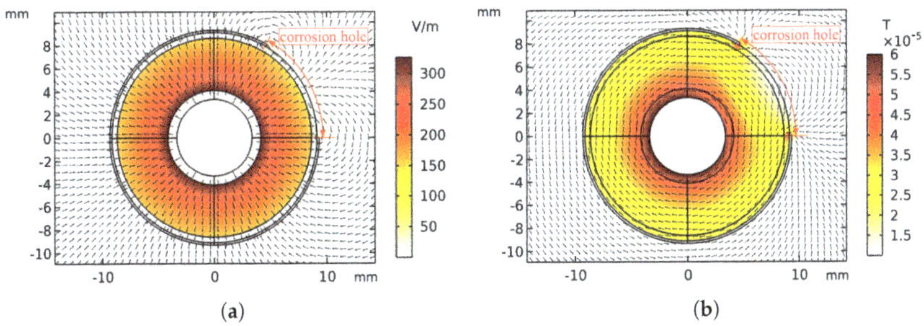

Figure 3. The distribution of the electric and magnetic fields in the cross-section of a cable defect. (**a**) The electric field; (**b**) the magnetic field. The arrows indicate the field directions.

Then, using the results of the electric and magnetic field distributions, the distributed parameters of shield hole defects, including R, L, G, and C, were calculated.

3.2. Simulation of Long Cables with Shield Hole Defects

The transfer function $H_c(\omega)$ of long cables with defects can be derived using network cascading. A cable with defects can be viewed as a cascade of two long healthy sections and a section with a defect. Therefore, the $H_c(\omega)$ can be derived as follows:

$$T_c = T_{p1} \cdot T_s \cdot T_{p2} \cdot T_o, \tag{16}$$

where T_c, T_{p1}, T_s, and T_{p2} represent the transmission matrices of the entire cable, the first healthy cable section, the defective section, and the second healthy cable section, respectively, whereas T_o denotes the transmission matrix of a high-impedance component that simulates an open terminal. $H_c(\omega)$ is the reflected coefficient spectrum, which can be obtained using T_c [20]. Therefore, the reflected signal $r(t)$ of a long cable with defects is calculated as

$$r(t) = \mathcal{F}^{-1}\{S(\omega) \cdot H_c(\omega)\}. \tag{17}$$

In the simulation, a 250 m long cable with shield hole defects located at the 50 m point was modeled. The shield hole angles ranged from 30° to 300°, and the lengths from 0.05 m to 0.5 m. A signal with a center frequency of 40 MHz, a bandwidth of 40 MHz, and t_0 = 100 ns was used as $s_1(t)$ for the first measurement [9]. For the second measurement, a signal with a center frequency of 80 MHz, with the same bandwidth and center time as $s_1(t)$, was used as $s_2(t)$. Some reflected signals $r(t)$ are shown in Figure 4. In the figure, the signals near 0 µs correspond to the injected signal, those around 0.7 µs are reflections from defects, and those near 3 µs result from reflections at the open end of the cable. Figure 4a shows that the peaks of the reflected signals are delayed and their magnitudes increase. Conversely, Figure 4b indicates that while the peak times of the reflected signals remain constant with the defect angle, their magnitudes increase with larger defect angles.

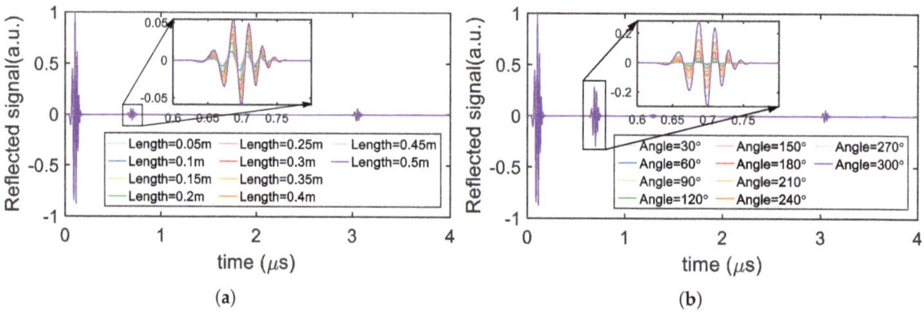

Figure 4. Reflected signals from shield hole defects of various sizes. (**a**) Results for defects with different lengths ($\theta = 150°$). (**b**) Results for defects with different angles ($l_c = 0.5$ m).

The defects in Figure 4 are evaluated using the improved dual-frequency TFDR method. The attenuation constant α_p of a healthy cable, calculated from (12), is 0.001867 m^{-1} at 40 MHz and 0.002831 m^{-1} at 80 MHz. The reference sets for Z_{ss} and M_{rs} are shown in Figure 5.

Using the two-step process, shield hole defects with various angles and lengths were estimated, and the corresponding errors are shown in Figure 6. The results indicate that as the defect angle increases, the estimation errors of both angle and length also increase, owing to the growing approximation error of (4). When the defect angle is 300° and the length is 0.05 m, the relative errors of angle and length estimation reach 5.76 % and 17.5 %, respectively. As the angle increases beyond 300°, the assumption $\Gamma_b \ll 1$ becomes invalid, and the method is no longer reliable.

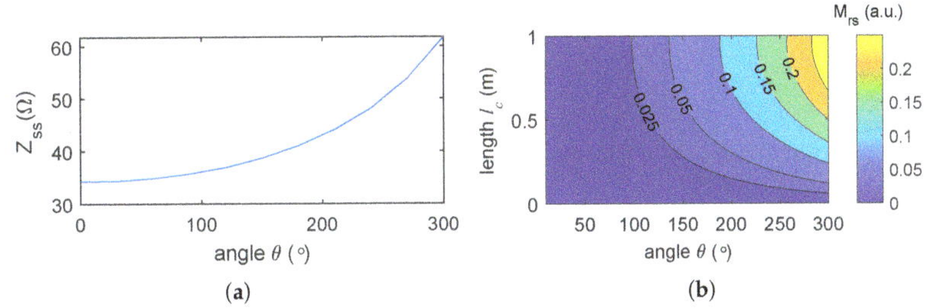

Figure 5. Reference values for assessing the size of shield hole defects in the 10 kV cable (shown in Table 1). (**a**) Reference characteristic impedance Z_{ss} for various angles θ (with $\theta = 0°$ representing a healthy cable); (**b**) reference damage ratio M_{rs} for various angles θ and lengths l_c.

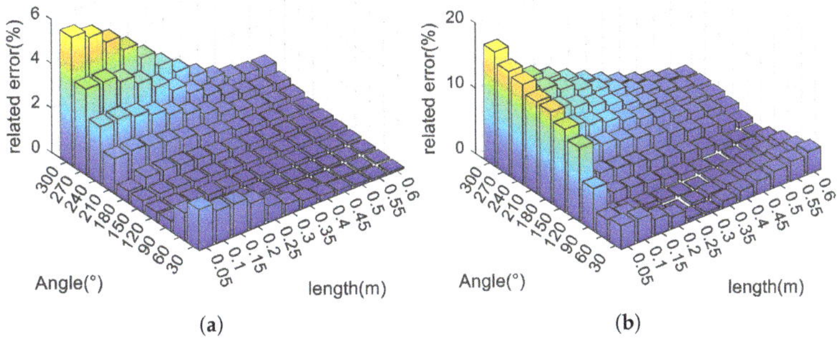

Figure 6. The estimation error of defect angle and length under various angles and lengths. (**a**) The estimation error of angles; (**b**) the estimation error of lengths. The colors also represent the magnitude of the error.

When using the two-step estimation process, the reference set has a significant influence on the estimated results. In practical measurements, there are differences between the tested cable and the reference cable, such as manufacturing tolerances. Therefore, the influence of cable parameter tolerances on the estimated results was analyzed. In the simulation, the core radius, shield radius, relative permittivity, and conductivity of the cable varied by ±5% to account for parameter tolerances. Using the Monte Carlo method, the estimated results and their 95 % confidence intervals were calculated and are shown in Figure 7.

Noise is another factor influencing estimation accuracy. To analyze its effect on the proposed method, we evaluated performance under various signal-to-noise ratios (SNRs). For each defect and SNR level, 1000 Monte Carlo simulations were performed. The mean estimates and their 95 % confidence intervals are presented in Figure 8.

As shown in Figure 8, length estimation is more sensitive to noise than angle estimation. With increasing SNR, the mean estimation errors of both length and angle decrease, the 95 % confidence intervals become narrower, and the estimation results become more accurate.

In practice, shield holes are normally irregular. To evaluate the robustness of the proposed method, 12 irregular shield hole defects were modeled and simulated. These irregular holes were generated by dividing a regular shield hole into multiple subsections and randomly perturbing their central angles. The structural parameters of the irregular shield holes are listed in Table 2, and an example of a typical irregular hole is shown in Figure 9.

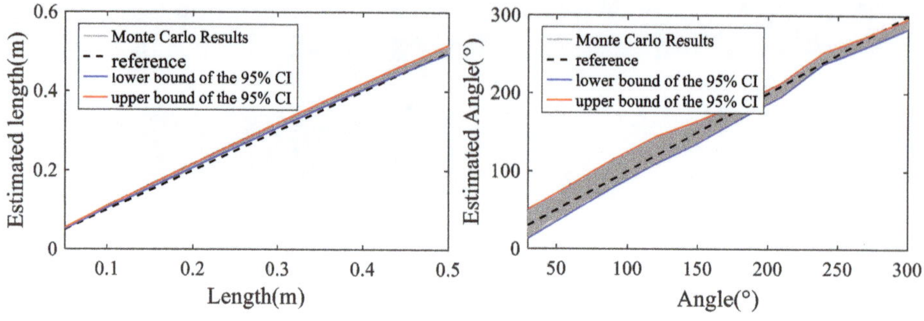

(a)

(b)

Figure 7. Estimated defect lengths and angles for a cable with ±5% manufacturing tolerance. (**a**) Defect angles are 150°, and (**b**) defect lengths are 0.5 m.

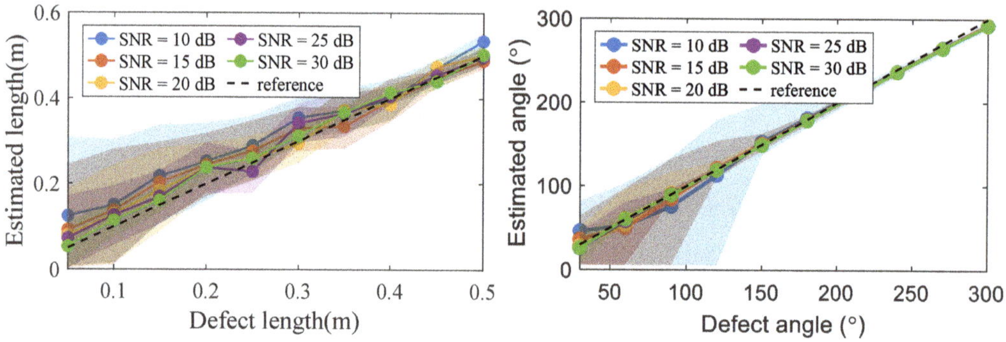

(a)

(b)

Figure 8. Size evaluation of shield hole defects under the signals with various SNRs. (**a**) Defect angles are 150°. (**b**) Defect lengths are 0.5 m.

Table 2. Angle and length parameters of the samples in the simulation.

Sample	θ_g (°)	$\Delta\theta_g$ (°)	l_c (m)	θ_{ave} (°)
1	90	±15	0.2	88.69
2	90	±45	0.2	85.97
3	120	±15	0.2	118.18
4	135	±45	0.2	136.62
5	150	±15	0.2	150.90
6	150	±60	0.2	148.72
7	180	±15	0.2	180.22
8	180	±60	0.2	182.16
9	180	±60	0.4	179.99
10	180	±60	0.6	178.85
11	210	±60	0.2	208.49
12	225	±45	0.2	228.01

The estimated sizes of the irregular shield holes are presented in Figure 10. Additionally, the estimated results are compared with the average geometric values of the corresponding holes. From the comparison, it can be observed that the estimated values are close to the average values and exhibit a consistent trend. Although the estimated angles and lengths for irregular shield holes show larger errors compared to those for regular holes, the deviations remain within an acceptable range. These results indicate that the proposed method is capable of providing a reliable evaluation of the sizes of irregular shield holes.

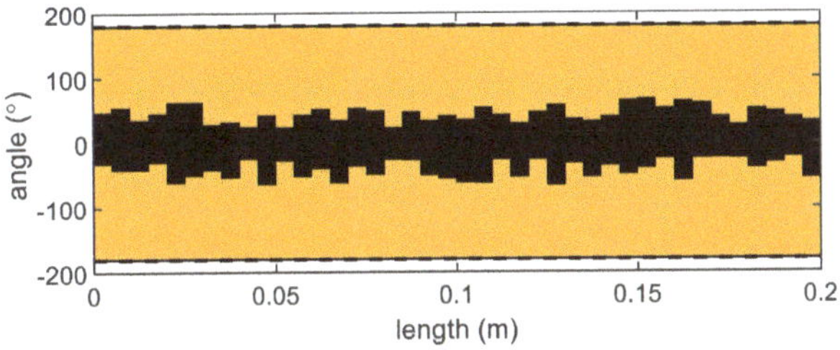

Figure 9. Irregular shield hole of sample 2 in Table 2.

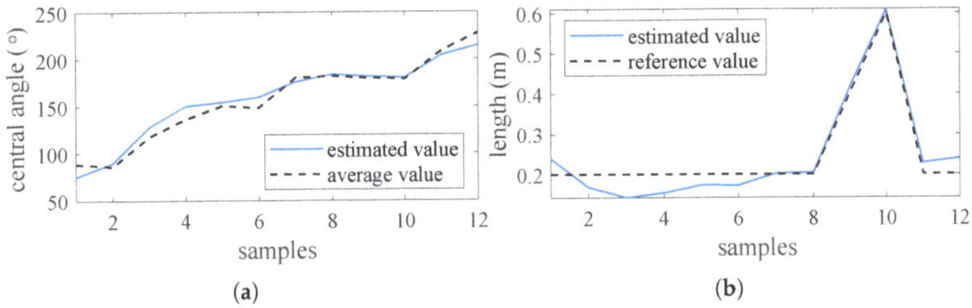

Figure 10. Size evaluation of irregular shield holes. (**a**) The estimated results of defect angles; (**b**) the estimated results of defect lengths.

4. Experiments and Results

In this section, an experiment was carried out to validate the proposed method. Direct validation on many long power cables is not feasible due to their high cost. Therefore, a scaled-model experiment was designed using a coaxial signal cable, which was chosen because of the structural similarity between signal cables and power cables. The experiment platform is shown in Figure 11.

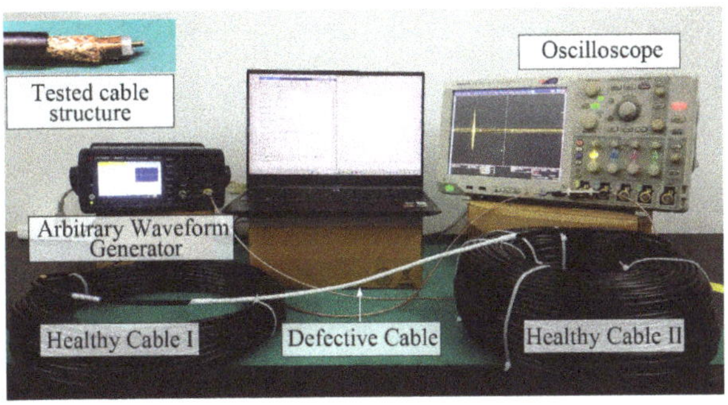

Figure 11. Experiment setup.

It consists of an arbitrary waveform generator to inject signals, an oscilloscope to capture the reflected signals, a computer to process and analyze the reflected signals, and a long cable under test. The cable under test consists of two healthy long sections, approximately 30 m and 150 m in length, and several 1 m long short cables with artificial defects. The short defective cables are inserted between the healthy sections using SMA connectors. The arbitrary waveform generator is Keysight 33621A (Keysight, Santa Rosa, CA, USA), and the oscilloscope is Tektronix MSO5204B (Tektronix, Beaverton, OR, USA). Both the output impedance of the waveform generator and the input impedance of the oscilloscope is 50 Ω. The cable under test is an SYV50 type with a 50 Ω characteristic impedance.

Shield hole defects were emulated by removing specific areas from the shield foil of short cable sections. To assess the performance of the proposed method, eleven cable samples were fabricated and tested. Among them, eight had regular shield holes, while three featured irregular ones. Considering both Type A uncertainty (from repeated measurements) and Type B uncertainty (from approximation of the method), the estimation results and their 95% confidence intervals are presented in Table 3, and the corresponding cable samples are shown in Figure 12.

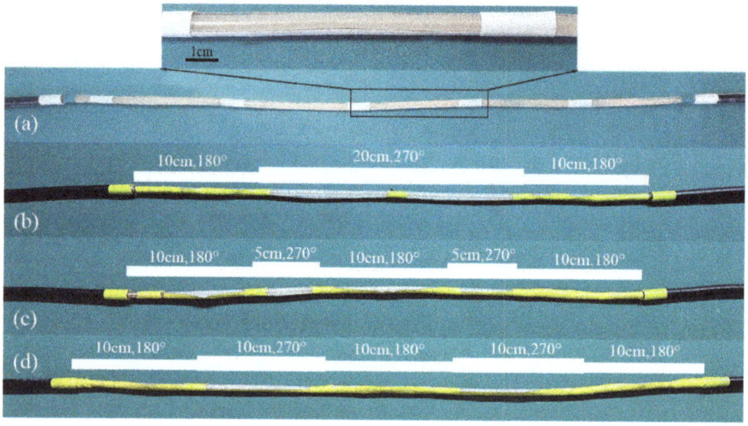

Figure 12. Cable sample mimicking shield hole defects. (**a**) A regular defect. (**b–d**) Irregular defects.

Figure 13 shows the measurement of reflected signals from defective cable samples. In the figure, the signals reflected by the defects are located between 0.25 µs and 0.35 µs. The analysis of the effect of the shield hole length and angle on the reflected signals reveals that the amplitude of the reflected signals increases with the length of the shield hole. Similarly, the amplitude also increases with the widening of the shield hole angles.

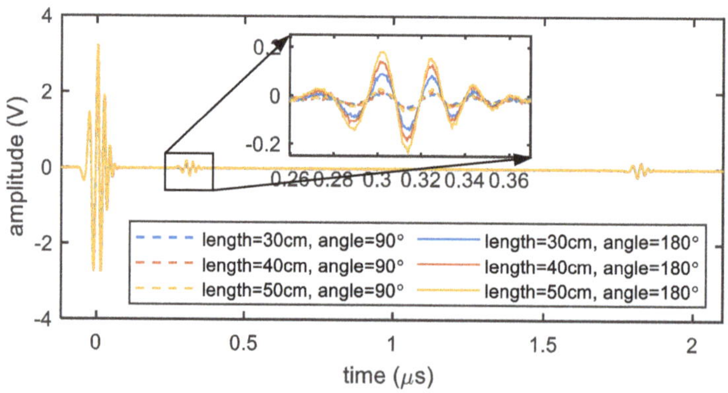

Figure 13. Effect of shield hole length l_c and angle θ on reflected signals.

The reference set Z_{ss} and M_{rs} were obtained by simulation. And the attenuation constant α_p of the SYV50 cable was measured using TFDR on a healthy cable. The attenuation α_p is $0.0072\,\mathrm{m}^{-1}$ at 40 MHz and $0.0085\,\mathrm{m}^{-1}$ at 80 MHz. To eliminate occasional factors and connector influence, 20 full connector disassembly and reassembly measurements were conducted for each sample. For each reassembly, the average of five measurements was taken as the measured value to reduce noise interference. The lengths and angles of the defective cable samples were evaluated using the two-step process, and the evaluated results were compared with the reference values in Table 3. The results in Table 3 demonstrate that the improved dual-frequency TFDR method is effective for evaluating the size of shield hole defects in cables. However, discrepancies between the evaluated and actual values remain, primarily due to noise and the precision of defect fabrication. For regular defects, discrepancies are generally smaller for defect angles than for defect lengths. This is because angle estimation relies on the ratio of measurements at different frequencies, which mitigates noise effects, whereas length estimation depends on angle estimation results, so any inaccuracies in angle measurement can further degrade length accuracy. For irregular defects, the method cannot capture detailed irregular shapes, but it still provides a reasonable evaluation of the overall defect size.

Table 3. Comparison between actual defect sizes and their estimation results.

Type	Defect Size	L_e and Confidence Interval (cm)	Relative Error (%)	θ_e and Confidence Interval (°)	Relative Error (%)
Regular defects	30 cm, 90°	32.54 ± 1.2	8.47	85.9 ± 2.19	4.56
	30 cm, 180°	32.38 ± 1.94	7.93	174.01 ± 4.57	3.33
	40 cm, 90°	44.27 ± 1.45	10.68	86.93 ± 2.21	3.41
	40 cm, 180°	39.06 ± 2.03	2.35	181.87 ± 3.55	1.04
	50 cm, 90°	53.95 ± 2.02	7.09	88.04 ± 1.48	2.18
	50 cm, 180°	54.38 ± 2.13	8.76	175.17 ± 4.16	2.68
	60 cm, 190°	66.50 ± 1.68	10.83	184.24 ± 4.29	3.03
	60 cm, 230°	55.19 ± 2.88	8.02	229.54 ± 6.14	0.2
Irregular defects	40 cm, 225°	37.26 ± 2.46	6.85	238.83 ± 7.4	6.15
	40 cm, 202.5°	42.46 ± 2.6	6.15	211.8 ± 5.16	4.59
	50 cm, 216°	54.35 ± 2.76	8.7	196.43 ± 3.95	9.06

The angle of irregular defects is the weighted average angle of the irregular shield hole, which is calculated as $\theta_{\text{ave}} = \frac{\sum_{i=1}^{N} \theta_i l_{ci}}{\sum_{i=1}^{N} l_{ci}}$.

To validate the effectiveness of the proposed method in multiple-defect scenarios, an additional experiment was conducted using a cable with two shield hole defects. The first defect was positioned at 30 m with varying sizes, while the second defect was fixed at 60 m with a length of 0.4 m and an angle of 180°. The reflected signals from the cable with two defects are shown in Figure 14.

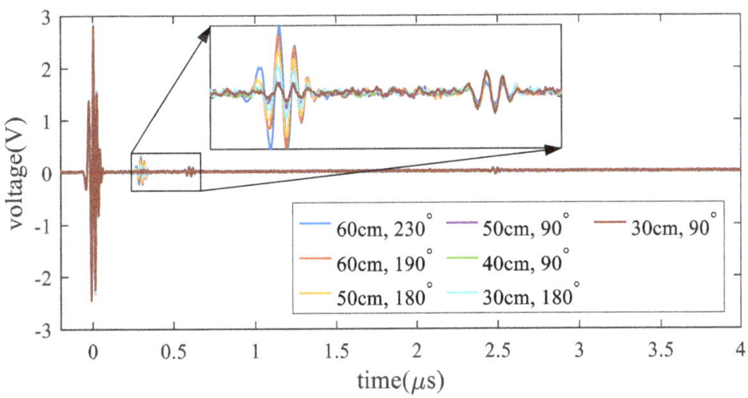

Figure 14. Reflected signals from cable with two shield hole defects.

Using the proposed method, the second defect was evaluated, and the results are presented in Table 4. A comparison of the estimated results with and without compensation for the first defect shows that compensation significantly improves the accuracy of the second defect's evaluation. The estimated results for multiple defects exhibit larger errors than those for single defects, primarily due to the accumulation of errors in evaluating multiple defects.

Table 4. Comparison the estimated results of the second defect with and without compensation for the first defect.

Algorithm	Defect 1	Defect 2 (40 cm, 180°)			
		L_e and Confidence Interval (cm)	Relative Error (%)	θ_e and Confidence Interval (°)	Relative Error (%)
No Comp.	30 cm, 90°	33.56 ± 1.85	16.1	145.38 ± 3.32	19.2
	30 cm, 180°	44.63 ± 2.23	11.58	126.74 ± 3.47	29.59
	40 cm, 90°	52.12 ± 2.17	30.3	138.97 ± 3.84	22.79
	50 cm, 90°	48.35 ± 2.18	20.87	125.83 ± 2.63	30.1
	50 cm, 180°	26.72 ± 1.78	33.2	105.41 ± 2.76	41.44
	60 cm, 190°	35.14 ± 2.01	12.15	90.25 ± 2.44	49.86
	60 cm, 230°	46.98 ± 2.37	17.45	88.76 ± 1.97	50.59
With Comp.	30 cm, 90°	42.52 ± 2	6.3	165.24 ± 4.09	8.2
	30 cm, 180°	35.43 ± 2.04	11.43	159.92 ± 3.85	11.16
	40 cm, 90°	35.75 ± 1.93	10.63	161.85 ± 3.49	10.08
	50 cm, 90°	43.23 ± 2.12	8.3	173.36 ± 3.38	3.69
	50 cm, 180°	34.23 ± 1.88	14.43	195.26 ± 4.11	8.48
	60 cm, 190°	33.28 ± 1.98	16.8	161.72 ± 3.29	10.16
	60 cm, 230°	44.15 ± 2.3	10.38	158.45 ± 2.91	11.97

5. Conclusions

In this work, we modeled cable shield hole defects and analyzed their effect on the propagation of TFDR signals. Building upon the analysis results, we proposed an improved dual-frequency TFDR method to evaluate the length and angle of the shield holes. A simulation and an experiment were conducted to demonstrate the improved dual-frequency TFDR method in detail and test its validity and accuracy. The main conclusions are summarized as follows:

1. In the long cable simulation results, the estimated error of defect angles and lengths increased with the increase in defect angle. When the defect angle is lower than 300°, the proposed method has a high accuracy and robustness.
2. In the experimental results, the maximum relative errors for regular defects were 11% in length estimation and 5% in angle estimation. For irregular defects, the proposed method was able to estimate both defect length and average angle, with errors of 8.7% and 9.06%, respectively. In multiple-defect scenarios, the method remained effective, yielding maximum relative errors of 14.43% for length and 11.97% for angle.
3. In practical measurements, the proposed method requires the signal source and data acquisition device to have a bandwidth of at least 100 MHz to ensure estimation accuracy. The on-site SNR is not lower than 20 dB, and excessive noise can lead to larger errors.

The estimated defect lengths and angles can be used to model defects in cable digital twins and to assess the operational state of the cable. By comparing these results with alarm criteria, appropriate maintenance plans can be formulated. Therefore, the proposed method can significantly enhance cable condition monitoring and increase the effectiveness of cable maintenance. The characteristic impedance of power cables is less strictly controlled than that of signal cables, which may affect the accuracy of the proposed method. As future work, we will investigate the impact of characteristic impedance variations on accuracy and perform experiments on power cables to further validate the effectiveness of the method.

Author Contributions: K.Z.: Conceptualization, Methodology, Writing—original draft. S.G.-T.: Supervision, Methodology, Writing—review and editing. P.M.: Supervision, Validation, Writing—review and editing. Y.Y.: Writing – review and editing. H.L.: Supervision, Writing—review and editing. All authors have read and agreed to the published version of the manuscript.

Funding: This research was founded by the China Postdoctoral Science Foundation (Certificate Number: 2024M762595).

Data Availability Statement: The original contributions presented in this study are included in the article. Further inquiries can be directed to the corresponding author.

Conflicts of Interest: The authors declare no potential conflicts of interests.

References

1. Tu, H.; Peng, C.; Chen, Y.; Li, L.; Deng, H.; Liu, G. Characteristic impedance modeling of nuclear power instrumentation and control cable shield breakage. *Energies* **2025**, *18*, 3008. [CrossRef]
2. Shao, Q.; Fan, S.; Zhang, Z.; Liu, F.; Fu, Z.; Lv, P.; Mu, Z. Artificial intelligence in cable fault detection and localization: Recent advances and research challenges. *Energies* **2025**, *18*, 3662. [CrossRef]
3. *IEEE 1617-2022*; IEEE Guide for Assessment, Mitigation, and Control of Corrosion of Metallic Shields for Extruded Dielectric Cables Rated 5 kv to 46 kv. IEEE: New York, NY, USA, 2022.
4. Liang, H.; Liu, Y.; Sheng, G.; Jiang, X. Fault-cause identification method based on adaptive deep belief network and time–frequency characteristics of travelling wave. *IET Gener. Transm. Distrib.* **2019**, *13*, 724–732. [CrossRef]
5. Jiang, K.; Zhou, K.; Ren, X.; Xu, Y. Frequency domain sampling optimization of cable defect detection and location method based on exponentially increased frequency reflection coefficient spectrum. *Energies* **2025**, *18*, 2428. [CrossRef]
6. Zhang, H.; Zhou, K.; Ren, X.; Xu, Y. Moisture localization and diagnosis method for power distribution cables based on dynamic frequency domain reflectometry. *Energies* **2025**, *18*, 2430. [CrossRef]
7. Abboud, L.; Cozza, A.; Pichon, L. A matched-pulse approach for soft-fault detection in complex wire networks. *IEEE Trans. Instrum. Meas.* **2012**, *61*, 1719–1732. [CrossRef]
8. Mu, H.B.; Zhang, H.T.; Zou, X.Y.; Zhang, D.N.; Lu, X.; Zhang, G.J. Sensitivity improvement in cable faults location by using broadband impedance spectroscopy with dolph-chebyshev window. *IEEE Trans. Power Deliv.* **2022**, *37*, 3846–3854. [CrossRef]
9. Song, E.; Shin, Y.J.; Stone, P.E.; Wang, J.; Choe, T.S.; Yook, J.G.; Park, J.B. Application of time-frequency domain reflectometry for detection and localization of a fault on a coaxial cable. *IEEE Trans. Electromagn. Compat.* **2009**, *54*, 2493–2500.
10. Lee, S.H.; Park, J.B.; Choi, Y.H. Wavelet-transform-based time-frequency domain reflectometry for reduction of blind spot. *Meas. Sci. Technol.* **2012**, *23*, 065004. [CrossRef]
11. Lim, H.; Kwon, G.Y.; Shin, Y.J. Fault detection and localization of shielded cable via optimal detection of time-frequency-domain reflectometry. *IEEE Trans. Instrum. Meas.* **2021**, *70*, 3521510. [CrossRef]
12. Lim, H.; Lee, Y.H.; Bang, S.S.; Shin, Y.J. Application of enhanced optimal-detection of time-frequency domain reflectometry on hts cable with high-resolution. *IEEE Trans. Appl. Supercond.* **2023**, *33*, 9000306. [CrossRef]
13. Zou, X.; Mu, H.; Wang, R.; Fan, K.; Cheng, Z.; He, Y.; Zhang, G. An efficient accuracy improvement method for cable defect location based on instantaneous filtering in time-frequency domain. *Measurement* **2024**, *226*, 114178. [CrossRef]
14. Manet, A.; Kameni, A.; Loete, F.; Genoulaz, J.; Pichon, L.; Picon, O. Equivalent circuit model of soft shield defects in coaxial cables using numerical modeling. *IEEE Trans. Electromagn. Compat.* **2017**, *59*, 533–536. [CrossRef]
15. Zou, X.Y.; Mu, H.B.; Zhang, H.T.; Qu, L.Q.; He, Y.F.; Zhang, G.J. An efficient cross-terms suppression method in time-frequency domain reflectometry for cable defect localization. *IEEE Trans. Instrum. Meas.* **2022**, *71*, 3511610. [CrossRef]
16. Lee, C.K.; Shin, Y.J. Detection and assessment of i&c cable faults using time-frequency R-CNN-based reflectometry. *IEEE Trans. Ind. Electron.* **2021**, *68*, 1581–1590.
17. Zhao, K.; Grivet-Talocia, S.; Manfredi, P.; Yan, Y.; Li, H. Size assessment of power cable shield corrosion defects using double-frequency time–frequency-domain reflectometry. *IEEE Trans. Instrum. Meas.* **2025**, *74*, 3513911. [CrossRef]
18. Papaleonidopoulos, I.C.; Karagiannopoulos, C.G.; Theodorou, N.J. Evaluation of the two-conductor hf transmission-line model for symmetrical indoor triple-pole cables. *Measurement* **2006**, *39*, 719–728. [CrossRef]

19. Yunusa-Kaltungo, A.; Sinha, J.K.; Elbhbah, K. An improved data fusion technique for faults diagnosis in rotating machines. *Measurement* **2014**, *58*, 27–32. [CrossRef]
20. Khan, A.S. *Microwave Engineering: Concepts and Fundamentals*, 1st ed.; CRC Press: Boca Raton, FL, USA, 2014.

energies

Article

Study on Partial Discharge Characteristics of Mixed Metal Particles Under Combined Power Frequency and Switching Impulse Voltage

Jiyun Ren [1], Yongfu Ma [1], Quanlei Qu [1], Zile Wang [1], Yuang Wang [2], Lili Wang [1], Xutao Han [2,*] and Xiaojie Yang [1]

[1] State Grid Qinghai Electric Power Research Institute, Xining 810001, China
[2] State Key Laboratory of Electrical Insulation and Power Equipment, Xi'an Jiaotong University, Xi'an 710049, China
* Correspondence: xutaohan@xjtu.edu.cn

Abstract: Under operating conditions, metallic particle contaminants inside Gas-Insulated Switchgears (GIS) represent a major threat that can initiate partial discharges (PD) and lead to insulation failure. To investigate the discharge patterns under combined AC and switching impulse voltages, this paper presents an experimental study conducted in SF6 gas on wire-shaped, spherical, and Mixed Metal Particles. By synchronously analyzing PD time-domain waveforms, Phase-Resolved Partial Discharge (PRPD) patterns, and high-speed motion camera recordings, the correlation between particle motion behavior and discharge signals was systematically examined. The results indicate that wire particles exhibit a significant discharge initiation delay under the combined voltage; however, intense, discrete discharges with large magnitudes occur during their vertical jumping phase. In contrast, spherical particles can be activated within the first power frequency cycle without delay, but the subsequent discharge magnitudes are limited. The characteristics of hybrid particles lie between these two types, demonstrating a staged evolution described as "spherical particles lead initiation, wire particles dominate discharge." Furthermore, under the sustained AC voltage, hybrid particles trigger a more dispersed and violent discharge process. These findings reveal the complex motion-discharge mechanism of Mixed Metal Particles, providing critical insights for fault mechanism analysis and insulation protection related to particle contamination in practical GIS equipment.

Keywords: AC-switching impulse superposition; partial discharge; motion characteristics; Gas-Insulated Switchgear (GIS); mixed metal particles

1. Introduction

Gas-Insulated Switchgear (GIS) serves as critical equipment in power systems, playing a vital role in power transmission and conversion. Its operational safety is directly linked to the stability and reliability of the power grid. In recent years, statistics from the State Grid Corporation of China indicate that the failure rate of ultra-high voltage and extra-high voltage GIS remains relatively high. In 2024 alone, 17 failures were recorded, 10 of which occurred during switching operations, frequently leading to busbar tripping or DC blockades. Failures triggered by switching operations have thus become a significant factor constraining the safe operation of GIS [1–3]. The operational conditions during GIS switching are severe, where mechanical vibrations and electrical transients can activate latent defects, with metallic particle contamination being particularly prominent. Reports

 https://doi.org/10.3390/en18215650

suggest that defects related to metallic particles account for approximately 60% of GIS failures, making them a primary cause of insulation degradation [4–6].

Consequently, the partial discharge (PD) characteristics initiated by metallic particles have long attracted widespread attention from researchers globally. Early studies primarily focused on the motion patterns and PD characteristics of particles under steady-state DC and AC voltages [7–9], exploring the influence of factors like size and shape on their movement behavior and discharge activity [10–12]. However, with the accumulation of field experience, it has been recognized that the mechanism behind particle-induced failures in actual GIS often cannot be explained by single steady-state voltage conditions alone. Instead, specific abnormal conditions are typically required to initiate sustained motion and discharge under steady-state voltage, eventually evolving into equipment failure [13,14]. In response, recent research has gradually shifted towards particle behavior under complex stresses, particularly combined with AC voltage. For instance, Shi Tianyi et al. investigated particle motion under the combined influence of AC voltage and mechanical vibration, finding that both transient and steady-state vibrations significantly excite particle movement. Furthermore, sustained vibration can enhance the discharge activity of smaller particles and increase the discharge repetition rate [15,16]. On the other hand, the research team led by Li Junhao focused on the combined AC and impulse voltages encountered during switching operations. They revealed the PD characteristics of particles under AC combined with lightning impulse voltage, noting that particle motion is more readily initiated when a negative lightning impulse superimposes on the negative half-cycle of the AC voltage, leading to enhanced subsequent PD activity [17,18]. Overall, research on metallic particle motion and discharge has expanded from single steady-state voltage conditions to complex operational scenarios and combined stress conditions, providing theoretical support for analyzing and preventing GIS failures in practical operating environments. Deng Jinyi et al. established a simulation model for 110 kV GIS equipment based on finite element simulation, simulating the impact of different shapes and multiple metal particle adhesion defects on the electric field of GIS equipment. The results showed that the electric field distortion at the location of elongated particles increases with their elongation; the electric field distortion is more severe at the sharp corners of spiny particles at an angle of 60°; and the thinner the circular flaky particles, the more severe the electric field distortion [19].

Although existing studies have demonstrated the significant excitatory effect of combined AC and impulse voltage on particle motion and discharge, related research remains exploratory and has limitations. Firstly, current work predominantly concentrates on AC combined with lightning impulse voltage, with insufficient attention paid to AC combined with switching impulse voltage. Secondly, most existing studies utilize spherical or wire-shaped particles of a single morphology. In contrast, particles in actual equipment are often a mixture of various shapes, and research on such complex contaminants remains limited.

Based on this, this paper takes the condition of AC superimposed with switching impulse voltage as the research object and conducts experimental research on the partial discharge characteristics involving a mixture of wire-shaped and spherical metallic particles. Differences between hybrid particles and single-type particles are compared and analyzed. By synchronously observing particle motion behavior and partial discharge signals, the motion-induced discharge mechanism of Mixed Metal Particles under combined voltage is revealed. The findings aim to provide new theoretical references for defect diagnosis and enhancing the operational reliability of GIS under complex working conditions.

2. Experimental Setup

2.1. AC-Switching Impulse Combined Voltage Test System

For the specific condition of AC combined with switching impulse voltage, the experimental setup established is illustrated in Figure 1. The setup comprises an AC test system, a switching impulse voltage circuit with a rated voltage of 100 kV, a sphere-gap synchronous triggering system, a partial discharge measurement circuit, and protective devices. The phase-shift triggering system acquires the voltage signal across the test object via a capacitive divider (C_2). Utilizing a control module and optical fibers, it synchronously triggers the main sphere gap (g_1) and the isolation/protection sphere gap (g_2). Nanosecond-level synchronous timing control technology ensures the simultaneous operation of the two sphere gaps, thereby guaranteeing phase consistency and reliability of the voltage waveform superimposed at the defective end.

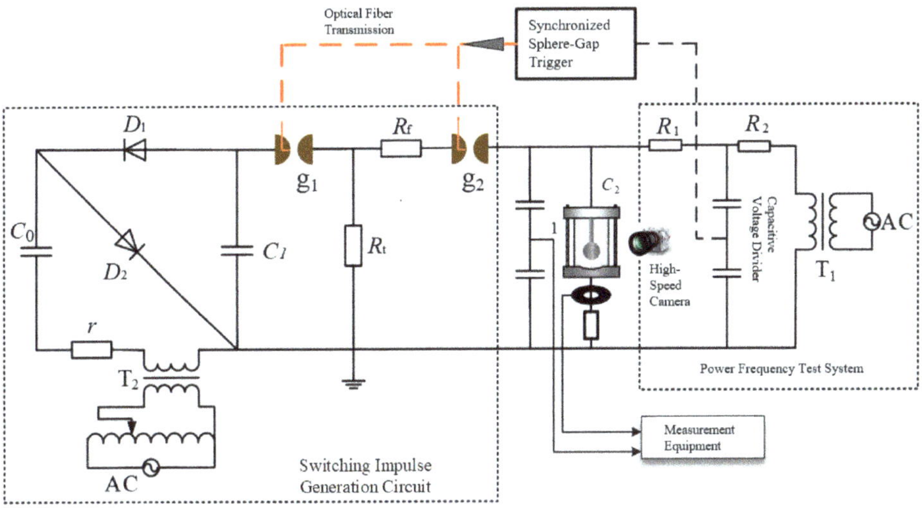

Figure 1. The AC and switching impulse superposition experiment platform.

The constructed partial discharge (PD) detection system consists of a high-frequency current transformer (HFCT), a partial discharge detector (DDX 9121b, Ruixuan Electronic Technology Co., Ltd., Shanghai, China), and a high-resolution digital oscilloscope (Tektronix HDO 9104, TeledyneLeCroy, New York, NY, USA). For detecting PD from metallic particles under the combined voltage, the HFCT (frequency range: 800 kHz to 102 MHz) coupled with the HDO 9104 oscilloscope was primarily used, while the PD detector was simultaneously employed to record Phase-Resolved Partial Discharge (PRPD) patterns.

The experiment utilized a phase-shift trigger to control the phase angle at which the switching impulse voltage was superimposed onto the AC voltage waveform. This enabled precise superposition of the impulse at different predefined phases of the AC cycle. Figure 2 shows the resulting voltage waveform output by the experimental platform, demonstrating a switching impulse successfully superimposed at the 270-degree point of the AC voltage cycle.

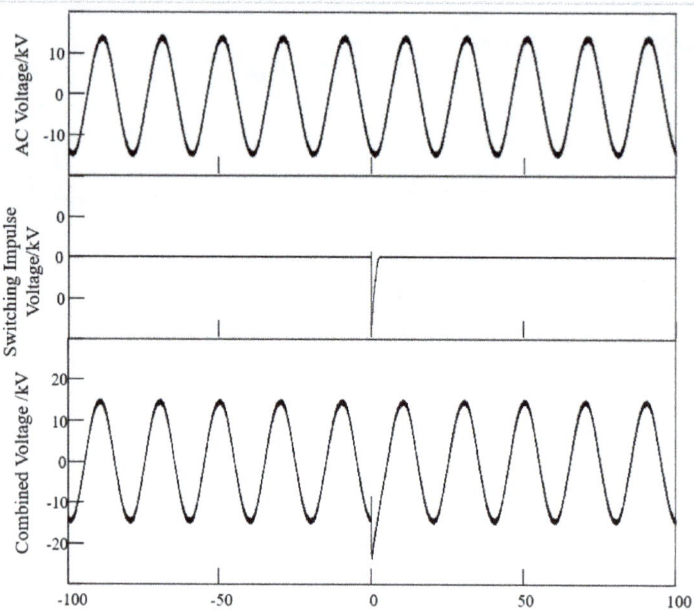

Figure 2. AC superposition switching impulse voltage waveform.

2.2. Metallic Particle Defect Model

Based on the dimensional proportions of internal components within a 252 kV GIS coaxial cylindrical electrode, a metallic particle defect model was designed, as illustrated in Figure 3. The upper electrode is an aluminum electrode with a radius of 17 mm, while a concentric grounded aluminum spherical-base electrode (base radius of 46 mm) is configured at the bottom. By controlling the electric field distribution, the field non-uniformity factor is maintained at 1.54, effectively replicating the quasi-uniform electric field characteristics of an actual GIS coaxial cylindrical electrode.

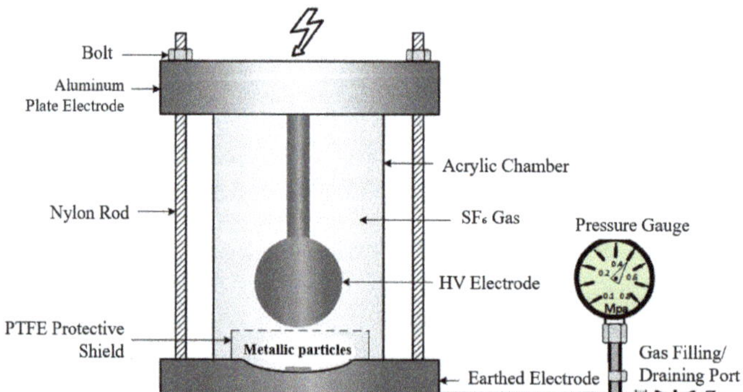

Figure 3. Metal particle defect test model.

To thoroughly analyze the partial discharge characteristics of Mixed Metal Particles, this study selected wire-shaped particles, known for causing the most severe electric field distortion and posing the greatest threat to insulation integrity, and combined them with representative spherical metallic particles for investigation. The wire-shaped particles had dimensions of 0.4 mm in diameter and 1 mm in length, while the spherical particles had a diameter of 1.5 mm. All the metal particles mentioned in the article are made of aluminum.

2.3. Correlative Analysis System for Partial Discharge Detection and Motion Capture

To correlate the discharge characteristics of metallic particles with their motion under combined AC and switching impulse voltage, a synchronized analysis system for partial discharge detection and motion capture was designed, as shown in Figure 4. The particle motion capture system employs a high-speed camera(High-Speed Camera M230M/C from Hefei Zhongke Junda Vision Technologies Co., Ltd., Hefei, China) with supplemental lighting. This camera can capture up to 3000 frames per second at a resolution of 1920 × 1080, achieving a temporal resolution of 0.3 ms, which meets the requirement for precise tracking of particle motion. The high-speed camera is externally triggered by the oscilloscope, ensuring synchronization between the application of the combined AC-switching impulse voltage and the image acquisition. This enables the simultaneous capture of partial discharge signals and particle motion behavior. By comparing the time-domain PD signals following the switching impulse with the corresponding transient frame images from the motion capture system, the correlation between specific discharge events and particle motion is established, allowing for a coupled analysis of motion behavior and partial discharge characteristics.

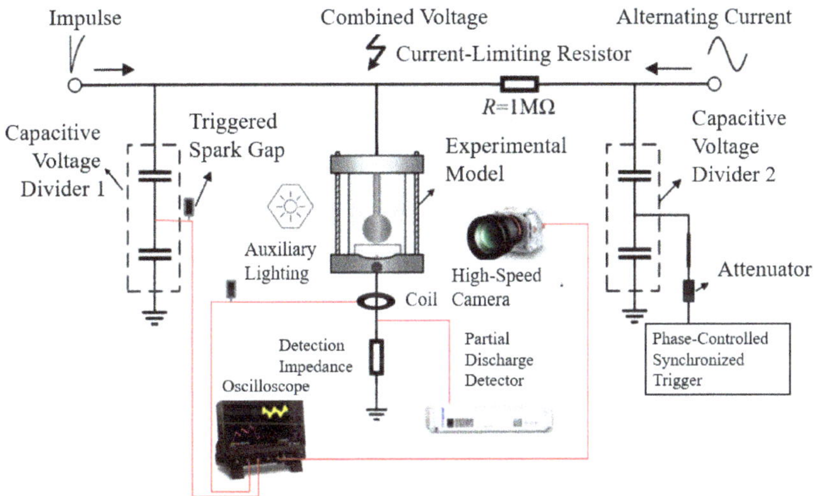

Figure 4. Partial discharge detection and motion capture correlation analysis system.

2.4. Experimental Methodology

The experiments on metallic particles under combined AC and switching impulse voltage were conducted in two stages: (1) measurement of the Partial Discharge Inception Voltage (PDIV), and (2) synchronized detection of discharge characteristics and motion behavior of metallic particles under the superimposed switching impulse voltage.

After determining the PDIV, the AC voltage was set to 95% of the PDIV, and a standard switching impulse with a peak value of 2 times the PDIV was applied to the defect. This was performed to investigate the excitatory effect of the superimposed impulse on the subsequent partial discharge activity of the particles. During the experiments, while the pulse current method was used to acquire the time-domain PD signals, the high-speed camera was externally triggered to capture the motion characteristics of the metallic particles. The experimental results were analyzed by comparing the state of the metallic particles before and after the application of the switching impulse voltage.

3. Partial Discharge Characteristics of Single-Type Metallic Particles Under Combined Voltage

3.1. Partial Discharge Characteristics of a Single Wire-Shaped Particle

Under AC voltage alone, the PDIV for the wire-shaped particle was 9.5 kV. Following the experimental procedure, a 19 kV switching impulse was superimposed onto a 9 kV AC background voltage (below the PDIV). The resulting time-domain PD waveform is shown in Figure 5. The experimental results indicate that no PD signals were detected when only the AC voltage below the inception level was applied. However, after the application of the switching impulse, sustained PD activity appeared in the subsequent AC cycles. A noticeable delay (Δt) of approximately 7 AC cycles was observed between the application of the switching impulse and the emergence of significant PD signals.

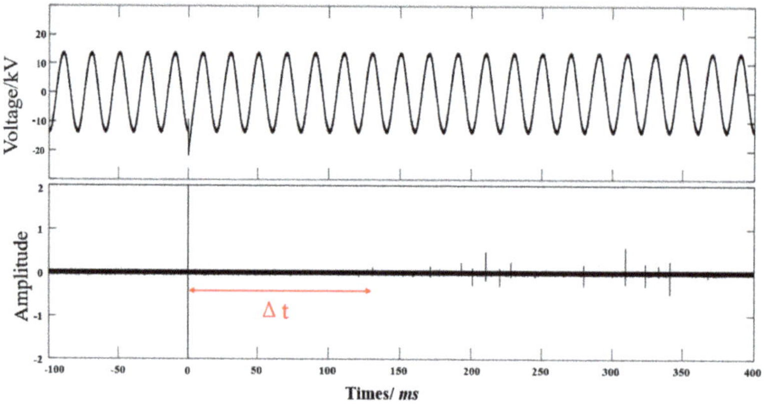

Figure 5. Partial discharge signals of linear particles under the impulse voltage of AC superposition switching.

Furthermore, the motion process of the wire-shaped particle was captured using the high-speed camera to observe its movement characteristics and discharge behavior following the application of the combined AC and switching impulse voltage. The experiments revealed that the excited motion of a single wire-shaped particle can be divided into two distinct stages.

In the first stage, initially excited by the transient switching impulse, the particle exhibited a slight lift-off at its end with the larger radius of curvature within approximately 1 ms. Subsequently, the entire particle entered a "seesaw-like" motion, as shown in Figure 6a. The positions indicated by the arrows in the figure represent the lifted and fallen ends of the metal particles. In this mode, one end of the particle rises while the other end falls, forming a reciprocating alternating lifting motion. The middle part remains relatively fixed, and when one end rises, the lifting angle changes between 0 and 90 degrees. In this mode, one end of the particle lifts while the opposite end descends, creating a reciprocating, alternating lifting motion with the central part remaining relatively fixed, resembling the action of a seesaw.

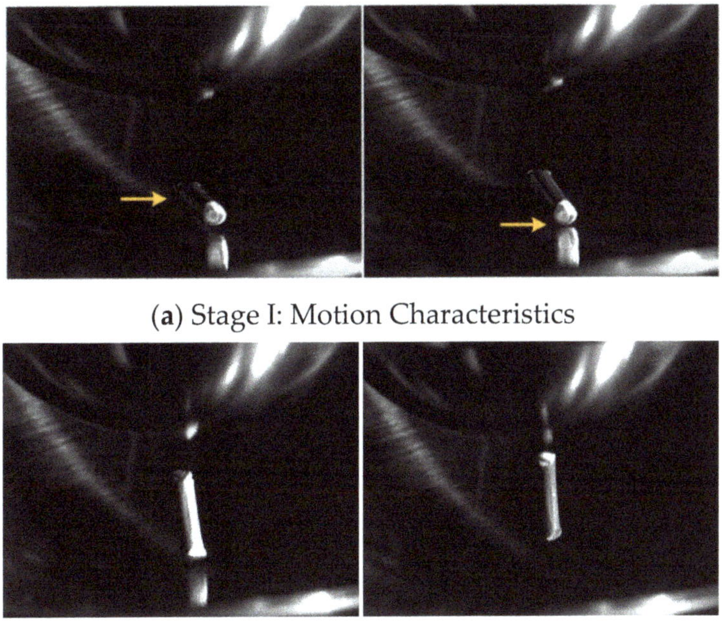

(a) Stage I: Motion Characteristics

(b) Stage II: Motion Characteristics

Figure 6. Motion characteristics of linear particles.

Upon transitioning to the second stage, the lift-off height at both ends of the particle gradually increased. Eventually, the end with the larger radius of curvature, experiencing a stronger electric field distortion, was vertically launched. At this point, the particle adopted a motion pattern characterized by vertical, reciprocating jumps accompanied by rotation, as illustrated in Figure 6b. During this process, the particle sometimes landed and then lifted off again, re-entering the vertical jumping state.

It is noteworthy that the motion amplitude in the first stage was relatively small and insufficient to initiate measurable partial discharge signals. In contrast, significant PD activity was gradually excited during the second stage, coinciding with the intensified motion and enhanced electric field distortion. This evolutionary process corresponds precisely to the time delay (Δt) observed in Figure 5, explaining the approximately 7 AC cycles required between the voltage application and the emergence of observable PD signals.

Under the subsequent AC voltage, sustained discharge activity was observed via the partial discharge detector, with its phase distribution characteristics shown in Figure 7. The Phase-Resolved Partial Discharge (PRPD) pattern indicates that the partial discharges initiated by the wire-shaped particle not only exhibit the typical feature of metallic particle discharges distributed across the entire AC cycle but are also accompanied by discrete, large-magnitude discharge pulses. These discrete discharges primarily originate from transient discharge signals excited when the wire particle jumps towards and makes contact with (or approaches) the high-voltage electrode during its motion. Consequently, wire-shaped metallic particles often demonstrate stronger discharge activity and pose a greater potential hazard in practical operation. Such high-magnitude discharges are more likely to initiate insulation breakdown or flashover, presenting a serious threat to equipment operational safety.

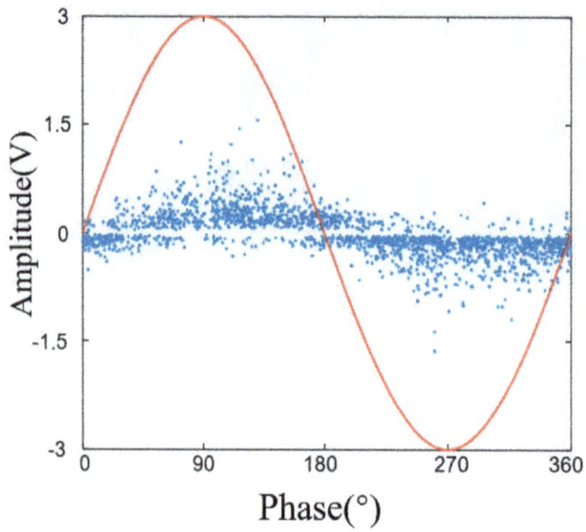

Figure 7. PRPD pattern of line particles under superimposed switching impulse.

3.2. Partial Discharge Characteristics of a Single Spherical Metallic Particle

The PDIV for the spherical metallic particle under AC voltage was 7.6 kV, which is lower than that of the wire-shaped particle. The time-domain waveform obtained after applying a 7.2 kV AC voltage superimposed with a 15.2 kV switching impulse voltage (i.e., 2 times PDIV) is shown in Figure 8. In contrast to the wire-shaped particle, the spherical particle generated significant partial discharge signals within the first AC cycle following the superimposed impulse, exhibiting no observable time delay phenomenon as seen in Figure 5.

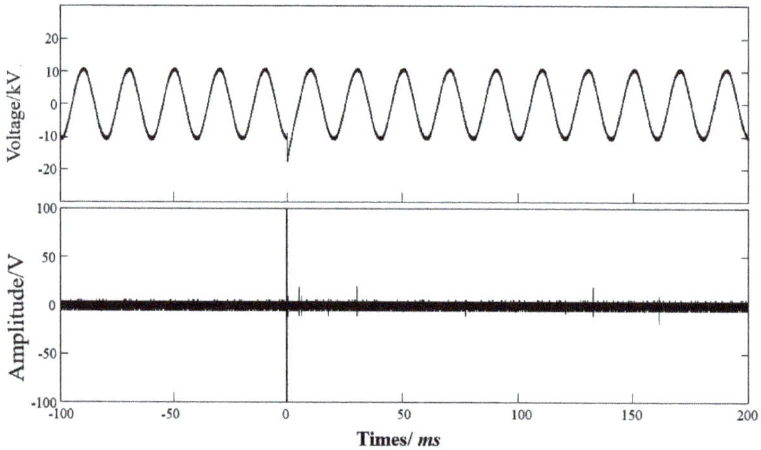

Figure 8. Partial discharge signals of circular particles under the impulse voltage of AC superposition switching.

Furthermore, as shown in Figure 9, synchronous observation of its motion behavior revealed that the spherical particle rapidly gains energy under the combined voltage and begins to bounce significantly. Subsequently, it triggers continuous discharge signals during this bouncing motion, which explains the absence of a discharge delay. This indicates that, compared to the wire-shaped particle, the spherical particle is more readily excited and can initiate partial discharge more rapidly under the combined AC and switching impulse voltage condition.

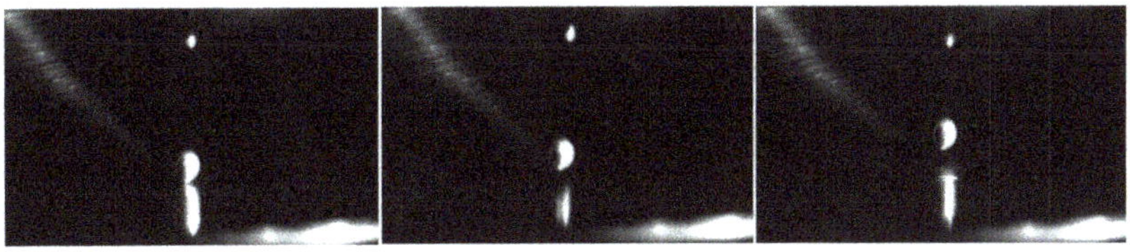

Figure 9. Motion characteristics of circular particles.

However, regarding the subsequent discharge behavior, the discharge intensity of the spherical particle was significantly weaker. As shown in Figure 10, the Phase-Resolved Partial Discharge (PRPD) pattern acquired under the AC voltage shows that while the discharges are distributed across the entire phase cycle, no discrete, large-magnitude discharge pulses are present. This is attributed to the limited electric field distortion caused by the spherical particle during its motion, which prevents significant energy accumulation. Furthermore, the bouncing amplitude is insufficient for the particle to reach (or closely approach) the high-voltage electrode, thus failing to generate the intense discharge activity observed with the wire-shaped particle.

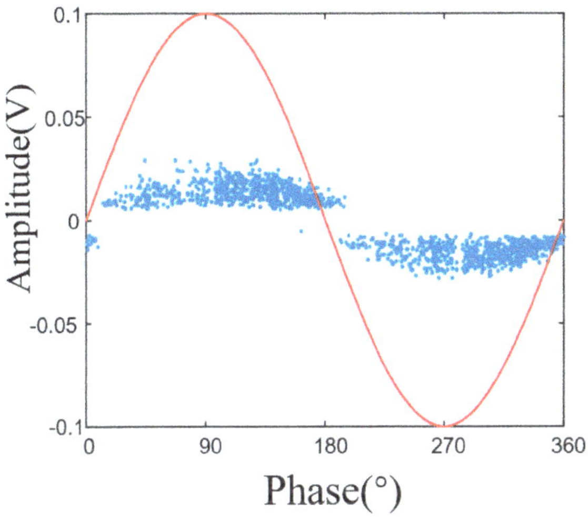

Figure 10. PRPD pattern of circular particles under superimposed switching impulse.

In summary, significant differences exist in the partial discharge characteristics between wire-shaped and spherical metallic particles under combined AC and switching impulse voltage: spherical particles are more readily excited and initiate discharge rapidly after the impulse, but their limited motion amplitude and weaker electric field distortion result in lower subsequent discharge intensity. In contrast, although the discharge initiation of wire-shaped particles exhibits a time delay, once they enter the vertical jumping phase, their discharge activity becomes more intense, often accompanied by large-magnitude discrete discharges, thereby posing a greater threat to the insulation safety of the equipment.

4. Partial Discharge Characteristics of Mixed Metal Particles Under Combined Voltage

Experiments were conducted by combining the previously studied spherical and wire-shaped metallic particles. The results show that the PDIV for the hybrid particles under AC voltage is 8.6 kV, which lies between that of the single spherical particle (7.6 kV) and the single wire-shaped particle (9.5 kV). This result suggests that the hybrid particle configuration has a certain homogenizing effect on the electric field distribution.

After applying an 8.1 kV AC voltage superimposed with a 17.2 kV switching impulse voltage, the obtained time-domain PD waveform is shown in Figure 11. It can be observed that in the initial stage following the superimposed impulse, PD signals of relatively small amplitude (not exceeding 50 mV) appear first, lasting approximately 450 ms, as shown in Figure 11. Subsequently, the discharge signals gradually intensify, with pulses reaching amplitudes up to about 150 mV. This result indicates that the partial discharge under the influence of Mixed Metal Particles exhibits an evolutionary process from weak to strong, demonstrating a unique transitional discharge characteristic. The hybrid particles inherit the ease of excitation under combined voltage seen in spherical particles, while also exhibiting a trend of intensified discharge in the later stage, similar to wire-shaped particles.

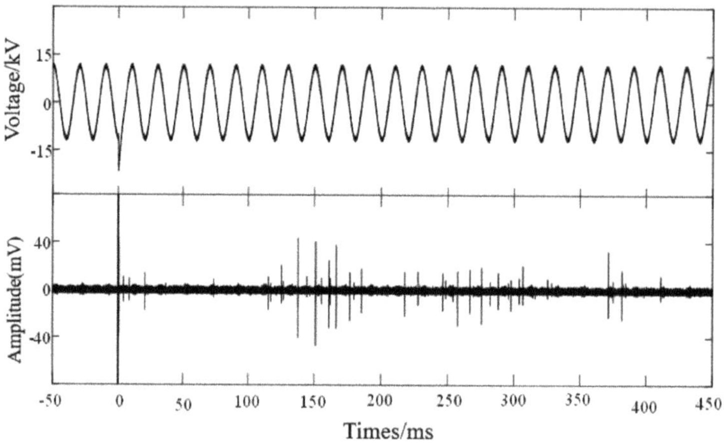

Figure 11. Partial discharge signals of mixed metallic particles under power-frequency superimposed operating impulse voltage.

The reason for these discharge characteristics lies in the more complex motion behavior of the Mixed Metal Particles. By synchronously observing the PD signals and particle motion, a clear correlation between the motion process and discharge features can be established. After the combined voltage is applied, the spherical particle jumps first while the wire-shaped particle remains stationary, as shown in Figure 12a. The yellow arrow shows spherical metal particles leaping from the stationary ground electrode into the air. Subsequently, as the spherical particle fully lifts off and collides with the wire-shaped particle (Figure 12b), the wire-shaped particle is excited into the "seesaw-like" motion stage; however, the overall discharge amplitude remains relatively small at this point. This process corresponds to the initial low-amplitude discharge stage. Further, with continued inter-particle collisions and the gradual lifting and eventual vertical jumping of the wire-shaped particle (Figure 12c), the discharge amplitude increases significantly, superimposed with smaller discharge signals generated by the spherical particle. Therefore, the discharge evolution of the hybrid particles exhibits a characteristic of "spherical particles lead the initiation, wire-shaped particles dominate the discharge".

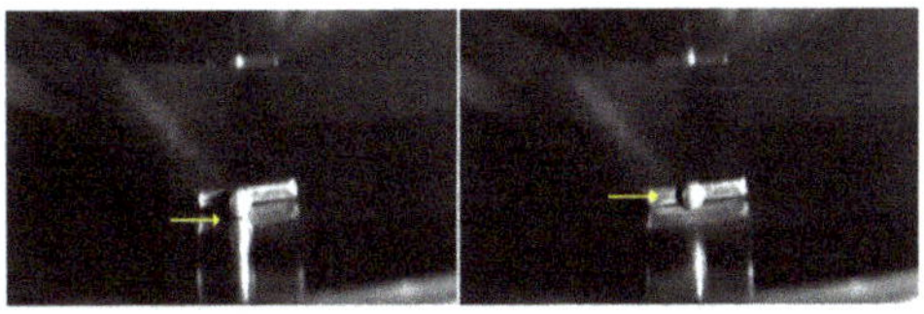

(**a**) Stage of Isolated Spherical Particle Bouncing

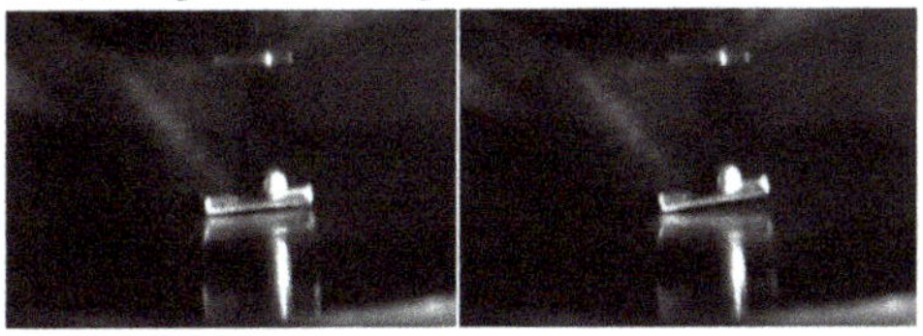

(**b**) Stage of Wire-shaped Particle Excitation and Motion

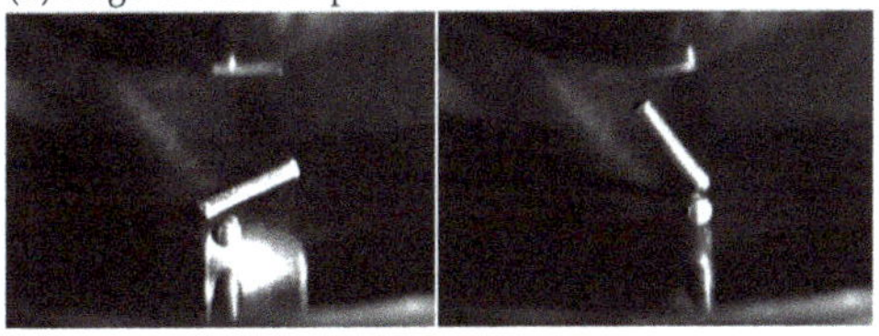

(**c**) Stage of Concurrent Particle Motion and Wire-shaped Particle Lifting

Figure 12. Motion characteristics of mixed metallic particles.

Under the subsequent sustained AC voltage, the obtained PRPD pattern for the Mixed Metal Particles is shown in Figure 13. Compared to the discharge characteristics of single-type particles, the discharge distribution of the hybrid particles is more dispersed, not only spanning the entire AC phase but also accompanied by a large number of high-amplitude discharge pulses. This indicates that under sustained voltage, the interaction between spherical and linear particles significantly enhances discharge activity, and the reasons behind this are diverse, including mechanical collisions, electromagnetic energy exchange, and so on, which lead to more intense and complex partial discharge processes. This indicates that under sustained voltage, the interaction between the spherical and wire-shaped particles significantly enhances the discharge activity, leading to a more intense and complex partial discharge process.

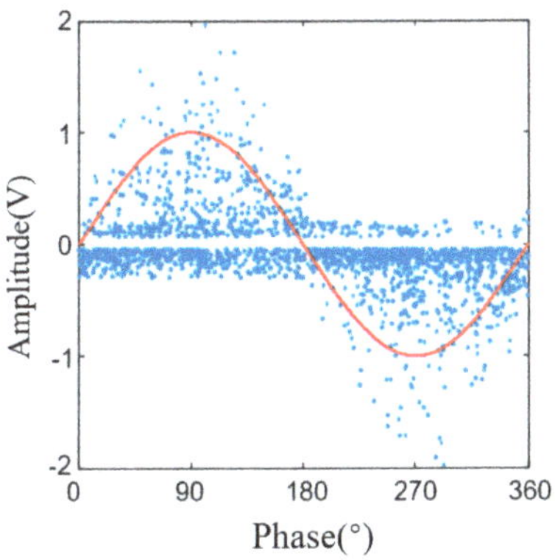

Figure 13. PRPD patterns of mixed metallic particles under superimposed operating impulse.

5. Conclusions

This paper investigated the partial discharge characteristics of wire-shaped, spherical, and Mixed Metal Particles in SF6 gas under combined AC and switching impulse voltage. The main conclusions are as follows:

(1) The switching impulse voltage can excite latent metallic particle motion under AC voltage and induce partial discharge. Wire-shaped particles exhibit a significant discharge initiation delay under the combined voltage; however, once they enter the vertical jumping stage, the discharge intensity increases rapidly and is often accompanied by large-magnitude discrete discharges. Spherical particles, in contrast, can be excited within the first AC cycle after the impulse without delay, but due to a weaker electric field distortion, their subsequent discharge amplitude is limited, without intense large-magnitude pulses.

(2) Hybrid particles exhibit a staged evolution characteristic under the combined voltage: initially, the spherical particle jumps and triggers low-amplitude discharges; subsequently, collisions from the spherical particle excite the wire-shaped particle into "seesaw-like" and vertical motion, leading to a significant increase in discharge amplitude. Ultimately, a more dispersed PRPD pattern accompanied by numerous large-amplitude pulses is observed, demonstrating the typical characteristic of "spherical particles lead initiation, wire-shaped particles dominate discharge", maintaining a more intense discharge state under the subsequent AC voltage.

(3) This study reveals the synergistic mechanism of power frequency voltage and switching impulse in triggering partial discharge: power frequency voltage (even if below the threshold) provides a "pre-excitation" field and charge preparation for discharge, while switching impulse plays a crucial "trigger" role by injecting high energy to drastically change the spatial charge distribution, thereby significantly reducing the threshold of subsequent power frequency discharge. Linear particles, due to their significant geometric asymmetry (tip effect), are the core cause of strong electric field distortion and delayed intense discharge. The study infers that an increase in the amplitude of the switching impulse accelerates this triggering process and shortens the discharge delay time. This mechanism indicates that in actual GIS operation, the risk of transient overvoltage "activating" latent metal particle defects is far more lethal than that of steady-state AC voltage itself.

Compared to single-type particles, hybrid particles are more commonly found in practical GIS interiors. The findings of this study indicate that Linear particles exhibit significant discharge triggering delay under superimposed voltage, but strong, discrete, large-amplitude discharge occurs during the vertical jump phase; spherical particles can be excited in the first power frequency cycle without any delay in discharge, especially under complex combined voltage conditions, exhibiting characteristics of being easily excited and producing high-amplitude discharges. The threat posed by complex contaminant combinations to GIS insulation safety is significantly higher than that from single contaminants.

Author Contributions: J.R.: funding acquisition, data curation, formal analysis; Y.M.: data curation; Q.Q.: formal analysis; Z.W.: investigation; Y.W.: methodology; L.W.: writing—original draft preparation; X.H.: writing—review and editing; X.Y.: validation. All authors have read and agreed to the published version of the manuscript.

Funding: This research was funded by Science and Technology Project of State Grid Qinghai Electric Power Company, grant number 522807240001. And the APC was funded by Science and Technology Project of State Grid Qinghai Electric Power Company.

Data Availability Statement: The raw data supporting the conclusions of this article will be made available by the authors on request.

Conflicts of Interest: Authors Jiyun Ren, Yongfu Ma, Quanlei Qu, Zile Wang, Lili Wang and Xiaojie Yang were employed by the State Grid Qinghai Electric Power Research Institute. The remaining authors declare that the research was conducted in the absence of any commercial or financial relationships that could be construed as a potential conflict of interest.

References

1. Song, X.; Shao, M. Health Assessment Method for Gas-Insulated Switchgear Based on Fault Tree Analysis. In Proceedings of the 2025 IEEE 3rd International Conference on Power Science and Technology (ICPST), Kunming, China, 16–18 May 2025.
2. Lu, W.; Yang, H.; Zhang, J.; Yang, C.; Wang, Y.; Wang, Z. Review on Mechanism and Suppression Strategies of Surface Flashover Induced by Metal Particles in DC GIS/GIL. *High Volt. Eng.* **2025**, *51*, 1010–1024.
3. Li, J.; Han, X.; Liu, Z.; Li, Y. Review of Partial Discharge Detection Techniques for Electrical Equipment. *High Volt. Eng.* **2015**, *41*, 2583–2601.
4. Jin, Z.; Du, B.; Xiao, M.; Liang, H.; Yao, H.; Zhang, W. Optimization of Permittivity Graded Spacer for Suppressing Metal Particles in AC-GIS. In Proceedings of the 2025 IEEE 5th International Conference on Electrical Materials and Power Equipment (ICEMPE), Harbin, China, 3–6 August 2025; pp. 1–4.
5. Ji, H.; Li, C.; Pang, Z.; Ma, G.; Cui, X.; Zeng, Z.; Rong, Z. Moving behaviors and harmfulness analysis of multiple linear metal particles in GIS. *IEEE Trans. Dielectr. Electr. Insul.* **2016**, *23*, 3355–3363. [CrossRef]
6. Li, Y.; Peng, Z.; Chen, C.; Xu, D.; Su, Y.; Lei, Y.; Li, Q. Influence of metal particles with different morphologies on the electric field distribution and discharge characteristics of basin-type insulators in GIS. In Proceedings of the 2025 10th Asia Conference on Power and Electrical Engineering (ACPEE), Beijing, China, 15–19 April 2025; pp. 2737–2741.
7. Zhang, C.; Li, W.; Yang, X.; Li, M.; Wang, M.; Luo, L. Digital Deduction Method for Metal Particle Size in Gas Insulated Metal-enclosed Switchgear. *J. Shanghai Jiaotong Univ.* **2025**, 1–20. [CrossRef]
8. Hu, Z. Flying Motion and Discharge Characteristics of Linear Particle Swarm and Cooperative Suppression Methods. Master's Thesis, North China Electric Power University, Beijing, China, 2024.
9. Sha, W.; Luo, Y.; Ma, P.; Qi, L. Research on Surface Discharge and Electric Field Characteristics of GIS Basin Insulator under Different Metal Particle Defects. *Insul. Mater.* **2023**, *56*, 89–93.
10. Li, X.; Hu, X.; Xu, H.; Jiang, P.; Wu, M.; Niu, B.; Li, Z.; Zhang, Q. Multi-Particle-in-Series Phenomenon and Discharge Properties Induced by Multiple Free Particles in GIS. *IEEE Trans. Power Deliv.* **2023**, *38*, 4039–4048. [CrossRef]
11. Ma, F.; Niu, B.; Zhang, T.; Tian, Y.; Cao, W. Study on the Motion Process of Foreign Metal Particles in GIS under High Voltage. *High Volt. Appar.* **2021**, *57*, 35–41.
12. Ji, H. Motion Behavior and Discharge Characteristics of Metal Particles in GIS Under AC Operating Voltage. Ph.D. Thesis, North China Electric Power University, Beijing, China, 2017.
13. Hu, X.; Li, X.; Wang, S.; Li, Z.; Hao, Y.; Zhang, Q. Motion of Free Metal Particles in GIS under Impact Vibration Excitation and Characteristics of Induced Gap Breakdown. *High Volt. Eng.* **2024**, *50*, 339–347.
14. Song, Y.; Zheng, Z.; Yu, J.; Sun, Y.; Zhang, S.Y.; Hou, J.C.; Wang, H.T.; Han, X.T.; Li, J.H. Study on Motion Behavior and Partial Discharge Characteristics of Metal Particles in SF_6 Gas under Electro-Mechanical Combined Action. *J. Xi'an Jiaotong Univ.* **2023**, *57*, 74–83.
15. Shi, T.; Zhang, Z.; Li, L.; Huang, H.; Niu, B.; Han, X.; Zhang, X.; Yao, C.; Li, J. Movement Behavior and Discharge Characteristics of Submillimeter Metal Particles Under AC Voltages Superimposed Mechanical Vibration. *IEEE Trans. Dielectr. Electr. Insul.* **2024**, *32*, 2460–2468. [CrossRef]
16. Niu, B.; Bai, T.; Shi, T.; Wu, X.; Han, X.; Liu, W. Initial Characteristics of Submillimeter Metal Particles in GIS under Impact Vibration. *Energies* **2024**, *17*, 5100. [CrossRef]
17. Liu, H.; Han, W.; Fan, Z.; Sun, J.; Qin, W.; Liao, X.; Wang, S.; Ma, G. Acoustic Emission and Propagation Characteristics of Metal Particles in GIS. *IEEE Trans. Dielectr. Electr. Insul.* **2025**, *32*, 20–27. [CrossRef]
18. Han, X.; Wu, X.; Zhao, Y.; Niu, B.; Wang, H.; Li, J. Partial Discharge and Motion Characteristics of Linear Metal Particles in SF_6 under AC Superimposed Lightning Impulse Voltage. *High Volt. Eng.* **2024**, *50*, 5406–5414.
19. Deng, J.; Gao, K.; Ren, M.; Huang, H.; Jin, L. Electric field analysis of different forms of metal particle adhesion defects in 110 kV GIS. *J. Electr. Eng.* **2024**, *19*, 316–324.

energies

Article

Research on High-Frequency Impedance Characteristics of Damaged Circuit Breaker Closing Resistance

Ce Zhang [1], Bo Niu [1], Feiyue Ma [1], Lingjun Yin [2], Shangpeng Sun [1] and Xutao Han [2,*]

1 Electric Power Research Institute, State Grid Ningxia Electric Power Co., Ltd., Yinchuan 750011, China
2 School of Electrical Engineering, Xi'an Jiaotong University, Xi'an 710049, China
* Correspondence: xutaohan@xjtu.edu.cn

Abstract: The closing resistor in a circuit breaker are prone to damage during operation due to extreme factors such as over-voltage, over-current, and mechanical shock, which alter their high-frequency impedance characteristics. Comparing impedance before and after damage can indicate the severity of degradation. However, the high-frequency impedance properties of damaged closing resistors remain insufficiently understood. This study investigates three classic damage types through simulation and external testing on a physical circuit breaker, validating the accuracy of the simulation results. Further high-frequency impedance measurements inside the tank examine the characteristics under varying damage degrees. Results show that external testing reflects the intrinsic impedance changes in the resistor string, exhibiting primarily resistive and inductive traits, with negligible capacitive influence. In contrast, internal measurements are affected by the tank's capacitance, leading to a resonance point in the high-frequency range. Different damage degrees cause noticeable shifts in the resonance frequency and a gradual increase in impedance magnitude. These findings offer practical guidance for field inspection of circuit breaker closing resistor conditions using high-frequency impedance techniques.

Keywords: circuit breakers; pre-insertion resistors; high-frequency impedance; damage diagnosis

1. Introduction

Against the backdrop of large-scale application of ultra-high voltage and extra-high voltage power grids, switch-gear such as gas-insulated switch-gear and dead-tank circuit breakers at 750 kV and above voltage levels equipped with closing resistor stacks are widely employed due to their compact structure and high reliability [1–4].

The closing resistor inside the circuit breaker is connected into the circuit via auxiliary contacts prior to the main closing contacts. It utilizes the damping effect of the resistor to suppress electromagnetic oscillations, reduce over-voltage amplitude, and limit the over-voltage generated when energizing unloaded transmission lines. These unique operational characteristics of the closing resistor dictate that during long-term operation, it must withstand extreme mechanical-electrical-thermal coupled stresses such as impact loads from the operating mechanism, over-voltages, and over-currents [5–10]. This easily leads to irreversible failures of the resistor discs, including blackening, charring, cracking, and breakage [11–15], resulting in insulation degradation or the presence of conductive contaminants, which seriously threatens the overall operation of the circuit breaker.

Currently, regarding the damage issues of closing resistors, researchers primarily rely on actual fault cases. Post-fault operation, problems are identified through disassembly inspections, which involve removing all closing resistors for individual examination. Combined with simulation calculations, the causes of equipment failure triggered by closing resistor damage are analyzed and discussed. For instance, Ma et al. uses EMTDC/PSCAD to establish a simulation model, the effect of pre-insertion resistor, pre-insertion resistor insertion duration and closing target angle on closing inrush current and energy absorption by pre-insertion resistor is calculated [16]. Bhatt et al. analyzing the effect of statistical scatter of mechanical operating time (MOT), tolerance of auxiliary contact used as feedback for a controlled switching device, and variation in mechanical insertion time (MIT) and electrical insertion time (EIT) of the PIR-CB [17]. Wang et al. established an impulse current test platform to simulate high-current impact scenarios during switching operations. By applying microsecond-level impulse currents to resistor discs, the time-domain evolution of closing resistance was calculated and analyzed [18].

In recent years research on the detection methods for the damage of closing resistors in circuit breakers has involved multiple fields. Reference [18] utilize ultrasonic sensing to detect the dynamic resistance characteristics of closing resistors. Reference [19] employ wavelet decomposition and the TLS-ESPRIT algorithm to analyze transient and instantaneous current data during the circuit breaker switching process, thereby assessing the operational state of the closing resistor. Although the use of non-contact sensors for detecting closing resistors eliminates the need for switching operations on the circuit breaker and significantly reduces maintenance time, the measurement accuracy remains relatively low.

Although existing research provides some guidance for explaining the causes of circuit breaker equipment failures due to closing resistor damage, it lacks reliable methods for effectively detecting closing resistor damage. Currently, to prevent damaged closing resistors from affecting equipment safety, substations often perform irregular tank-opening inspections on circuit breakers. This approach is highly blind and significantly increases the workload of field equipment operation and maintenance [20–25].

Frequency-sweeping technology has developed rapidly. Ref. [26] proposed a frequency modulation fuze anti-sweeping jamming method based on outlier reconstruction, which utilizes Iterative Filter Decomposition (IFD) and Long Short-Term Memory (LSTM) networks to suppress sweeping jamming. Another application of frequency-sweeping detection technology has been in identifying the coupling error coefficients of micro-electromechanical gyroscopes. This work derived the identification method for coupling error coefficients and analyzed the impact of amplitude and phase detection errors on identification accuracy through numerical simulation [27]. Furthermore, a machine learning-based method for correcting distortions in OCT images has been proposed, demonstrating the use of machine learning techniques to improve the outcomes of frequency-sweeping processes [28].

In theory, when minor microscopic defects occur in the closing resistor, the high-frequency impedance characteristics of the entire resistor stack change. Measuring these changes in high-frequency impedance characteristics could enable diagnostic analysis of its damage condition. However, current domestic attention to high-frequency impedance detection for circuit breaker closing resistors is insufficient. The frequency-sweep impedance method is mostly used in the field of transformer winding deformation detection. Therefore, conducting high-frequency impedance tests on damaged circuit breaker closing resistors and further researching the high-frequency impedance characteristics measured both outside and inside the tank for closing resistors under different damage conditions hold significant practical engineering and scientific research importance.

Based on this, this paper investigates the high-frequency impedance characteristics of closing resistors with varying degrees of damage by leveraging the frequency-sweeping impedance method, combined with simulation analysis and field tests. In the context of UHV systems, the 800-kV circuit breaker is currently the most prevalent and widely deployed type equipped with a pre-insertion resistor assembly. To address the fault issues caused by damage to the pre-insertion resistors, internal tank detection of the closing resistor is performed on an actual 800 kV circuit breaker to further study the variations in its high-frequency impedance characteristics under such conditions. This approach allows for the identification of damaged closing resistor positions based on their characteristic changes before deciding on removal, thereby shortening maintenance downtime. Approximately three-quarters of the time can be saved in the disassembly step of the pre-insertion resistor string. The study aims to provide reference and guidance for the field application of high-frequency impedance-based detection of closing resistor damage in intact circuit breakers.

2. Simulation Results of the Damaged Closing Resistor

2.1. Simulation Model

Currently, the closing resistors used in power equipment mainly employ Al-Si-C series resistor materials. The constituent phases of Al-Si-C series closing resistors can be categorized into three types: ceramic phase, pores, and graphite. Among these, both the ceramic phase and the pores are insulating phases, while graphite is a conductive phase (semi-metal). Therefore, its electrical equivalent circuit can be represented by a parallel connection of a series resistive-inductive conductive path and a capacitive insulating path.

The closing resistor within the circuit breaker has a series stacked structure. It is assembled on an insulating rod, with compression springs ensuring reliable electrical contact between resistor discs. Both ends are connected via terminal plates in series with metal plates to achieve uniform electrical signal transmission. Thus, when the resistor disc material is homogeneous, each resistor disc can be considered equipotential and equivalently represented as a single RLC circuit in the simulation model. For the entire closing resistor stack, referring to the actual number of resistor discs, it can be viewed as a series connection of numerous identical resistor discs. In the model, this is represented by a series connection of numerous identical RLC circuits. The established equivalent RLC network circuit model is shown in Figure 1.

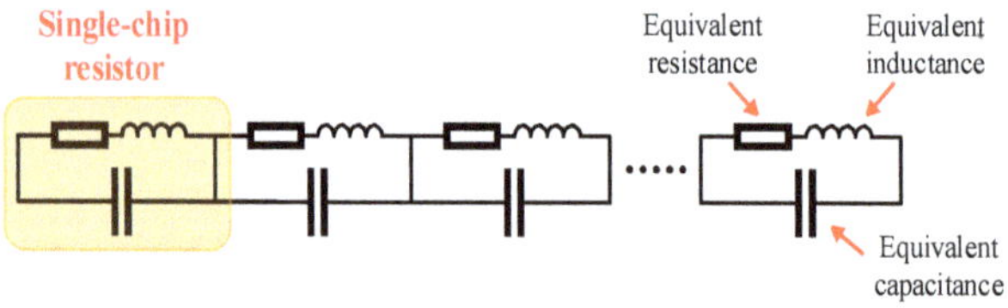

Figure 1. Equivalent circuit model of the closing resistor string.

This paper selects three typical types of damage for simulation: discoloration and charring damage, through-crack damage, and edge breakage damage. Among these, discoloration and charring damage and edge breakage damage only require modifying the parameter values of the equivalent circuit of the damaged resistance sheet according to the actual damage conditions. In contrast, through-crack damage necessitates a change in the circuit topology.

When through-crack damage occurs in the closing resistor, a gas gap forms at the crack location. For simulation, the equivalent circuit topology is altered by representing the RLC circuit shown in Figure 1 as the parallel RLC circuit illustrated in Figure 2. The crack is treated as a medium with the same dielectric constant as the gas inside the circuit breaker tank, and its equivalent capacitance is calculated accordingly.

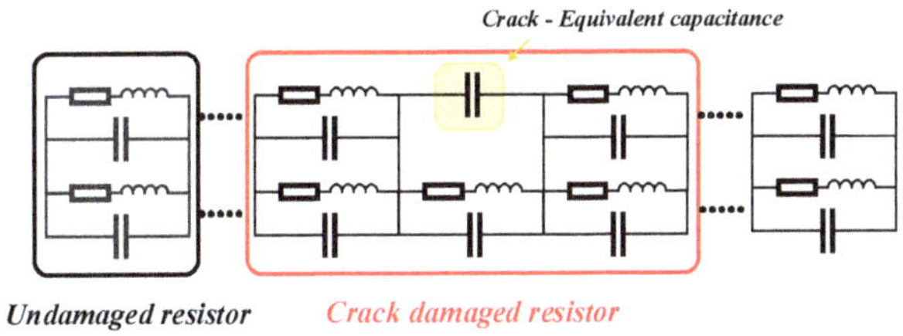

Figure 2. Equivalent circuit model of the crack damaged closing resistor string.

2.2. Edge Drop Damage

Based on the measured variations in the equivalent parameters of the closing resistor string under different damage conditions, as shown in Table 1, the resistance and inductance parameters in the equivalent circuit model illustrated in Figure 1 were modified to reflect the electrical characteristics of the closing resistor under varying degrees of damage.

Table 1. Resistance and inductance values of resistance strings with different degrees of damage.

Degree of Damage	Resistance/Ω	Inductance/μH
Undamaged	156.3	13.30
Minor damage	157.8	13.45
Severe damage	159.3	13.71

Using the previously established simulation model of the closing resistor string, high-frequency impedance tests were conducted on single resistor strings under different damage scenarios. The resulting curves obtained from the simulation results are shown in Figure 3, where the horizontal axis is plotted on a logarithmic scale, S-U represents the comparison between the severely damaged curve and the undamaged curve, while M-U represents the comparison between the mildly damaged curve and the undamaged curve.

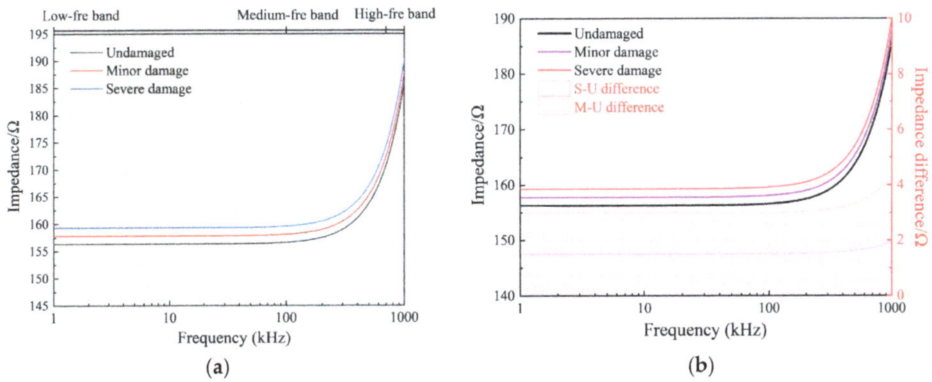

Figure 3. Simulation results of edge drop damage. (**a**) Impedance characteristic curve. (**b**) The difference in impedance curves before and after damage.

As can be seen from the test results in Figure 3, the impedance of the closing resistor string generally exhibits an increasing trend with the rise in the applied frequency, indicating the presence of a significant parasitic inductance effect in addition to its inherent resistive characteristics. Its impedance behaviour shows distinct differences across various frequency bands. In the low-frequency range of 1 kHz to 100 kHz, the inductive effect is not yet pronounced, and the closing resistor string primarily exhibits resistive characteristics, with its impedance value being close to the DC resistance measured directly by a multi-metre. In the mid-frequency range of 100 kHz to 700 kHz, the parasitic inductance gradually becomes influential, leading to a slow increase in impedance with rising frequency. However, at this stage, the inductive effect remains smaller than the resistive component, so the impedance curve does not show a distinct linear upward trend. In the high-frequency range of 700 kHz to 1 MHz, the inductive effect dominates, and the impedance of the closing resistor string increases approximately linearly with frequency.

For the difference in impedance curves before and after damage, in the low-frequency and mid-frequency bands, it is mainly affected by the change in resistance value, and the difference is almost a straight line. In the high-frequency band, it gradually increases due to the influence of parasitic inductance, and the curve difference shows a gradually increasing trend.

2.3. Entire Crack Damage

The crack damage of the closing resistor reduces the effective cross-sectional area for current flow, increases the path length, and leads to an increase in resistance. At the same time, the cracks damaged the integrity of the resistor sheet, weakened the eddy current cancellation effect, intensified the magnetic flux change, and led to an increase in inductance.

Simulations were conducted on closing resistors with full-section cracks of varying damage degrees, and the obtained high-frequency impedance curves along with their difference comparisons are shown in Figure 4.

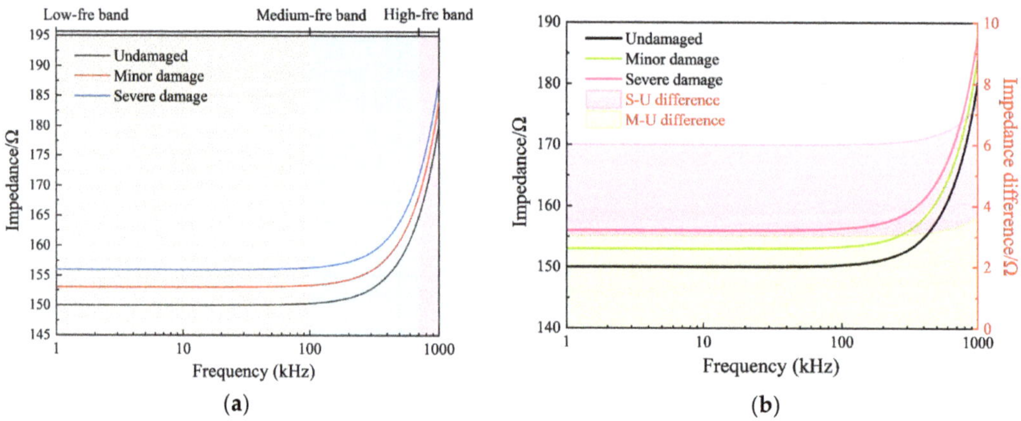

Figure 4. Simulation results of entire crack damage. (**a**) Impedance characteristic curve. (**b**) The difference in impedance curves before and after damage.

The results indicate that the variation trend of the impedance curve under full-section crack damage is similar to that observed in edge loss damage. Notably, the difference in the impedance curve between the damaged and undamaged conditions exhibits a gradual increase only in the high-frequency range.

2.4. Blackened and Charred Damage

When a closing resistor suffers from blackening and charring damage due to factors such as mechanical shock and energy injection, black particles like carbon and metal oxides appear on the material surface. These substances possess significantly lower electrical conductivity than the resistor material, resulting in a permanent and irreversible increase in the resistivity of the resistor string. In severe cases, the material surface may exhibit bulging and peeling, while the internal structure becomes loose and porous, leading to an overall increase in the inductance of the resistor chip.

Simulations were performed on closing resistors with varying degrees of blackening and charring damage, and the obtained high-frequency impedance curves along with their difference comparisons are shown in Figure 5.

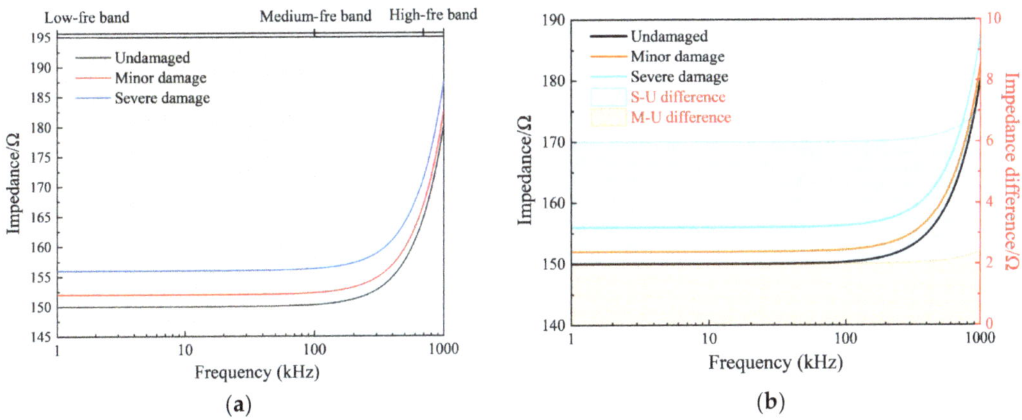

Figure 5. Simulation results of blackened and charred damage. (**a**) Impedance characteristic curve. (**b**) The difference in impedance curves before and after damage.

As can be seen from the figure, the variation trend of the impedance curve and the difference curve before and after damage is almost identical to that of the first two damage types.

For the three aforementioned types of closing resistor damage, the variation trends of the high-frequency impedance curves under different damage levels are generally consistent. In the low-frequency range, the curve approximates a straight line, with the impedance value increasing slightly as the frequency rises. In the mid-frequency range, influenced by parasitic inductance, the curve exhibits a noticeable upward trend, while it shows a linear increase in the high-frequency range. Regarding the difference in curves before and after damage, the variation remains largely constant across the tested frequency band, with only a slight increase observed in the high-frequency region.

It can be concluded that the closing resistor overall exhibits resistive-inductive characteristics, while its inherent capacitance has a negligible effect on its high-frequency impedance properties and can be disregarded.

3. On-Site Test Results of the Damaged Closing Resistor

3.1. Wiring Method for Entity Detection

This paper conducts experiments on an 800 kV circuit breaker with a parallel double-break interrupter unit structure in He Nan, China. As shown in Figure 6, the circuit breaker features a symmetrical layout on both sides, with the interrupting chambers located at each end to ensure electrical and mechanical balance. To improve the transient characteristics during the closing operation, closing resistor stacks are installed on both sides, effectively suppressing switching overvoltages and enhancing operational stability.

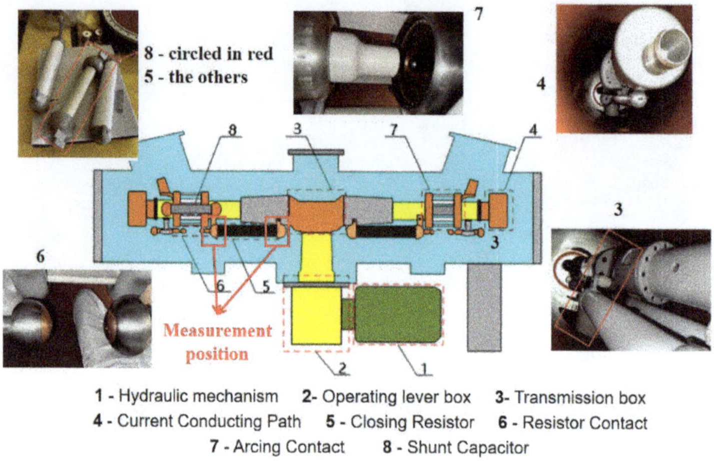

Figure 6. The structure of the test circuit breaker body.

This circuit breaker employs a parallel configuration, and its equivalent circuit is depicted in Figure 7. In the circuit, the closing resistor is connected in series with the auxiliary break and is paralleled with the main break and the grading capacitor to form the complete circuit. Each closing resistor stack consists of 30 resistor discs, with a total resistance of 150 ± 7.5 Ω; the capacitance of the grading capacitor is 700 ± 21 pF.

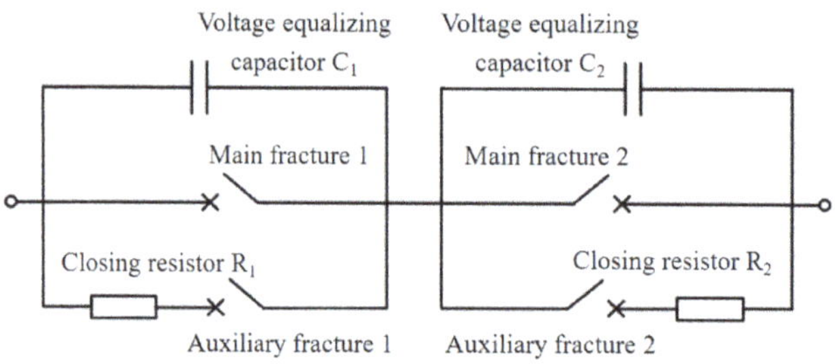

Figure 7. Simplified circuit of the test circuit breaker.

Tests were performed on the closing resistor stacks both outside and inside the tank using a frequency-sweep impedance analyzer to investigate their high-frequency impedance characteristics.

During the testing process, frequent disassembly and replacement of the closing resistors were required. To facilitate operation, the bushings were not installed on the circuit breaker. Consequently, suitable measurement points had to be identified inside the tank for the internal testing.

Resistance measurements were taken at multiple potential locations within the tank using a multimeter. The results indicated that a stable resistance value, approximately 156.7 Ω, was only obtainable at the holes in the insulating rods located in the central area of the circular support bases at both ends of the closing resistor stack. Compared to the nominal resistance of the closing resistor (150 ± 7.5 Ω), the measured value is essentially consistent with the theoretical value. Considering the presence of additional resistance from components such as the aluminum tabs of the resistor discs and the support bases themselves, this result is entirely reasonable within the expected error margin. More importantly, this measurement point provides a reliable conductive path while avoiding measurement failure caused by the insulating coating, satisfying the requirements from both the physical mechanism and testing feasibility perspectives.

Based on the above factors, this location was ultimately selected as the measurement point for the inside-the-tank tests. When testing a single closing resistor stack inside the tank, the measurement points were chosen at both ends of the stack. The connection method is illustrated in Figure 8.

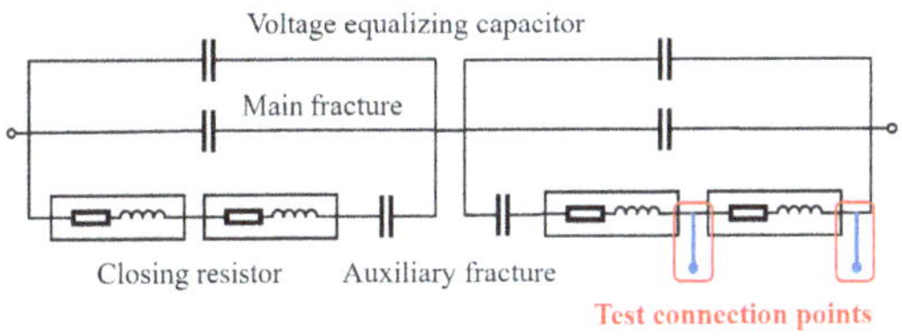

Figure 8. Schematic diagram of the wiring method for closing resistor detection inside the tank.

3.2. Simulation of the Damage Condition of the Closing Resistor

During circuit breaker operation, the most common forms of damage to closing resistors primarily stem from two sources: firstly, the intense mechanical impact generated by opening and closing operations, and secondly, the fragmentation and dislodgment of resistor discs caused by improper assembly. Among these, edge fragmentation, spalling, and subsequent dislodgment of resistor discs represent the most typical and significantly hazardous damage mode. Based on this, this study selects edge fragment loss as the representative damage mode for simulation experiments.

The morphology of the edge fragment loss under different damage severity levels is shown in Figure 9. This study uses the lost volume fraction of the resistor disc as the evaluation metric for damage severity: loss of approximately 1/8 of the volume is defined as slight damage, while loss of approximately 1/4 is defined as severe damage. To more closely simulate actual conditions, small fragments were additionally introduced at the breakage locations on the South and North Non-mechanism-side resistor stacks to simulate fragmentation caused by switching impacts or assembly defects.

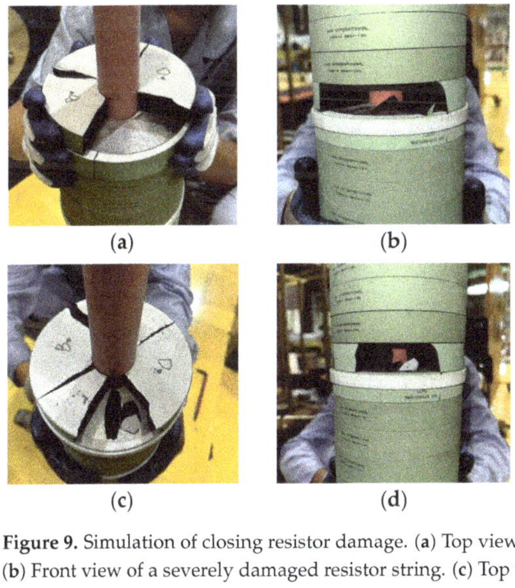

Figure 9. Simulation of closing resistor damage. (**a**) Top view of severely damaged resistance string. (**b**) Front view of a severely damaged resistor string. (**c**) Top view of minimally damaged resistance string. (**d**) Front view of a minimally damaged resistor string.

After assembling the resistor stacks, the overall resistance of the stacks under different damage states was measured using a multi-metre; the results are presented in Table 2. It can be observed that under both slight and severe damage conditions, the overall resistance of the closing resistors exhibited only minor changes. Relying solely on resistance measurement often proves difficult for effectively distinguishing the degree of damage.

Table 2. The measured resistance value of the resistance series.

Degree of Damage	M-South	M-North	N-South	N-North
Undamaged	156.4 Ω	155.6 Ω	156.5 Ω	156.3 Ω
Minor damage	157.7 Ω	157.6 Ω	157.8 Ω	157.8 Ω
Severe damage	158.4 Ω	158.8 Ω	158.6 Ω	159.3 Ω

The test subjects are the four closing resistor stacks inside the circuit breaker. Since the hydraulic mechanism is located on the right side of the breaker, the resistor stacks on the right are designated as the mechanism-side closing resistors, and those on the opposite side as the non-mechanism-side closing resistors. Based on their relative positions, they are specifically named M-south, M-north, N-south, and N-north.

The traditional method for diagnosing closing resistor damage, which relies solely on whether the resistance value change exceeds ±5%, may overlook certain cases where damage is already significant. The frequency-sweeping impedance testing method proposed in this study can, to some extent, address this limitation.

3.3. Test Results of Closing Resistance

Field tests were conducted on single closing resistor strings under different damage conditions using a frequency-sweeping impedance analyzer, and the resulting curves are shown in Figure 10.

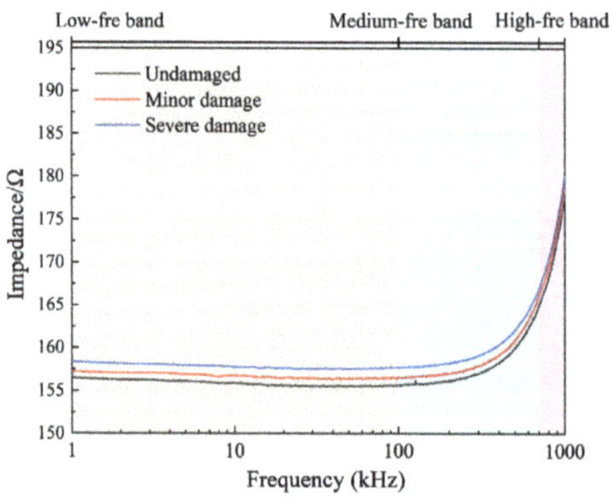

Figure 10. On-site test results.

As can be seen from the figure, resistor strings with different degrees of damage still exhibit resistive-inductive characteristics, meaning the overall impedance of the closing resistor string increases with rising frequency. A comparison of the impedance curves reveals that changes in the degree of damage are primarily manifested in two aspects:

Firstly, in the low-frequency range, the impedance of the closing resistor shows a certain increase as the damage becomes more severe. This is attributed to the reduction in the effective conductive cross-sectional area caused by chipping at the edges of the resistor discs, which narrows the current path and thereby leads to an increase in resistance. Although the magnitude of this change is relatively limited, it still demonstrates a discernible distinction between slight and severe damage.

Secondly, in the high-frequency range, the slope of the impedance curve varies with the degree of damage, reflecting changes in the equivalent inductance of the closing resistor. The underlying reason is that the loss of resistor disc fragments alters the current path and the loop area, resulting in variations in the parasitic inductance. This manifests in the frequency response as differences in the slope of the impedance-frequency curve within the approximately linear region.

The above analysis indicates that the closing resistor string can be electrically modelled as a series combination of a resistor and an inductor. The impedance magnitude of this resistive-inductive element is given by:

$$|Z(j\omega)| = \sqrt{R^2 + (\omega L)^2} \tag{1}$$

where $Z(j\omega)$ is the impedance of the closing resistor string, R is its equivalent resistance, L is its equivalent parasitic inductance, and ω is the angular frequency. Based on Equation (1), the following linear expression can be derived:

$$|Z(j\omega)|^2 = \omega^2 L^2 + R^2 \tag{2}$$

The relationship curve between the square of the impedance and the square of the angular frequency for the closing resistor string, plotted from experimental test data, is shown in Figure 11.

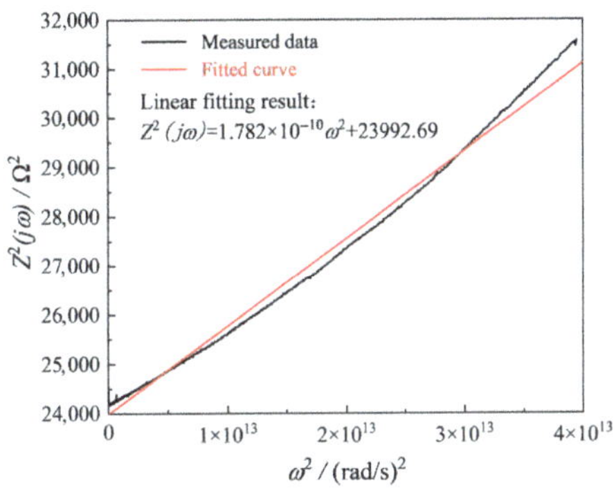

Figure 11. The relationship curve between the square of the string impedance and the square of the angular frequency of a single-string resistor.

The results from Figure 11 confirm that the characteristics of the closing resistor string align with the resistive-inductive model. With a goodness-of-fit greater than 0.999, the fitted equation for the curve is:

$$|Z(j\omega)|^2 = 1.782 \times 10^{-10}\omega^2 + 23992.69 \tag{3}$$

By combining this with Equation (2), the inductance value for the single string of 30 series-connected closing resistors in this study is calculated to be 13.30 μH, and the resistance value is 154.89 Ω. This resistance value is consistent with the result obtained using a multimeter.

Applying the calculation principle of Equation (2), the resistance and inductance values for closing resistor strings with different damage degrees can be obtained, as summarized in Table 3.

Table 3. The resistance and inductance values of resistance strings with different degrees of damage calculated based on the fitting curve results.

Degree of Damage	Resistance/Ω	Inductance/μH
Undamaged	154.89	13.30
Minor damage	155.72	13.42
Severe damage	156.90	13.51

Due to the inherent limitations of the measurement method, imperfect environmental conditions, instrument errors, and other influencing factors, the combined standard uncertainty in the determined resistance parameter $u_{0\Omega}$ is approximately 0.224%, while that for the inductance parameter u_{0L} is approximately 0.447%. The comparison of Tables 2 and 3 indicates that while the resistance values differ slightly, the inductance values are almost identical, which validates the rationality and accuracy of the simulation analysis.

As can be seen from Table 3, after the closing resistor is damaged, both its overall equivalent resistance and inductance increase. Although the change in resistance appears larger in magnitude, this level of variation is often difficult to measure effectively in practical applications. The reasons are twofold: firstly, a 2 Ω increase represents only a minor disturbance relative to the nominal resistance of several hundred ohms; secondly, field testing conditions are complex, where fluctuations in contact resistance and external interference often mask such subtle changes, making damage diagnosis based solely on resistance values unreliable.

In contrast, although the absolute change in inductance (0.2 μH) is relatively small, its effect is significantly amplified in high-frequency testing due to the frequency-sensitive nature of inductance. When the frequency rises to several hundred kHz or higher, the inductive component becomes dominant in the overall impedance, leading to observable differences in the slope of the impedance curve. Compared to low-frequency resistance measurement, this characteristic not only offers better measurability but also exhibits stronger noise immunity under high-frequency conditions. Therefore, the variation in inductance at high frequencies can serve as a more effective indicator for identifying damage in closing resistors. Consequently, establishing a damage diagnosis criterion based on high-frequency impedance characteristics is more feasible and valuable for engineering applications than relying solely on DC or low-frequency resistance measurements.

A comparison between the simulation curve under undamaged conditions and the field test curve, along with the calculated difference between them, is shown in Figure 12.

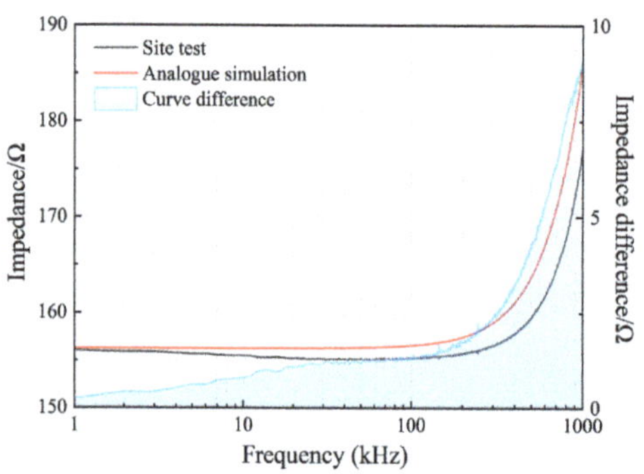

Figure 12. Comparison of simulation results with test results.

As observed in the figure, the discrepancy between the curves is relatively small in the low-frequency range. As the frequency increases, the difference gradually grows, showing a significant and linearly increasing trend in the mid- and high-frequency ranges, the discrepancy between the curves increases gradually from 0.5% at 1 kHz to 5.6% at 1 MHz. The discrepancy in the low-frequency region is primarily attributed to the difference in resistance values, while the difference in the high-frequency region is mainly influenced by the variation in inductance values. Since the reactance variation is linearly proportional to the frequency, the curve discrepancy exhibits a linear growth trend. Influenced by factors such as inherent measurement errors of the sweep-frequency impedance system (e.g., test cables, clip positioning), structural complexity of the test object, and environmental interference, the field results exhibit a descending trend in the mid-frequency range.

3.4. Test Results of the Closing Resistor in the Tank

The circuit breaker used in this study has two closing resistor stacks arranged on the same side, connected by a link located near the centre of the breaker. In a practical internal tank test, measuring only a single resistor stack would require manual entry into the tank or partial extraction of the internal structure to connect one end of the test circuit directly to this linking piece. This procedure is not only time-consuming and labour-intensive but also increases the uncertainty and operational risk of the experiment. In light of this, for testing damaged closing resistors, this study selected both resistor stacks on the same side as the test subject.

Figure 13 shows the impedance curves obtained from in-tank tests performed on two closing resistor stacks on the same side under different damage conditions. As can be observed from the figure, resonance points appear in the high-frequency band for the impedance curves under all damage conditions. A comparison of these curves reveals that the variation in the degree of damage is reflected across all three frequency bands.

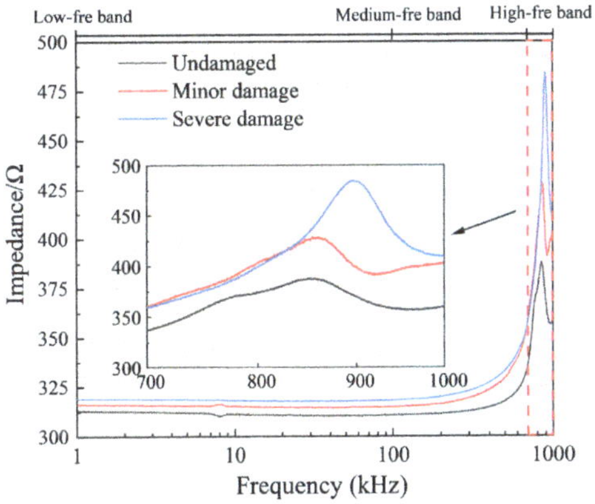

Figure 13. The impedance curves of resistance string inside the tank with different degrees of damage.

Firstly, in the low-frequency band, the change in the impedance value of the closing resistors is consistent with the impedance characteristic changes observed during the out-of-tank tests on a single resistor stack under different damage conditions. As the severity of the damage increases, the impedance shows a certain degree of increase. Furthermore, the magnitude of this change exhibits distinguishable differences between slight and severe damage states.

Secondly, in the medium-frequency band, the rate of rise in the impedance curve varies with the extent of damage. The more severe the damage to the resistor stack, the faster the rise in the curve in this band. The contrast between the curves for damaged and undamaged conditions is more pronounced in the in-tank test compared to the out-of-tank test. The reason for this is that the capacitive effects of various components within the circuit breaker cause the rising trend, typically observed in the high-frequency band during out-of-tank tests, to manifest earlier.

Finally, in the high-frequency band, significant changes are observed at the resonance points of the impedance curves for different damage levels, with the resonant frequencies and corresponding impedance listed in Table 4. Measured values presented in Table 4 are averaged from five repeated measurements, and the resonance point positions show negligible variation across tests. The impedance at the resonance point of the curve changes with the damage severity; more severe damage results in a higher resonant point impedance.

Table 4. The resonant point of the same side closing resistor string tested in the tank.

Degree of Damage	Undamaged	Minor Damage	Severe Damage
Resonant frequency	869 kHz	854 kHz	897 kHz
Resonant point impedance	421.44 Ω	478.84 Ω	534.39 Ω

The reason for the resonant peak in the impedance curve during the internal tank test is as follows. As shown in Figure 7, when testing a single closing resistor stack inside the breaker, both the main and auxiliary breaks are in the open state, primarily exhibiting capacitive effects. They can be considered as capacitors, similar to the grading capacitor. Therefore, the measurement circuit simplifies to a parallel combination of the single resistor stack and a series circuit of another resistor stack and a capacitor. Since an independent closing resistor stack exhibits resistive-inductive characteristics, the introduction of capacitance into the loop creates a resonant frequency point.

These results indicate that when detecting closing resistors inside the circuit breaker tank, the influence of damage-induced inductance changes on the high-frequency impedance characteristics is amplified. The frequency range where the inductive component dominates the impedance shifts towards lower frequencies. The distinguishing feature of the curves evolves from a change in slope in the out-of-tank test to a more pronounced and easily identifiable shift in the resonance point in the in-tank test. This provides stronger evidence supporting the use of inductance variation characteristics at high frequencies as an effective indicator for assessing the damage state of closing resistors.

Based on the changing trends of the sweep frequency curves of the closing resistors with different damage degrees obtained from the above image, we have summarized a set of damage assessment standards for circuit breaker closing resistors for subsequent reference. This standard is defined by three judgement criteria: Initial value change ratio r, Resonance point frequency shift ratio r_1, Resonance point impedance shift ratio r_2.

The average value a of the low-frequency band is calculated from N1 data points. This value is then compared with the baseline average b obtained under undamaged conditions to determine the initial value shift ratio of the curve $r = |a - b|/b$. This ratio serves as an indicator for assessing the extent of closing resistor degradation.

Furthermore, the resonant frequency f and its corresponding impedance Z at the resonance point within the high-frequency band are identified. These parameters are compared with their baseline counterparts, the resonant frequency f_0 and impedance Z_0 from the undamaged state, to calculate the frequency shift ratio r_1 and the impedance shift ratio r_2 at the resonance point, where $r_1 = |f - f_0|/f_0$ and $r_2 = |Z - Z_0|/Z_0$. The degradation level of the closing resistor is subsequently evaluated based on these calculated ratios. The extent of degradation can be determined upon the fulfilment of any one of these specified ratio-based criteria. The specific criteria for judgement corresponding to different degrees of injury are shown in Table 5.

Table 5. The basis for judging the degree of damage to the closing resistor.

Degree of Damage	Resonance Point Shift Ratio r_1, r_2 Initial Value Change Ratio r
Severe damage	$r_1 \geq 5\%$ or $r_2 \geq 20\%$ or $r \geq 15\%$
Moderate damage	$2\% \leq r_1 \leq 5\%$ or $10\% \leq r_2 \leq 20\%$ or $10\% \leq r \leq 15\%$
Minor damage	$1\% \leq r_1 \leq 2\%$ or $5\% \leq r_2 \leq 10\%$ or $7.5\% \leq r \leq 10\%$
Undamaged	$r_1 \leq 1\%$ and $r_2 \leq 5\%$ and $r \leq 7.5\%$

4. Conclusions

This study employs the frequency-sweep impedance testing method to conduct simulation analysis and field tests on the high-frequency impedance characteristics of closing resistors in real tank-type circuit breakers. The main conclusions are as follows:

(1) For the three classical damage types of the closing resistor, the variation trends of the curves under different damage degrees and the trends in the impedance differences before and after damage are fundamentally consistent, with all cases exhibiting overall resistive and inductive characteristics. When damage occurs, both the equivalent resistance and inductance increase, further demonstrating that the more severe the damage, the more significant the increase.

(2) The frequency-impedance curve obtained from measurements performed on the closing resistor assembly inside the tank exhibits resistive characteristics in the low-frequency segment and reactive characteristics in the high-frequency segment. Moreover, variations in the inductance properties of the closing resistor caused by different degrees of damage demonstrate superior sensitivity and significance compared to changes in resistance. Therefore, inductive characteristics can serve as an effective indicator for identifying the damage state of closing resistors, providing an experimental basis and methodological support for reliable damage detection in subsequent research.

(3) Based on the variation patterns observed in the sweep-frequency impedance curve, we have developed a foundational approach for establishing a standard to diagnose the degradation level of closing resistors using the sweep-frequency impedance method. This approach is based on a combined assessment of the mean value shift in the low-frequency band, and the resonance point shift in the high-frequency band.

The sweep-frequency impedance method employed in this study has contributed to reduced maintenance time and improved detection accuracy for circuit breakers. In future developments of this technique, the trends in frequency-impedance curve variations under different degradation levels summarized in this paper can be leveraged. By incorporating approaches such as machine learning, it would be possible to predict the degradation of closing resistors and further optimize the proposed method.

Author Contributions: Conceptualization, S.S. and X.H.; Methodology, C.Z., B.N. and X.H.; Software, C.Z.; Validation, C.Z. and L.Y.; Formal analysis, B.N.; Investigation, B.N.; Resources, F.M. and X.H.; Data curation, B.N., F.M. and X.H.; Writing—original draft, L.Y.; Writing—review & editing, C.Z.; Visualization, F.M. and L.Y.; Supervision, F.M. and S.S.; Project administration, X.H.; Funding acquisition, S.S. All authors have read and agreed to the published version of the manuscript.

Funding: This research was funded by Science and Technology Project of State Grid Ningxia Electric Power Co., Ltd., grant No. 5229DK24000S.

Data Availability Statement: The original contributions presented in this study are included in the article. Further inquiries can be directed to the corresponding author.

Conflicts of Interest: The authors declare that this study received funding from Electric Power Research Institute, State Grid Ningxia Electric Power Co., Ltd., Yinchuan 750011, China. The funder had the following involvement with the study: Ce Zhang, Bo Niu, Feiyue Ma, Shangpeng Sun.

References

1. Niu, B.; Ma, F.; Xiang, Z.; Wu, H.; Sun, S.; Wen, Q. Research on operation reliability and fault diagnosis technology of pre-insertion resistors for circuit breakers. *Ningxia Electr. Power* **2021**, *1*, 23–35.
2. Legate, A.C.; Brunke, J.H.; Ray, J.J. Elimination of closing resistors on EHV circuit breakers. *IEEE Trans. Power Deliv.* **1988**, *3*, 223–231. [CrossRef]
3. Huang, H.; Li, W.; Lu, W.; Zhang, C.; Chen, W.; Zhou, H. Simulation of radial breakdowns on the breaker grading capacitor of 500 kV AC filter. *High Volt. Appar.* **2018**, *54*, 56–62.
4. Ji, R.; Yi, Q.; Su, F.; Sun, K.; Chen, J.; Zhou, H. Study on applicability of circuit breaker closing resistance in EHV and UHV AC systems. *Power Syst. Technol.* **2011**, *35*, 18–25.
5. Chen, X.; He, C.; Deng, J.; Zhu, X.; Zhang, G.; Ni, H.; Ma, F.; Chen, L. Energization transient suppression of 750 kV AC filters using a pre-insertion resistor circuit breaker with a controlled switching device. *IEEE Trans. Power Deliv.* **2022**, *37*, 3381–3390.
6. Awad, E.A.; Badran, E.A.; Youssef, F.M.H. Mitigation of temporary over-voltages in weak grids connected to DFIG-based wind farms. *J. Electr. Syst.* **2014**, *10*, 431–444.
7. Sun, Q.; Tu, Y.; Fu, J.; Wang, X.; Liu, X.; Zhou, J. Fracture analysis of fastening bolt of closing resistance for 1100 kV GIS circuit breaker. *Zhejiang Electr. Power* **2023**, *42*, 68–74.
8. Shu, S.; Ruan, J.; Huang, D.; Wu, G.; Liu, B.; Zhang, K. Dynamic voltage-sharing measures for vacuum circuit breakers with triple breaks. *Power Syst. Technol.* **2012**, *36*, 252–259.
9. Liu, S.; Lv, D.; Zhang, F. Study on the suppression efficacy of filter closing inrush currents by shunt dynamic closing resistor of circuit breaker. *Power Syst. Technol.* **2024**, *48*, 3931–3937.
10. Yang, R.; Zhang, H.; Zhang, G.; Liu, K.; Wang, Y.; Gao, L. Adaptability of cancellation of the closing resistance of circuit breaker for 330 kV transmission line. *Insul. Surge Arresters* **2023**, *6*, 165–171.
11. Kumar, R.; Bhargava, P. Fabrication of low specific resistance ceramic carbon composites by colloidal processing using glucose assoluble carbon source. *Bull. Mater. Sci.* **2017**, *40*, 1197–1202. [CrossRef]
12. Chen, L.; Chen, X.; Wei, Y.; Wu, C.; Deng, J.; Ma, F. Study of controlled switching with pre-insertion resistor during energization of AC filters. In Proceedings of the 22nd International Symposium on High Voltage Engineering (ISH 2021), Hybrid Conference, Xi'an, China, 21–26 November 2021; pp. 1682–1687.
13. Lazimov, T.; Saafan, E.A.; Babayeva, N. Transitional processes at switching-off capacitor banks by circuit-breakers with pre-insertion resistors. In Proceedings of the 2015 Modern Electric Power Systems (MEPS), Wroclaw, Poland, 6–9 July 2015; pp. 1–4.
14. Hedman, D.E.; Johnson, I.B.; Titus, C.H.; Wilson, D.D. Switching of Extra-High-Voltage Circuits II-Surge Reduction with Circuit-Breaker Resistors. *IEEE Trans. Power Appar. Syst.* **1964**, *83*, 1196–1205. [CrossRef]
15. Niu, B.; Chen, L.; Wu, X.; Ni, H.; Zhang, Q.; Wei, Y. Fault analysis of pre-insertion resistors for circuit breakers used in 750kV ACF. In Proceedings of the 16th IET International Conference on AC and DC Power Transmission (ACDC 2020), Online Conference, 2–3 July 2020; pp. 283–289.
16. Ma, F.; He, C.; Chen, X.; Zhu, X.; Deng, J.; Ni, H. Suppression Methods of 750 kV AC Filter Switching Inrush Surge for Circuit Breakers Having Pre-insertion Resistors. In Proceedings of the 2020 4th International Conference on HVDC (HVDC), Xi'an, China, 6–9 November 2020; pp. 1207–1212.
17. Bhatt, K.A.; Bhalja, B.R.; Parikh, U.B. Evaluation of controlled energisation of an unloaded power transformer for minimizing the level of inrush current and transient voltage distortion using PIR-CBs. *IET Gener. Transm. Distrib.* **2018**, *12*, 2788–2798. [CrossRef]
18. Wang, X.; Chen, Z.; Wu, Y.; Bao, Y.; Wu, X.; Qiu, Z.; Li, S.; Guo, L. Study on the Dynamic Resistance Characteristic of Circuit Breaker Closing Resistor Under Impulse Current. In Proceedings of the 2025 7th International Conference on Energy Systems and Electrical Power (ICESEP), Wuhan, China, 20–22 June 2025; pp. 1206–1211.
19. Ma, F.; Li, A.; Niu, B.; Ni, H.; Liu, B.; Deng, J. Extra/Ultra High Voltage AC Filter Circuit Breaker Pre-Insertion Resistors State Evaluation Based on Wavelet Decomposition and TLS-ESPRIT. In Proceedings of the 2024 4th International Conference on Energy Engineering and Power Systems (EEPS), Hangzhou, China, 9–11 August 2024; pp. 1016–1019.
20. Wu, D.; Zhou, Y.; Li, W.; Li, H.; Xu, Y.; Xia, H. Analysis on a fault of grading capacitor for 500 kV circuit breaker. *High Volt. Appar.* **2016**, *52*, 195–199.
21. Patcharoen, T.; Ngaopitakkul, A. Transient inrush current detection and classification in 230 kV shunt capacitor bank switching under various transient-mitigation methods based on discrete wavelet transform. *IET Gener. Transm. Distrib.* **2018**, *12*, 3718–3725. [CrossRef]
22. Meng, W.; Yin, F.; Chang, Y.; Tang, W.; Wei, J. Research of closing resistor in 800 kV SF6 tank circuit breaker. *High Volt. Appar.* **2020**, *56*, 109–113.

23. Han, B.; Chen, W.; Chen, X.; Ban, L.; Xiang, Z.; Ge, D. Study on Eliminating Closing Resistor of Circuit Breaker by Using Controllable Surge Arrester in UHV System. In Proceedings of the 2018 International Conference on Power System Technology (POWERCON), Guangzhou, China, 6–8 November 2018; pp. 2839–2844.
24. Sun, J.; Dai, H.; Chen, W.; Cui, B.; Yang, F.; Yao, X. Energy Tolerance and Failure Characteristics of Carbon-ceramic Closing Resistor for Circuit Breaker. *High Volt. Eng.* **2024**, *50*, 2590–2600.
25. Liu, K.; Wen, D.; Bao, Y.; Qiu, Z.; Li, S.; Ji, S. Stress distribution numerical simulation and influencing factor analysis of stress under operating shock of pre-insertion resistor of 750 kV dead tank circuit breaker. *High Volt. Appar.* **2023**, *59*, 180–185.
26. Duan, L.; Hao, X.; Chen, Q. Anti-sweep jamming method for FM fuze based on outlier reconstruction. *J. Beijing Univ. Aeronaut. Astronaut.* **2025**, *51*, 1377–1384.
27. Zheng, X.; Xia, C.; Wang, X.; Tong, W. Coupling error coefficient identification based on swept-frequency measurement in MEMS gyroscopes. *J. Chin. Inert. Technol.* **2024**, *32*, 170–179.
28. Ma, C.; Chen, P.; Yang, L.; Han, T.; Tang, Y.; Chen, C.; Ding, Z. A machine learning approach to correct imaging distortions in swept-source optical coherence tomography. *Acta Phys. Sin.* **2025**, *74*, 120–129. [CrossRef]

2. Wide-Band Modeling of Electromagnetic Current Transformer

Under the power frequency (50 Hz) condition, the equivalent circuit model of the electromagnetic current transformer mainly includes the leakage inductance, leakage resistance, and excitation impedance of the primary and secondary sides. The corresponding equivalent circuit is shown in Figure 1 [13], where R_1 and L_1 are the primary winding resistance and leakage inductance; R_2 and L_2 are the secondary winding resistance and leakage inductance; and R_m and L_m represent the core loss resistance and magnetizing inductance, respectively.

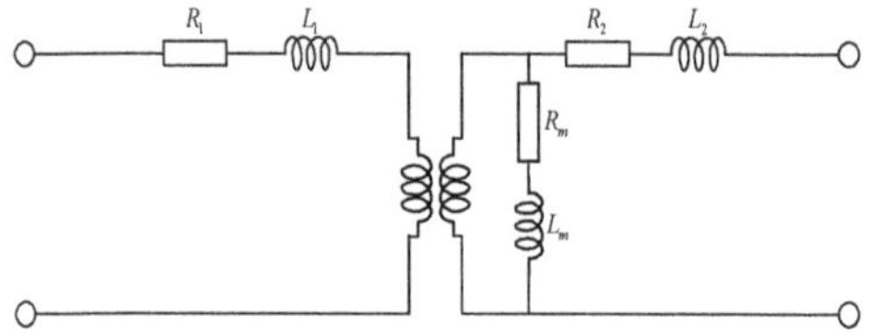

Figure 1. CT power frequency equivalent circuit.

In the power frequency operation state, the magnetic coupling effect of the electromagnetic coil inside the CT is dominant [14]. At this time, the stray capacitance between the windings and the stray capacitance to the ground are very small, and the influence on the operation of the equipment can be neglected. However, as the current frequency gradually increases, the influence of the ground capacitance of the primary and secondary windings of the CT and the stray capacitance between the windings on the measurement accuracy is more and more significant, which cannot be ignored. The primary current of the instrument's current transformer is at the ampere level, the secondary current is at the milliampere level, and the number of secondary side turns is more than 2000 turns. At this time, the distributed capacitance between the secondary coils is about tens to hundreds of micrometers [15]. In the low frequency band of 1~200 Hz, the distributed capacitance can be ignored, but in the middle frequency and high frequency band, its influence cannot be ignored, which will have a significant effect on the measurement. Considering that the number of turns of the primary side winding is relatively limited, its influence can be ignored. Based on the power frequency equivalent circuit, the inter-layer capacitance is added to characterize the inter-winding capacitance effect, and the high-frequency equivalent model shown in Figure 2 is constructed.

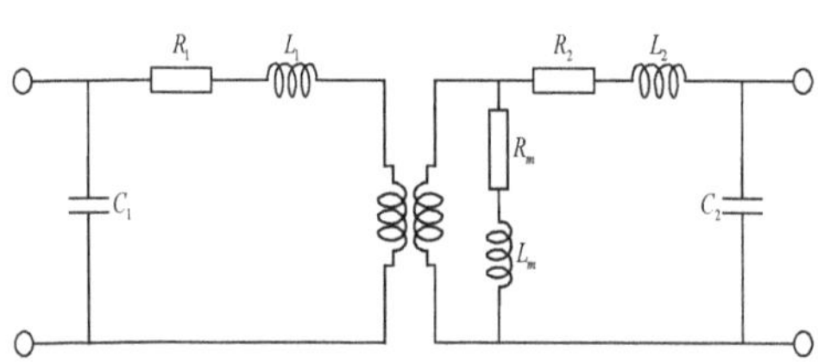

Figure 2. CT harmonic equivalent circuit.

Considering the small number of turns and thick wire diameter of the CT primary side coil, its distributed capacitance is ignored [16]. The simplified CT harmonic equivalent circuit is converted to the primary side, and the converted broadband equivalent circuit is shown in Figure 3.

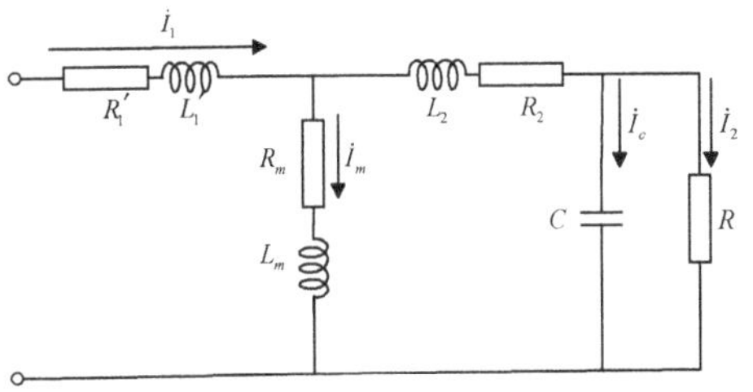

Figure 3. CT simplified harmonic equivalent circuit.

In Figure 3, L_1' and R_1' are the primary winding leakage inductance and DC resistance (converted to the secondary side), L_2 and R_2 are the secondary winding leakage inductance and DC resistance, R_m and L_m are the excitation resistance and magnetizing inductance of CT, C is the secondary side stray capacitance, and R is the equivalent load of the current transformer.

In the designed electromagnetic current transformer, L_1, L_2, R_1, and R_2 are all fixed constants. However, the equivalent resistance R_m, which is related to the core loss, and the magnetizing inductance L_m, which is affected by the excitation characteristics of the ferromagnetic material, are not constant but fluctuate with the change in the saturation degree of the core [17]. Therefore, the key to obtaining the component parameters of the broadband equivalent circuit of the current transformer is to obtain the broadband parameters of the excitation impedance.

3. Study on the Frequency Characteristics of Complex Permeability of Ferromagnetic Materials

When analyzing the broadband response characteristics of current transformers, researchers often focus on the influence of distributed capacitance but pay less attention to the dynamic characteristics of excitation impedance changing with frequency. In fact, in broadband high-precision measurement applications, the frequency dependence of the excitation impedance is a key factor that cannot be ignored. This nonlinear characteristic is essentially derived from the inherent properties of the excitation process of ferromagnetic materials. The analysis of the broadband characteristics of the current transformer needs to focus on two key parameters. One is that the permeability μ is a key parameter to measure the magnetic properties of ferromagnetic materials, which reflects the magnetization ability and magnetic flux conduction efficiency of the material under the action of the magnetic field. The second is the core loss angle θ, which refers to the phase delay of the magnetic induction intensity B relative to the magnetic field intensity H due to the hysteresis effect and eddy current effect of the ferromagnetic material during the alternating magnetization process. This angle reflects the energy loss characteristics of the core. Both H and B can be described by Formulas (1) and (2).

$$H = H_{\mathrm{m}} \sin \omega t = H_{\mathrm{m}} \mathrm{e}^{\mathrm{j}\omega t} \tag{1}$$

$$B = B_{\mathrm{m}} \sin(\omega t - \theta) = B_{\mathrm{m}} \mathrm{e}^{\mathrm{j}(\omega t - \theta)} \tag{2}$$

In the formula, ω is the angular frequency, H_m is the magnetic field intensity amplitude, and B_m is the magnetic flux density amplitude.

The complex permeability of ferromagnetic materials in a period can be expressed as follows:

$$\mu = \frac{B}{H} = \frac{B_{\mathrm{m}}}{H_{\mathrm{m}}} \mathrm{e}^{-\mathrm{j}\theta} = \mu' - \mathrm{j}\mu'' \tag{3}$$

In the formula, the real and imaginary parts of the complex permeability are, respectively,

$$\mu' = \frac{B_m \cos\theta}{H_m} \tag{4}$$

$$\mu'' = \frac{B_m \sin\theta}{H_m} \tag{5}$$

The tangent of the core loss angle is

$$\mathrm{tg}\theta = \frac{\mu''}{\mu'} \tag{6}$$

At present, the core of the electromagnetic transformer is a laminated structure, and its thickness is far less than the width and length, so only the one-dimensional flow of the current is considered, and the following equation can be obtained:

$$\frac{\partial^2 H_Z}{\partial x^2} = j\omega\mu_0\mu_r(\sigma + j\omega\varepsilon)H_z = k^2 H_Z \tag{7}$$

where ω is the angular frequency of the excitation signal μ_r; is the relative permeability; ε is the complex permittivity; σ is the core conductivity; H_Z is the component of the magnetic field intensity in the z direction; μ_0 is the vacuum permeability; and k is the propagation constant, and its calculation formula is

$$k = (1+j)\sqrt{\frac{\omega\sigma\mu_0\mu_r}{2}} \tag{8}$$

In the formula, σ is the core conductivity.

The calculation formula of the magnetic field intensity on the core surface is

$$H = \frac{k\Phi\cosh(k\frac{b}{2})}{2\mu_0\mu_r h\sinh(k\frac{b}{2})} \tag{9}$$

In the formula, b represents the thickness of the silicon steel sheet; H represents the magnetic field intensity on the surface of the iron core; h is the width of the lamination; and ∅ denotes the magnetic flux with a cross-sectional area of bh. Therefore, the spatial average magnetic flux density B in the z direction is related to the magnetic flux passing through the cross-sectional area, and the relationship is

$$\mu = \frac{\overline{B}}{H} = k_{Fe}\mu_r\frac{\tanh(\frac{kb}{2})}{\frac{kb}{2}} \tag{10}$$

In the formula, k_{Fe} is the core lamination coefficient, $k_{Fe} = b/(b+\Delta)$; Δ is the thickness of the insulating layer. The relative permeability μ_r is close to the initial permeability μ_i of the iron core, and μ_i is defined as the relative permeability at zero magnetic field strength, which is

$$\mu_r \approx \mu_i = \frac{1}{\mu_0}\lim_{H\to 0}\frac{B}{H} \tag{11}$$

There are

$$\mu = \frac{\overline{B}}{H} = k_{Fe}\mu_i\frac{\tanh(\frac{kb}{2})}{\frac{kb}{2}} \tag{12}$$

Based on the simulation model, the real part μ' imaginary part μ'', and its amplitude $|\mu|$ of the relative complex permeability of the core and the curve of the relationship between the loss angle of the core and the frequency change are shown in Figures 4–6, and the specific parameters used in the simulation model are shown in Table 1.

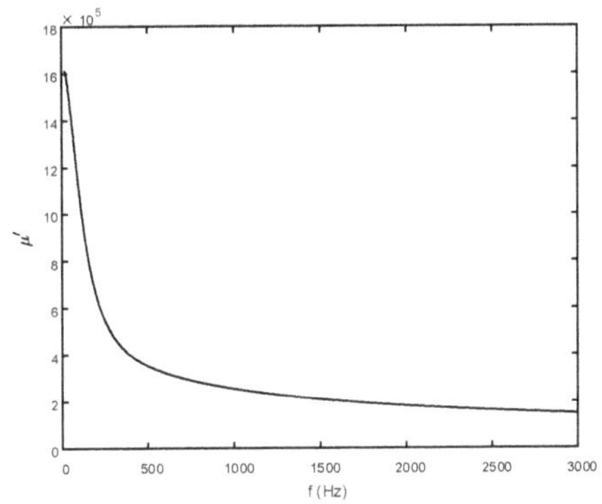

Figure 4. μ′-f frequency response curve.

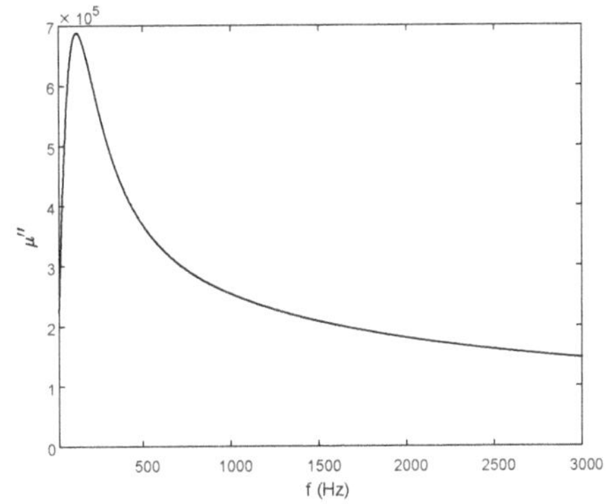

Figure 5. μ″-f frequency response curve.

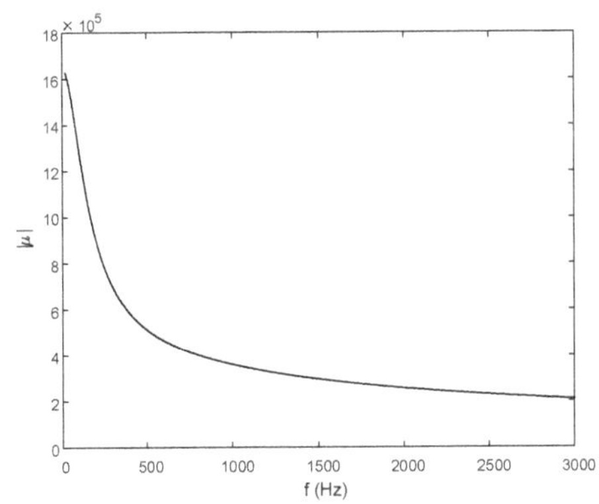

Figure 6. | μ |-f frequency response curve.

From the simulation results of Figures 4–6, it can be seen that the real part and amplitude of the relative permeability decrease with the increase in frequency. The imaginary part of the relative permeability increases with the increase in frequency in the range of 0~1000 Hz and decreases with the increase in frequency when it exceeds 1000 Hz.

Table 1. Core parameters.

Parameter Name and Symbol	Numerical Value
Core thickness b/m	0.0634
Insulation layer thickness Δ/mm	1.7
Core resistivity $\rho/(\Omega\cdot\text{m})$	1.72×10^{-7}
Conductivity of iron core $\sigma/(\text{S}\cdot\text{m}^{-1})$	5.6×10^{7}
The relative permeability of iron core $\mu_i/(\text{H}\cdot\text{m})$	0.03212
Magnetic flux path length L/M	0.78
Primary side winding thickness d_1/mm	0.62
Secondary winding thickness d_2/mm	12.58
Winding height h/mm	620
The average winding turn length p/mm	100
Vacuum permeability $\mu_0/(\text{H}\cdot\text{m}^{-1})$	1.26×10^{-6}

4. Frequency Characteristic Analysis of Current Transformer

4.1. Mathematical Model of the Frequency Characteristics of Current Transformers

Usually, the error of the current transformer mainly covers the ratio difference and the phase difference. The following will elaborate on the definition of these two types of errors.

The so-called ratio difference refers to the error caused by the difference between the actual current ratio and the rated current ratio in the measurement process of the current transformer. The so-called phase difference refers to the phase difference between the primary and secondary current phasors of the current transformer [18].

Ratio error GB1208-87: The definition of the ratio error is as follows [19]:

$$e\% = \frac{KI_s - I_p}{I_p} \times 100 \tag{13}$$

In the formula, I_s and I_p are the primary and secondary side currents, respectively, and the rated ratio $K = N_s/N_p$; N_p and N_s are the number of turns of the primary and secondary windings, respectively.

The definition of the phase error in GB1208-87 is as follows:

$$\varepsilon = \alpha_s - \alpha_p[\text{deg}] \tag{14}$$

In the formula, α_s is the secondary current phase angle, and α_p is the primary current phase angle.

According to the CT broadband equivalent circuit established above (as shown in Figure 3), the transfer function expression of CT can be derived:

$$G(s) = \frac{I_2(s)}{I_1(s)} = \frac{Z_m}{Z_m + Z_{2L}} \cdot \frac{1/sC}{1/sC + R} \tag{15}$$

In the formula, $Z_{2L} = ({}^{1}/{}_{sC}// R) + (R_2 + sL_2)$ is substituted into Equation (15) and simplified to obtain the transfer function of CT:

$$G(s) = \frac{R_m + sL_m}{(RCs + 1)[(L_m + L_2)s + (R_m + R_2)] + R} \tag{16}$$

When the CT material and winding mode are unchanged, the stray capacitance of the primary and secondary sides and the capacitance of the secondary side winding to the ground are fixed in the frequency range of the research current [19]. Therefore, as long as the excitation impedance parameters of the transformer are obtained, the amplitude–frequency characteristic curve can be obtained by bringing them into Equation (18). According to the amplitude–frequency curve, the ratio difference and angle difference in the transformer can be obtained.

In the calculation of transformer parameters, the value of excitation impedance is related to the real and imaginary parts of complex permeability [20], and the corresponding relationship is shown in Equations (17) and (18):

$$Lm = \frac{N^2\mu' S}{l} \tag{17}$$

$$Rm = \frac{\omega N^2\mu'' S}{l} \tag{18}$$

According to the simulation model, the frequency response curve of the excitation impedance is drawn as shown in Figures 7 and 8.

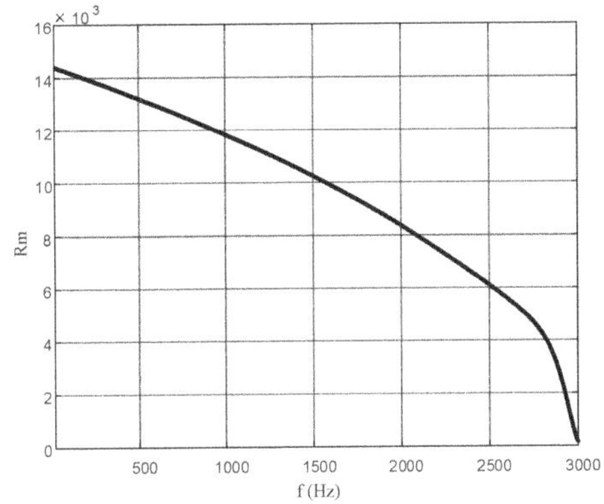

Figure 7. Rm-f frequency response curve.

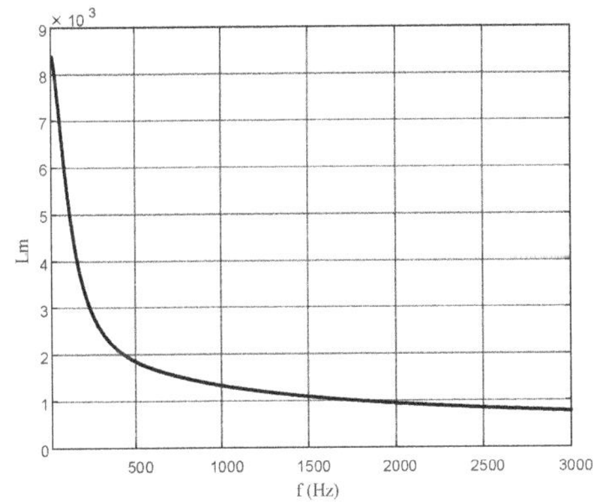

Figure 8. Lm-f frequency response curve.

As the frequency increases, the eddy current effect will increase, resulting in a decrease in the excitation inductance in the transformer coil. At the same time, at high frequencies, the imaginary part of the complex permeability decreases faster than the frequency increase rate, resulting in a decrease in the excitation resistance as the frequency increases. The equivalent circuit parameters of the current transformer are shown in Table 2.

Table 2. Parameters of current transformer equivalent circuit.

Parameter	Values
Transformer ratio	1:300
Secondary distributed capacitance C/pF	60
Secondary leakage inductance L_2/H	0.73
Secondary leakage resistance R_2/Ω	0.08
Accuracy level	0.2
Rated load	20

According to the corresponding calculation formulas of Equations R_m and L_m, the broadband response curve of the transformer at 50–3000 Hz can be obtained by introducing the transfer function Equation (16) derived from the broadband equivalent circuit, as shown in Model 1.

The accuracy level of the current transformer in this study is 0.2, which, according to the standard (GB 1208-2006), means its maximum permissible ratio error is ±0.2% at the rated current and burden. From the above analysis, it can be seen that when the frequency is in the range of 50–650 Hz, the ratio difference is controlled within −0.2%, which meets the accuracy level requirements. In the frequency band of more than 1000 Hz, the error increases significantly. When the frequency reaches 3000 Hz, the ratio difference has reached −0.61%, and the angle difference has reached −5.3°, which is much larger than the error level under the working condition. At this time, the ratio difference and angle difference cannot meet the accuracy requirements, indicating that the current transformer is difficult to accurately measure the high-frequency current.

4.2. Comparative Analysis of the Experiment

Based on the above theoretical analysis, in order to study the variation law of the current transformer ratio difference and phase angle under different frequency signals, the electromagnetic current transformer, which is the same as the simulation model, is selected as the test object, and the programmable power supply was used to apply a primary current signal, the amplitude and frequency of which could be accurately adjusted (current resolution: 0.1%, frequency resolution: 0.1 Hz). In order to obtain accurate reference datum, the standard inductive voltage divider (ratio error $<\pm 2 \times 10^{-6}$, calibration uncertainty: 5×10^{-6}, and k = 2) and the electromagnetic current transformer to be tested are connected in parallel to the test circuit. In the process of experimental data acquisition, a high-precision digital lock-in amplifier is used to record the output characteristics of the measured transformer under different working conditions in real time so as to calculate the ratio difference and angle difference in the CT to be tested (the specific test results are shown in Table 3). To quantitatively assess the reliability of the measurement results, we conducted five repeated measurements at each test frequency point. Based on the dispersion of the measured values, the experimental standard deviation was calculated using the Class A evaluation method. The lengths of the error bars in Model 1 correspond to the measurement standard deviations at each frequency point, reflecting the repeatability of the measurement results, the field experiment diagram is shown in Figure 9.

When analyzing the broadband characteristics of current transformers, the existing models often ignore the variation in excitation impedance with frequency [21]. However, for high-precision current transformers with broadband characteristics, the frequency characteristics of the excitation impedance need to be taken into account.

Table 3. Measured test data of ratio error and phase error at different frequencies.

Current Frequency/Hz	Ratio Difference/(%)	Angle Difference/(°)
50	−0.1785	−0.1457
100	−0.1786	−0.1693
150	−0.1744	−0.2085
200	−0.1684	−0.2556
250	−0.1639	−0.2948
300	−0.1577	−0.3576
350	−0.1487	−0.4361
400	−0.1438	−0.5145
450	−0.1475	−0.5694
500	−0.1568	−0.6479
550	−0.1683	−0.7028
600	−0.1775	−0.7891
700	−0.1879	−0.9225
800	−0.1909	−1.0716
900	−0.1954	−1.1814
1000	−0.2022	−1.2991
2000	−0.4107	−2.3584
3000	−0.6032	−3.5432

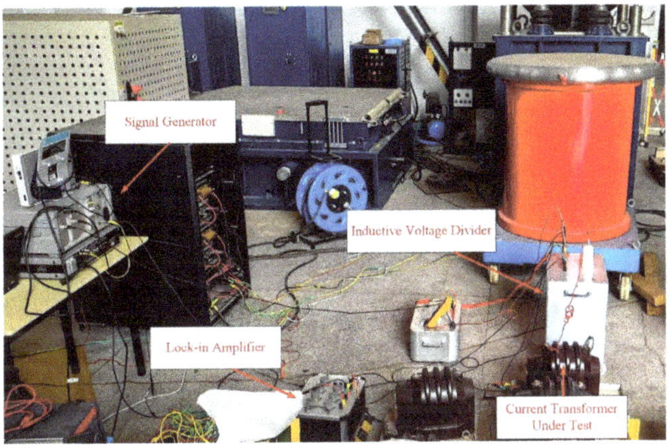

Figure 9. Field experiment diagram.

The derived error frequency curve (Model 1) and the existing model error curve (Model 2) are compared with the measured data, as shown in Figures 10 and 11. The results show that, in the low frequency range of 50–300 Hz, the trend of the three curves is basically consistent. In the low frequency range of 50–300 Hz, the maximum deviation of the ratio difference between Model 1 and the measured value is 0.03%, and the angle difference deviation is 0.3°, which is better than 0.06% and 0.5° of Model 2, which verifies the accuracy of the complex permeability modeling. When the frequency exceeds 300 Hz, the deviation between Model 2 and the measured value increases significantly, while the deviation between Model 1 and the measured data is still controlled within 0.05% because of the broadband characteristics of the core hysteresis loss. This shows that the existing model does not take into account the complex characteristics of permeability at high frequencies, resulting in the insufficient prediction accuracy of high-frequency error, while this model is closer to the actual operating conditions.

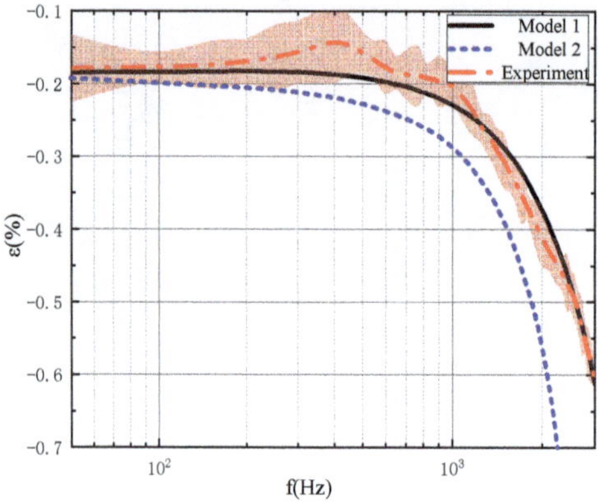

Figure 10. Comparison of ratio errors from simulations and tests. (The shaded band represents the measurement uncertainty range calculated based on the experimental standard deviation and an expansion factor of k = 2).

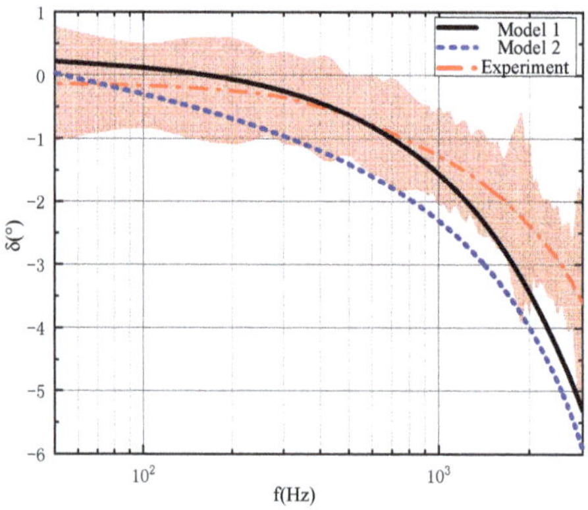

Figure 11. Comparison of phase errors from simulations and tests. (The shaded band represents the measurement uncertainty range calculated based on the experimental standard deviation and an expansion factor of k = 2).

5. Frequency Characteristic Improvement and Frequency Band Broadening

In actual operating conditions, some parameters change due to the influence of the current transformer manufacturing process, winding process, and material selection difference [22–25]. Through the transfer function expression of the current transformer, it can be seen that its frequency characteristics are closely related to the excitation impedance and the secondary side coil parameters. These influencing factors are reflected in the above relationship, and the measurement error of the circuit transformer can be compensated by adjusting the relevant parameters so as to optimize its performance.

The influence of changing the excitation impedance value on the frequency characteristics of the current transformer is shown in Figures 12 and 13. Through in-depth analysis of the experimental data of Figures 12 and 13, it is found that reducing the excitation impedance parameter value can significantly expand the amplitude–frequency characteristic bandwidth and effectively improve the amplitude–frequency response performance. However, this same adjustment yields only a marginal improvement in the phase–frequency characteristic. Consequently, while the amplitude error is greatly reduced across a broader band, the phase error remains largely unchanged and becomes the dominant factor limiting the overall measurement accuracy at high frequencies.

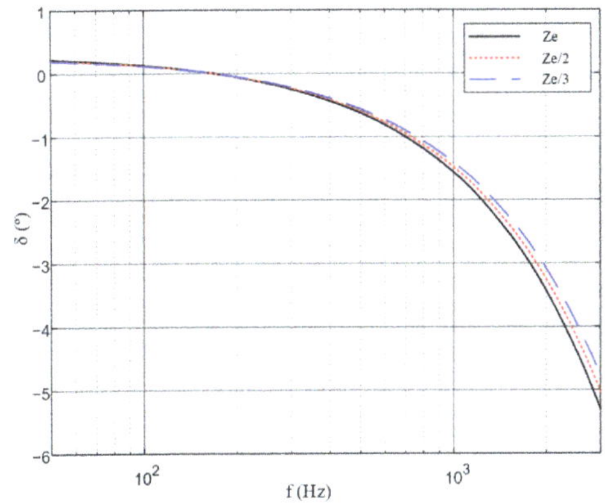

Figure 12. Influence of excitation impedance variation on phase error characteristics.

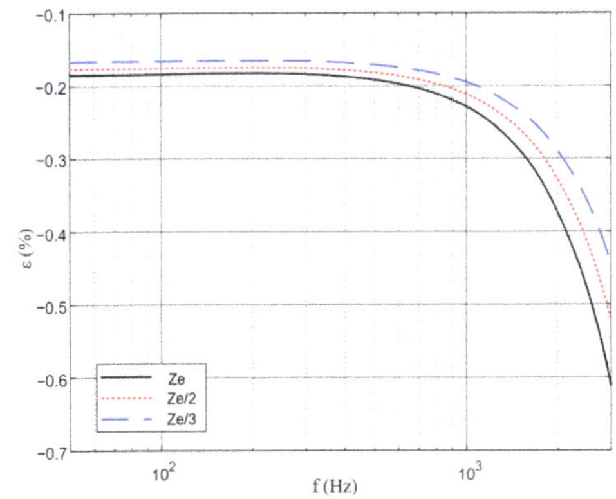

Figure 13. Influence of excitation impedance variation on ratio error characteristics.

Figures 14 and 15 show the influence of the secondary winding parameter Z_2 on the frequency characteristics and working frequency band of the current transformer. The analysis results show that reducing the excitation impedance value has a significant effect on expanding the frequency band of amplitude–frequency characteristics and optimizing the amplitude–frequency characteristics but has a slight effect on the phase–frequency characteristics.

Affected by differences in the CT manufacturing process, winding process, and material properties in actual production, some parameters may change [26]. Through the analysis of the simplified high-frequency equivalent circuit of CT, it is found that the capacitance C of the secondary winding to the ground is the key factor affecting the accuracy of the CT measurement. In order to explore the influence of stray capacitance change on the CT ratio difference and angle difference, the simulation calculation is carried out. The results are shown in Figures 16 and 17. The research shows that the influence of the secondary side winding on the ground capacitance change contrast difference and angle difference is consistent. With the increase in the capacitance parameter, the CT ratio difference and angle difference will increase accordingly. In addition, the higher the current frequency, the more significant the impact on the measurement error.

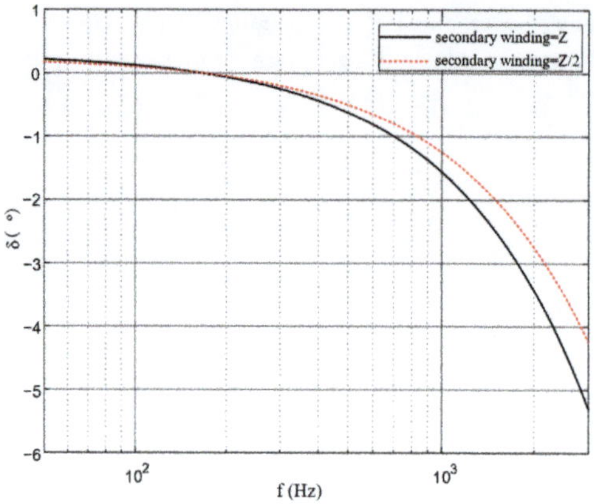

Figure 14. Influence of secondary impedance variation on phase error characteristics.

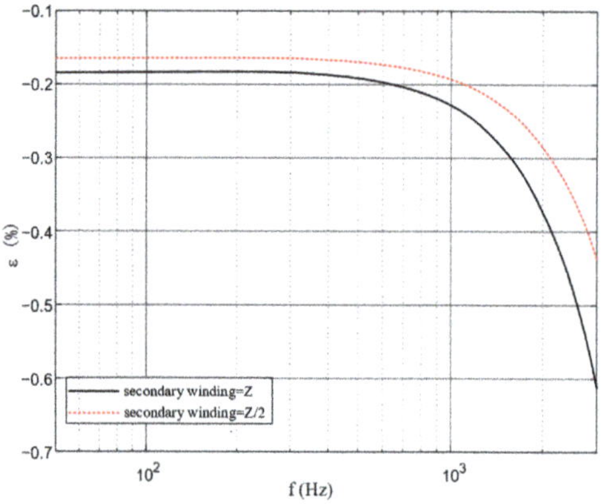

Figure 15. Influence of secondary impedance variation on ratio error characteristics.

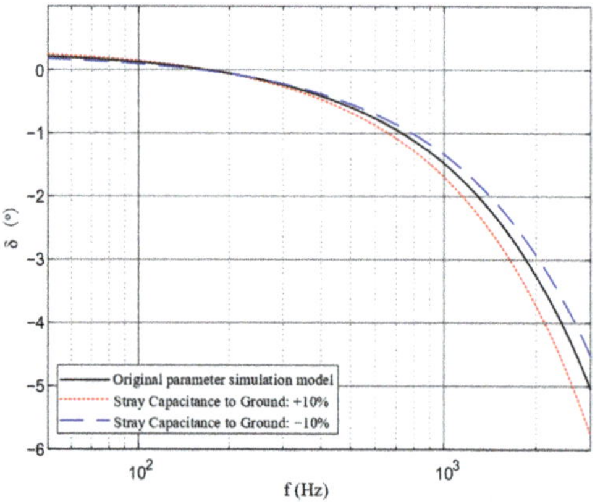

Figure 16. Influence of stray capacitance variation on phase error characteristics.

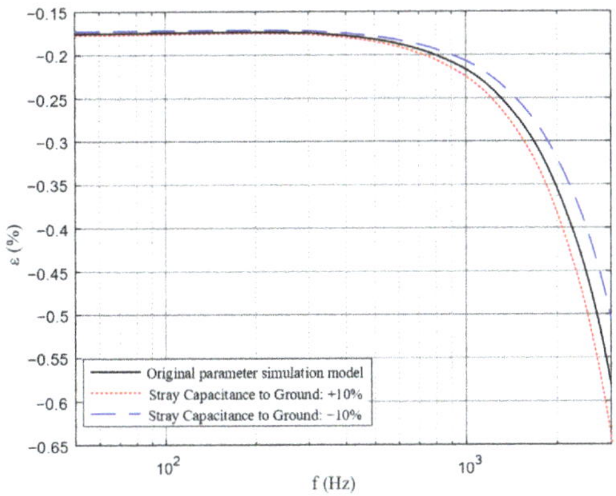

Figure 17. Influence of stray capacitance variation on ratio error characteristics.

Through the above analysis, we can improve the frequency response performance of the current transformer and extend its effective working frequency band from many aspects. At the material and structural level, high permeability materials are selected to reduce the excitation impedance. By optimizing the parameters of the secondary side coil, the coil impedance is reduced. At the same time, the appropriate winding process is adopted to reduce the ground capacitance of the secondary side winding. In engineering application practice, the accurate adjustment of parameters can be realized by means of optimized design scheme. For example, the use of iron-based nanocrystalline alloy as the core material has the advantage of high initial permeability and is the ideal core material; optimizing the parameters of the secondary side coil can improve the winding process and adjust the wire diameter [27–30]. In addition, it can also consider building a compensation network, using hardware compensation, and using algorithm compensation to further improve its frequency characteristic performance.

6. Conclusions

Based on an in-depth analysis, the broadband equivalent circuit of the electromagnetic current transformer was established and effectively simplified. Utilizing this simplified model, the mechanisms governing the generation of ratio and phase errors in CT were thoroughly investigated, leading to the development of an accurate simulation model for analyzing its wide-frequency behavior.

The core of the proposed model lies in the precise representation of the excitation impedance, which is intrinsically linked to the complex permeability of the core material. The frequency-dependent values of the excitation impedance were calculated analytically, and their response curves were obtained via simulation, forming the basis for determining the CT's error characteristics across the 50–3000 Hz frequency range.

Simulation results demonstrate that the CT maintains its specified 0.2 accuracy class within the 50–650 Hz range, with the ratio error effectively contained within −0.2%. However, beyond 1000 Hz, both the ratio and phase errors exhibit a pronounced increasing trend. At 3000 Hz, the ratio error deteriorates to −0.61% and the phase error reaches $-5.3'$, values that significantly exceed the permissible limits for accurate measurement, thereby highlighting the CT's inherent limitation in measuring high-frequency currents.

The validity and advantage of the proposed model were unequivocally confirmed through experimental measurements. A direct comparison between the simulated error trends and the measured data revealed close agreement, proving that the model more accurately reflects actual operating conditions, particularly for high-order harmonics where conventional CTs exhibit substantial errors.

Furthermore, the parameter sensitivity analysis conducted with the simulation model provided critical insights for performance improvement. It was found that, while reducing the excitation impedance effectively broadens the amplitude–frequency bandwidth, its effect on improving the phase–frequency characteristic is limited. Consequently, a comprehensive optimization strategy—simultaneously reducing the secondary winding impedance and stray capacitance—is identified as essential for mitigating the dominant phase error and achieving optimal broadband performance. This systematic analysis offers valuable guidance for the error compensation and structural design of future wide-band CT.

Author Contributions: Validation, C.Y.; Writing—original draft, X.W.; Writing—review & editing, Z.L. (Zhenhua Li), Z.L. (Zhenxing Li) and L.Z. All authors have read and agreed to the published version of the manuscript.

Funding: This research was funded by the National Natural Science Foundation of China (52311530337) and in part by the National Natural Science Foundation of China (52277012).

Data Availability Statement: The original contributions presented in this study are included in the article. Further inquiries can be directed to the corresponding author.

Conflicts of Interest: The authors declare no conflicts of interest.

References

1. Danquah, O.J.; Mandava, S. Machine learning-based current transformer saturation detection for medium voltage in power grids system. *Measurement* **2025**, *253*, 117636. [CrossRef]
2. Yang, Y.; Zhu, B.; She, X.; Wang, K.; Liu, A. A Family of Voltage Balancers with Fault-Tolerance and Interpolar Voltage Self-Balancing Ability for Bipolar DC Power Systems. *IEEE Trans. Transp. Electrif.* **2025**, *11*, 9864–9876. [CrossRef]
3. Mahmoud, A.R. Digital protection scheme based on Durbin Watson and Pearson similarity indices for current signals practically applied to power transformers. *Sci. Rep.* **2025**, *15*, 12214. [CrossRef] [PubMed]
4. Wang, A.; Lin, S.; Zhou, Q.; Wu, G.; Li, X.; Liu, J.; Cheng, H. Frequency analysis and calculation of transformer neutral DC caused by metro stray currents. *Int. J. Electr. Power Energy Syst.* **2025**, *166*, 110531. [CrossRef]
5. Rom, M.; Brom, D.V.E.H.; Houtzager, E.; van Leeuwen, R.; van der Born, D.; Rietveld, G.; Muñoz, F. Measurement System for Current Transformer Calibration from 50 Hz to 150 kHz Using a Wideband Power Analyzer. *Sensors* **2025**, *25*, 5429. [CrossRef]
6. Yao, L.; Xu, L.J.; Li, S. Research on traceability technology of broadband standard current transformer based on digital comparison method. *Electr. Meas. Instrum.* **2024**, *61*, 97–103. (In Chinese)
7. Kaczmarek, M.; Blus, K. Analytical Investigation of Primary Waveform Distortion Effect on Magnetic Flux Density in the Magnetic Core of Inductive Current Transformer and Its Transformation Accuracy. *Sensors* **2025**, *25*, 4837. [CrossRef]
8. Zhang, D.Q.; Yang, L.; Xiao, Y. Transfer characteristics of wide-range current transformer based on zero-flux theorem. *South. Power Syst. Technol.* **2024**, *18*, 150–158. (In Chinese)
9. Feizbakhsh, A.; Naimi, M.H.; Vakilian, M. An accurate method for detection and compensation of current transformer saturation using unscented Kalman filter. *Electr. Power Syst. Res.* **2025**, *249*, 112015. [CrossRef]
10. Lu, Y.Q.; Fan, S.T.; Li, X.Q. A new method for measuring frequency characteristics of industrial current transformers. *J. Beijing Inst. Petrochem. Technol.* **2009**, *17*, 45–48. (In Chinese)
11. Cataliotti, A.; Cara, D.D.; Emanuel, E.A.; Nuccio, S. A novel approach to current transformer characterization in the presence of harmonic distortion. *IEEE Trans. Instrum. Meas.* **2009**, *58*, 1446. [CrossRef]
12. *GB/T 17215.302-2013*; National Technical Committee 104 on Electricity Metering Equipment (SAC/TC 104). AC Electricity Metering Equipment—Particular Requirements—Part 2: Static Harmonic Meters for Active Energy (Classes 0, 2 S, 0, 5 S, 1 and 2). China Standards Press: Beijing, China, 2013.
13. Li, Z.H.; Cui, J.X.; Lu, H.P.; Zhou, F.; Diao, Y.; Li, Z. Prediction method for instrument transformer measurement error: Adaptive decomposition and hybrid deep learning models. *Measurement* **2025**, *253*, 117592. [CrossRef]
14. Yang, N.; Xiong, Z.; Ding, L.; Liu, Y.; Wu, L.; Liu, Z.; Shen, X.; Zhu, B.; Li, Z.; Huang, Y. A Game-Based Power System Planning Approach Considering Real Options and Coordination of All Types of Participants. *Energy* **2024**, *312*, 133400. [CrossRef]

15. Li, Z.H.; Cui, J.X.; Wei, Y.K.; Huang, Y.-H.; Xie, H.-C.; Yao, W.-F. Calculation study of ion flow field in ±800 kV transmission lines under the effects of air pressure, temperature, humidity and improvement measures. *Electr. Power Syst. Res.* **2024**, *234*, 110574. [CrossRef]
16. Yang, N.; Shen, X.; Liang, P.; Ding, L.; Yan, J.; Xing, C.; Wang, C.; Zhang, L. Spatial-temporal Optimal Pricing for Charging Stations: A Model-Driven Approach Based on Group Price Response Behavior of EVs. *IEEE Trans. Transp. Electrif.* **2024**, *10*, 8869–8880. [CrossRef]
17. Yu, G.; Liu, C.; Tang, B.; Chen, R.; Lu, L.; Cui, C.; Hu, Y.; Shen, L.; Muyeen, S. Short term wind power prediction for regional wind farms based on spatial-temporal characteristic distribution. *Renew. Energy* **2022**, *199*, 599–612. [CrossRef]
18. Li, Z.H.; Cui, J.X.; Lu, H.P.; Zhou, F.; Diao, Y.-L.; Li, Z.-X. Prediction model of measurement errors in current transformers based on deep learning. *Rev. Sci. Instrum.* **2024**, *95*, 044704. [CrossRef]
19. Zhao, X.Z.; Wu, H.D.; Yang, D.Q. Research on eliminating residual magnetism in large power transformers using energy storage oscillation. *IEEE Trans. Power Deliv.* **2025**, 1–10, *early Access*.
20. Zhao, X.Z.; Wu, G.; Yang, D.Q.; Xu, G.; Xing, Y.; Yao, C.; Abu-Siada, A. Enhanced detection of power transformer winding faults through 3D FRA signatures and image processing techniques. *Electr. Power Syst. Res.* **2025**, *242*, 111433. [CrossRef]
21. Lou, W.; Lu, B.; Pan, Y.; Han, Z.; Wang, L. Numerical Analysis of Electric Field in Oil-Immersed Current Transformer with Metallic Particles Inside Main Insulation. *Energies* **2025**, *18*, 3628. [CrossRef]
22. Kaczmarek, M. Operating Properties of the Inductive Current Transformer and Evaluation of Requirements for Its Compliance with the IEC 61869-1 WB2 Class Extension for Frequency up to 20 kHz. *Energies* **2025**, *18*, 2595. [CrossRef]
23. Odinaev, I.; Aty, A.H.A.; Pazderin, A.; Safaraliev, M.; Matrenin, P.; Senyuk, M.; Menaem, A.A.; Kanan, M. Restoration of the current transformer secondary current under core saturation conditions based on ANN. *Heliyon* **2024**, *10*, e37960. [CrossRef] [PubMed]
24. Zhang, J.; Yin, X.; Wang, Q.; Peng, C.; Chen, H.; Lu, B.; Zhao, Q.; Tong, T.; Ding, C. Measurement equipment and result analysis of B-H curve of current transformer core material under mixing frequency. *Rev. Sci. Instrum.* **2024**, *95*, 084710. [CrossRef] [PubMed]
25. Liu, G.; Xiong, X.F.; Liao, R.J. Effect and analysis of leakage current on error characteristics of current transformer. *Trans. China Electrotech. Soc.* **2018**, *33*, 697–704.
26. Qi, X.; Wan, B.; Zhang, J.; Wang, Q.; Chen, H.; Zhao, Q.; Wang, B. Active on-site calibration technology of current transformer based on non-rated frequency current injection. *Rev. Sci. Instrum.* **2024**, *95*, 074705. [CrossRef]
27. You, H.; Gan, M.; Duan, D.; Zhao, C.; Lv, F.; Yuan, J. Approach of obtaining the error of current transformers through the measurement of turns correction factor and impedance. *Sens. Actuators A Phys.* **2024**, *376*, 115566. [CrossRef]
28. Li, Y.M.; Jiang, J.P.; Jian, Y.W. Research on the influence of harmonics on the measurement of electromagnetic transformers. *Electr. Technol.* **2021**, *22*, 36–41. (In Chinese)
29. Wang, S.Q.; Jiang, C.Y.; Nie, G.N. Research on wideband error calibration system of current transformer. *High. Volt. Eng.* **2024**, *50*, 5281–5288. (In Chinese)
30. Zhang, Z.; Chen, B.; Tian, C.; Chen, Y.; Wang, Y. A novel measurement-protection-integrated current transformer based on hybrid core and magnetic field sensor. *IET Electr. Power Appl.* **2024**, *18*, 739–755. [CrossRef]

energies

Article

A New Method for PMU Deployment Based on the Preprocessed Integer Programming Algorithm

Hanyuan Dan [1,*], Zhenhua Li [1,2] and Hongda Dou [1]

[1] School of Electrical Engineering and New Energy, China Three Gorges University, Yichang 443002, China
[2] Provincial Engineering Research Center of Intelligent Energy Technology, China Three Gorges University, Yichang 443002, China
* Correspondence: danhy00@163.com

Abstract: To further enhance the deployment efficiency of synchronous phasor measurement units and rationally select deployment configuration schemes, an improved configuration method based on integer programming algorithms is proposed. Based on the existing deployment method of integer programming algorithm, on the one hand, the special conditions of end nodes and zero injection nodes are taken into consideration. By analyzing the corresponding node model matrix, special nodes are given priority processing, achieving the condition simplification of the algorithm model. On the other hand, the evaluation indicators for the construction of schemes with the same minimum deployment quantity in the solution set obtained from the iterative solution of the algorithm are further analyzed and compared, so as to screen out a more reasonable deployment method. After conducting simulation tests on the IEEE-14, IEEE-30, and NE-39 power node systems using the MATLAB platform, the depth-first search algorithm and the improved simulated annealing algorithm were compared with the improved method. Eventually, this method had fewer deployments in a similar number of deployments and less deployment time in a similar number of deployments. The results verified the superiority of this method in terms of time and deployment quantity for PMU deployment problems.

Keywords: phasor measurement unit; deployment optimization; integer programming; electric power system; constraint conditions

1. Introduction

In recent years, the development of smart grid construction has been constantly advancing. Therefore, achieving effective and efficient full-network status monitoring to ensure the safe and stable operation of the entire system is particularly crucial. The stability of grid operation is related to detection equipment such as PMUs and transformers [1–3]. Nowadays, there are many different detection schemes for equipment in power systems from the past, such as the use of image recognition mentioned in [4,5] to detect the status of transformers. In terms of the measurement accuracy of the equipment, references [6,7] further ensure the accuracy of sensor measurement by establishing a more precise model.

In power systems, the deployment methods of measuring equipment have gradually become a hot topic of research. This article mainly focuses on the deployment research of PMUs [8–10]. The essence of this research is actually a kind of objective optimization problem, which achieves the optimization of device deployment by combining optimization algorithms [11–13]. The descriptions in references [14–16] reveal the impact of PMU deployment on the state estimation of power systems.

PMU is a high-precision measuring device in power grids, which operates relying on satellite timing signals [17–19]. As a component of the WAMS system, it plays a crucial role in the overall operation, monitoring and control of the distribution network, and is a key link in the current smart grid construction [20,21]. The PMU can detect in real time the voltage phasor, current phasor, frequency, active power, reactive power, and other data in the power network, and promptly transmit the data to the dispatching center [22–24]. The existing research on the optimal deployment of PMU mainly focuses on two aspects: the efficiency of the algorithm and its related constraints [25–27]. Due to the real-time nature of PMU and the synchronization of all network devices, it is bound to gradually replace the role of SCADA systems in state cognition and become an important support for the development of smart grids. Therefore, how to further achieve the efficient and reasonable deployment of PMUs to realize the full visibility of the power grid is of great significance.

Typically, the deployment configuration schemes of synchronous vector measurement units may vary due to different considerations, such as climatic factors and power consumption factors [28,29]. If all bus nodes in the entire network are equipped with PMUs, the entire system will be directly and completely observable, and the power flow equation of the entire system does not require any iterative calculation and solution. However, due to the high cost of PMU equipment, it is necessary to implement how to rationally deploy PMUs so as to not only reduce the number of PMUs deployed but also ensure the observability of the entire system. Taking into account the changes in the global economy and the demand situation in power production, current research in this area is a point worthy of attention, and it is also a research topic with certain production significance [30–32].

Research on the optimal configuration of synchronous phasor measurement units has become a hot topic in the field of power system equipment measurement, and certain research results have been achieved. Reference [33] employs an improved adaptive genetic algorithm, which can calculate crossover probability and mutation probability. However, when the maximum fitness value of the population is equal to the average fitness value, it may fall into local optimum. Reference [34] employed the simulated annealing algorithm. The actual effect of this method is related to the iterative conditions of the algorithm, and the result may not be the optimal situation. Reference [35] adopted an algorithm based on binary particle swarm to achieve deployment optimization, but the algorithm is generally prone to getting stuck in local optimal solutions. Heuristic algorithms have a certain degree of randomness when solving problems, and they may not have a good effect on the final optimization result. Moreover, since their solution effect is related to the initial algorithm setting, the obtained solution may not be the global optimal solution. The integer programming algorithm is a number theory method. It has better adaptability in solving problems under complex conditions and can also achieve good results for complex constraints.

Reference [36] uses the integer programming algorithm to solve the PMU deployment problem, but the setting of its constraint conditions is rather complex. Reference [37] employed an improved integer programming method to address the configuration issue of PMUs, optimizing the algorithm by simplifying the constraint conditions. However, it failed to take into account the special nodes of the power system, leaving room for further improvement in practical application fields. Reference [38] took zero-injection nodes into consideration, but its method has deficiencies in the setting of constraint conditions for zero-injection nodes in terms of condition setting. In the problem of PMU optimization configuration, most of the literature does not take into account the situation of minimizing the number of analog channels.

In this paper, aiming at the complete observability of the power system state and the minimization of the number of configured PMUs, an improved PMU configuration method is proposed based on the existing integer programming algorithm. The optimal deployment of PMUs is achieved by setting preprocessing conditions and evaluation indicators. On the one hand, the special conditions of the end nodes and the zero injection nodes are taken into consideration. By analyzing the corresponding node model matrix and preprocessing the special nodes, the conditions of the algorithm model are simplified and the time complexity of the algorithm is reduced. On the other hand, a multi-scheme solution is achieved by improving the solution method, and evaluation indicators are constructed for the solution set. Through further analysis and comparison, a more reasonable deployment method is selected. This study conducted simulation calculations on the IEEE-14, IEEE-30, and NE-39 node systems, compared the algorithm proposed in this paper with different algorithms, and verified the effectiveness and rationality of the method proposed in this paper.

2. Materials and Methods

2.1. Power System and Its Observability

2.1.1. Observability Requirements

The prerequisite for the deployment of PMU in the power grid is to meet the observability of the system. For each node, it has state quantities such as amplitude and phase Angle. If the state of each node in a power system can be obtained through direct or indirect observation, then the power system is overall observable [39]. The observability conditions of power systems are usually described from two aspects: "algebraic observability" and "topological observability" [40].

(1) The algebra is considerable

From an algebraic perspective, the observability of a power system with n nodes and m measurements can be expressed by Equation (1).

$$z = Hx + v, \tag{1}$$

Among them, z is the m-dimensional measurement vector; H is the Jacobian matrix of $m \times (2n - 1)$ dimensional measurements; x is a $2n - 1$ dimensional voltage state vector; v is the m-dimensional measurement error vector, and the elements in the vector conform to the normal distribution of $N(0, {\sigma_i}^2)$.

When the Jacobian matrix H is measured to be of full Rank and in a good state, that is, it satisfies the condition $Rank(H) = 2n - 1$, then it is determined that the system satisfies algebraic observability.

(2) The topology is considerable

From the perspective of graph theory, a power system can be regarded as a graph composed of n vertices and m edges, $G = (V, E)$, where V represents the set of vertices corresponding to the busbars within the system and E represents the set of edges corresponding to the branches within the system. The deployment of the measurement network constitutes a measurement subgraph $G' = (E', V')$, which corresponds to the conditions of $V' \in V$ and $E' \in E$. When the subgraph G' contains all the points of graph G, it is determined that the system satisfies topological observability.

2.1.2. Configuration Rules Under Observable Conditions

In the power system, there exist zero-injection nodes, which refer to nodes without generators or load connection. The net injection power of these nodes is zero, and they are usually used as transmission nodes, thus having certain particularities.

For a zero-injection node, it and its adjacent nodes can jointly form an N-node set. If N − 1 nodes are observable, then all nodes can be observable through calculation, as shown in Figure 1.

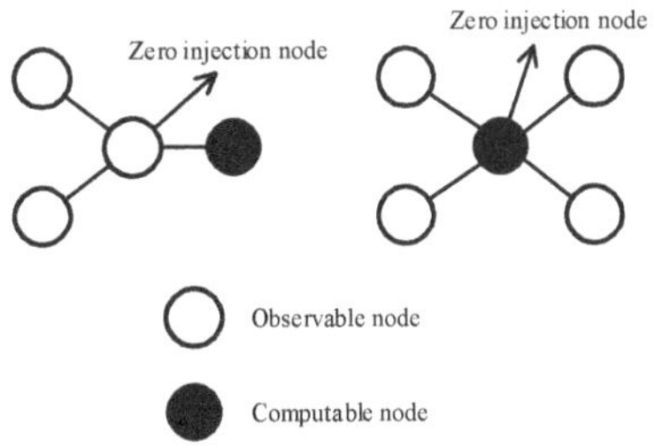

Figure 1. Zero injection node sizable situation conditions.

According to the observability requirements of nodes, the following rules can be obtained:

(1) When node i is configured with a PMU, the node itself can directly measure and achieve observability.
(2) When node i is configured with a PMU, the state quantities of its adjacent nodes can be calculated, and the implementation of its adjacent nodes is considerable.
(3) When node i is injected into a node with zero, in the set of N nodes formed by itself and adjacent nodes, as long as N − 1 nodes achieve observability, the remaining node can be calculated, and all N nodes achieve observability.

2.1.3. Average Channel Index

The analog channels of PMU can be divided into voltage analog channels and current analog channels. Typically, the voltage analog channel is used to measure the voltage phasor at the installation point, while the current analog channel is used to measure the current phasor at the adjacent branch of the installation point. For each PMU, if the number of its connection channels can be kept as few as possible, the entire system will have better stability. Therefore, for the same node system, even if the same number of PMUs are used in different deployment schemes, the deployment locations of different schemes are also worth exploring. In this paper, the final optimal deployment method is selected by constructing the average channel index of PMU, that is: the average number of connected channels = the total number of adjacent channels connected to PMU/the number of PMU installations.

2.2. Improve the Integer Programming Deployment Method After Preprocessing

Compared with heuristic algorithms, integer programming algorithms are deterministic algorithms, and the results they obtain are deterministic optimal solutions. Under normal circumstances, mathematical models are simple and intuitive. To meet actual needs, only the objective function and constraints need to be modified. The traditional deployment method that combines planning algorithms generally takes the observability premise and associated parameters of all nodes within the system as the objective function and constraint conditions. For the target of the power system with the fewest observable PMU devices installed throughout the network and the fewest average number of connected channels, the basic form of the target equation is shown in Equation (2). Its fundamental main objective is to minimize the funds required for the deployment plan as much as possible.

$$\begin{cases} \min \sum\limits_{j=1}^{n} x_j \\ \\ \min \dfrac{P_w}{P_n} \end{cases}, \tag{2}$$

In this formula: n represents the total number of nodes in the system, $x_j = 1$ indicates that a PMU is installed at node j, $x_j = 0$ indicates that a PMU is not installed, P_w represents the total number of connection channels of the PMU, and P_n represents the number of PMUs installed.

The constraint equation that satisfies the observable condition of non-zero injection nodes is shown in Equation (3).

$$F(i) = \sum_{j \in i} a_{ij} x_j \geq 1, \tag{3}$$

In this formula, when $F(i) \geq 1$, it indicates that the non-zero injection node i can be observed; a_{ij} is the corresponding element in the node adjacency matrix A. The node adjacency matrix A represents the connection status among various nodes in the power system, and matrix A is an n × n matrix. When node i is connected to node j, $a_{ij} = 1$; when node i is not connected to node j, $a_{ij} = 0$. That is shown in Equation (4) below.

$$a_{ij} = \begin{cases} 1 & i = j \\ 1 & i \text{ is connected to } j \\ 0 & i \text{ is not connected to } j \end{cases}, \tag{4}$$

The representation of the adjacency matrix A is shown in Equation (5) below.

$$A = \begin{bmatrix} A_1 \\ A_2 \\ A_3 \\ \vdots \\ A_n \end{bmatrix} = \begin{bmatrix} a_{11} & a_{12} & a_{13} & \cdots & a_{1n} \\ a_{21} & a_{22} & a_{23} & \cdots & a_{2n} \\ a_{11} & a_{11} & a_{11} & \cdots & a_{3n} \\ \vdots & \vdots & \vdots & \vdots & \vdots \\ a_{n1} & a_{n2} & a_{n3} & \cdots & a_{nn} \end{bmatrix}, \tag{5}$$

Considering the characteristics of the end node and the zero injection node, certain preprocessing is carried out on the solution model in this paper. From the perspective of the cost of PMU installation, the end node is usually not regarded as an installation node in the power system. When the adjacent nodes meet certain observability conditions, the state parameters of the end node can be obtained through calculation. Therefore, in this paper, it is considered to first identify the end nodes, and then, respectively, identify the situations where the end nodes are connected to regular nodes or zero-injection nodes. For the case of connecting to regular nodes, pre-deployment is set. For the case of connecting to zero-injection nodes, deployment conditions are set in combination with the characteristics of zero-injection nodes, thereby reducing the overall number of constraint conditions.

In terms of algorithmic solution, for the solution set of the deployed schemes, calculate the average number of connection channels for each scheme, and ultimately select the scheme with the smallest average number of channel connections. The overall method process is shown in Figure 2.

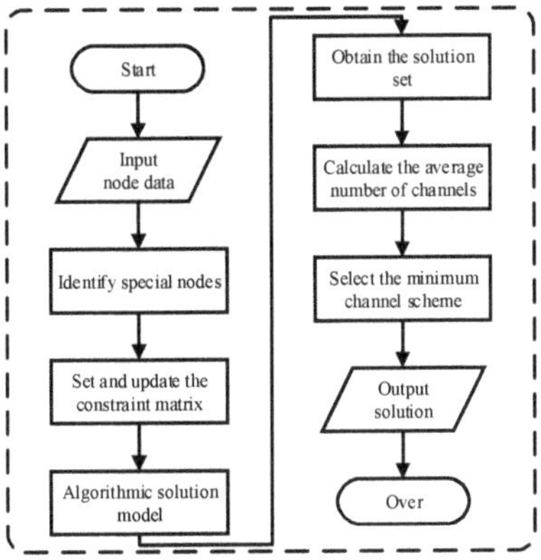

Figure 2. Deployment flowchart.

3. Simulation and Results

The simulation in the text runs in the MATLAB R2023b environment. The IEEE-14 node, IEEE-30 node and NE-39 node were, respectively, compared and simulated for verification.

3.1. IEEE-14

IEEE-14 node system as shown in Figure 3.

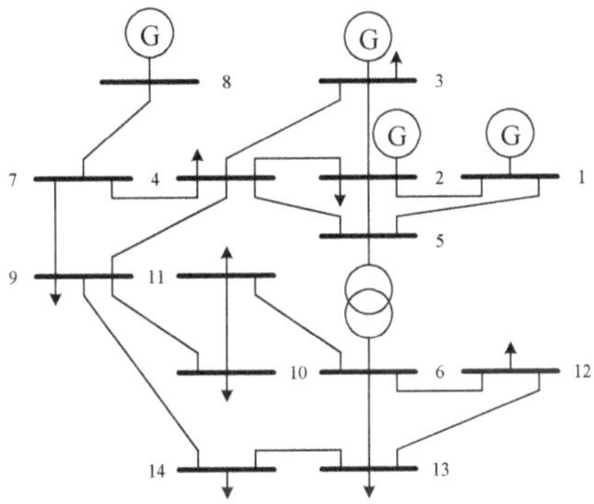

Figure 3. IEEE-14 node model.

According to node data, the zero-injection node in the IEEE-14 node system is node 7, and the terminal node within the system is node 8. During the preprocessing stage, according to the deployment rules, when all the adjacent nodes of node 7 except node 8 are observable, nodes 7 and 8 meet the observability condition. Therefore, the node association parameter matrix is set to zero for node 8 at this time, and the association matrix is updated. For other nodes, generate constraint conditions based on the node association matrix:

$$\begin{cases} F(1) = x_1 + x_2 + x_5 \geq 1 \\ \vdots \\ F(7) = x_4 + x_7 + x_9 \geq 1 \\ F(9) = x_4 + x_7 + x_9 + x_{10} + x_{14} \geq 1 \\ \vdots \\ F(14) = x_9 + x_{13} + x_{14} \geq 1 \end{cases}, \tag{6}$$

After solving, the preferred deployment nodes are 2, 6, and 9 nodes. Table 1 shows the comparison of the deployment results of the algorithm in this paper with those of other algorithms.

Table 1. Comparison of IEEE-14 node deployment algorithms.

Algorithm	PMU Deployment Quantity	PMU Deployment Node Location	Average Algorithm Time
Improved algorithm in the text	3	2\6\9	0.02 s
Depth-first search algorithm	6	2\4\6\8\10\14	0.004 s
Improve the simulated annealing algorithm	4	2\6\7\9	0.03 s

It can be seen from the results in Table 1 that under the IEEE-14 node system conditions, the number of PMUs required for deployment in this paper is the least and the average completion time of the algorithm is relatively short.

3.2. IEEE-30

IEEE-30 node system as shown in Figure 4.

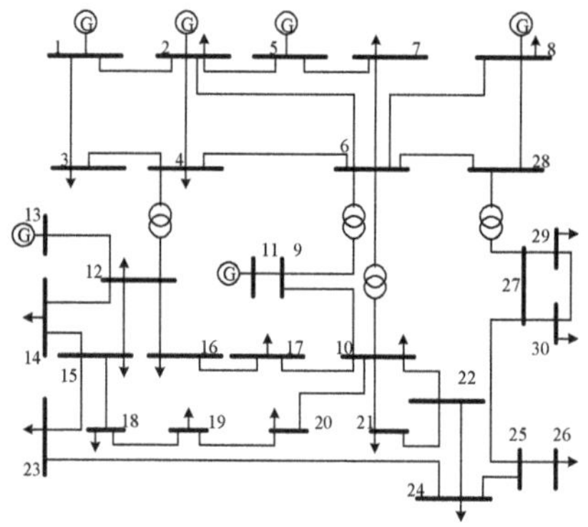

Figure 4. IEEE-30 node model.

According to the node data, the zero injection nodes in the IEEE-30 node system are nodes 6, 9, 22, 25, 27, and 28, and the terminal nodes existing in the system are nodes 11, 13, and 26. In the preprocessing stage, according to the deployment rules, for the classification processing of end nodes, the nodes connected to the zero injection node are node 11 and node 26, and the corresponding connected end nodes are node 9 and node 25. Therefore, node 12 is a node that must be deployed, and thus the constraint conditions for nodes 4, 12, 13, 14, 15, and 16 can all be omitted. At this point, the node association parameter matrix sets nodes 11 and 26 to zero and updates the association matrix.

For the remaining other nodes, generate corresponding constraint conditions based on the node association matrix:

$$\begin{cases} x_{12} = 1 \\ F(1) = x_1 + x_2 + x_3 \geq 1 \\ F(2) = x_1 + x_2 + x_4 + x_5 + x_6 \geq 1 \\ F(3) = x_1 + x_3 + x_4 \geq 1 \\ F(5) = x_2 + x_5 + x_7 \geq 1 \\ \vdots \\ F(30) = x_{27} + x_{29} + x_{30} \geq 1 \end{cases}, \tag{7}$$

After solving, it is found that under the condition of the same average number of PMU channels, the preferred deployment nodes are 3, 7, 10, 12, 19, 24, and 29 nodes, and the average number of channels is 2.571.

In Table 2, the partial solution sets obtained from the solution and the corresponding average number of channels are listed. Table 3 shows the comparison of the deployment of the method proposed in this paper with other algorithms.

Table 2. Comparison of different scenarios for IEEE-30 node system.

Plan	PMU Deployment Location (First Line)/ Number of PMU Connection Channels (Second Line)	Average PMU Channel
1	3\7\10\12\19\24\29 2\2\5\4\1\2\2	2.571
2	1\2\10\12\19\24\29 2\4\5\4\1\2\2	2.857
3	2\4\10\12\19\24\30 4\3\5\4\1\2\2	3.000

Table 3. Comparison of IEEE-30 node deployment algorithms.

Algorithm	PMU Deployment Quantity	PMU Deployment Node Location	Average Algorithm Time
Improved algorithm in the text	7	3\7\10\12\ 19\24\29	0.02 s
Depth-first search algorithm	10	1\5\6\10\11\12\ 18\24\26\27	0.004 s
Improve the simulated annealing algorithm	7	1\5\10\12\ 18\24\27	0.1 s

As can be seen from Table 2, even for different schemes with the same number of PMU deployments, the average number of channels varies. From Table 3, it can be observed that the improved method proposed in this paper has different advantages and disadvantages over other algorithms in terms of deployment quantity and time. Under the same number of deployments, the average completion time of the algorithm is shorter.

3.3. NE-39 Nodes

NE-39 node system as shown in Figure 5.

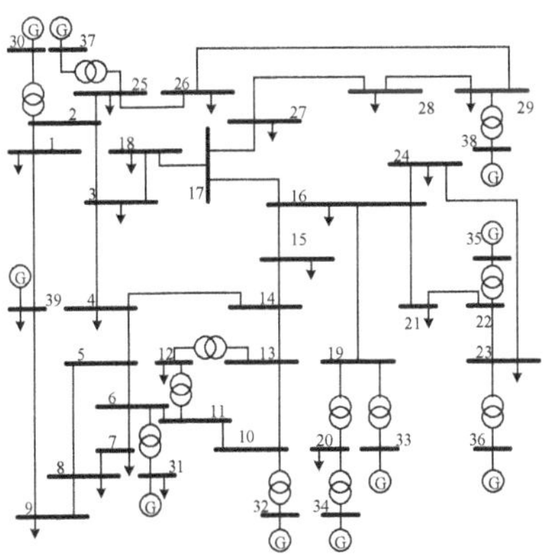

Figure 5. NE-39 node model.

According to the node data, the zero injection nodes in the NE-39 node system are nodes 2, 5, 6, 10, 11, 13, 14, 17, 19, and 22, and there are end nodes 30, 31, 32, 33, 34, 35, 36, 37, and 38 within the system. During the preprocessing stage, according to the deployment rules, the end nodes are classified. The end nodes connected to the zero injection nodes are 30, 31, 32, 33, and 35, and the corresponding zero injection nodes connected are 2, 6, 10, 19, and 22. Therefore, nodes 20, 23, 25, and 29 are the nodes that must be deployed. Therefore, the constraints on itself and the adjacent nodes of the corresponding connection can be omitted. At this time, the node association parameter matrix is set to zero for nodes 30, 31, 32, 33, and 35, and the association matrix is updated. For the remaining other nodes, generate constraint conditions based on the node association matrix:

$$\begin{cases} x_{20} = 1 \\ x_{23} = 1 \\ x_{25} = 1 \\ x_{29} = 1 \\ \vdots \\ F(18) = x_3 + x_{17} + x_{18} \geq 1 \\ F(21) = x_{16} + x_{21} + x_{22} \geq 1 \\ \vdots \\ F(29) = x_{26} + x_{28} + x_{29} + x_{38} \geq 1, \end{cases} \quad (8)$$

After solving, the obtained deployment nodes are 3, 6, 13, 16, 20, 23, 25, 29, and 39 nodes. Table 4 shows the comparison of the deployment results of the algorithm in this paper with those of other algorithms.

Table 4. Comparison of NE-39 node deployment algorithms.

Algorithm	PMU Deployment Quantity	PMU Deployment Node Location	Average Algorithm Time
Improved algorithm in the text	9	3\6\13\16\20\23\25\29\39	0.03 s
Depth-first search algorithm	16	2\4\6\8\10\12\16\18\20\22\26\33\36\37\38\39	0.004 s
Improve the simulated annealing algorithm	12	2\8\10\11\16\22\26\34\36\37\38\39	0.2 s

It can be seen from the results in Table 4 that under the NE-39 node system condition, the number of PMUs required for deployment in this paper is the least and the average completion time of the algorithm is shorter.

4. Discussion

This study shows that by preprocessing the deployment conditions, the efficiency of the deployment algorithm can be improved to a certain extent. Meanwhile, adjusting the algorithm's solution mode can make the deployment plan more rational. This research represents a further alteration to the traditional algorithm deployment problem, making the issues to be addressed more organized. These results are of certain help for the study of the PMU deployment field, providing a new approach to solving the PMU deployment optimization problem, rather than being limited to simply replacing different algorithms to change the optimization effect of the PMU deployment problem.

This study provides some new research ideas for the deployment of PMUs, but it also has certain limitations. The current research mainly focuses on the improvement of deployment algorithms under static distributed network conditions. However, in actual application scenarios, there may be situations where the node attributes change due to the variation in switch states. Therefore, future research in this field can consider further in-depth exploration and analysis of deployment issues under complex conditions.

5. Conclusions

This paper mainly studies the optimization problem of PMU deployment. Traditional deployment algorithms have certain room for improvement in the optimization methods of PMU deployment. This paper makes certain improvements on the traditional deployment algorithm. Starting from the observability of the power system, a deployment method that comprehensively considers the preprocessing conditions and the number of PMU connection channels is proposed. On the one hand, compared with traditional methods, this method reduces the overall constraints through preprocessing conditions, thereby improving the performance of the deployment algorithm. On the other hand, the solution method of the algorithm has been changed. For multiple deployment methods, the average number of connection channels is used as the condition for screening the final deployment plan, making the final plan more economical.

(1) In the test case of the IEEE-14 node model, it can be seen that compared with other deployment algorithms, the improved algorithm only takes 0.02 s to obtain the solution, and the number of deployed PMUs is only 3 to meet the global observability condition.
(2) In the test case of the EIEE-30 node model, it can be seen that there are multiple solutions obtained by the improved algorithm. It only takes 0.02 s to obtain the optimal solution, and the number of deployed PMUs is only 7 to meet the global observability condition.
(3) In the test case of the NE-39 node model, it can be seen that the improved algorithm only takes 0.03 s to obtain the solution, and the number of deployed PMUs only needs 9 to meet the global observability condition.

In conclusion, through the comparison of simulation tests, it can be seen that the deployment algorithm, through certain pre-processing of data and the setting of algorithm strategies, can ultimately make the improved method have certain advantages in the solution time and deployment quantity of the PMU deployment optimization scheme.

Author Contributions: Conceptualization, Z.L. and H.D. (Hanyuan Dan); methodology, Z.L. and H.D. (Hanyuan Dan); software, Z.L. and H.D. (Hanyuan Dan); validation, Z.L. and H.D. (Hongda Dou); formal analysis, Z.L. and H.D. (Hanyuan Dan); resources, Z.L.; data curation, H.D. (Hanyuan Dan); writing—original draft preparation, H.D. (Hanyuan Dan) and H.D. (Hongda Dou); writing—review and editing, Z.L.; supervision, Z.L.; project administration, Z.L.; funding acquisition, Z.L. All authors have read and agreed to the published version of the manuscript.

Funding: This research received no external funding.

Data Availability Statement: Data model of this study is based on the IEEE node, the data content available in MATPOWER database or online sites such as https://icseg.iti.illinois.edu/power-cases (accessed on 15 September 2024).

Acknowledgments: The author expresses sincere gratitude to all those who have supported this research.

Conflicts of Interest: The authors declare no conflicts of interest.

Abbreviations

The following abbreviations are used in this manuscript:

IEEE	Institute of Electrical and Electronic Engineers
NE	New England
PMU	Phasor Measurement Unit
SCADA	Supervisory Control and Data Acquisition
WAMS	Wide Area Measurement System

References

1. Xu, Y.; Liu, X.; Li, Z.; Lv, M. Influence of PMU quasi-real-time data on robust estimation of active distribution network. *Electr. Power Autom. Equip.* **2020**, *40*, 15–22. [CrossRef]
2. Li, Z.; Cui, J.; Lu, H.; Zhou, F.; Diao, Y.; Li, Z. Prediction method for instrument transformer measurement error: Adaptive decomposition and hybrid deep learning models. *Measurement* **2025**, *253*, 117592. [CrossRef]
3. Almasabi, S.; Mitra, J. A fault-tolerance based approach to optimal PMU placement. *IEEE Trans. Smart Grid* **2019**, *10*, 6070–6079. [CrossRef]
4. Zhao, X.; Wu, G.; Yang, D.; Xu, G.; Xing, Y.; Yao, C.; Abu-Siada, A. Enhanced detection of power transformer winding faults through 3D FRA signatures and image processing techniques. *Electr. Power Syst. Res.* **2025**, *242*, 111433. [CrossRef]
5. Zhao, X.; Wu, H.; Yang, D.; Wei, J.; Li, Z.; Guo, Z.; Zhang, S. Residual Magnetism Elimination Method for Large Power Transformers Based on Energy Storage Oscillation. *IEEE Trans. Power Deliv.* **2025**, *40*, 2759–2768. [CrossRef]
6. Li, Z.; Cui, J.; Wei, Y.; Huang, Y.; Xie, H.; Yao, W. Calculation study of ion flow field in ±800 kV transmission lines under the effects of air pressure, temperature, humidity and improvement measures. *Electr. Power Syst. Res.* **2024**, *234*, 110574. [CrossRef]
7. Li, Z.; Cui, J.; Lu, H.; Zhou, F.; Diao, Y.; Li, Z. Prediction model of measurement errors in current transformers based on deep learning. *Rev. Sci. Instrum.* **2024**, *95*, 044704. [CrossRef]
8. Carvajal, G.; Carrión, D.; Jaramillo, M. Planning Scheme for Optimal PMU Location Considering Power System Expansion. *Energies* **2025**, *18*, 3283. [CrossRef]
9. Ogbogu, C.E.; Thornburg, J.; Okozi, S.O. Smart Grid Fault Mitigation and Cybersecurity with Wide-Area Measurement Systems: A Review. *Energies* **2025**, *18*, 994. [CrossRef]
10. Zhang, Y.; Cao, S.; Cai, Y.; Hui, Q.; Xu, J. Power system situation awareness method based on trend analysis. *Electr. Meas. Instrum.* **2024**, *61*, 99–106. [CrossRef]
11. Liu, C.; Han, K.; Wang, Y.; Ye, Z. Multi-objective optimization of distributed multi-energy system considering shared energy storage. *Electr. Meas. Instrum.* **2024**, *61*, 88–96. [CrossRef]
12. Zheng, J. The current situation and prospect of frontier electric power technologies. *China Electr. Power* **1999**, *10*, 11–16. [CrossRef]
13. Wang, Z.; Ge, F.; Dai, C. Power system modeling and parameter identification based on PMU. *Relay* **2007**, *S1*, 378–382. (In Chinese)
14. Kong, X.; Wang, Y.; Yuan, X.; Yu, L. PMU optimal configuration considering multiple topological observability of distribution network based on customized genetic algorithm. *Electr. Power Autom. Equip.* **2020**, *40*, 66–72. [CrossRef]
15. Al-Odienat, A.I.; Malahmeh, B.; Tarawneh, A.R. The optimal PMU placement in the power systems for the enhancement of state estimation. In Proceedings of the 2020 International Conference on Electrical, Communication, and Computer Engineering (ICECCE), Istanbul, Turkey, 12–13 June 2020; pp. 1–6.
16. Mishra, A.; de Callafon, R.A. Algebraic approach to PMU placement for minimum variance linear state estimation in power networks. *IEEE Trans. Power Syst.* **2022**, *38*, 4381–4390. [CrossRef]
17. Zhang, W.; Qiu, Y.; Zheng, X. GPS and its application in power system. *Power Grid Technol.* **1996**, *05*, 38–40+45. (In Chinese)
18. Xu, W.; Xue, Y.; Chen, S.; Ge, F. Difficulties and prospects in extracting knowledge from measured trajectories. *Autom. Electr. Power Syst.* **2009**, *33*, 1–7. [CrossRef]
19. Chai, J.; Li, S.; He, H. Parameter estimation of transmission grid based on PMU and multiple time sections. *Autom. Electr. Power Syst.* **2009**, *33*, 49–52. [CrossRef]
20. Zhu, X.; Wen, M.H.; Li, V.O.; Leung, K.-C. Optimal PMU-communication link placement for smart grid wide-area measurement systems. *IEEE Trans. Smart Grid* **2018**, *10*, 4446–4456. [CrossRef]
21. Matsukawa, Y.; Watanabe, M.; Mitani, Y.; Othman, M.L. Multi-objective PMU placement optimization considering the placement cost including the current channel allocation and state estimation accuracy. *Electr. Eng. Jpn.* **2019**, *207*, 20–27. [CrossRef]
22. Chen, X.; Sun, L.; Chen, T.; Sun, Y.; Rusli; Tseng, K.J.; Ling, K.V.; Ho, W.K.; Amaratunga, G.A. Full coverage of optimal phasor measurement unit placement solutions in distribution systems using integer linear programming. *Energies* **2019**, *12*, 1552. [CrossRef]

23. Zheng, M. Research on Optimal Allocation of PMU Based on 0-1 Integer Programming. Master's Degree, Tianjin University, Tianjin, China, 2014.
24. Zhang, B. PMU-Based Distribution Network State Estimation and PMU Configuration Optimization. Master's Degree, Zhejiang University, Hangzhou, China, 2020.
25. Liu, X.; Jiang, Q.; Cao, Y. Optimal configuration method of PMU without loss of observability under N-1 condition. *Proc. CSEE* **2009**, *29*, 47–51. [CrossRef]
26. He, K.; Zhang, Q.; Zheng, M. Optimal configuration method of PMU with minimum number of analog channels. *Power Syst. Prot. Control* **2015**, *43*, 14–18.
27. Zhao, Y.; Yuan, P.; Ai, Q.; Lv, T. Optimal configuration of PMU considering topological constraints and using improved genetic algorithm. *Power Grid Technol.* **2014**, *38*, 2063–2070. [CrossRef]
28. Tang, L.; Han, Y.; Zalhaf, A.S.; Zhou, S.; Yang, P.; Wang, C.; Huang, T. Resilience enhancement of active distribution networks under extreme disaster scenarios: A comprehensive overview of fault location strategies. *Renew. Sustain. Energy Rev.* **2024**, *189*, 113898. [CrossRef]
29. Citroni, R.; Mangini, F.; Frezza, F. Efficient integration of ultra-low power techniques and energy harvesting in self-sufficient devices: A comprehensive overview of current progress and future directions. *Sensors* **2024**, *24*, 4471. [CrossRef]
30. Niu, S.; Jia, Y.; Wang, W.; He, R.; Hu, L.; Liu, Y. Electricity consumption and human development level: A comparative analysis based on panel data for 50 countries. *Int. J. Electr. Power Energy Syst.* **2013**, *53*, 338–347. [CrossRef]
31. Zhao, C.; Dong, K.; Wang, K.; Nepal, R. How does artificial intelligence promote renewable energy development? The role of climate finance. *Energy Econ.* **2024**, *133*, 107493. [CrossRef]
32. Luo, K.; Lee, C.-C.; Zhuo, C. A pathway to coordinated regional development: Energy utilization efficiency and green development-Evidence from China's major national strategic zones. *Energy Econ.* **2024**, *131*, 107402. [CrossRef]
33. Ren, Z.; San, Y. Improvement of adaptive genetic algorithm and its application in system identification. *J. Syst. Simul.* **2006**, *18*, 41–43+66. [CrossRef]
34. Dongjie, X.; Renmu, H.; Peng, W.; Tao, X. Comparison of several PMU placement algorithms for state estimation. In Proceedings of the 2004 Eighth IEE International Conference on Developments in Power System Protection, Amsterdam, The Netherlands, 5–8 April 2004; pp. 32–35.
35. Wang, X.; Liu, Y.; Sheng, G.; Jiang, X.; Luo, L.; Zhang, X.; Liu, Z. A new method for PMU optimal configuration based on improved BPSO algorithm. *Guangdong Electr. Power* **2018**, *31*, 62–67.
36. Jiang, Z.; Cao, Y.; Sun, W. Multi-objective optimal PMU configuration algorithm based on 01 integer programming. *Power Syst. Prot. Control* **2008**, *21*, 12–17. [CrossRef]
37. Yan, L.; Liu, Z.; Ai, Y.; Luo, C.; Chen, B. A new method for optimal configuration of PMU measurement points based on 0-1 integer programming algorithm. *Power Syst. Prot. Control* **2017**, *45*, 101–106.
38. Cao, P.; Liu, M. PMU optimization configuration method based on improved integer programming method combined with zero injection node. *Power Syst. Prot. Control* **2021**, *49*, 143–150. [CrossRef]
39. Monticelli, A. *State Estimation in Electric Power Systems: A Generalized Approach*; Springer Science & Business Media: Berlin, Germany, 2012.
40. Krumpholz, G.; Clements, K.; Davis, P. Power system observability: A practical algorithm using network topology. *IEEE Trans. Power Appar. Syst.* **2007**, *PAS-99*, 1534–1542. [CrossRef]

energies

MDPI

Review

A Comprehensive Review of Reliability Analysis for Pulsed Power Supplies

Xiaozhen Zhao [1,2,*], Haolin Tong [1], Haodong Wu [1], Ahmed Abu-Siada [3], Kui Li [1,2] and Chenguo Yao [4]

[1] State Key Laboratory of Smart Power Distribution Equipment and System, Hebei University of Technology, Tianjin 300401, China; 202531402077@stu.hebut.edu.cn (H.T.); 202421401141@stu.hebut.edu.cn (H.W.); likui@hebut.edu.cn (K.L.)
[2] Key Laboratory of Electromagnetic Field and Electrical Apparatus Reliability of Hebei Province, Hebei University of Technology, Tianjin 300401, China
[3] Electrical and Computer Engineering Discipline, Curtin University, Perth, WA 6102, Australia; a.abusiada@curtin.edu.au
[4] State Key Laboratory of Power Transmission Equipment Technology, Chongqing University, Chongqing 400044, China; yaochenguo@cqu.edu.cn
* Correspondence: xzzhao@hebut.edu.cn

Abstract

Achieving high reliability remains the critical challenge for pulsed power supplies (PPS), whose core components are susceptible to severe degradation and catastrophic failure due to long-term operation under electrical, thermal and magnetic stresses, particularly those associated with high voltage and high current. This reliability challenge fundamentally limits the widespread deployment of PPSs in defense and industrial applications. This article provides a comprehensive and systematic review of the reliability challenges and recent technological progress concerning PPSs, focusing on three hierarchical levels: component, system integration, and extreme operating environments. The review investigates the underlying failure mechanisms, degradation characteristics, and structural optimization of key components, such as energy storage capacitors and power switches. Furthermore, it elaborates on advanced system-level techniques, including novel thermal management topologies, jitter control methods for multi-module synchronization, and electromagnetic interference (EMI) source suppression and coupling path optimization. The primary conclusion is that achieving long-term, high-frequency operation depends on multi-physics field modeling and robust, integrated design approaches at all three levels. In summary, this review outlines important research directions for future advancements and offers technical guidance to help speed up the development of next-generation PPS systems characterized by high power density, frequent repetition, and outstanding reliability.

Keywords: pulsed power supplies; wide bandgap semiconductor switch; capacitive storage; reliability

1. Introduction

Pulsed power technology is a field of science and engineering dedicated to generating high-power pulses. This is achieved by rapidly storing and releasing high-density energy (typically in the microsecond to nanosecond range), effectively compressing and converting the primary energy source [1]. The basic topology of a typical pulsed power supply system, illustrated in Figure 1, comprises essential stages: energy storage, pulse formation, and switch control. The system operates by charging an energy storage element, which then quickly releases its stored energy to the load through a fast-switching component,

https://doi.org/10.3390/en19020518

generating a high-power pulse [2]. Recent progress in power electronics, thin-film dielectrics, and insulation has greatly broadened pulsed power applications. This expansion ranges from national defense and military applications, such as electromagnetic launchers (EML) [3] and directed energy weapons [4] and scientific research including Z-pinches [5] and plasma sources [6], to a diverse set of industrial and civilian sectors, such as sewage purification [7,8], food sterilization [9,10], and biomedicine [11,12]. Consequently, pulsed power technology is fostering a series of groundbreaking applications.

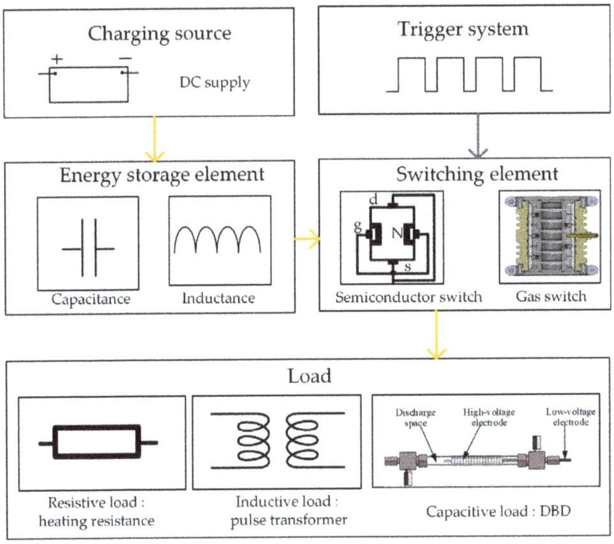

Figure 1. Schematic diagram of a typical pulsed power supply system structure [13–15].

Although pulsed power technology offers extensive practical applications, its implementation encounters considerable obstacles. In sectors such as national defense and heavy industry, these systems must sustain reliable, long-term performance under challenging circumstances, including frequent discharge-recharge cycles and complex environmental factors. As such, reliability is essential for the successful operation of pulsed power systems. Neglecting this aspect can negatively affect the mean time between failures (MTBF) and result in increased maintenance expenditures as well as prolonged periods of system inactivity.

This challenge arises from the operational mode of slow energy storage and fast release, rather than any fundamental power limitation. Figure 2 reveals the interaction and coupling pathways of the electro-thermal–magnetic–mechanical stress fields within the pulsed power supply. The Joule heating effect and the Lorentz force, acting at the core of the components, serve as the respective hubs for electro-thermal and magnetic–mechanical coupling. These coupled effects trigger a series of cascading failure reactions, including dielectric breakdown, electrode erosion, thermal fatigue, and structural deformation, which collectively restrict the system's overall reliability. Electrically, the high voltage rate dV/dt induces rapid changes in the electric field across the dielectric material, leading to a non-uniform electric field distribution. This results in localized energy concentration at dielectric interfaces and defect sites, where the transient electric field enhances displacement currents and initiates polarization dynamics that are out of equilibrium with the external field. These transient stresses give rise to thermal and electrical energy dissipation, which activate a series of physical processes, such as partial discharge, charge injection, and localized dielectric heating. Over time, this repeated energy injection causes microstructural

https://doi.org/10.3390/en19020518

damage, including the formation of Micropores, cracks, and space-charge accumulation. The cumulative effect of these degradation mechanisms induces critical parameter shifts, such as threshold voltage Vth drift, ultimately leading to dielectric breakdown or short-circuit failure. [16]. Thermally, high-repetition-rate pulsed operation inevitably induces thermal accumulation within the device. This sustained thermal cycling leads to excessive thermal stress, which subsequently promotes the formation of fatigue cracks on critical interfaces, such as the capacitor weld points and the switch surfaces [17]. Magnetically, the high electromagnetic forces inherent in the pulsed power supply leads to electrode surface erosion and morphological degradation. Importantly, when subjected to significant electromagnetic stress from high-voltage and high-current pulses, the contact resistance of the electrodes may exhibit substantial fluctuations, which can significantly undermine the overall stability of the system [18]. Mechanically, large mechanical stresses are inherently induced in certain demanding applications, such as electromagnetic ejection. This mechanical loading poses a direct threat to the structural integrity of the system, potentially resulting in deformation, structural failure, or fatigue damage [19]. The combined and interactive effects of these multi-physical factors elevate the reliability of the pulsed power supply to an interdisciplinary challenge. Addressing this issue necessitates a comprehensive analysis and the development of optimized design strategies spanning diverse fields, including electrical engineering, materials science, electromagnetics, and thermodynamics. Consequently, systematic research into the coupling mechanisms is paramount for achieving robust and sustainable pulse power system operation.

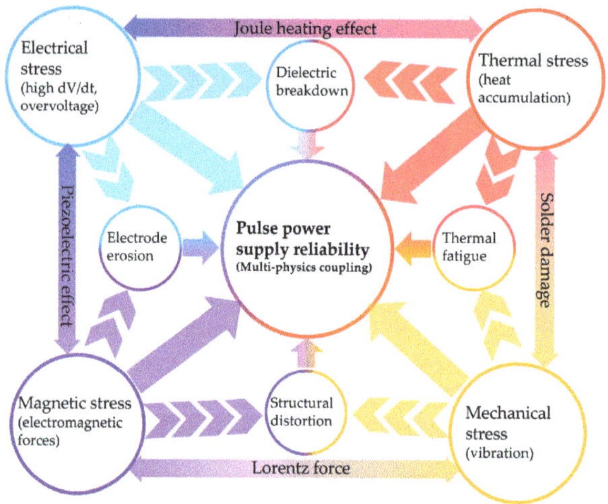

Figure 2. Schematic diagram illustrating the multi-physics coupling effects in a pulsed power supply, showing the inter-related stress fields and resulting failure modes.

Currently, several significant research gaps remain in the reliability of pulsed power systems. For example, there is a lack of efficient, high-speed fault diagnosis and dynamic isolation methods at the system level. In addition, current modeling and prediction capabilities for structural and long-term cumulative fatigue effects are inadequate. Importantly, while coupled models for specific devices exist, there is currently no widely accepted system-level quantitative model that can comprehensively account for the complex interactions among the four main stresses—electric, thermal, magnetic, and mechanical. As a result, without a comprehensive database of power failures, it is not possible to establish an accurate operational lifetime model for these systems.

https://doi.org/10.3390/en19020518

Given these challenges, improving pulsed power supply reliability is crucial for technological progress. This paper reviews recent studies, examining reliability issues and solutions from components to system integration and application scenarios. Its main aim is to outline a technical roadmap for developing durable, dependable pulsed power systems for stable, critical future uses.

2. Component Reliability

2.1. Energy Storage Elements

The core principle of pulsed power supply is the instantaneous release of energy stored over a significantly longer duration. Therefore, the features of the energy storage element are crucial because they essentially determine both the system's highest possible output and the range of applications it can support [20]. More specifically, the energy density of the storage element establishes the theoretical upper bound on the total energy delivered per single pulse, while its power density simultaneously governs the achievable peak pulse power output of the system. Equivalent parameters such as the inductance L_{eq} and resistance R_{eq} are crucial for shaping the output waveform, specifically by influencing the pulse width and the rising edge [21]. Furthermore, the charging frequency and efficiency directly govern the output repetition rate, while the energy density of the system imposes a direct constraint on the overall power supply volume and weight. However, under long-term, repetitive pulsed operation, the accumulation of electro-thermal stress causes the gradual degradation of the storage element's dielectric properties and internal structure. This degradation directly results in the drift of key performance parameters, which consequently manifests as output energy attenuation, waveform distortion, and a decline in system efficiency. Ultimately, this process causes the actual power supply performance to severely deviate from the initial design specifications [22]. Therefore, thoroughly investigating the reliability of energy storage components is essential to ensure the stable, long-term operation of pulsed power systems.

Table 1 outlines pulsed power system classifications by energy storage type, followed by a detailed description of each.

Table 1. Classification of Pulsed Power Energy Storage Elements based on Storage Medium.

Energy Storage Type	Energy Storage Form	Lifetime	Influencing Factors	References
Capacitive storage	Electric field	High cyclic stability and longevity	Overvoltage and overheating lead to material denaturation and lifetime reduction	[23,24]
Inductive storage	Magnetic field	Exceptional longevity and minimal inductive loss	High voltage and current handling requirements for switching elements.	[25–27]
Flywheel energy storage	Kinetic energy	Extended service life, requires periodic maintenance	Limited by bearing friction and rotor fatigue	[28–30]

2.1.1. Capacitive Storage

Persistent electro-thermal stress is inevitable during long-term repetitive pulsed operation. This sustained stress drives the gradual aging of the energy storage capacitor's dielectric material, resulting in the weakening of its internal polarization mechanism. Studies of the microstructure and overall electrothermal performance reveal that changes in the electrical structure of energy storage capacitors are not just due to cumulative channel damage or localized breakdowns caused by nanosecond, high-intensity pulses. These

https://doi.org/10.3390/en19020518

changes are also influenced by the combined effects of rising internal temperatures and distortions in the local electric field, both of which result from operations at high repetition rates [22,31]. The most widely used empirical model for capacitors is given by (1), which describes the influence of temperature and voltage stress.

$$L = L_0 \times \left(\frac{V}{V_0}\right)^{-n} \times \exp\left[\left(\frac{E_a}{K_B}\right)\left(\frac{1}{T} - \frac{1}{T_0}\right)\right] \tag{1}$$

where L and L_0 are the lifetime under the use condition and testing condition, respectively. V and V_0 are the voltage at use condition and test condition, respectively. T and T_0 are the temperature in Kelvin at use condition and test condition, respectively. E_a is the activation energy, K_B is Boltzmann's constant (8.62×10^{-5} eV/K), and n is the voltage stress exponent. Therefore, the values of E_a and n are the key parameters to be determined in the above model [32]. Due to variations in dielectric materials and failure mechanisms, the values of E_a and n differ significantly across capacitor types. For electrolytic capacitors, E_a typically ranges from 0.6 to 0.95 eV, with n approximately 3–5 [33]. In contrast, polymer film capacitors exhibit E_a values of 0.4–1.5 eV and n values of 5–9 [34]. while ceramic capacitors demonstrate E_a in the range of 0.8–1.5 eV and n between 2 and 6 [35]. Based on (1), Figure 3a,b illustrates the variation trends of capacitor lifetime with respect to temperature and voltage stress, respectively. These plots demonstrate that the lifetime exhibits a strong non-linear sensitivity to operating conditions.

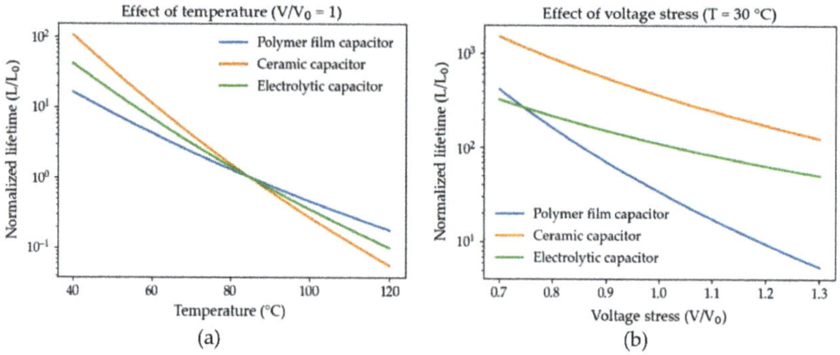

Figure 3. Theoretical analysis of capacitor durability under different operating conditions: (**a**) Influence of temperature on normalized lifetime degradation at unity voltage stress, (**b**) Impact of voltage stress on normalized lifetime degradation at 30 °C [32].

Furthermore, under pulsed conditions, each type of energy storage capacitor exhibits unique failure mechanisms, such as dielectric aging in ceramic capacitors, electrical breakdown in polymer film capacitors, and electrolyte dry-out in electrolytic capacitors. Although these mechanisms depend on the specific material, Table 2 highlights their shared foundation: changes in microstructure and declines in performance caused by the combined impact of electro-thermal stress and charge injection. Parameter drifts, influenced by differing media characteristics and architectural aspects of the storage element, can result in substantial deviations of the power output's energy and waveform from intended specifications. As a result, such drift constitutes a significant constraint on both the reliability and operational longevity of the pulsed power system.

https://doi.org/10.3390/en19020518

Table 2. Summary of characteristic failure mechanisms for various capacitor types.

Capacitor Type	Characteristic Failure Mechanisms	References
Polymer film capacitors	Dielectric breakdown Thermal accumulation causes dielectric aging and breakdown	[36–38]
Ceramic capacitors	Dielectric breakdown Thermal shock cracking	[39–41]
Electrolytic capacitors	Electrolyte evaporation Oxide film degradation and breakdown	[42,43]

Researchers, addressing the failure mechanisms of energy storage capacitors under pulsed conditions, have recently focused on mitigation strategies such as material modification, structural innovation, and system-level management to improve reliability, lifespan, and performance. As the energy density and repetition frequency of pulsed power systems continue to advance, the inherent limitations of traditional energy storage capacitors—specifically regarding dielectric strength, thermal shock resistance, and operational lifetime—have become increasingly acute. This situation has served as a major impetus for the accelerated development of novel dielectric materials.

The energy storage capacitor is often the main reliability limitation in pulsed power supplies due to its restricted dielectric strength, poor thermal shock resistance, and reduced lifespan under high energy and frequent use. Researchers have developed a multi-level reliability framework that innovates in material properties, internal structure, and packaging. Enhancing material properties aims to improve dielectric thermal stability and functionality, meeting stricter requirements for high-temperature and high-frequency use. For polymer film capacitors, traditional biaxially oriented polypropylene (BOPP) has been progressively substituted by polymers possessing superior glass transition temperatures and exceptional thermal stability, such as polyethylene terephthalate (PET), polyethylene naphthalene (PEN), and even advanced materials like polyphenylene sulfide (PPS) and polyether ether ketone (PEEK). These novel material systems ensure stable dielectric strength and energy density in environments up to 150–250 °C [44,45]. Furthermore, the introduction of inorganic nano-fillers with high dielectric constants, such as $BaTiO_3$, Al_2O_3, and BN, via nanocomposite technology to form functional composite films, has yielded multiple benefits [46,47]. This not only enhances the dielectric's overall polarization energy storage capacity but also significantly reduces its sensitivity to partial discharge and improves breakdown resistance at the dielectric mechanism level.

Reliability for ceramic capacitors centers on improving grain boundaries and DC bias performance. Researchers have successfully mitigated oxygen vacancy migration and grain boundary charge accumulation through fine element doping Zr, Mn and texturing techniques [48,49]. This suppression is vital for slowing aging and maintaining capacitance stability under high direct current bias, thereby greatly extending the effective device lifetime.

Significant improvements in the reliability of electrolytic capacitors depend on creating innovative electrolyte systems [50,51]. These new electrolytes effectively inhibit the evaporation and gasification of conventional electrolytes under high temperature and pulsed conditions. The resulting benefits include a significant reduction in equivalent series resistance (ESR), broader operating temperature ranges, and substantially improved capacitor pulse durability.

Beyond intrinsic material innovation, the internal structure and packaging technology of the capacitor also play a pivotal role in determining overall system performance. At the structural level, optimizing the stacking or winding configuration aims to reduce

 https://doi.org/10.3390/en19020518

the effective current loop area. This is a key approach to minimizing equivalent series inductance (ESL) and inductive coupling [52,53]. Concurrently, the internal electrode connection method is being upgraded from traditional spot welding to ultrasonic welding or more refined contact architectures. The objective is to reduce contact resistance, enhance the mechanical and electrical stability of the connection, and minimize localized heat loss [54].

High-reliability packaging materials and designs, including flame-retardant epoxy resins, ceramics, and metal enclosures, protect internal components against humidity, vibration, and extreme temperatures. This reduces risks like moisture absorption, electrode corrosion, and mechanical fatigue [55–57].

During high-repetition-rate, large-current pulsed operation, energy storage capacitor failure is mainly caused by electro-thermal coupling stress. As shown in Figure 4, the process begins with rapid dV/dt and overvoltage, leading to electric field distortion and charge injection. Continued energy dissipation and rising temperatures accelerate electrode migration, dielectric carbonization, and terminal corrosion, resulting in reduced capacitance, increased ESR, and ultimately diminished output energy and waveform distortion of the power module [58,59].

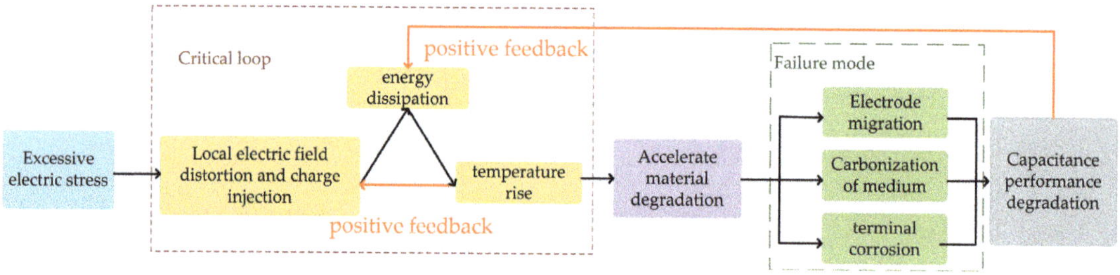

Figure 4. Key steps in the degradation and failure mechanism of the pulsed power capacitor.

Additionally, to achieve higher energy storage capacity, capacitors in pulsed power supplies are typically deployed in cascaded configurations, particularly in Marx generators and pulse forming network (PFN). In energy storage modules comprising multiple capacitors connected in series and parallel, a single-point failure has the potential to trigger a severe cascading failure. In a series branch, a short-circuit failure effectively bypasses the faulty capacitor. Consequently, the total branch voltage is redistributed across the remaining intact capacitors, subjecting them to voltage stresses exceeding their rated limits. Without a rapid voltage balancing mechanism, this leads to sequential breakdown due to overvoltage. Conversely, in a parallel group, a shorted capacitor creates an extremely low-impedance path. This causes adjacent healthy capacitors to discharge rapidly through this path within microseconds or even nanoseconds. The resulting transient current generates excessive electromagnetic forces and Joule heating, potentially damaging the electrode connections or internal structures of neighboring components [60,61]. Furthermore, localized heating from the failed unit propagates via thermal conduction, accelerating dielectric aging or inducing direct thermal breakdown in adjacent capacitors, thereby initiating a positive feedback loop of thermal runaway.

To effectively mitigate this degradation from the source, a dual, parallel strategy is typically employed. On the one hand, at the topology level, multi-level and soft-charging control strategies are implemented to constrain the peak dV/dt and charging stress on individual capacitors. Furthermore, cross-cell voltage equalization and redundant bypass are realized via equalization circuits to minimize the impact of single-point failures on the module [62,63]. On the other hand, the operational level focuses on online monitoring.

https://doi.org/10.3390/en19020518

This is achieved by implementing real-time estimation of capacitance, ESR, and leakage current, coupled with an electro-thermal degradation model, which supports residual life estimation and predictive maintenance decisions [64–66]. The synergistic linkage between these topology and online prediction strategies allows for a significant extension of the capacitor module's operational lifetime and a reduction in unexpected downtime risk, all without substantially compromising power density.

2.1.2. Inductive Storage

In pulsed power supplies, inductive energy storage offers a significantly higher energy density compared to capacitive storage, making it particularly suitable for storing enormous energy above the megajoule level [21]. Furthermore, pulsed power systems based on inductive storage can typically achieve ultra-high transient response and extremely fast current rise rates (di/dt) using magnetic or semiconductor switches. This capability makes it irreplaceable in applications requiring instantaneous large current shocks, such as driving particle accelerators, high-energy lasers, or electromagnetic emission systems [67,68].

The combination of high energy density and high di/dt in inductive energy storage creates significant reliability challenges. During pulse discharge, coils face intense mechanical stress from Lorentz forces [69,70], while high di/dt causes eddy current losses and transient voltages that speed up insulation aging [71]. These stresses reduce pulsed power supply reliability over time. Additionally, switching elements must handle sudden turn-off under high current, with transient voltage dominated by $L \cdot di/dt$, placing strict demands on switch design [72]. Thus, the overall reliability of the system heavily depends on the switch's performance.

2.1.3. Flywheel Energy Storage

In a Flywheel Energy Storage System (FESS), electrical energy is converted to kinetic energy and stored by spinning a flywheel at high speeds. FESS operates in three main modes: charging, discharging, and holding. During charging, the flywheel rotor absorbs external energy, storing it as kinetic energy while accelerating to a set speed. In discharging mode, generators draw kinetic energy from the flywheel rotor, convert it back to electrical energy, and supply current and voltage to electrical devices with the help of power control systems. As the flywheel releases energy, it slows down until it reaches a specified minimum speed. In holding mode, once the flywheel reaches its target speed, it neither gains nor loses energy; if energy losses are ignored, its energy remains unchanged. These modes allow the flywheel system to efficiently control energy input, output, and storage [73].

Flywheel Energy Storage is well-suited for pulse generators due to its high-power density, fast response capability, longevity, and high efficiency [74,75]. It can continuously absorb low-power input from the grid and release enormous energy (MJ-level) to the generator load within a very short duration, thereby significantly enhancing the system's instantaneous power output capacity [76]. However, achieving high- power density necessitates extremely high-speed operation, which generates immense centrifugal forces. This, in turn, causes complex resonance and fatigue accumulation among the flywheel rotor, bearings, and stator [77], accelerating the mechanical wear of the assembly and amplifying material defects in any part, which may ultimately result in catastrophic failure [29].

2.2. Switching Elements

The switching element regulates energy flow from storage to load, determining when and how much energy is released. Its performance parameters are decisive in determining the quality of the output pulse and the upper limit of the overall system capacity in pulsed power systems, this energy involves high voltage and large currents, so the switch must

https://doi.org/10.3390/en19020518

endure significant electro-thermal stress. Its failures are often sudden and severe, making reliability crucial for stable system operation [78,79].

Switching elements in pulsed power systems fall into three main categories based on their technical principles: mechanical, semiconductor, and plasma switches [80]. Table 3 outlines the specific failure mechanisms corresponding to each of the three categories, followed by a comprehensive explanation for each.

Table 3. Comparative Analysis of Characteristic Failure Mechanisms and Operational Lifetimes for Major Pulsed Power Switch Categories.

Switch Category	Representative Devices	Characteristic Failure Mechanisms	References
Plasma switch	Gas spark gap switch Thyratron	Electrode melting and vaporization Particle splashing	[81–83]
Semiconductor switch	SiC MOSFETs, GaN HEMTs Photoconductive switches	High dV/dt Pulsed short-circuit Severe overcurrent Irreversible electrode degradation Excessive or uneven optical excitation energy	[84–86]
Vacuum switch	Triggered vacuum switch	Degradation of Vacuum Integrity Contact Erosion and Wear	[87,88]

2.2.1. Gas Spark Gap Switch

The failure mechanism in gas spark gap switches is predominantly derived from the synergistic effect of multiple physical and chemical processes. Firstly, during each discharge event, the combination of high current density and short-duration thermal input causes partial melting or vaporization of the electrode surface. This action results in electrode mass loss, noticeable melting traces, and material erosion. This entire process is intricately linked to the electrode material's thermophysical properties (such as thermal conductivity, melting point, and specific heat) and the single-pulse transferred charge [81].

Secondly, with the accumulation of multiple discharges, structural damage—including cracks, pitting, bulges, and splash spheres—gradually develops on the electrode surface. These formations do not only increase the electrode roughness but also concentrate the local electric field, thereby aggravating partial discharge or premature breakdown [89].

Thirdly, particle sputtering and migration between electrodes, with subsequent deposition in the gap or on walls, contaminate the insulation path and create localized conductive channels. This process undermines the stability and reliability of the self-breakdown voltage V_{SB} [90].

Ultimately, these physical and chemical effects degrade key switch parameters, causing instability or reduction in V_{SB}, lower trigger reliability, and eventual switch failure after repeated cycles.

Prolonged, repetitive pulse discharge affects the reliability of gas spark gap switches primarily because of electrode damage caused by intense high current, high voltage, and frequent switching. The main failure arises when the electrode material melts and vaporizes, which leads to particle splashing and deposition. To address these issues, research has progressed from modeling and analyzing fundamental damage mechanisms to optimizing electrode materials and structural design. In exploring these mechanisms, researchers have used detailed numerical models to better understand how material is removed during operation.

The volume of fluid (VOF) method is used to construct a multiphase flow removal numerical model and compare it with the experimental data. The results show that reducing

 https://doi.org/10.3390/en19020518

the wear caused by melting or vaporization can be achieved by two main ways: one is to increase the polar pressure drop, and the other is to select materials with better thermal properties, thereby extending the effective working life of the switch [91].

Based on the understanding of the mechanism, the improvement of the material system has become the key to improving the durability of the switch. Researchers have carried out a wide range of pulse discharge experiments on various high-performance alloy electrodes, such as high-pulse current discharge tests on alloy electrodes such as W-Ni-Fe and W-Cu. By measuring their surface morphology and erosion degree, it is clear that material alloying and microstructure control are strategic methods to effectively reduce melting and vaporization damage [92].

The structural design of the electrode has a critical influence on the dispersion and concentration of thermal stress. Studies show that optimizing the electrode geometry alongside appropriate material selection can significantly mitigate thermal stress concentration, thereby minimizing localized damage and ensuring the switch's long-term stable operation [93]. Long-term reliability depends not just on initial design, but also on operational conditions and maintenance. Analysis under conditions like 0.3 MPa pressure, 30 Hz repetition rate, 8.5 kA peak current, and −35 kV voltage shows electrode spatter affects switch lifespan. Using materials prone to less spatter and regular maintenance can extend switch life [94].

Periodic electrode replacement is considered a feasible strategy to extend the switch's operational lifetime. To prevent or minimize thermal damage to the new electrode during this process, theoretical studies leveraging the Holm model, Hertz formula, and heat conduction theory have demonstrated that manufacturing electrode brackets from metals or alloys with superior thermal resilience or those incorporating trace amounts of rare earth elements can effectively mitigate replacement-induced thermal damage [95].

Ultimately, the dynamic assessment of the switch's internal environment offers crucial theoretical support. By applying the Joule heat accumulation method to analyze the resistance characteristics of the spark channel, research revealed the pressure and temperature evolution law within the switch chamber: this typically features an initial rapid growth, followed by a slower increase, and ultimately stabilizing. Notably, the increase in pulse current amplitude significantly enhances heat deposition. This understanding provides a vital thermal management theoretical basis for designing highly reliable, long-life gas spark gap switches [96,97].

2.2.2. Wide Bandgap Semiconductor Switch

Silicon-based devices have traditionally dominated the field of semiconductor switching technology. However, their lower bandgap, limited breakdown electric field, and restricted carrier mobility impose fundamental constraints that increasingly hinder advancements in power density and efficiency for pulsed power systems [98]. Wide-bandgap (WBG) semiconductor materials, specifically SiC and GaN present a robust solution to these limitations, owing to their superior intrinsic properties. As depicted in Figure 5 both SiC and GaN demonstrate significant advantages over conventional silicon in several critical material parameters: their larger bandgaps support stable operation at elevated temperatures; higher breakdown electric fields allow for more compact device architectures and improved voltage blocking capabilities; and enhanced electron saturation velocity and thermal conductivity facilitate rapid switching and effective heat dissipation. These characteristics collectively result in greater power density, minimized switching losses, and improved resilience to severe electro-thermal stress in switching devices [99]. Therefore, the following section will focus on the reliability aspects of WBG semiconductor switches.

https://doi.org/10.3390/en19020518

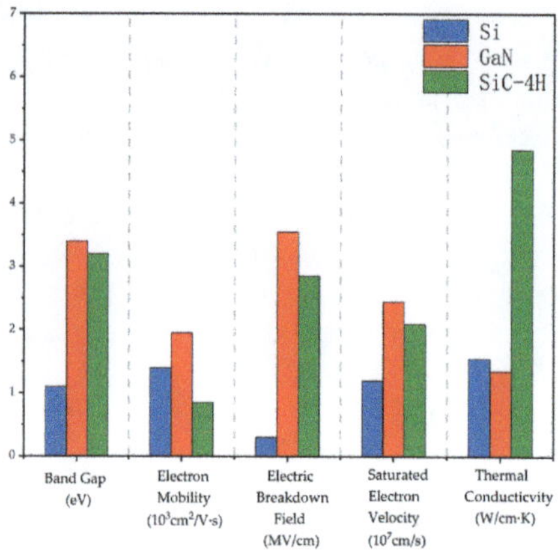

Figure 5. Comparative analysis of key material properties for Silicon (Si), Gallium Nitride (GaN), and Silicon Carbide (SiC) [99].

Wide-bandgap semiconductor switches are widely adopted in pulsed power supply applications due to their high switching speeds, precise controllability, and extended operational lifespans. Nevertheless, their reliability under strenuous electro-thermal-mechanical multi-field conditions remains a critical limiting factor in overall system performance. Comprehensive reliability analysis of these semiconductor devices is essential for advancing pulsed power technologies. The varied failure mechanisms observed in these switches primarily originate from uncontrolled energy dynamics and material degradation, driven by the interactions across extreme electro-thermal-mechanical fields.

Firstly, electrical overstress represents the primary threat: when the switch is rapidly turned off, the circuit's parasitic inductance generates an induced voltage. This induced voltage superimposes onto the direct current bus voltage, resulting in a transient overvoltage that significantly exceeds the device's rated value and subsequently leads to localized or complete device failure. In the case of MOSFETs, this failure mode is typically initiated by the device entering an avalanche state [86,100].

Secondly, thermal failure is often triggered by electrical overstress: excessive electrical stress intensifies the thermal effect, which subsequently affects the device's internal material structure and current distribution. This process causes the current carrying capability to concentrate locally, leading to the formation of a 'hot spot'. As illustrated in Figure 6, within a repetitive pulse sequence, each pulse induces a transient temperature spike at the device's core. This initiates a positive thermal runaway feedback loop in the affected region: increased temperature reduces resistance, which elevates the carrier concentration, thereby boosting the current flow. This vicious cycle ultimately leads to device destruction via fusing or secondary breakdown [101,102]. Cyclic thermal stress gradually degrades the interlayer dielectric, which can lead to source metallization melting and a partial short-circuit between the gate and source. The fatigue life associated with this thermal failure mechanism typically follows the Coffin–Manson model [103]. The relationship between the number of thermal cycles and device failure under various electro-thermal stresses is expressed by (2) [104].

$$N_f = K\Delta T_j^{\beta_1} e^{\left(\frac{\beta_2}{T_{j-min}+273}\right)} t_{on}^{\beta_3} I^{\beta_4} \tag{2}$$

where N_f is the number of cycles to failure, t_{on} is the duration of the device on time during each cycle, ΔT_j is the junction temperatures swing, $T_{j\text{-}min}$ is the minimum temperature of the device, and I is the current per bond wire. The model parameters K and β_1-β_4 are determined by least squares curve fitting of the test data.

 https://doi.org/10.3390/en19020518

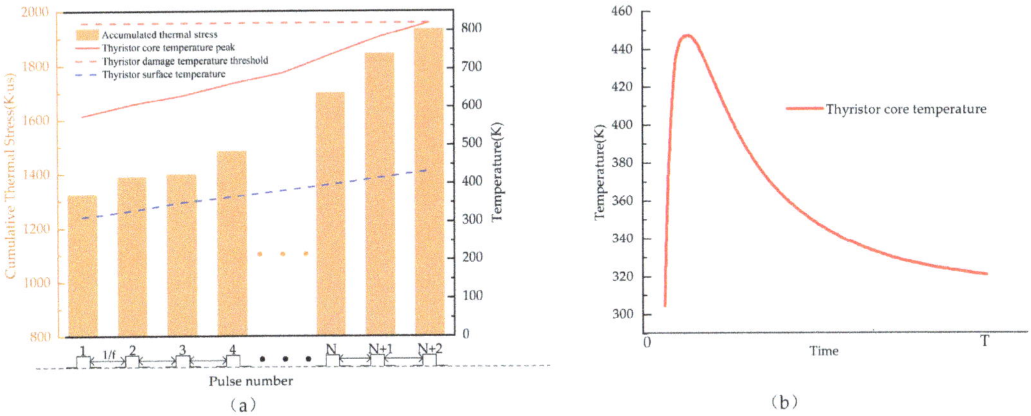

Figure 6. Thermal behavior of the semiconductor switch core under pulsed operation [105]: (**a**) Comparison of the core and back temperature profiles during repetitive pulse sequences. (**b**) Transient temperature trend of the switch core under a single pulsed event.

In summary, the synergistic coupling of electro-thermal-mechanical stresses leads to physical failure modes categorized into two distinct dimensions: chip-level and package-level [106]. At the chip level, degradation primarily manifests as gate oxide failure and body diode failure. Specifically, gate oxide failures are typically induced by electrical overstress and high temperatures, which trigger interfacial charge accumulation or dielectric breakdown. Conversely, body diode failures predominantly stem from forward voltage bias stress [107]. To quantify these degradation mechanisms, the key parametric indicators include the threshold voltage (Vth), on-state resistance (R_{Dson}), critical avalanche energy (E_{AV}), and gate leakage current (I_{GSS}). The threshold voltage (Vth) instability is a complex phenomenon primarily driven by bias temperature instability (BTI) and charge trapping in the gate oxide. Two distinct failure modes arise from this instability: A negative drift, typically induced by hole trapping, is critical as it erodes the noise margin. Once the threshold drops below a safety limit, it can trigger spurious turn-on events and catastrophic shoot-through. Conversely, a positive drift induced by electron trapping increases the on-state resistance (R_{DSon}) by reducing the effective overdrive voltage, thereby degrading switching speed and increasing conduction losses [108,109]. Functionally, a failure is defined when the on-state resistance increases by 15% [104]. Furthermore, monitoring the magnitude of the gate leakage current (I_{GSS}) serves as a robust method for failure detection. Under healthy operating conditions, I_{GSS} typically remains in the nano-ampere (nA) range; however, upon device failure, this current can escalate drastically to the milli-ampere (mA) level. Moreover, driven by cumulative physical degradation, the evolution of I_{GSS} is often non-linear. The onset of an exponential growth pattern in leakage current is widely identified as a critical precursor to device failure [103]. In terms of energy tolerance, the critical avalanche energy (E_{AV}) defines the device's thermodynamic limit. When the dissipated avalanche energy causes the junction temperature to exceed the melting point of the source metallization (such as Aluminum), it operates outside its safe operating area (SOA). This thermal breakthrough serves as a deterministic criterion for catastrophic avalanche failure [110,111].

Common package-level failures predominantly comprise bond-wire lift-off and solder layer fatigue. The critical parameters for characterizing these degradation processes are the on-state voltage (V_{Dson}) and the thermal impedance (Zth) between the chip and the copper substrate. According to the AQG 324 guideline, the failure criteria are explicitly defined: a device is deemed to have reached its end-of-life (EOL) when the on-state voltage

 https://doi.org/10.3390/en19020518

increases by more than 5% relative to its initial value, or when the thermal impedance rises by 20% [112].

To effectively mitigate the failures arising from electro-thermal field coupling in pulsed environments, researchers have developed and substantiated a broad spectrum of strategies.

Firstly, in terms of device process and structure optimization, the core objective is to enhance the durability and avalanche tolerance of the gate dielectric. By establishing a unit-level behavior model for short-circuit and avalanche breakdown faults, researchers can clearly describe the thermal failure progression of these two events. This enables designers to intuitively and rapidly evaluate and improve device characteristics, thereby identifying weak links at the design origin [113]. Reliability assessments of gate oxide layers; mainly conducted via precise time-dependent dielectric breakdown (TDDB) tests highlight how the lifetimes of various MOSFET structures, such as planar and asymmetric trench designs, depend on electric field stress from bias. For example, planar oxides demonstrate superior lifespans when subjected to negative bias, whereas trench oxides perform better under positive bias [114]. Moreover, innovative device architectures greatly enhance overall performance. A newly developed deep trench sub-gate super-junction 4H-SiC MOSFET offers reduced specific on-resistance and switching loss while preserving high breakdown voltages by eliminating complicated multi-epitaxial processes [115]. Additionally, the SiC dual-channel MOSFET, featuring a lower-barrier diode and an L-shaped gate-source embedded in the gate trench, efficiently suppresses body diode reverse conduction. Its split gate design enables faster switching speeds, surpassing conventional structures in both voltage drop and switching performance [116].

Secondly, research at the chip and packaging levels focuses on reducing parasitic inductance and improving thermal management, which are fundamental ways to manage transient overshoot and thermal stress. Novel discrete device packaging, such as structures employing stepped etching active metal brazed substrates and flexible printed circuits, achieves a seamless combination of high thermal conductivity, full electrical isolation, and low stray inductance, significantly enhancing the SiC MOSFET's electrical performance [113]. Innovative ultra-high thermal reliability embedded liquid-cooled SiC 3D packaging structures, achieved by stacking embedded micro-pin fin SiC substrates on copper cold plate manifolds, demonstrate superior thermal stability, heat dissipation efficiency, and temperature uniformity over traditional DBC substrate SiC power modules [117]. In connection technology, laser micro-welding has been implemented for bonding the source contact of Ag-sintered, copper-bonded top plate-enhanced SiC MOSFETs. This method demonstrates reduced sensitivity to surface properties and improved reliability when compared to conventional ultrasonic copper wire bonding techniques [118].

Thirdly, optimizing the gate drive and controlling dV/dt are effective in suppressing spurious turn-on due to Miller coupling and reducing turn-off overshoot. Gate drive design has formed two primary optimization strategies. The first is the active intelligent control scheme, which utilizes detection and feedback loops to achieve accurate dynamic control over the switching transient. The dV/dt control method in the active gate driver (AGD), for instance, actively modulates the turn-off gate resistance to realize voltage balancing in stacked devices during the turn-off transient [119]. This method is further refined by using digital open-loop control AGD [120] or by introducing negative feed-back [121], aiming to suppress inrush voltage and shorten the switching transient time. Another AGD variant, the IS-AGD (independent suppression AGD) [122], focuses on optimizing the switching trajectory through independent suppression of overshoot and oscillation, achieving finer control. As illustrated in Figure 7 [123], the circuit utilizes a resistive divider network (R_1, R_2, R_6, R_7) to sense the switching status. Subsequently, the comparators (OP_3, OP_6) and logic gates ($AND_{1\text{–}4}$) accurately determine the optimal timing for active intervention.

https://doi.org/10.3390/en19020518

Based on these logic outputs, the auxiliary MOSFETs (MOS_1, MOS_2) are activated to regulate the gate current through shunting or injection. This specific configuration allows for a significant reduction in electromagnetic interference (EMI), with experimental data showing a decrease in current and voltage overshoots of 39–53% and 48–54%, respectively, while maintaining the original switching speed. The second strategy explores more concise and efficient Passive Resonance Assistance schemes. This approach avoids complex active components, inherently lowering drive circuit complexity and cost. A typical example, the resonant auxiliary drive circuit shown in Figure 8, the capacitor C_q and inductor L_r placed between the gate and the source constitute a resonant loop. Simultaneously, R_p, R_q, C_p, and C_q constitute a voltage division network, while VD_1, VD_2, VD_3, and VD_Z function as a voltage clamping mechanism. This ingeniously achieves a smooth, zero-voltage transition of the gate voltage during the turn-off period. The resulting mechanism not only effectively suppresses bidirectional crosstalk but also yields faster switching speeds and lower switching losses [124]. Thus, these passive schemes prioritize topological simplicity and efficiency alongside high performance.

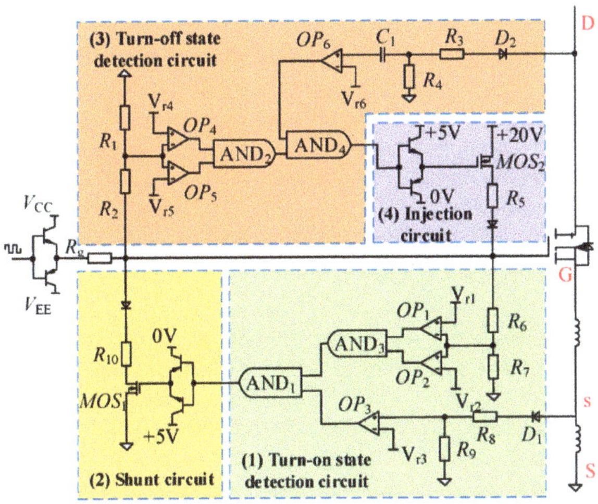

Figure 7. Schematic of the AGD for dynamically regulating the drive current [123].

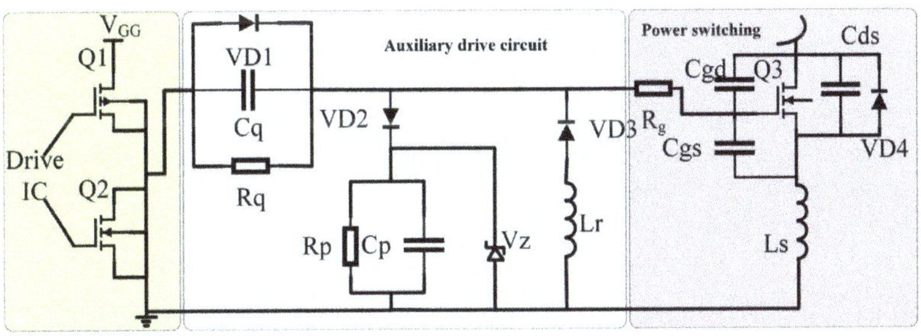

Figure 8. Resonant auxiliary drive circuit [124].

Finally, the absorption and energy feedback circuit is adopted, with the optimized layout or design, to reduce the overvoltage caused by parasitic inductance and improve the single pulse avalanche and transient tolerance of the device. For example, researchers have proposed a SiC MOSFET overvoltage and oscillation suppression circuit (OVSC) that integrates switching loss optimization and clamping energy feedback characteristics.

https://doi.org/10.3390/en19020518

The circuit not only achieves excellent suppression performance for overvoltage and oscillation, but also significantly reduces the switching loss of the device by cleverly using clamping capacitors and mode-locked capacitors, achieving the unity of performance and efficiency [125]. At the same time, in-depth theoretical analysis also provides the basis for the design of the buffer circuit. By establishing the research model of the parasitic parameters of the device, the parameters of the buffer component are accurately selected and the circuit performance is optimized [126,127].

2.2.3. Photoconductive Switch

Photoconductive Switches (PCSS), as a representative solid-state switch, are widely employed in pulsed power systems, owing to their ultra-fast response, low jitter, and high peak current capability. However, despite these advantages, PCSS is less prevalent than other switch categories, and its practical application is limited by numerous coupling failure paths [85]. Specifically, the device is prone to forming self-sustaining avalanches and hot spot current channels under the challenging conditions of high electric field and low optical excitation. Furthermore, instantaneous, concentrated energy injection can induce irreversible electrode damage [128]. Moreover, the high current density resulting from light triggering generates significant Joule heating. The inability to dissipate this heat quickly leads to localized melting and material denaturation. Furthermore, uneven or excessive optical excitation accelerates cumulative damage, consequently limiting the device lifetime [129,130]. In summation, the failure of the PCSS is a complex electro-optical-thermal multi-field coupling process, characterized by both transient, catastrophic breakdown and long-term cumulative degradation.

Researchers have developed a multi-dimensional optimization framework to improve PCSS performance and reliability by controlling electrode contact, internal electric fields, and external triggers.

The primary objective is to ensure efficient current injection and extraction while suppressing surface failure under high fields. Studies confirm a competitive interplay between PCSS surface flashover breakdown and bulk current channel formation. Analysis of back-illuminated trigger electrode structures provides a vital empirical basis for optimizing electrode geometry and layout to mitigate surface electric field concentration [131]. For achieving high-performance ohmic contact in WBG systems, advanced techniques are paramount. In AlGaN/GaN structures, forming a good ohmic contact to the high-concentration two-dimensional electron gas (2DEG); produced by spontaneous and piezoelectric polarization, is achieved through metal deposition and annealing, which improves peak current and device reliability [130]. Moreover, research suggests fundamentally improving ohmic contact quality by filling the electrode–substrate gap with a 100 nm p-type epitaxial layer and Ti/Pt/Au metal electrodes [132].

The core objective in device structure and electric field regulation is to optimize the internal E-field distribution to enhance the switch's voltage withstand capability. Researchers introduced a design strategy utilizing a low resistance region (LRR) structure to modulate the PCSS electric field and mitigate local peak E-field concentration. This method successfully improves the device's breakdown voltage under high fields without compromising the light absorption area or increasing the on-state resistance [133]. Furthermore, analysis of the trigger position revealed the structure's sensitivity to conduction performance, showing that cathode triggering achieves a shorter delay time and lower on-resistance compared to anode triggering under identical conditions. This provides a clear strategic direction for subsequent trigger mechanism and structure optimization [134].

https://doi.org/10.3390/en19020518

In terms of external triggers, expanding the trigger region width facilitates rapid carrier density multiplication and the formation of avalanche ionization domains. Consequently, this shortens the PCSS delay and switching times, effectively minimizing trigger jitter [135].

2.3. Impact of Component Reliability on System Performance

The reliability of energy storage and switching elements within pulsed power systems is interconnected, as each component's performance affects the others. The overall system output depends heavily on the reliability of individual elements. For example, changes in the parameters of the energy storage capacitor impact both the pulse energy and the waveform delivered to the load. At the same time, deterioration in the switching device can further disrupt the pulse waveform and compromise timing accuracy [136,137]. When both components degrade together, the system's output may significantly stray from its intended design specifications. Therefore, future designs for individual devices must incorporate sufficient safety margins to ensure that a localized failure does not propagate to other system components.

3. System Integration Reliability

Studies show that even if each part is designed for maximum performance, the whole system may still fall short of its required reliability standards. Having reliable components is only one piece of the puzzle but it doesn't guarantee the entire system will work as intended. When these components interact after being put together, new potential failures can arise, holding back overall effectiveness. Therefore, to truly improve reliability in pulsed power supplies, it is essential to shift focus from just individual parts to the entire system, using structured approaches for designing, evaluating, and managing reliability.

3.1. System-Level Thermal Management

As illustrated in Figure 1, the fundamental unit of the pulsed power module comprises a charging power supply, an energy storage element, a switching element, and a load. The system realizes slow energy storage and rapid discharge through the modulation of the switch, typically resulting in the output of an instantaneous high-power square wave. However, this high instantaneous power inherently generates significant energy losses, such as switching losses and ESR losses within the capacitor [38]. These losses inevitably accumulate as heat within the energy storage and switching components. Ultimately, component failure is induced, which consequently precipitates the failure of the entire pulsed power supply. As illustrated in Figure 9, thermal management design must be integrated across all hierarchical levels of the power supply system. By implementing a rational combination of component-level, board-level, and system-level thermal design, the complexity of the cooling system is reduced, and the overall heat dissipation effectiveness is significantly enhanced. At the component level, the design objective is to efficiently extract heat from the power chip core to the package housing via high-performance thermal interface materials (TIMs). At the board level, the focus shifts to employing effective cooling technologies to address multi-chip layouts and complex thermal coupling effects. At the system level, a final cooling design must be chosen with careful attention to electromechanical integration. The thermal system is shaped by environmental conditions and the system's operating profile, aiming to meet safety, reliability, and longevity requirements throughout its lifecycle.

 https://doi.org/10.3390/en19020518

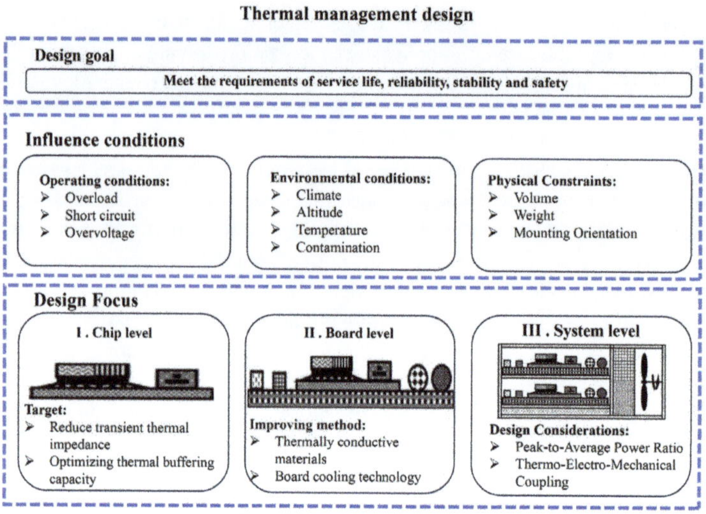

Figure 9. Overall framework for thermal management design across different hierarchical packaging levels [138].

Cooling technologies are broadly categorized into four main methods: air cooling, solid-state cooling, single-phase liquid cooling, and phase-change cooling [139]. From a thermal engineering perspective, the objective of the cooling system is to minimize the total thermal resistance R_{conv} from the junction to the ambient environment. The convective thermal resistance is governed by Newton's Law of Cooling, as shown in (3).

$$q = hA(T_s - T_\infty) \tag{3}$$

where q denotes the heat flux, h represents the heat transfer coefficient, A is the effective heat transfer area, T_s is the surface temperature, T_∞ refers to the ambient fluid temperature. The definition of thermal resistance is provided in (5). Consequently, as derived in (4), the thermal resistance is inversely proportional to the heat transfer coefficient and the effective heat transfer area [140].

$$R_{conv} = \frac{T_s - T_\infty}{q} \tag{4}$$

$$R_{conv} = \frac{1}{hA} \tag{5}$$

Equations (3)–(5) govern the fundamental heat rejection mechanism for fluid-based systems. Although Solid-state cooling operates via active conduction mechanics, its system-level performance is ultimately constrained by the convective heat rejection efficiency at its hot side. Consequently, maximizing h or A remains the universal design goal.

Air cooling typically involves adding fans to heat-generating components or the entire assembly for temperature reduction, or employing advanced dual-piezoelectric cooling jets (DCJ), which disturb the boundary layer over thermal elements to enhance heat dissipation [141].

Solid-state cooling encompasses thermal conduction techniques and thermoelectric cooling technology (TECT). Thermal conduction technologies rely on conduction mechanisms to enhance heat transfer efficiency by expanding the contact surface area between the semiconductor casing and the cooling medium [142]. Thermoelectric cooling technology exploits the Peltier effect in semiconductor materials, whereby heat dissipates when current flows through two dissimilar materials [143].

https://doi.org/10.3390/en19020518

Liquid cooling technology is broadly divided into microchannel cooling and immersion cooling. The underlying principle of microchannel cooling involves heat transfer from the device, through the solid material, to the microchannel wall, where the coolant then absorbs the heat via convective transfer as it flows through the channel [144]. To further enhance microchannel performance, researchers have developed various structural optimizations, including the compact layered manifold microchannel design [145] and the incorporation of cylinder structures within the channel to promote fluid mixing and boost cooling effectiveness [146]. Immersion cooling involves completely immersing the entire circuit or a specific component in a dielectric liquid, so that heat is transferred directly to the liquid. For example, in the LTD pulse power supply, the heat dissipation mainly depends on the conduction heat dissipation of the transformer oil-immersed environment, and the magnetic core with a thin strip laminated structure is used to reduce the switching and core losses [147].

Phase change cooling technology has established itself as the dominant solution for high power density applications, primarily due to its superior heat dissipation efficiency, primarily including two-phase immersion cooling and the application of solid phase change materials (PCMs). The two-phase immersion system absorbs the heat load via the coolant, undergoing a liquid-to-gas phase transition. The resulting vapor is then condensed by a condenser, returning to the liquid state for circulation within the system [148]. Researchers have proposed structural innovations, such as a compact sealed two-phase immersion cooling scheme where the entire PCB board is submerged in a sealed dielectric fluid. This utilizes the highly efficient liquid–gas transition process for heat dissipation while simultaneously providing electrical insulation [149]. More innovatively, a new packaging structure for SiC MOSFET without DBC (direct bonded copper) or bonding wires has been developed. This structure forms a sealed cavity that can be filled with dielectric fluid to realize two-phase immersion cooling, thereby comprehensively enhancing the SiC MOSFET's performance regarding volume, weight, and thermal management, offering an effective strategy for high power density systems [150]. This high-efficiency cooling technology also significantly extends system limits: applying two-phase immersion cooling to the avalanche transistor Marx circuit successfully increased the maximum repetition rate from 80 kHz to 260 kHz and reduced the pulse drift from 900 ps to 200 ps, effectively resolving the contradiction between high repetition rate and high stability [151]. Furthermore, integrating a novel three-component composite PCM into the copper substrate of the MOSFET power module, coupled with active heat dissipation via single-line water transfer in the bottom plate, is an effective method for addressing the dynamic thermal management challenge of SiC MOSFET modules [152].

Thermal management for pulsed power supplies is recognized as a quintessential multi-level problem, spanning the entire three-tier hierarchy of component, board, and system scales. Figure 10 illustrates the typical heat dissipation characteristics of the four cooling architectures. Consequently, the selection of the cooling architecture necessitates a comprehensive trade-off analysis across several key metrics, including peak heat load, heat flux density, system volume, energy efficiency, compactness, and cost. To facilitate this selection, Table 4 provides a systematic comparison of these methods, summarizing their typical heat transfer coefficients (h), applicable levels, and respective advantages and disadvantages. Among these options, phase-change cooling is particularly well-suited for next-generation high-voltage pulse generators operating at elevated power levels and repetition rates. This suitability stems from its capability to rapidly handle thermal transients and severe high heat flux concerns [153]. In addition to choosing the best cooling technique, using materials that can withstand high temperatures during the manufacturing and packaging of components is a keyway to address thermal issues. This focus on durable

https://doi.org/10.3390/en19020518

materials greatly reduces the risk of heat-related damage and naturally boosts the reliability of the pulsed power supply.

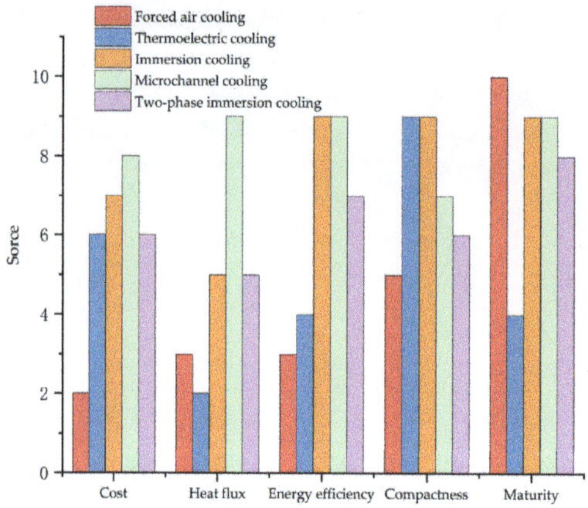

Figure 10. Key considerations and trade-off metrics for selecting the optimal cooling technology in pulsed power systems [154].

Table 4. Suitability of Various Thermal Management Technologies for Different Pulsed Power Topologies [140,153,155,156].

Cooling Method	Suitable for Hierarchical Levels	Core Strengths	Limitations	Heat Transfer Coefficient (W/m^2K)	High Heat Flux (W/cm^2)
Forced air cooling	System-level	Low cost, Simple structure, Large heat dissipation surface area	Limited heat dissipation capacity, High noise Dust accumulation	2~25	10~35
Thermoelectric cooling	Device-level Board-level Compact system-level	Precise temperature control, High reliability	Low efficiency, Self-generated heat degrades output parameters	Not applicable	1~10
Immersion cooling	Device-level Board-level	High heat dissipation capacity, Compact structure, Low noise	System complexity and risk of leakage	50~1000	50~70
Microchannel cooling	Device-level Board-level	Compact volume Rapid transient response	High cost High pressure drop	100~20,000	100~1000
Two-phase immersion cooling	Device-level Board-level System-level	Excellent heat dissipation, Excellent temperature uniformity, Strong environmental adaptability, Low noise	High cost Complex structure Poor Maintainability	2500~100,000	50~100

3.2. Multi-Module Synchronous Triggering

To increase voltage and current output, pulse power systems commonly connect multi-stage Marx, LTD, and PFN modules in series and parallel, boosting their output capabilities.

 https://doi.org/10.3390/en19020518

Ideally, each module's rise time should be perfectly synchronized so their output matches in phase and amplitude at the load. The ideal output is expressed as (6).

$$V_{total}(t) = \sum_{i=1}^{N} V_i(t) \quad (6)$$

when accounting for jitter (δ_i), the output is represented by (7)

$$V_{total}(t) = \sum_{i=1}^{N} V_i(t - \delta_i) \quad (7)$$

where V_{total} represents the total output voltage, $V_i(t)$ denotes the output voltage of a single module, and t is the time variable.

Therefore, a larger jitter degrades the rising edge of the output waveform and attenuates the output peak voltage. Asynchronous switching results in non-uniform voltage and current sharing among modules. Prematurely triggered switches endure disproportionate electrical and thermal stresses, whereas delayed switches are susceptible to reverse voltage or severe transient overstress [157]. Under high-repetition-rate conditions, this cumulative stress imbalance accelerates device degradation and compromises system longevity. If the distributed switch array's trigger jitter exceeds 5 ns, it may cause peak reduction and waveform distortion, and exceeding 20 ns results in unacceptable performance, potentially damaging them or creating hazardous hot spots within components [158]. Therefore, the system's required synchronization accuracy directly affects both its transient responses and long-term reliability. Making synchronization a primary design goal is crucial from both engineering and scientific perspectives.

In multi-module pulsed power systems, synchronization faces severe challenges as the number of modules scales up. Research indicates that electromagnetic coupling and structural variances lead to a non-linear accumulation of timing errors, a phenomenon referred to as "synergistic jitter deterioration." Reference [159] clearly illustrates this trend: while a single module operates with a manageable jitter of 4.7 ns, parallel operation of two modules causes a significant increase to the 4.79–9.5 ns range due to mutual interference—an escalation of 20–43%. in a three-module configuration, the combined effects of asymmetric discharge channels and coupling exacerbate the system jitter to 6.38–10.98 ns. This more than twofold increase demonstrates that simple parallel triggering strategies are insufficient for large-scale arrays, necessitating advanced isolation and timing control mechanisms. However, achieving superior synchronization often incurs a steep penalty in terms of system complexity and cost. To resolve the conflict between synchronization precision and system complexity, researchers have developed a spectrum of triggering architectures, extending from robust electrical topologies to ultra-precise optical controls. To elucidate the performance boundaries and applicability of these methodologies, Table 5 presents a comprehensive comparative analysis focusing on their jitter characteristics, scalability, and technical trade-offs.

As indicated in Table 5, current research on triggering technology has shifted significantly from gas and corona-stabilized switches toward solid-state architectures, incorporating advanced FPGA timing control to minimize intrinsic jitter. However, at the system level, mitigating inconsistent switching caused by signal jitter and delay requires more than just device optimization; it necessitates robust isolation mechanisms (such as magnetic or optical isolation) and low-inductance isochronous transmission lines. Fundamentally, the core technical challenge arises from an irreconcilable conflict between the demand for extremely high synchronization accuracy and the system's inherent parasitic effects and EMI.

https://doi.org/10.3390/en19020518

Table 5. Comparative analysis of technical characteristics, jitter precision, and scalability for different triggering methods.

Triggering Method	Technical Characteristics	Jitter Precision	Scalability	Cost	Reference
Electrical triggering	Based on corona stabilized triggered switch	3.1–4.8 ns	Low	Low	[159]
	Cascaded amplification architecture: comprises an 8-channel digital delay generator (DDG), a pulse generator (PG), and a compact low-inductance Marx generator.	<2 ns	High	Middle	[160]
	FPGA timing control and magnetic ring transformers	Jitter depends on FPGA clock	High	Middle	[161]
	All-solid-state design, primary windings on individual cores; secondary winding encloses all cores, functioning as the Marx circuit magnetic switch	0.64–2.53 ns	High	Middle	[162]
	FPGA timing control combined with a compensation circuit consisting of an active gated integrator and an output comparator	Picosecond level	High	high	[163]
Optical triggering	Cascaded amplification architecture and high-power ultraviolet laser	<1 ns	High	Very high	[164]
	Combining DC bias avalanche photoconductive semiconductor switch (PCSS) and spark gap switch	<3 ns	High	Middle	[165]

3.3. Electromagnetic Interference

The generation of EMI is fundamentally driven by extreme electrical transients during pulse discharge. Regarding conduction coupling, the interference spectrum is heavily dependent on the switching topology and parasitic elements. Quantitative analysis of a 2.4 kV pulsed power system demonstrates that conducted EMI exhibits distinct characteristics across frequency bands. In the low-frequency domain (typically 20 kHz to 1 MHz), Differential mode (DM) noise dominates, driven by the fundamental switching frequency and its harmonics. the circulation of these high-amplitude DM currents through circuit resistive elements (particularly the ESR of capacitors) significantly exacerbates joule heating Conversely, in the high-frequency domain (> 1 MHz), Common mode (CM) noise becomes prevalent. This transition is fundamentally driven by the parasitic capacitance (C_p) between the switching devices and the heatsink, which provides a low-impedance path for high dV/dt currents. Furthermore, unmitigated stray inductances in the circuit layout often interact with these capacitances to induce significant resonance peaks in the 10–30 MHz range. Crucially, this CM interference activates two distinct failure mechanisms: firstly, the high-frequency voltage stress across the insulation interfaces promotes partial discharge and dielectric breakdown; secondly, the coupling of CM transients into sensitive control loops induces signal corruption. This can precipitate false triggering events, subjecting the power switches to catastrophic shoot-through faults or cumulative electro-thermal overstress [166]. However, this interference generates electromagnetic radiation characterized by an extremely wide spectrum. Via conduction coupling, it affects the internal control circuitry, accelerates the aging of internal switching elements, and interferes with external measurement equipment. This interference notably disrupts the normal operation of oscilloscopes, hindering system analysis by personnel, and also negatively impacts adjacent circuitry [167]. The EMI problem in pulsed power systems fundamentally adheres to the classic source–path–receptor model (interference source–coupling path–sensitive device), as illustrated in Figure 11. Conduction coupling transfers interference directly via

 https://doi.org/10.3390/en19020518

physical connections, while radiation coupling propagates as electric and magnetic field energy through free space and surrounding media.

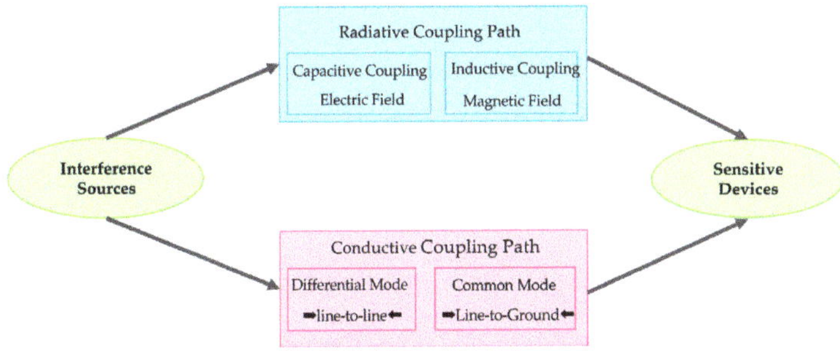

Figure 11. The fundamental EMI model illustrating the source–path–receptor relationship [168].

Firstly, at the packaging level, the primary strategy focuses on minimizing parasitic capacitance and inductance through structural innovation. Studies indicate that common-mode (CM) current can be effectively curtailed by replacing conventional direct bonded copper (DBC) ceramics with low-permittivity materials (such as epoxy) and by minimizing the copper area at the DBC's base. This optimization yields a 52.5% reduction in CM capacitance and a noise attenuation of 4–5 dB during steady-state operation, all while maintaining thermal stability [169]. Furthermore, for high-frequency switching modules, integrating a π-type common-mode filter (CMF) directly within the GaN module helps mitigate the impact of parasitic inductance in the interference path, achieving a significant noise reduction of up to 50 dB across the 10–100 MHz frequency range [170].

Secondly, regarding filtering, the objective is to actively cancel parasitic parameters and suppress broadband noise. One innovative technique involves magnetically coupling two inductors onto a single core to generate negative capacitance. This effectively neutralizes the inductor's equivalent parallel capacitance (EPC), reducing interference by 10 dB at frequencies above 5 MHz [171]. Moreover, to address wideband noise, a hybrid suppression strategy combining an active common-mode canceler (ACMC) with passive CMFs has been proposed. Experimental results demonstrate that this synergistic approach suppresses total conducted emissions by over 30 dB, ensuring signal levels remain well below standard regulatory limits [172].

Thirdly, material-level optimization aims to enhance EMC through advanced polymer nanocomposites. Unlike traditional metals, emerging composites reinforced with conductive fillers (such as graphene, CNTs, and MXenes) prioritize an absorption-dominated mechanism, effectively reducing secondary electromagnetic pollution. For instance, novel segregated structures facilitate the formation of efficient conductive networks at low filler loadings, attaining electrical conductivities exceeding 40 S/m and a Shielding effectiveness (SE) of 50–70 dB in the X-band (8.2–12.4 GHz) [173]. Similarly, conducting polymers—such as polyaniline, polypyrrole, and polythiophene—exhibit superior EMI shielding capabilities, realizing an SE of 60.2 dB in the 8–12 GHz range. Concurrently, the development of low-k dielectric materials further enhances signal integrity by minimizing capacitive coupling [174].

In summation, the EMI problem is characterized by its pervasive and latent nature. Fundamentally, the strategy involves segmenting the primary space into multiple distinct regions and implementing optimized shielding designs to maximally suppress the coupled interference currents, electric fields, and magnetic fields.

https://doi.org/10.3390/en19020518

3.4. Reliability Characteristics of Typical Topologies

While fundamental failure mechanisms—electrical, thermal, magnetic, and mechanical—are universal, the hierarchy of dominant stresses and mitigation strategies is intrinsically linked to the specific topology of the pulsed power supply (PPS). For instance, a design optimized for a Marx generator may be incompatible with a PFN. Consequently, analyzing reliability from a topological perspective is a prerequisite for effective system design. Table 6 presents a comparative analysis of the reliability characteristics of four representative PPS topologies across four dimensions: dominant stresses, critical vulnerable components, typical failure modes, and targeted design strategies.

Table 6. Comparison of reliability characteristics of four representative PPS topologies [175–180].

PPS Architecture	Operational Characteristics	Critical Components	Main Failure Mechanisms	Mitigation Strategies
Marx generator	High-voltage pulsed output High efficiency Flexible parameter adjustment	Switch Energy storage elements	Dielectric aging Switch Electrode Erosion Synchronization-induced Overvoltage	Electrode material and cooling optimization Low-jitter trigger system
Linear transformer driver (LTD)	High efficiency High power density Modular scalability	Magnetic Core Winding Switch Capacitance	Magnetic core loss and heating Dielectric breakdown Winding deformation	High-frequency low-loss magnetic material and liquid cooling Optimization of winding mechanical fixation
Pulse Forming Network (PFN)	Precise pulse waveform control Low system complexity High load adaptability	Inductance Capacitance Switch	Breakdown of capacitors and inductors Component overvoltage Breakdown	Overvoltage and Overcurrent protection Improved thermal management strategies
Blumlein Pulse Forming Line (BPFL)	Excellent pulse quality High voltage efficiency	Switch Blumlein transmission lines	Overcurrent-induced Switch breakdown Dielectric breakdown of transmission line	Addition of switch snubber circuit

Switching element failures are a critical bottleneck for high-voltage impulse generators. For Marx generators, the dominant issue is switch electrode erosion due to material ablation, whereas Blumlein lines are vulnerable to catastrophic overcurrent-induced switch breakdown caused by high current densities. Beyond switching, insulation and passive component degradation are universal challenges for all topologies in Table 6. While Marx generators suffer primarily from capacitor dielectric aging, LTDs, PFNs, and Blumlein lines exhibit similar vulnerabilities, including capacitor or inductor failure and dielectric breakdown in modules and transmission lines. Notably, LTDs encounter unique thermal stresses arising from magnetic core loss; without effective thermal management, this heating significantly accelerates insulation degradation. Regardless of topologies, common critical requirements remain. Rigorous thermal management and EMC design are paramount to minimizing failure risks. Fundamentally, reliability is not an intrinsic, static constant; it is a dynamic characteristic that evolves under the unique stress profiles of each topology.

https://doi.org/10.3390/en19020518

3.5. Synergistic Impact of System Integration on Reliability

This section reviews common reliability challenges in integrating pulsed power supply systems. Reliability depends on complex interactions between electrical, thermal, and magnetic fields, not just on individual parts. For example, in a solid-state Marx generator, microchannel cooling introducing parasitic parameters and EMI, complicate system design. Achieving low-jitter synchronization also increases EMI and adds complexity to EMC considerations [157]. Moreover, the switch's junction temperature directly influences its switching characteristics, meaning an unstable thermal field compromises switch synchronization and induces jitter [181]. These intertwined phenomena demonstrate that system-level challenges—including thermal management, EMC, trigger synchronization and topological characteristics—are not isolated, but rather interwoven into a tightly coupled challenge network. Furthermore, the distinct reliability profiles of different PPS topologies arise from how they configure and stress the basic components. The choice of topology dictates the primary stress vectors, which in turn prioritizes different hierarchies of the aforementioned system challenges, making a universal optimization approach unrealistic.

Therefore, achieving high reliability in pulsed power systems necessitates a component-driven, system-aware design philosophy that explicitly accounts for cross-domain interactions. The properties of components define the system-level constraints, but these constraints are entangled. Future design must evolve from independent, layered optimization to a holistic co-design framework, which requires multi-physics modeling and co-simulation tools that can capture the electro-thermal–magnetic coupling both within components and across system integration choices, enabling designers to navigate the inherent trade-offs and preemptively decouple failure pathways for robust, long-term operation.

4. Reliability Issues and Solutions in Extreme Environments

The above review provides a comprehensive analysis of reliability challenges and mitigation strategies for pulsed power supplies at both the component and system integration levels. However, pulsed power supplies seldom function within ideal laboratory conditions; rather, their ultimate reliability depends greatly on specific application scenarios. Variations in performance criteria, operational modes, and external environments cause each scenario to introduce distinct stressors to the power system, resulting in different failure modes and reliability requirements.

4.1. Aerospace

Leveraging its high-power density and adjustable frequency, the pulsed power supply is extensively utilized in the aerospace domain for critical applications such as satellite radar systems and plasma thrusters. However, space presents a comprehensive extreme environment—distinct from terrestrial conditions. Radiation effects, vacuum exposure, potentially leading to internal component failure, mission termination, and irreparable financial loss. In aerospace applications, the vacuum environment precludes convective heat transfer, leading to thermal accumulation and elevated junction temperatures under repetitive pulse operations. Concurrently, low ambient pressure and material outgassing compromise dielectric strength. When exacerbated by the high dV/dt characteristic of power switches, these factors significantly increase the susceptibility to surface flashover [182]. Furthermore, the radiation environment exerts a synergistic effect with the electric field. High-energy particles can trigger Single event burnout (SEB) in power devices, particularly during the off-state. Given that SEB sensitivity exhibits an exponential dependence on operating voltage, the implementation of strict voltage derating is imperative to mitigate

https://doi.org/10.3390/en19020518

these stochastic failure modes [183,184]. Consequently, mitigation strategies prioritize decoupling these environmental-electrical synergies. Insulation degradation is mitigated by employing radiation-hardened materials and robust packaging [185]. Furthermore, the exponential risk of SEB is suppressed through strict voltage derating and the integration of protection circuits [184].

4.2. Navigation

Leveraging its high energy density and rapid response capability, the pulsed power supply exhibits immense potential in maritime applications, including ship integrated power systems, directed energy weapons, and underwater detection. The marine environment creates major reliability challenges due to corrosion from salt and humidity, ongoing temperature changes, and strong mechanical stress from seawater. Figure 12 schematically illustrates the environmental stresses encountered by pulsed power supplies in the marine environment. The presence of salt mist and high humidity facilitates the formation of conductive aqueous films on PCB surfaces. Under continuous electric bias, this environment induces electrochemical migration (ECM), where metal ions dissolve at the anode and migrate toward the cathode to form dendritic structures. This process progressively degrades the surface insulation resistance (SIR), eventually leading to short-circuit failures [186]. Simultaneously, the system is subjected to thermal cycling arising from both environmental fluctuations and pulsed load operations. Due to the mismatch in coefficients of thermal expansion (CTE) among semiconductor chips, solder layers, and substrates, cyclic shear stress is generated at bonding interfaces. This stress induces the accumulation of plastic strain within solder joints, driving crack propagation and resulting in eventual fatigue fracture [187]. Therefore, the mitigation strategy focuses on blocking the electrochemical reaction path and alleviating interfacial stress. To prevent ECM, conformal coatings and hermetic housings are applied to isolate the circuit from moisture and chloride ions, effectively cutting off the electrolyte formation. To mitigate thermal fatigue, the design employs CTE-matched baseplates and advanced silver sintering technologies. These materials reduce the shear strain range experienced by the interconnection layers, thereby extending the fatigue life under cyclic thermal loads [186,188].

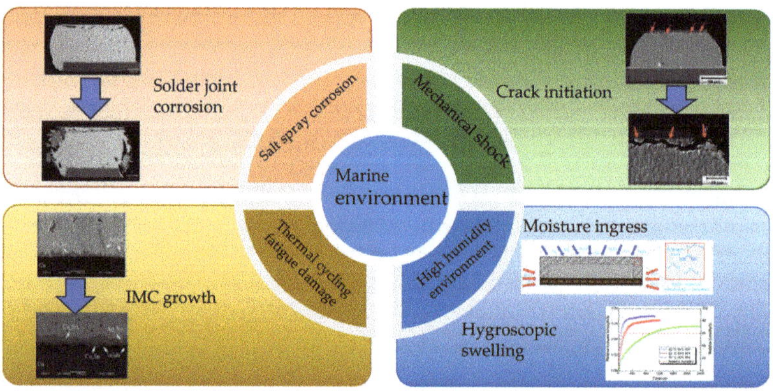

Figure 12. Environmental stressors and reliability challenges for pulsed power systems deployed in marine applications [189–192].

4.3. Geological Exploration and Resource Exploration

Leveraging its potent instantaneous electromagnetic field excitation capability, the pulsed power supply has become central to geological exploration equipment, such as those used for deep subsurface surveying and oil resource identification. However, deployment

https://doi.org/10.3390/en19020518

in extreme environments (such as deserts and plateaus) imposes a dual constraint: the system must maintain mobility while ensuring reliable operation.

In desert environments, system reliability is fundamentally challenged by the disruption of thermal equilibrium. The primary stressor is the high ambient temperature induced by prolonged solar radiation, which accelerates the degradation of internal materials and reduces convective heat transfer efficiency, ultimately resulting in an increase in ESR. Furthermore, windblown sand tends to accumulate on heat dissipation surfaces, forming a thermal insulation layer. This accumulation creates a positive feedback loop with the high-temperature environment, significantly increasing the system's thermal resistance Rth [193]. Consequently, internal heat dissipation is impeded, causing the junction temperature of power devices to rise continuously until system failure occurs. Therefore, mitigation strategies must focus on maintaining thermal balance and implementing sand-resistant designs. The risk of thermal aging is minimized by employing high-temperature tolerant materials such as gold [194], high-temperature tolerant devices and high-efficiency cooling methods. Simultaneously, the ingress of windblown sand is prevented through the use of sealed enclosures or filtration systems.

In high-altitude plateau applications, the reliability of pulsed power supplies is strictly constrained by the coupling of hypobaric and cryogenic conditions. The primary threat is the degradation of air dielectric strength due to the rarefied atmosphere. According to equation 5, Paschen's Law, the breakdown voltage of the air gap decreases significantly at lower pressures, drastically increasing the risk of corona inception and surface flashover across high-voltage components [195].

$$V_B = \frac{B \cdot p \cdot d}{\ln(A \cdot p \cdot d) - \ln(\ln(1 + \frac{1}{\gamma_{se}}))} \tag{8}$$

where V_B denotes the gas breakdown voltage, p is the gas pressure, and d represents the inter-electrode distance. A and B denote Townsend's first and second ionization coefficients, while γ_{se} represents the secondary electron emission coefficient. Furthermore, extreme low temperatures cause the electrolyte in energy storage capacitors to freeze. This leads to a sharp rise in ESR and a reduction in capacitance, which distorts the output pulse waveform and degrades energy transfer efficiency [196]. Therefore, mitigation strategies must address both corona discharge hazards and the adverse effects of low temperatures. Key measures include the utilization of corona-resistant materials, increasing inter-electrode spacing, and selecting low-temperature tolerant devices.

4.4. Comparative Analysis and Mitigation Strategies

In conclusion, while specific environmental manifestations vary significantly, the fundamental reliability challenge for pulsed power supplies remains consistent: the coupling of environmental physics with high-voltage electrical stress. Environmental factors serve as catalysts that significantly accelerate the degradation of components under electrical load. External stressors do not act in isolation but exacerbate intrinsic failure mechanisms such as thermal aging and insulation breakdown. Table 7 provides a comprehensive summary of these environmental-physical interactions and corresponding mitigation strategies. This systematic mapping underscores that in future, ensuring the reliability of pulsed power systems in extreme scenarios requires a transition from empirical retrofitting to physics-based predictive design, fundamentally aiming to decouple sensitive components from these degradation pathways.

https://doi.org/10.3390/en19020518

Table 7. Overview of environmental stressors, dominant failure mechanisms, and targeted mitigation strategies for pulsed power systems.

Scenario	Environmental Stressors	Dominant Failure Mechanism	Mitigation Strategy
Aerospace	Vacuum High-energy radiation	Surface flashover Thermal accumulation Oxide layer damage	Radiation-hardened design Derating usage
Navigation	High humidity High salinity Thermal cycling	Reduction in insulation resistance Solder joint fatigue	Coating design Advanced soldering technology
Desert	Long-term high temperature Sand interference	Thermal feedback loop caused by dust blockage and solar heat	High-temperature resistant design Hermetic sealing

5. Conclusions and Prospects

Serving as the fundamental technological foundation for transient energy release and ultra-high-power output, the reliability of pulsed power supplies is critical in assessing both system engineering readiness and sustained operational capacity. This paper systematically reviews recent progress at three levels, component technologies, system integration, and adaptability to harsh environments, identifying a distinct "layered optimization" trend in current technologies. At the component level, innovations in dielectric modification, composite films, porous electrodes, and wide-bandgap semiconductors have significantly enhanced energy density, temperature stability, and switching performance. Concurrently, advanced packaging has improved the management of internal electro-thermal stresses under high-repetition pulses. At the system level, solutions for thermal management, multi-module synchronization, and EMI suppression are summarized. At the application level, protective designs tailored for extreme environments are discussed to ensure survivability.

However, a fundamental bottleneck remains in this layered approach. Failures in PPS operation are intrinsically the result of dynamic coupling, transmission, and evolution of multi-physics stresses (electrical–thermal–magnetic–mechanical) across the "Component–System–Environment" hierarchy. Existing research largely addresses static stresses in isolation or considers simple unidirectional impacts, failing to establish a framework that bridges these layers to describe the dynamic synergistic evolution of failures. This lack of systemic methodology means that high reliability at the component level does not necessarily translate into system-level longevity. This has resulted in notable voids within the current body of knowledge regarding PPS reliability, particularly at three key levels:

1. Current research predominantly focuses on failure analysis of individual components or interactions within limited physical fields. There is a notable lack of investigation into the performance interaction and degradation correlation among multiple components under coupled electrical–thermal–magnetic–mechanical stresses. Consequently, a widely accepted system-level prediction model capable of describing the cross-component propagation and evolution of failures has yet to be established.
2. Current reliability verification experiments are predominantly conducted under standardized or idealized conditions. They fail to adequately replicate the dynamic coupling of electrical–thermal–magnetic–mechanical stresses and environmental factors inherent in real-world applications. Consequently, there is a significant discrepancy between laboratory data and actual field performance.
3. Current condition monitoring technologies are primarily focused on reactive fault detection and alarming. There is a distinct lack of comprehensive capabilities for real-time health assessment and prognostic prediction.

https://doi.org/10.3390/en19020518

In conclusion, this systematic review highlights the critical need to transcend the limitations of traditional layered design. Future research must prioritize bridging gaps in multi-component synergistic modeling, realistic system verification, and intelligent health management. By adopting multi-scale modeling and simulation, designing coupled multi-stress accelerated life testing protocols, and employing hybrid intelligent algorithms combining data-driven and model-driven approaches, a full-lifecycle reliability assurance system can be established. This framework will be pivotal in facilitating the deployment of next-generation high-performance and high-reliability PPS.

Funding: This project was supported by the Tianjin Municipal Education Commission Scientific Research Project (Natural Science, Grant No. 2023KJ301) and the Youth Fund for Scientific Research Projects of Higher Education Institutions in Hebei Province of China (No. QN2025339). The Article Processing Charge (APC) was funded by the Youth Fund for Scientific Research Projects of Higher Education Institutions in Hebei Province of China.

Data Availability Statement: No new data were created or analyzed in this study.

Conflicts of Interest: The authors declare no conflict of interest.

References

1. Martin, J.C. Nanosecond pulse techniques. *Proc. IEEE* **1992**, *80*, 934–945. [CrossRef]
2. Jiang, S.; Hu, H.; Wang, S.; Li, L.; Peng, T. Design of pulsed power supply for repetitive pulsed high magnetic field for water electrolysis. *Rev. Sci. Instrum.* **2021**, *92*, 114708. [CrossRef] [PubMed]
3. Dong, J.; Gui, Y.; Li, J.; Zhang, J. Design of Pulsed Power Supply for Electromagnetic Launch. *High Volt. Eng.* **2007**, *33*, 105–107.
4. Novac, B.; Istenic, M.; Luo, J.; Smith, I.; Brown, J.; Hubbard, M.; Appelgren, P.; Elfsberg, M.; Hurtig, T.; Möller, C.; et al. A 10-GW pulsed power supply for HPM sources. *IEEE Trans. Plasma Sci.* **2006**, *34*, 1814–1821. [CrossRef]
5. Qiu, H.; Wang, S.H.; Sun, F.J.; Wang, Z.G.; Zhang, N.M. Transient Electromagnetic Field Analysis for the Single-Stage Fast Linear Transformer Driver with Two Different Configurations Using the Finite-Element Method and Finite Integration Technique. *IEEE Trans. Magn.* **2020**, *56*, 7515805. [CrossRef]
6. Borthakur, S.; Talukdar, N.; Neog, N.K.; Borthakur, T.K.; Kumar, R.; Verma, R.; Shyam, A. 200 kJ Pulsed Power System for Pulsed Plasma Device. *IEEE Trans. Plasma Sci.* **2017**, *45*, 1769–1775. [CrossRef]
7. Jang, S.-R.; Ryoo, H.-J.; Ok, S.-B. Pulsed-Power System for Leachate Treatment Applications. *J. Power Electron.* **2011**, *11*, 612–619. [CrossRef]
8. Tseng, S.Y.; Wu, T.F.; Wu, M.W. Bipolar narrow-pulse generator with energy-recovery feature for liquid-food sterilization. *IEEE Trans. Ind. Electron.* **2008**, *55*, 123–132. [CrossRef]
9. Katiyo, W.; Yang, R.; Zhao, W. Effects of combined pulsed electric fields and mild temperature pasteurization on microbial inactivation and physicochemical properties of cloudy red apple juice (Malus pumila Niedzwetzkyana (Dieck)). *J. Food Saf.* **2017**, *37*, e12369. [CrossRef]
10. Ferraz, L.; Silva, E. Pulsed electric field-assisted extraction techniques for obtaining vegetable oils and essential oils: Recent progress and opportunities for the food industry. *Sep. Purif. Technol.* **2025**, *354*, 128833. [CrossRef]
11. Golberg, A.; Yarmush, M. Nonthermal Irreversible Electroporation: Fundamentals, Applications, and Challenges. *IEEE Trans. Biomed. Eng.* **2013**, *60*, 707–714. [CrossRef] [PubMed]
12. Jiang, C.; Davalos, R.; Bischof, J. A Review of Basic to Clinical Studies of Irreversible Electroporation Therapy. *IEEE Trans. Biomed. Eng.* **2015**, *62*, 4–20. [CrossRef] [PubMed]
13. Jiang, H.; Jiang, X.; Wang, Z.; Sun, F.; Wei, H.; Lou, C.; Qiu, A. Discharge characteristics of trigger brick gas switches for 12-stage linear transformer driver. *High Power Laser Part. Beams* **2024**, *36*, 115001.
14. Lv, Y.; Wang, H.; Liu, X.; Lu, S.; Lian, H. A High Frequency Bipolar Nanosecond Pulse Generator with High Voltage Gain Based on Inductive Discharge for DBD Application. *IEEE Trans. Power Electron.* **2025**, *40*, 17998–18010. [CrossRef]
15. Pandit, P.P.; Jency, J.G.; Babu, M.; Kumar, K.; Nirmal, D. Fabrication of ultra flexible super capacitor using polyvinylidene fluoride. In Proceedings of the 2017 International Conference on Innovations in Electrical, Electronics, Instrumentation and Media Technology (ICEEIMT), Coimbatore, India, 3–4 February 2017.
16. Liu, W.; Sun, X.; Yan, X.; Gao, Y.; Zhang, X.; Wang, K.; Ma, Y. Review of Energy Storage Capacitor Technology. *Batteries* **2024**, *10*, 271. [CrossRef]
17. He, Y.; Wang, F.; Du, G.; Pan, L.; Wang, K.; Gerhard, R.; Plath, R.; Rozga, P.; Trnka, P. Revisiting the thermal ageing on the metallised polypropylene film capacitor: From device to dielectric film. *High Volt.* **2023**, *8*, 305–314. [CrossRef]

 https://doi.org/10.3390/en19020518

18. Luo, C.; Cong, P.; Zhang, T.; Zhong, X.; Luo, W. Prediction and experimental verification of erosion resistance of gas switch electrode materials. *AIP Adv.* **2021**, *11*, 055206. [CrossRef]
19. Sun, H.; Liu, W.; Liu, K. Summary of research on control technology of pulsed power supply in electromagnetic launch system. In Proceedings of the 16th Annual Conference of China Electrotechnical Society: Volume II, Beijing, China, 24–26 September 2021.
20. Bluhm, H. *Pulsed Power Systems: Principles and Applications*, 1st ed.; Springer: Berlin/Heidelberg, Germany, 2006; pp. 7–54.
21. Aghmadi, A.; Mohammed, O. Energy Storage Systems: Technologies and High-Power Applications. *Batteries* **2024**, *10*, 141. [CrossRef]
22. Chen, Z.; Sun, J.; Dong, Z.; Ren, J.; Jia, W.; Guo, F.; Xie, L.; Wu, W.; Yao, X. Failure behavior of dielectric films for peaking capacitor subjected to surface flashover under nanosecond pulses. *AIP Adv.* **2023**, *13*, 125203. [CrossRef]
23. Xu, Y.; Xiang, Y.; Wan, Q.; Yang, R.; Xiao, H.; Ding, H.; Li, L. A Dual-Capacitors Type Energy Recovery Power System for Repetitive Pulsed High Magnetic Fields. *J. Low Temp. Phys.* **2013**, *170*, 488–495. [CrossRef]
24. Wang, J.; Gao, D.-Q.; Shen, W.-Z.; Yan, H.-B.; Mao, L.-J. Reliability of DC-link capacitor in pulsed power supply for accelerator magnet. *Nucl. Sci. Tech.* **2024**, *35*, 151. [CrossRef]
25. Kolobov, V.; Barannik, M.; Selivanov, V.; Kuklin, D. A current source with an inductive energy storage for measuring pulse impedances of grounding connections. *Instrum. Exp. Tech.* **2014**, *57*, 572–578. [CrossRef]
26. Liebfried, O. Review of Inductive Pulsed Power Generators for Railguns. *IEEE Trans. Plasma Sci.* **2017**, *45*, 1108–1114. [CrossRef]
27. Perez, J.; Sugai, T.; Jiang, W.; Tokuchi, A.; Horie, M.; Ohshio, Y.; Ueno, K. High current pulse forming network switched by static induction thyristor. *Matter Radiat. Extremes* **2018**, *3*, 261–266. [CrossRef]
28. Vazquez, S.; Lukic, S.M.; Galvan, E.; Franquelo, L.G.; Carrasco, J.M. Energy Storage Systems for Transport and Grid Applications. *IEEE Trans. Ind. Electron.* **2010**, *57*, 3881–3895. [CrossRef]
29. Kondoh, J.; Funamoto, T.; Nakanishi, T.; Arai, R. Energy characteristics of a fixed-speed flywheel energy storage system with direct grid-connection. *Energy* **2018**, *165*, 701–708. [CrossRef]
30. Mousavi, G.S.M.; Faraji, F.; Majazi, A.; Al-Haddad, K. A comprehensive review of Flywheel Energy Storage System technology. *Renew. Sustain. Energy Rev.* **2017**, *67*, 477–490. [CrossRef]
31. Huang, Y.; Wang, L.; Zhang, D.; Chen, L.; Yuan, J.; Xie, W.; Deng, M.; Kong, Y. Electrical and thermal aging analysis of all-film pulsed capacitor. *High Power Laser Part. Beams* **2024**, *36*, 025020.
32. Wang, H.; Blaabjerg, F. Reliability of Capacitors for DC-Link Applications in Power Electronic Converters-An Overview. *IEEE Trans. Ind. Appl.* **2014**, *50*, 3569–3578. [CrossRef]
33. Chun, J.; Heo, J.; Lee, K.; Ye, B.U.; Kang, B.S.; Yoon, S.-H. Thermal activation energy on electrical degradation process in $BaTiO_3$ based multilayer ceramic capacitors for lifetime reliability. *Sci. Rep.* **2024**, *14*, 616. [CrossRef]
34. Ye, X.; Hu, Y.; Zheng, B.; Chen, C.; Feng, R.; Liu, S.; Zhai, G. Reliability assessment of film capacitors oriented by dependent and nonlinear degradation considering three-source uncertainties. *Microelectron. Reliab.* **2021**, *126*, 114277. [CrossRef]
35. Liu, H.; Qiu, J.; Zhang, W.; Zhang, M.; Dou, Z.; Chen, L. Lifetime prediction and reliability analysis for aluminum electrolytic capacitors in EV charging module based on mission profiles. *Front. Electron.* **2023**, *4*, 1226006. [CrossRef]
36. Umran, H.; Wang, F.; He, Y. Ageing: Causes and Effects on the Reliability of Polypropylene Film Used for HVDC Capacitor. *IEEE Access* **2020**, *8*, 40413–40430. [CrossRef]
37. Tortai, J.; Bonifaci, N.; Denat, A.; Trassy, C. Diagnostic of the self-healing of metallized polypropylene film by modeling of the broadening emission lines of aluminum emitted by plasma discharge. *J. Appl. Phys.* **2005**, *97*, 053304. [CrossRef]
38. Gupta, A.; Yadav, O.; DeVoto, D.; Major, J. A review of degradation behavior and modeling of capacitors. In Proceedings of the ASME International Technical Conference and Exhibition on Packaging and Integration of Electronic and Photonic Microsystems, San Francisco, CA, USA, 27–30 August 2019.
39. Choi, J.S.; Kim, D.C.; Shin, H.S.; Yeo, D.H.; Lee, J.H. Effect of oxygen vacancies on dielectric property and reliability of anti-ferroelectric PLZT applicable to EV-MLCC. *Bull. Mater. Sci.* **2024**, *47*, 39. [CrossRef]
40. Laadjal, K.; Cardoso, A.J.M. Multilayer Ceramic Capacitors: An Overview of Failure Mechanisms, Perspectives, and Challenges. *Electronics* **2023**, *12*, 1297. [CrossRef]
41. Zheng, B.; Ma, R.; Yuan, S. Transient failure mode of $(Ba,Ca)(Ti,Zr)O_3$ pulsed power capacitor based on electromechanical coupling mechanism. *Ceram. Int.* **2019**, *45*, 7654–7660. [CrossRef]
42. David, J.; Ciufo, P.; Elphick, S.; Robinson, D. Preliminary Evaluation of the Impact of Sustained Overvoltage on Low Voltage Electronics-Based Equipment. *Energies* **2022**, *15*, 1536. [CrossRef]
43. Tu, C.; Chai, M.; Yu, Y.; Guo, Q.; Chen, J.; Xiao, B.; Long, L. Method for monitoring ESR and capacitance based on discharge rule of aluminum electrolytic capacitors. *Electr. Power Autom. Equip.* **2020**, *40*, 108–113.
44. Cao, L.; Xi, R.; Zhou, C.; He, G.; Yang, F.; Xu, L.; Li, H. Polymer Capacitor Films with Nanoscale Coatings for Dielectric Energy Storage: A Review. *Coatings* **2024**, *14*, 1193. [CrossRef]
45. Tan, D.Q.; Wu, X. A case study of high-temperature polyetherimide film capacitor fabrication. *Mater. Today Energy* **2022**, *30*, 101167. [CrossRef]

https://doi.org/10.3390/en19020518

46. Cheng, R.; Wang, Y.; Men, R.; Lei, Z.; Song, J.; Li, Y.; Guo, M. High-energy-density polymer dielectrics via compositional and structural tailoring for electrical energy storage. *Iscience* **2022**, *25*, 104837. [CrossRef] [PubMed]
47. Jian, G.; Jiao, Y.; Feng, L.; Meng, Q.; Yang, N.; Zhu, S.; Lu, M.; Wong, C.-P. High energy density of BaTiO3@ TiO_2 nanosheet/polymer composites via ping-pong-like electron area scattering and interface engineering. *NPG Asia Mater.* **2022**, *14*, 4. [CrossRef]
48. Zhang, W.; Cheng, X.; Zhen, Y.; Cao, X.; Fu, Z.; Zhang, H.; Wang, X. Enhanced electrical reliability of Zr-doped BaTiO3 nanoceramics: First-principle calculations and experimental studies. *J. Am. Ceram. Soc.* **2025**, *108*, e20273. [CrossRef]
49. Wang, J.; Shen, Z.; Liu, R.; Shen, Y.; Chen, L.; Liu, H.; Nan, C. Texture Engineering Modulating Electromechanical Breakdown in Multilayer Ceramic Capacitors. *Adv. Sci.* **2023**, *10*, 2300320. [CrossRef]
50. Bahaa, A.; Alhammadi, A.; Lethesh, K.; Susantyoko, R.; Bamgbopa, M. Ionic liquid electrolyte selection for high voltage supercapacitors in high-temperature applications. *Front. Chem.* **2024**, *12*, 1349864. [CrossRef]
51. Castillo, J.; Santiago, A.; Judez, X.; Garbayo, I.; Clemente, J.; Morant-Miñana, M.; Villaverde, A.; González-Marcos, J.; Zhang, H.; Armand, M.; et al. Safe, Flexible, and High-Performing Gel-Polymer Electrolyte for Rechargeable Lithium Metal Batteries. *Chem. Mater.* **2021**, *33*, 8812–8821. [CrossRef]
52. Feng, M.; Feng, Y.; Zhang, T.; Li, J.; Chen, Q.; Chi, Q.; Lei, Q. Recent Advances in Multilayer-Structure Dielectrics for Energy Storage Application. *Adv. Sci* **2021**, *8*, 2102221. [CrossRef]
53. MacDougall, F.; Ennis, J.; Yang, X.; Cooper, R.; Gilbert, J.; Bates, J.; Naruo, C.; Schneider, M.; Keller, N.; Joshi, S.; et al. High energy density capacitors for pulsed power applications. In Proceedings of the 2009 IEEE Pulsed Power Conference, Washington, DC, USA, 28 June–2 July 2009.
54. Hunstig, M.; Schaermann, W.; Broekelmann, M.; Holtkaemper, S.; Siepe, D.; Hesse, H.J. Smart Ultrasonic Welding in Power Electronics Packaging. In Proceedings of the 11th International Conference on Integrated Power Electronics Systems, Berlin, Germany, 24–26 March 2020.
55. Xia, J.; Cao, F.; Chen, X.; Lin, Z.; Song, Y.; Chen, Y.; Yan, S.; Wang, G. Investigation on Discharge Behavior of Antiferroelectric Multilayer Ceramic Capacitors. *IEEE Trans. Dielectr. Electr. Insul.* **2023**, *30*, 643–648. [CrossRef]
56. Peng, Z.; Shi, Q.; Zhang, F.; Liu, J.; Nie, X.; Wang, J.; Xu, S.; Wu, D.; Yang, Z.; Chao, X. A new family of high temperature stability and ultra-fast charge-discharge KNN-based lead-free ceramics. *J. Mater. Sci.* **2022**, *57*, 9992–10002. [CrossRef]
57. Li, Z.; Li, H.; Qiu, T.; Lin, F.; Wang, Y. Reliability criterion of film capacitor based on moisture diffusion in encapsulation. *High Volt.* **2023**, *8*, 1242–1252. [CrossRef]
58. Li, H.; Huang, X.; Li, Z.; Li, H.; Wang, W.; Wang, B.; Zhang, Q.; Lin, F. Modeling of ESR in metallized film capacitors and its implication on pulse handling capability. *Microelectron. Reliab.* **2015**, *55*, 1046–1053. [CrossRef]
59. Lv, C.; Liu, J.; Zhang, Y.; Yin, J.; Zhang, X. A High-Resolution Analytical Thermal Modeling Method of Capacitor Bank Considering Thermal Coupling and Different Cooling Modes. *IEEE Trans. Power Electron.* **2023**, *38*, 7674–7684. [CrossRef]
60. Rezaei, M.A.; Fathollahi, A.; Akbari, E.; Saki, M.; Khorgami, E.; Teimouri, A.R.; Chronopoulos, A.T.; Mosavi, A. Reliability Calculation Improvement of Electrolytic Capacitor Banks Used in Energy Storage Applications Based on Internal Capacitor Faults and Degradation. *IEEE Access* **2024**, *12*, 13146–13164. [CrossRef]
61. Zhang, W.; Li, Y.; Xiao, W.; Zhang, S.; Xiao, X. Identification of electrical breakdown faults in shunt capacitor bank elements using instantaneous negative sequence current. *Int. J. Electr. Power Energy Syst.* **2025**, *170*, 110962. [CrossRef]
62. Mansourizadeh, H.; Hosseinpour, M.; Seifi, A.; Shahparasti, M. A 13-level switched-capacitor-based multilevel inverter with reduced components and inrush current limitation. *Sci. Rep.* **2025**, *15*, 290. [CrossRef]
63. Khan, M.; Barzegarkhoo, R.; Siwakoti, Y.; Khan, S.; Li, L.; Blaabjerg, F. A new switched-capacitor multilevel inverter with soft start and quasi resonant charging capabilities. *Int. J. Electr. Power Energy Syst.* **2022**, *135*, 107412. [CrossRef]
64. Ou, S.; Hassanifar, M.; Votava, M.; Sangwongwanich, A.; Sahoo, S.; Langwasser, M.; Liserre, M.; Blaabjerg, F. Monolithic Data-Driven Condition Monitoring Strategy for MMC Considering C and ESR. *IEEE Trans. Power Electron.* **2025**, *40*, 11339–11354. [CrossRef]
65. Zhao, J.; Zhou, Y.; Zhu, Q.; Song, Y.; Liu, Y.; Luo, H. A remaining useful life prediction method of aluminum electrolytic capacitor based on wiener process and similarity measurement. *Microelectron. Reliab.* **2023**, *142*, 114928. [CrossRef]
66. Zhao, X.; Wu, H.; Yang, D.; Wei, J.; Li, Z.; Guo, Z.; Zhang, S. Residual Magnetism Elimination Method for Large Power Transformers Based on Energy Storage Oscillation. *IEEE Trans. Power Delivery* **2025**, *40*, 2759–2768. [CrossRef]
67. Smirnov, A.; Ivanov, E.; Pronikov, A.; Kutsaev, S. High-Voltage Pulsed Power Generator for Beam Injection Systems. *Electronics* **2025**, *14*, 535. [CrossRef]
68. Spielman, R.B.; Froula, D.H.; Brent, G.; Campbell, E.M.; Reisman, D.B.; Savage, M.E.; Shoup, M.J., III; Stygar, W.A.; Wisher, M.L. Conceptual design of a 15-TW pulsed-power accelerator for high-energy-density-physics experiments. *Matter Radiat. Extrem.* **2017**, *2*, 204–223. [CrossRef]
69. Kumar, A.; Jeyan, J.V.M.L.; Agarwal, A. Numerical analysis on 10 MJ solenoidal high temperature superconducting magnetic energy storage system to evaluate magnetic flux and Lorentz force distribution. *Phys. C-Supercond.* **2019**, *558*, 17–24. [CrossRef]

https://doi.org/10.3390/en19020518

70. Zhao, X.; Wu, G.; Yang, D.; Xu, G.; Xing, Y.; Yao, C.; Abu-Siada, A. Enhanced detection of power transformer winding faults through 3D FRA signatures and image processing techniques. *Electr. Power Syst. Res.* **2025**, *242*, 111433. [CrossRef]
71. Yuan, Q.; Deng, Z.; Ding, W.; Wang, Y.; Wu, J. New advances in solid-state pulse generator based on magnetic switches. *Rev. Sci. Instrum.* **2022**, *93*, 051501. [CrossRef]
72. Go, T.; Kanesawa, K.; Yamazaki, N.; Mukaigawa, S.; Takaki, K.; Fujiwara, T. Energy efficiency of inductive energy storage system pulsed power generator using fast recovery diode. *IEEJ Trans. Fundam. Mater.* **2009**, *129*, 23–29. [CrossRef]
73. Xu, K.; Guo, Y.; Lei, G.; Zhu, J. A Review of Flywheel Energy Storage System Technologies. *Energies* **2023**, *16*, 6462. [CrossRef]
74. Tao, X.; Liu, K. Pulse shaping method for compulsator. *High Power Laser Part. Beams* **2018**, *30*, 72–76.
75. Wang, H.; Liu, K.; Zhu, B.; Feng, J.; Ao, P.; Zhang, Z. Analytical Investigation and Scaled Prototype Tests of a Novel Permanent Magnet Compulsator. *IEEE Trans. Magn.* **2015**, *51*, 8203309. [CrossRef]
76. Sun, Z.; Liu, Y.; Hao, X. Present status and development trend of pulsed power supplies for railguns. *Adv. Technol. Electr. Eng. Energy* **2022**, *41*, 49–63.
77. Hu, D.; Dai, X.; Xie, B.; Li, W.; Yu, H.; Chen, H. Fatigue Life of Flywheel Energy Storage Rotors Composed of 30Cr2Ni4MoV Steel. *Energies* **2024**, *17*, 3730. [CrossRef]
78. Manivasagam, R.; Richard, S. Review on recent developments in pulsed high power systems. In Proceedings of the 3rd International Conference on Engineering Facilities Maintenance and Management Technologies (EFM2T'21), Palapye, Botswana, 28–29 October 2021.
79. Sun, L.; Zhang, Y.; Zu, Y.; Guo, J.; Yin, H.; Song, Q.; Tang, X. A solid-state pulse power sub-nanosecond SiC DSRD-based generator with high-voltage and high repetition frequency for pulse discharge water treatment. *Environ. Res.* **2024**, *252*, 119053. [CrossRef] [PubMed]
80. Panda, S.K.; Mitra, S.; Mondal, J.; Roy, A. Comparative Study of Different High Voltage Switches Used in Pulsed High Voltage Application. *Power Res. J. CPRI* **2024**, *19*, 109–115. [CrossRef]
81. Yao, X.; Zeng, Z.; Chen, J. Electrode erosion of a high energy impulse spark gap switch. *Plasma Sci. Technol.* **2005**, *7*, 3157–3160. [CrossRef]
82. Man, L.; Liang, Y.; Zhang, N.; Cao, N.; Wang, Y.; Xue, P. Research on Self-breakdown Performance of Gas Spark Gap Switch. *High Volt. Eng.* **2021**, *57*, 146–151.
83. Li, X.; Pei, Z.; Zhang, Y.; Liu, X.; Li, Y.; Zhang, Q. Electrode erosion properties of gas spark switches for fast linear transformer drivers. *Phys. Plasmas* **2017**, *24*, 122108–1–122108-13. [CrossRef]
84. Rukin, S.N. Pulsed power technology based on semiconductor opening switches: A review. *Rev. Sci. Instrum.* **2020**, *91*, 011501. [CrossRef]
85. Meyers, V.; Voss, L.; Flicker, J.D.; Rodriguez, L.G.; Hjalmarson, H.P.; Lehr, J.; Gonzalez, N.; Pickrell, G.; Ghandiparsi, S.; Kaplar, R. Photoconductive Semiconductor Switches: Materials, Physics, and Applications. *Appl. Sci.* **2025**, *15*, 645. [CrossRef]
86. Wang, J.; Chung, H.S.-H.; Li, R.T.-H. Characterization and Experimental Assessment of the Effects of Parasitic Elements on the MOSFET Switching Performance. *IEEE Trans. Power Electron.* **2013**, *28*, 573–590. [CrossRef]
87. Wu, H.; He, Z.; Guo, R.; Ma, J.; Zhao, H.; Chen, X.; Chen, W.; Xu, T. Characterization of the high current triggered vacuum switch. *Vacuum* **2012**, *86*, 1911–1915. [CrossRef]
88. Scholfield, D.W.; Butcher, M.D.; Hilko, B.; Dorr, G. Vacuum switch performance in a 1.2 MJ pulse forming network. *Rev. Sci. Instrum.* **2008**, *79*, 024703. [CrossRef] [PubMed]
89. Xie, C.; Tan, X.; Du, T.; Tang, B.; Shang, S.; Li, R.; Li, Z. Electrode erosion research of gas spark gap. *High Power Laser Part. Beams* **2014**, *26*, 131–139.
90. Zhong, W.; Liu, Y.; Xu, A.; Shang, S.; Jin, D. Experimental investigation of electrode erosion of triggered spark gap. In Proceedings of the 27th International Symposium on Discharges and Electrical Insulation in Vacuum (ISDEIV), Suzhou, China, 18–23 September 2016.
91. Sun, Q.; Zhou, Q.; Yang, W.; Dong, Y.; Zhang, H.; Song, M.; Zhou, M.; Yang, W. Modeling of Electrode Erosion Process in Gas Spark Switch. *Plasma Chem. Plasma Process.* **2023**, *43*, 1195–1215. [CrossRef]
92. Wu, J.; Han, R.; Ding, W.; Qiu, A.; Tang, J. Note: Erosion of W-Ni-Fe and W-Cu alloy electrodes in repetitive spark gaps. *Rev. Sci. Instrum.* **2018**, *89*, 026103. [CrossRef]
93. Silakhori, K.; Neghabi, M.; Torabi, R. Investigation on electrode erosion effects in high frequency rotary spark gap switches. *Optik* **2021**, *237*, 166763. [CrossRef]
94. Song, F.; Li, F.; Zhang, B.; Zhu, M.; Li, C.; Wang, G.; Gong, H.; Gan, Y.; Jin, X.; Novac, B.M.; et al. Electrode Erosion and Lifetime Performance of a Compact and Repetitively Triggered Field Distortion Spark Gap Switch. *IEEE Trans. Plasma Sci.* **2020**, *48*, 212–218. [CrossRef]
95. Li, L.; Li, H.; Li, C.; Qi, X.; Lin, F. Analysis on Electrode Replacement of Spark-Gap Switches with Graphite Electrodes. *IEEE Trans. Plasma Sci.* **2011**, *39*, 1874–1880. [CrossRef]

 https://doi.org/10.3390/en19020518

96. Zhang, B.; Gan, Y.; Li, C.; Li, F.; Jin, X.; Song, F. Research on channel resistance and thermal effect characteristics of gas-filled spark gap switch. *High Power Laser Part. Beams* **2024**, *36*, 134–139.
97. Zhong, W.; Liu, Y.; Wang, L.; Jin, D.; Tan, X. Statistical analysis of particulate concentration from electrode erosion of gas discharge. *J. Phys. D Appl. Phys.* **2017**, *50*, 015202. [CrossRef]
98. Akbar, G.; Di Fatta, A.; Rizzo, G.; Ala, G.; Romano, P.; Imburgia, A. Comprehensive Review of Wide-Bandgap (WBG) Devices: SiC MOSFET and Its Failure Modes Affecting Reliability. *Physchem* **2025**, *5*, 10. [CrossRef]
99. Ma, C.-T.; Gu, Z.-H. Review on Driving Circuits for Wide-Bandgap Semiconductor Switching Devices for Mid- to High-Power Applications. *Micromachines* **2021**, *12*, 65. [CrossRef]
100. Mao, H.; Wu, B.; Lan, X.; Xia, Y.; Chen, J.; Tang, L. Research on Improving the Avalanche Current Limit of Parallel SiC MOSFETs. *Electronics* **2025**, *14*, 2502. [CrossRef]
101. Grome, C.A.; Ji, W. A Brief Review of Single-Event Burnout Failure Mechanisms and Design Tolerances of Silicon Carbide Power MOSFETs. *Electronics* **2024**, *13*, 1414. [CrossRef]
102. Wu, R.; Blaabjerg, F.; Wang, H.; Liserre, M.; Iannuzzo, F. Catastrophic Failure and Fault-Tolerant Design of IGBT Power Electronic Converters—An Overview. In Proceedings of the 39th Annual Conference of the IEEE Industrial-Electronics-Society (IECON), Vienna, Austria, 10–14 November 2013.
103. Oliveira, J.; Reynes, J.-M.; Morel, H.; Frey, P.; Perrotin, O.; Allirand, L.; Azzopardi, S.; Piton, M.; Coccetti, F. Test Methodology for Short-Circuit Assessment and Safe Operation Identification for Power SiC MOSFETs. *Energies* **2024**, *17*, 5476. [CrossRef]
104. Abbaszada, M.A.; Gajanur, N.; Mazumder, S.K. Assessing the reliability of SiC MOSFET through inverter-like accelerated test vs. power cycling test. *Microelectron. Reliab.* **2025**, *165*, 115589. [CrossRef]
105. Duan, G.; Vainshtein, S.N.; Kostamovaara, J.T. Modified High-Power Nanosecond Marx Generator Prevents Destructive Current Filamentation. *IEEE Trans. Power Electron.* **2017**, *32*, 7845–7850. [CrossRef]
106. Wang, Y.; Ding, Y.; Yin, Y. Reliability of Wide Band Gap Power Electronic Semiconductor and Packaging: A Review. *Energies* **2022**, *15*, 6670. [CrossRef]
107. Huang, T.; Singh, B.P.; Liu, Y.; Norrga, S. Failure Characterization of Discrete SiC MOSFETs under Forward Power Cycling Test. *Energies* **2024**, *17*, 2557. [CrossRef]
108. Lelis, A.J.; Green, R.; Habersat, D.B.; El, M. Basic Mechanisms of Threshold-Voltage Instability and Implications for Reliability Testing of SiC MOSFETs. *IEEE Trans. Electron. Devices* **2015**, *62*, 316–323. [CrossRef]
109. Volosov, V.; Bevilacqua, S.; Anoldo, L.; Tosto, G.; Fontana, E.; Russo, A.-L.; Fiegna, C.; Sangiorgi, E.; Tallarico, A.N. Positive Bias Temperature Instability in SiC-Based Power MOSFETs. *Micromachines* **2024**, *15*, 872. [CrossRef]
110. Weckbrodt, J.; Ginot, N.; Batard, C.; Azzopardi, S. Monitoring of Gate Leakage Current on SiC Power MOSFETs: An Estimation Method for Smart Gate Drivers. *IEEE Trans. Power Electron.* **2021**, *36*, 8752–8760. [CrossRef]
111. Ouaida, R.; Berthou, M.; Leon, J.; Perpina, X.; Oge, S.; Brosselard, P.; Joubert, C. Gate Oxide Degradation of SiC MOSFET in Switching Conditions. *IEEE Electron. Device Lett.* **2014**, *35*, 1284–1286. [CrossRef]
112. *AQG 324*; Qualification of Power Modules for Use in Power Electronics Converter Units in Motor Vehicles. ECPE European Center for Power Electronics e.V.: Nuremberg, Germany, 2021.
113. Janabi, A.; Shillaber, L.; Ying, W.; Mu, W.; Hu, B.; Jiang, Y.; Iosifidis, N.; Ran, L.; Long, T. Substrate Embedded Power Electronics Packaging for Silicon Carbide mosfets. *IEEE Trans. Power Electron.* **2024**, *39*, 9614–9628. [CrossRef]
114. Wang, L.; Zhang, T.; Yang, F.; Ma, D.; Zhao, C.; Pei, Y.; Gan, Y. Cu Clip-Bonding Method with Optimized Source Inductance for Current Balancing in Multichip SiC MOSFET Power Module. *IEEE Trans. Power Electron.* **2022**, *37*, 7952–7964. [CrossRef]
115. Ma, R.; Wang, R.; Fang, H.; Li, P.; Zhao, L.; Wu, H.; Huang, Z.; Tao, J.; Hu, S. A Novel Deep-Trench Super-Junction SiC MOSFET with Improved Specific On-Resistance. *Micromachines* **2024**, *15*, 684. [CrossRef] [PubMed]
116. Ou, Y.; Lan, Z.; Hu, X.; Liu, D. Novel SiC Trench MOSFET with Improved Third-Quadrant Performance and Switching Speed. *Micromachines* **2024**, *15*, 254. [CrossRef] [PubMed]
117. Zhang, X.; Zhao, X.; Li, W.; Wang, Z.; Liao, A.; Song, Y.; Wang, Y.; Zhang, L. Ultra-thermostable embedded liquid cooling in SiC 3D packaging power modules of electric vehicles. *Energy Convers. Manag.* **2023**, *276*, 116499. [CrossRef]
118. Pavlicek, N.; Mohn, F. Laser bonding of copper ribbons and clips on SiC power MOSFETs with sintered copper bond buffers. In Proceedings of the 8th IEEE Electronics System-Integration Technology Conference (ESTC), Vestfold, Norway, 15–18 September 2020.
119. Lee, I.; Yao, X. Active dv/dt Control with Turn-off Gate Resistance Modulation for Voltage Balancing of Series Connected SiC MOSFETs. In Proceedings of the 2021 IEEE Applied Power Electronics Conference and Exposition (APEC), Phoenix, AZ, USA, 14–17 June 2021.
120. Takayama, H.; Okuda, T.; Hikihara, T. Digital active gate drive of SiC MOSFETs for controlling switching behavior-Preparation toward universal digitization of power switching. *Int. J. Circ. Theor. Appl.* **2022**, *50*, 183–196. [CrossRef]
121. Shao, T.; Sun, Y.; Bai, Z.; Zheng, T.Q.; Zhang, Y.; Jia, P. Active Gate Drive Based on Negative Feedback for SiC MOSFETs to Suppress Crosstalk Parasitic Oscillation and Avoid Decreased Efficiency. *Electronics* **2025**, *14*, 2100. [CrossRef]

 https://doi.org/10.3390/en19020518

122. Li, Q.; Yang, Y.; Wen, Y.; Zhang, G.; Xing, W. Active Gate Driver with the Independent Suppression of Overshoot and Oscillation for SiC MOSFET Modules. *IEEE Trans. Ind. Electron.* **2025**, *72*, 2325–2335. [CrossRef]
123. Xu, C.; Miao, Y. SiC MOSFET Active Gate Drive Circuit Based on Switching Transient Feedback. *Energies* **2024**, *17*, 1997. [CrossRef]
124. Huang, Y.; Zhang, J.; Wang, N. A Resonant Auxiliary Drive Circuit for SiC MOSFET to Suppress Crosstalk. *Trans. China Electrotech. Soc.* **2022**, *37*, 3004–3015.
125. Yang, C.; Pei, Y.; Wang, L.; Yu, L.; Zhang, F.; Ferreira, B. Overvoltage and Oscillation Suppression Circuit with Switching Losses Optimization and Clamping Energy Feedback for SiC MOSFET. *IEEE Trans. Power Electron.* **2021**, *36*, 14207–14219. [CrossRef]
126. Liang, M.; Chen, J.; Jia, P. Analysis of DC-Side Snubbers for SiC Devices Application. *Electronics* **2022**, *11*, 3874. [CrossRef]
127. Xue, P.; Hoene, E.; Davari, P. An RC snubber design method to achieve optimized switching noise-loss trade-off of cascode GaN HEMTs. *IET Power Electron.* **2024**, *17*, 1583–1593. [CrossRef]
128. Liu, C.; Xu, M.; Chen, S.; Sun, L.; Zhang, L.; Li, Q.; Wang, J. Analysis of carrier dynamics and thermal effect of the GaAs photoconductive semiconductor switch in lock-on mode. *J. Phys. D Appl. Phys.* **2024**, *57*, 495101. [CrossRef]
129. Sun, Y.; Hu, L.; Li, Y.; Zhu, L.; Dang, X.; Hao, Q.; Li, X. Research on the thermal failure mechanism of an opposed-contact gallium arsenide photoconductive semiconductor switch in avalanche mode. *J. Phys. D Appl. Phys.* **2022**, *55*, 215103. [CrossRef]
130. Xu, J.; Jiang, L.; Cai, P.; Feng, C.; Xiao, H.; Wang, X. Properties Investigation and Damage Analysis of GaN Photoconductive Semiconductor Switch Based on SiC Substrate. *Micromachines* **2024**, *15*, 1178. [CrossRef]
131. Chu, X.; Xun, T.; Wang, L.; Liu, J.; Yang, H.; He, J.; Zhang, J. Breakdown Behavior of GaAs PCSS with a Backside-Light-Triggered Coplanar Electrode Structure. *Electronics* **2021**, *10*, 357. [CrossRef]
132. Zong, J.; Shi, Y.; Qian, G.; Wang, J.; Wei, Z.; Kong, Y.; Zhang, J.; Chen, T. Performance Enhancement of Planar GaAs Photoconductive Semiconductor Switches by Introducing p-Type Epitaxial Layer. *Photonics* **2025**, *12*, 152. [CrossRef]
133. Cai, P.; Jiang, L.; Xu, J.; Xiao, H.; Feng, C.; Wang, Q.; He, T.; Zhou, M.; Wang, X. Design of a lateral photoconductive semiconductor switch with a low resistivity region on semi-insulating GaN to enhance breakdown characteristics. *Opt. Commun.* **2024**, *555*, 130332. [CrossRef]
134. Sun, Y.; Hu, L.; Dang, X.; Zhu, L.; Yang, X.; Huang, J.; Li, Y.; Li, X. Investigation on the Mechanism of Triggering Efficiency of High-Power Avalanche GaAs Photoconductive Semiconductor Switch. *IEEE Electron. Device Lett.* **2021**, *42*, 1646–1649. [CrossRef]
135. Chen, H.; Wei, J.; Zeng, F.; Jia, C.; Fu, Z.; Li, S.; Qian, B. Influence of the width of triggering region on output characteristics of GaAs photoconductive semiconductor switch. *High Power Laser Part. Beams* **2023**, *35*, 154–159.
136. Pokryvailo, A.; Carp, C.; Scapellati, C. High Performance, High-Power Capacitor Charging: Focus on Pulse-to-Pulse Repeatability. In Proceedings of the 25th Annual IEEE Applied Power Electronics Conference and Exposition, Palm Springs, CA, USA, 21–25 February 2010.
137. Oomori, H.; Omura, I. An experimental study on switching waveform design with gate charge control for power MOSFETs. *Power Electron. Devices Compon.* **2023**, *6*, 100043. [CrossRef]
138. Tong, X.C. *Advanced Materials for Thermal Management of Electronic Packaging*, 1st ed.; Springer Science & Business Media: New York, NY, USA, 2011; pp. 1–57.
139. Wang, Y.; Ren, L.; Yang, Z.; Shen, S.; Deng, Z.; Yuan, Q.; Ding, W.; Ding, Z. Heat management technology for solid-state high voltage and high repetitive pulse generators: Towards better effects and reliability. *High Volt.* **2024**, *9*, 2–21. [CrossRef]
140. Bergman, T.L. *Fundamentals of Heat and Mass Transfer*; John Wiley & Sons: Hoboken, NJ, USA, 2011; pp. 5–138.
141. de Bock, H.P.; Chamarthy, P.; Jackson, J.L.; Whalen, B. Investigation and Application of an Advanced Dual Piezoelectric Cooling Jet to a Typical Electronics Cooling Configuration. In Proceedings of the 13th IEEE InterSociety Conference on Thermal and Thermomechanical Phenomena in Electronic Systems (ITherm), San Diego, CA, USA, 30 May–1 June 2012.
142. Sauciuc, I.; Yamamoto, R.; Culic-Viskota, J.; Yoshikawa, T.; Jain, S.; Yajima, M.; Labanok, N.; Amoah-Kusi, C. Carbon Based Thermal Interface Material for High Performance Cooling Applications. In Proceedings of the 14th InterSociety Conference on Thermal and Thermomechanical Phenomena in Electronic Systems (ITherm), Orlando, FL, USA, 27–30 May 2014.
143. Cai, Y.; Wang, Y.; Liu, D.; Zhao, F.-Y. Thermoelectric cooling technology applied in the field of electronic devices: Updated review on the parametric investigations and model developments. *Appl. Therm. Eng.* **2019**, *148*, 238–255. [CrossRef]
144. Orville, T.; Tajwar, M.; Bihani, R.; Saha, P.; Hannan, M.A. Enhancing Thermal Efficiency in Power Electronics: A Review of Advanced Materials and Cooling Methods. *Thermo* **2025**, *5*, 30. [CrossRef]
145. Back, D.; Drummond, K.P.; Sinanis, M.D.; Weibel, J.A.; Garimella, S.V.; Peroulis, D.; Janes, D.B. Design, Fabrication, and Characterization of a Compact Hierarchical Manifold Microchannel Heat Sink Array for Two-Phase Cooling. *IEEE Trans. Compon. Packag. Manuf. Technol.* **2019**, *9*, 1291–1300. [CrossRef]
146. Wang, H.; Xia, X.; Wang, P.; Zhou, B.; Chen, S.; Shi, Y. Study on the pillars structure impacts on the embedded microchannel heat sink. In Proceedings of the 23rd International Conference on Electronic Packaging Technology (ICEPT), Dalian, China, 10–13 August 2022.

 https://doi.org/10.3390/en19020518

147. Sporer, B.J.; Shah, A.P.; Dowhan, G.V.; Shapovalov, R.V.; Packard, D.A.; Wisher, M.; Leckbee, J.J.; Hendricks, K.J.; Hoff, B.W.; Lau, Y.Y.; et al. Multicavity linear transformer driver facility for Z-pinch and high-power microwave research. *Phys. Rev. Accel. Beams* **2021**, *24*, 100402. [CrossRef]
148. Zhang, C.; Wang, H.; Huang, Y.; Zhang, L.; Chen, Y. Immersion liquid cooling for electronics: Materials, systems, applications and prospects. *Renew. Sustain. Energy Rev.* **2025**, *208*, 114989. [CrossRef]
149. Ristic-Smith, A.; Rogers, D.J. Compact Two-Phase Immersion Cooling with Dielectric Fluid for PCB-Based Power Electronics. *IEEE Open J. Power Electron.* **2024**, *5*, 1107–1118. [CrossRef]
150. Tang, G.; Wai, L.C.; Feng, H. Development and Demonstration of a Novel Immersion Two Phase Cooling High Power SiC Power Module. In Proceedings of the IEEE 73rd Electronic Components and Technology Conference (ECTC), Orlando, FL, USA, 30 May–2 June 2023.
151. Ren, L.; Wang, Y.; Yang, Z.; Ding, W. High power and high repetition frequency sub-nanosecond pulse generator with two-phase immersion cooling technique. *Rev. Sci. Instrum.* **2021**, *92*, 034716. [CrossRef] [PubMed]
152. Manepalli, V.V.; McAfee, R.C.; Randriambololona, A.M.; Graham, S.; Agonafer, D. Dynamic Thermal Management of a MOSFET Power Module Using a Novel Three-Component Phase Change Material. *J. Heat Transf.* **2025**, *147*, 102401. [CrossRef]
153. Aleinikov, E.; Barzdenas, V. Comparative Analysis of High-Voltage High-Frequency Pulse Generation Techniques for Pockels Cells. *Appl. Sci.* **2025**, *15*, 10830. [CrossRef]
154. Laloya, E.; Lucía, O.; Sarnago, H.; Burdío, J. Heat Management in Power Converters: From State of the Art to Future Ultrahigh Efficiency Systems. *IEEE Trans. Power Electron.* **2016**, *31*, 7896–7908. [CrossRef]
155. Gorecki, K.; Posobkiewicz, K. Cooling Systems of Power Semiconductor Devices-A Review. *Energies* **2022**, *15*, 4566. [CrossRef]
156. Dhumal, A.R.; Kulkarni, A.P.; Ambhore, N.H. A comprehensive review on thermal management of electronic devices. *J. Eng. Appl. Sci.* **2023**, *70*, 140. [CrossRef]
157. Shi, F.; Chen, P.; Jiang, S.; Zhuang, J.; Rao, J. A Solid-State Marx Generator with Prevention of through Current for Rectangular Pulses. *Electronics* **2024**, *13*, 101. [CrossRef]
158. Zhang, H.; Chen, J. Analysis of control signal jitter in the distributed synchronous trigger control system for pulse power combining. *J. Phys. Conf. Ser.* **2025**, *3079*, 012022. [CrossRef]
159. Geng, J.; Cheng, X.; Yang, J.; Yang, X.; Chen, R. Research on multi-switch synchronization based on single trigger generator. *AIP Adv.* **2018**, *8*, 055117. [CrossRef]
160. Damideh, V.; Btaiche, J.C.; Ho, A.; Spielman, R.B.; Lehr, J.M.; Mehlhorn, T.A.; Hassen, I.; Akoulov, A.; Aranfar, E.; McDonald, A.; et al. Experimental results of a 330 GW impedance-matched Marx generator. *Sci. Rep.* **2024**, *14*, 16889. [CrossRef]
161. Chen, L.; Zhu, C.; Zheng, J.; Qiu, J.; Zhao, H.; Liu, K. A Flexible Solid-State Marx Modulator Module Based on Discrete Magnetic Coupling Drivers. *Electronics* **2023**, *12*, 3831. [CrossRef]
162. Geng, J.-Y.; Yang, J.-H.; Cheng, X.-B.; Chen, R.; Shu, T. A compact, low jitter, high voltage trigger generator based on fractional-turn ratio saturable pulse transformer and its application. *Rev. Sci. Instrum.* **2022**, *93*, 084709. [CrossRef] [PubMed]
163. Wu, M.; Liu, G.; He, M.; Shu, W.; Jiao, Y.; Li, H.; Yao, W.; Liang, X. A Low-Jitter Delay Synchronization System Applied to Ti:sapphire Femtosecond Laser Amplifier. *Appl. Sci.* **2025**, *15*, 9424. [CrossRef]
164. Chen, L.; Zou, W.; Zhou, L.; Wang, M.; Liu, Y.; Liu, L.; Deng, M.; Liu, D.; Zhu, J.; Lian, K.; et al. Development of a fusion-oriented pulsed power module. *Phys. Rev. Accel. Beams* **2019**, *22*, 030401. [CrossRef]
165. Wang, J.; Wang, Z.; Wei, H.; Shen, Y.; Jiang, H. Fiber-optic controlled high voltage trigger generator for a ten-stage linear transformer driver module. *Rev. Sci. Instrum.* **2025**, *96*, 065205. [CrossRef]
166. Nia, M.S.S.; Shamsi, P.; Ferdowsi, M. EMC Modeling and Conducted EMI Analysis for a Pulsed Power Generator System Including an AC-DC-DC Power Supply. *IEEE Trans. Plasma Sci.* **2020**, *48*, 4250–4261. [CrossRef]
167. Muttaqi, K.; Haque, M. Electromagnetic interference generated from fast switching power electronic devices. *Int. J. Innov. Energy Syst. Power* **2008**, *3*, 19–26.
168. Yuan, L.; Zhang, J.; Liang, Z.; Hu, M.; Chen, G.; Lu, W. EMI challenges in modern power electronic-based converters: Recent advances and mitigation techniques. *Front. Electron.* **2023**, *4*, 1274258. [CrossRef]
169. Choi, S.; Choi, J.; Warnakulasooriya, T.; Shin, J.-W.; Imaoka, J.; Yamamoto, M. Improved Packaging of Power Module with Low-Permittivity Material for Low Common-Mode Noise and High Reliability. *IEEE Trans. Power Electron.* **2024**, *39*, 14261–14270. [CrossRef]
170. Jia, N.; Tian, X.; Xue, L.; Bai, H.; Tolbert, L.M.; Cui, H. Integrated Common-Mode Filter for GaN Power Module with Improved High-Frequency EMI Performance. *IEEE Trans. Power Electron.* **2023**, *38*, 6897–6901. [CrossRef]
171. Li, Y.; Wang, S.; Sheng, H.; Lakshmikanthan, S. Reduction and Cancellation Techniques for the Near Field Capacitive Coupling and Parasitic Capacitance of Inductors. In Proceedings of the EEE Symposium on Electromagnetic Compatibility, Signal Integrity and Power Integrity (EMC, SI & PI), Long Beach, CA, USA, 30 July–3 August 2018.

 https://doi.org/10.3390/en19020518

172. Xue, P.; Davari, P. Comparative Study on Temperature Dependency of dV/dt, dI/dt and EMI Generation for IGBTs, Si and SiC MOSFETs. In Proceedings of the 2025 International Symposium on Electromagnetic Compatibility—EMC Europe, Paris, France, 1–5 September 2025.
173. Bheema, R.K.; Gopu, J.; Bhaskaran, K.; Verma, A.; Chavali, M.; Etika, K.C. A review on recent progress in polymer composites for effective electromagnetic interference shielding properties—structures, process, and sustainability approaches. *Nanoscale Adv.* **2024**, *6*, 5773–5802. [CrossRef]
174. Shahapurkar, K.; Gelaw, M.; Tirth, V.; Soudagar, M.E.M.; Shahapurkar, P.; Mujtaba, M.A.; Kiran, M.C.; Ahmed, G.M.S. Comprehensive review on polymer composites as electromagnetic interference shielding materials. *Polym. Polym. Compos.* **2022**, *30*, 09673911221102127. [CrossRef]
175. Jiang, W. Review of solid-state linear transformer driver technology. *Matter Radiat. Extremes* **2018**, *3*, 159–164. [CrossRef]
176. Ranjan, A.; Abhishek, A.; Kumar, N.; Verma, B.K. Review of Nanosecond Pulse Generator with High-Power Switching Devices. *IEEE Trans. Electron. Devices* **2024**, *71*, 5165–5176. [CrossRef]
177. Mazarakis, M.G.; Fowler, W.E.; LeChien, K.L.; Long, F.W.; Matzen, M.K.; McDaniel, D.H.; McKee, R.G.; Olson, C.L.; Porter, J.L.; Rogowski, S.T.; et al. High-Current Linear Transformer Driver Development at Sandia National Laboratories. *IEEE Trans. Plasma Sci.* **2010**, *38*, 704–713. [CrossRef]
178. Li, M.J.; Kang, Q.; Tan, J.; Luo, M.; Xiang, F. A High-Power Pulse Generator Based on Pulse Forming Network and Linear Transformer. *Laser Part. Beams* **2021**, *2021*, e19. [CrossRef]
179. Shen, Y.; Ye, M.; Liu, Y.; Zhang, H.; Xia, L. Electric Field Enhancement Breakdown Caused by the Pulse Formation and Transmission of Parallel-Plate Blumlein Line. *IEEE Trans. Plasma Sci.* **2024**, *52*, 2281–2284. [CrossRef]
180. Zhou, Y.; Liu, B.-Y.; Xie, Y.-Z. Investigation of the waveform quality of a kA-level rectangular pulse generator based on pulse forming network. *Rev. Sci. Instrum.* **2025**, *96*, 124709. [CrossRef]
181. Mishra, R.K.; Joshi, D.; Agrawal, P.K. Effect on timing drift due to temperature variation in high voltage pulse power supply. In Proceedings of the 2020 International Conference on Power, Energy, Control and Transmission Systems (ICPECTS), Chennai, India, 10–11 December 2020.
182. He, Y.; Gao, R.; Ma, T.; Zhang, X.; Zhang, X.; Yang, Y. Total Ionizing Dose Effects on Lifetime of NMOSFETs Due to Hot Carrier-Induced Stress. *Electronics* **2025**, *14*, 2563. [CrossRef]
183. Pearton, S.J.; Haque, A.; Khachatrian, A.; Ildefonso, A.; Chernyak, L.; Ren, F. Review-Opportunities in Single Event Effects in Radiation-Exposed SiC and GaN Power Electronics. *ECS J. Solid State Sci. Technol.* **2021**, *10*, 075004. [CrossRef]
184. Zeng, Q.; Yang, Z.; Wang, X.; Li, S.; Gao, F. Research Progress on Radiation Damage Mechanism of SiC MOSFETs Under Various Irradiation Conditions. *IEEE Trans. Electron. Devices* **2024**, *71*, 1718–1727. [CrossRef]
185. Mao, J.; Wang, S.; Shi, Q.; Cheng, Y.; Chen, Y. Excellent Vacuum Pulsed Flashover Characteristics Achieved inDielectric Insulators Functionalized by Electronegative Halogen-Phenyl and Naphthyl Groups. *Langmuir* **2022**, *38*, 4129–4137. [CrossRef] [PubMed]
186. Zhao, Q.; Liu, X.; Wang, H.; Zhu, Y.; An, Y.; Yu, D.; Qi, J. Research Progress in Corrosion Protection Technology for Electronic Components. *Metals* **2023**, *13*, 1508. [CrossRef]
187. Xu, J.; Guo, Y.; Su, Y.; Tang, R.; Long, X. Growth kinetics of intermetallic compound in solder joints during thermal cycling: A review. In Proceedings of the 23rd Electronics Packaging and Technology Conference, Singapore, 1–3 December 2021.
188. Chen, X.; Zhou, G.; Wang, X.; Xu, H.; Wang, C.; Yao, Q.; Chi, J.; Fu, X.; Wang, Y.; Yin, X.; et al. Progress in semiconductor materials for photocathodic protection: Design strategies and applications in marine corrosion protection. *Chemosphere* **2023**, *323*, 138194. [CrossRef]
189. Pongvittayapanu, K.; Srisrual, A.; Fakpan, K. Effect of thermal cycling and vibration on cracking in Sn-3.0 Ag-0.5 Cu solder bump. *Mater. Today Proc.* **2022**, *52*, 2372–2376. [CrossRef]
190. Liu, B.; Lee, T.-K.; Liu, K.-C. Impact of 5% NaCl Salt Spray Pretreatment on the Long-Term Reliability of Wafer-Level Packages with Sn-Pb and Sn-Ag-Cu Solder Interconnects. *J. Electron. Mater* **2011**, *40*, 2111–2118. [CrossRef]
191. Tian, W.; Chen, X.; Zhang, G.; Chen, Y.; Luo, J. Delamination of Plasticized Devices in Dynamic Service Environments. *Micromachines* **2024**, *15*, 376. [CrossRef]
192. He, Y.; Kabiri, M. In-situ characterization of moisture absorption and hygroscopic swelling of an epoxy molding compound for electronic packaging. *J. Therm. Anal. Calorim.* **2022**, *147*, 5667–5675. [CrossRef]
193. Liu, X.; Wang, N.; Zhao, M.; Hu, X. Experimental Study on the Effect of Sand and Dust on the Performance of Photovoltaic Modules in Desert Areas. *Energies* **2024**, *17*, 682. [CrossRef]
194. Alshatnawi, F.; Enakerakpo, E.; Alhendi, M.; Abdelatty, M.; Umar, A.; Al-Haidari, R.; Shaddock, D.; Hoel, C.; Boyd, L.; Poliks, M.; et al. High stability and reliability additively manufactured metal-insulator-metal capacitors for high-temperature applications. *Mater. Today Commun.* **2024**, *39*, 108682. [CrossRef]

195. Montanari, G.C.; Myneni, S.B. Operation at Reduced Atmospheric Pressure and Concept of Reliability Redundancy for Optimized Design of Insulation Systems. *Energies* **2025**, *18*, 2371. [CrossRef]
196. Jeong, S.; Kareem, A.B.; Song, S.; Hur, J.-W. ANN-Based Reliability Enhancement of SMPS Aluminum Electrolytic Capacitors in Cold Environments. *Energies* **2023**, *16*, 6096. [CrossRef]

https://doi.org/10.3390/en19020518

MDPI AG
Grosspeteranlage 5
4052 Basel
Switzerland
Tel.: +41 61 683 77 34

Energies Editorial Office
E-mail: energies@mdpi.com
www.mdpi.com/journal/energies

www.ingramcontent.com/pod-product-compliance
Lightning Source LLC
Chambersburg PA
CBHW061603220926
51850CB00005B/408
9783725867400